Modern Thermodynamics

Modern Thermodynamics

Edited by
Nora Jonas

Willford Press
www.willfordpress.com

Published by Willford Press,
118-35 Queens Blvd., Suite 400,
Forest Hills, NY 11375, USA

ISBN: 978-1-68285-589-8

Cataloging-in-Publication Data

Modern thermodynamics / edited by Nora Jonas.
p. cm.
Includes bibliographical references and index.
ISBN 978-1-68285-589-8
1. Thermodynamics. 2. Heat-engines. 3. Heat engineering. I. Jonas, Nora.

TJ265 .M63 2019
621.402 1--dc21

For information on all Willford Press publications
visit our website at www.willfordpress.com

Willford Press

Printed in China.

Contents

Preface

The purpose of the book is to provide a glimpse into the dynamics and to present opinions and studies of some of the scientists engaged in the development of new ideas in the field from very different standpoints. This book will prove useful to students and researchers owing to its high content quality.

This book is a valuable compilation of topics, ranging from the basic to the most complex advancements in the field of thermodynamics. It explores the modern systems and principles introduced in this field. Thermodynamics is the study of the relation of heat and temperature with forms of energy. It is based on four principal laws that define and characterize physical quantities. The technological progress over the years has fueled the advancements in the field of thermodynamics. The prevalent thermodynamic principles and concepts have applications across a range of fields such as chemical engineering, industrial ecology, physical chemistry, etc. The data collated in this book provides in-depth knowledge about the modern theories of thermodynamics. The extensive content herein will be helpful for those who want to broaden the expanse of their knowledge in this area of study.

At the end, I would like to appreciate all the efforts made by the authors in completing their chapters professionally. I express my deepest gratitude to all of them for contributing to this book by sharing their valuable works. A special thanks to my family and friends for their constant support in this journey.

Editor

1

Assessing the Activity of Solid-Suported $SnCl_2$ Catalysts on the Oleic Acid Esterification for Biodiesel Production

Márcio José Da Silva* and Abiney Lemos Cardoso

Chemistry Department, Federal University of Viçosa, Viçosa, Minas Gerais State, Brazil

Abstract

The search for cleaner methodologies has forcing the chemical industry to seek environmentally benign acid catalysts. $SnCl_2.2H_2O$ is an affordable commercially Lewis acid, water tolerant and easy to handling, cheaper, and less corrosive. Recently, we have demonstrated that it is an efficient catalyst to produce biodiesel in homogeneous catalysis conditions. In this work, we have supported $SnCl_2.2H_2O$ over different solid matrix (i.e., such as SiO_2, Nb_2O_5 and ZrO_2) and evaluated its catalytic activity in FFA esterification reactions. In this case, the reaction occurs in heterogeneous phase bringing with itself all the pertinent advantages to this rational system as the best separation of products and reuse of the catalyst. It was appraised reaction parameters such as thermal treatment temperature nature of the support; leaching test was also accomplished.

Keywords: Esterification; Tin chloride; Free fatty acids; Heterogeneous catalysis

Introduction

The exhaustion of the fossil petrol reserves, in combination with the environmental impact generated by the green-house effect gas emission has stimulated the search by alternative and renewable feedstock for energy production [1]. Besides this crescent demand, the industry chemistry and academy in all parts of world are developing environment friendly technologies for the production of alternative fuels [2]. Nowadays, the biodiesel is a "green" alternative fuel that has arisen as an attractive option, mainly because it's less pollutant than fossil diesel and can be obtained from renewable sources. Currently, the biodiesel is manufactured from alkaline transesterification of edible or non-edible vegetable oils via a well-established industrial process [3,4]. However, there are serious drawbacks related to its conventional production process, such as a greater generation of effluents and salts from neutralization steps of the products and wastes, the use of non-recyclable catalysts and the incompatibility with acid raw material with high amount of FFA [4,5]. These important features have hardly affected the final cost to biodiesel production [6].

An interesting alternative for lower biodiesel price is produce it directly from low cost lipidic raw materials are normally rich in FFA [7]. It should be mentioned that even though traditional mineral acids are an inexpensive catalysts able to those processes, they are equally corrosive, are not reusable and results in a large generation of acid effluents which should be neutralized at the end of the reaction, leaving greater amount of salts to be disposed of into environment [7,8]. Therefore, to develop recyclable alternative catalysts for FFA esterification reactions presents in low cost feedstokes and food industry rejects can be an option strategically important, and undoubtedly can produce biodiesel at more competitive cost via a cleaner technology [9]. Recently, several works have been described the development of heterogeneous catalysts based on acids solids that appear to offer a good perspective to make the biodiesel production more environment friendly [10,11]. Nevertheless, serious technological drawbacks such as drastic conditions reaction, the rigorous control of raw material quality in relation to water content, beyond of the leaching catalyst provoked by presence of alcohol seems suggest that those process yet are hard to control.

Alternatively, to develop catalysts for the direct conversion into biodiesel of lipid wastes basically constituted of FFA, or yet for the pre-esterification of raw materials that have high acidity seem be also a challenge to be overcome [12]. Nowadays, the majority of the catalysts conventionally used in the FFA esterification reactions are Brønsted acids. However, features such the higher corrosive character besides greater generation of residues and salts compromises the use of these traditional catalysts [13]. Lewis acids can be interesting alternative catalysts for biodiesel production. However, their high cost, the manipulation difficult and the intolerance to water of compounds traditionally used such as BF3 and others common reagents of organic synthesis, also does not favor the use of these latters in FFA esterification at industrial scale [14].

Particularly, we have concentrating efforts in developing an alternative esterification processes based on tin chloride, an simple, easily handling, water tolerant and inexpensive Lewis acid [15,16]. On the other hand, the $SnCl_2$ is also most attractive, because it is are commercially available and easy to handle [17]. Moreover, its display remarkably low toxicity, is tolerant to water, economically cost effective, and can be used in recyclable processes [18]. Although Sn(II) catalysts have been used in homogeneous phase, they are though less corrosive than mineral acids [19]. In this work, the application of solid supported $SnCl_2$ catalysts on FFA esterification is described. Our intention was assess if supported tin(II) catalysts can be potentially used on direct conversion of FFA into biodiesel via ethanol esterification reactions. The reusability, the less corrosion, the mild condition reactions and the compatibility with ethanol, which is also a renewable raw material, comprise positive aspects of this process. Moreover, the stability and possibility of reuse of supported Sn(II) catalysts were investigate.

***Corresponding author:** Márcio José Da Silva, Chemistry Department, Federal University of Viçosa, Viçosa, Minas Gerais State, Brazil, E-mail: silvamj2003@ufv.br

Materials and Methods

Chemicals

All chemicals were purchased from commercial sources and used as received unless otherwise stated. Ethanol was acquired from Sigma-Aldrich (99.9% wt.). The tin(II) chloride dihydrate ($SnCl_2·2H_2O$; 99.5% wt.), niobium(V) oxide (Nb_2O_5), silicon(IV) oxide (SiO_2) and zirconium(IV) oxychloride octahydrate ($ZrOCl_2·8H_2O$) were purchased from Sigma Aldrich. Oleic acid (99% wt.) was obtained from Modern Chemistry (São Paulo, SP, Brazil).

Synthesis of supported tin(II) catalysts

The supported tin catalysts were prepared by impregnation method. $SnCl_2$⊠$2H_2O$ alcoholic solutions were prepared using 10 ml of solution per gram of supported catalyst to obtain a concentration of $SnCl_2$ 10, 30 and 50% w/w. The support (SiO_2, Nb_2O_5 or ZrO_2) was added to the solution resulting in a suspension. The suspension was kept under constant stirring and evaporated at 80 degrees until complete dryness. Then the solid obtained was pulverized and dried at 100 degrees in an oven for 3 hours. The supported tin samples were divided into two portions and each was thermally treated at 200 or 300 degrees for 4 h.

Catalytic runs

Catalytic tests were conducted in a 50 mL three-necked glass flask, equipped with sampling system, a reflux condenser, in thermostatic bath and under magnetic stirrer. In all runs, ethanol was used in large excess for to shift the equilibrium towards the esters formation. In a typical experiment, oleic acid (*ca.* 1.00 mmol) was dissolved in 7 ml (*ca.*155.0 mmol) of ethanol, which were heated to the reflux temperature. After the addition of catalyst (*ca.* 50.0 mg) the reaction was started.

Reaction monitoring and identification of products reaction

The reaction progress was continuously monitored taking aliquots at regular intervals of 30 min and analyzing them via gas chromatography. The temperature program was as follows: 150°C for 1 min, 10°C/min up to 240°C, hold time of 5 min. Both injector and FID detector were kept at 250 and 280°C, respectively. The GC was fitted with a DB5 capillary column (0.25 μm × 0.25 mm × 30 m) and reaction yields were calculated by matching the areas of the esters GC peaks into the corresponding calibration curves. Analyses were carried out in a Shimadzu GC gas chromatograph coupled with a MS-QP 2010A mass spectrometer (Tokyo, Japan), fitted with a DB5 capillary column (0.25 μm × 0.25 mm × 30 m), with helium as the carrier gas at 2 mL/min. The temperature program was as follows: 150°C for 1 min, 10°C/min up to 250°C, hold time of 5 min. The GC injector and MS ion source temperatures were kept at 250 and 260°C, respectively. The MS detector was operated in the EI mode at 70 eV, with a scanning range of *m/z* 50-400.

Recovery and reuse of heterogeneous catalysts

After reaction end, the solid catalyst was filtered, washed and dried in oven. Immediately, it was reused in another run without any previous activation.

Determination of the Sn content

The acid digestion and measurements of tin content was done in according with literature [18]. Typically, a mix of HNO_3/H_2O_2 solutions was added to 0.1 gram of sample drop by drop. Afterward, the sample was gently swirled for a complete homogenization. After transfer to a tube, the digestion was performed at microwave oven. The digestion parameters as following: 1000 Watt; 80°C/ 5 min; 10°C min^{-1} to 200°C; time hold by 20 min. After cooling the rotor until the solution reaches room temperature, open the vessel and transfer the solution to a volumetric flask. The concentration of tin was determined by AAS analysis (Varian Spectra A- 200 model), in a spectrophotometer equipped with broker-absorbing substance. The samples were pumped into flame: acetylene-nitrous oxide. The parameters used were λ (235.5 nm), flame (C_2H_2/N_2O) and detection limit (1 ppm). A calibration curve was built based on a stock standard solution of Tin (Merck, Germany).

Results and Discussion

General aspects

Initially, the catalytic activity of the stannous chloride was compared with the activity of others typical tin catalysts. It can be seen (Figure 1) that although a large excess of ethanol has been used, the reactions achieved poor conversions when the catalysts with low activity were present. Indeed, the insoluble oxide catalysts (i.e., SnO, SnO_2) were the less active. These commercial oxides have low surface area and consequently are few active. For this reason, they are generally used supported on the matrix with higher surface area [20].

However, it should be noted the significant activity of tin(II) sulfate, which was almost insoluble; literature suggests that sulfate ion exert an electron withdrawing effect on the tin increasing its Lewis acidity [21]. Conversely, although totally soluble, the tin(II) acetate was almost inactive. Probably, the presence of bidentate ligand on tin(II) hampered the activation of oleic acid carbonyl group, making difficult its attack by the hydroxyl group of ethyl alcohol. Notably, the $SnCl_2$ was the most active catalyst. This result is in agreement with literature [22]. This support the choice of $SnCl_2$ as dopant to synthesizes heterogeneous catalysts. In next sections, the results of oleic acid esterification with ethyl alcohol over different supports (i.e., ZrO_2, Nb_2O_5 and SiO_2) treated thermally at temperatures of 100, 220 and 300 degrees are displayed. In each case are also presented the results of the reaction without catalyst, only in the presence and support catalyzed by $SnCl_2$ in homogeneous phase.

Esterification of oleic acid over $SnCl_2$/ ZrO_2: Effect of the thermal treatment

The temperature of thermal treatment was the parameter selected for an adequate comparison between the catalytic activities of different tin(II) composites. Temperatures higher than 300°C may reduce the surface area of support. For this reason, and because this parameter will not discuss herein, the temperatures selected were 100, 200 and 300°C. The kinetic curves are displayed in Figure 2. Even being carried

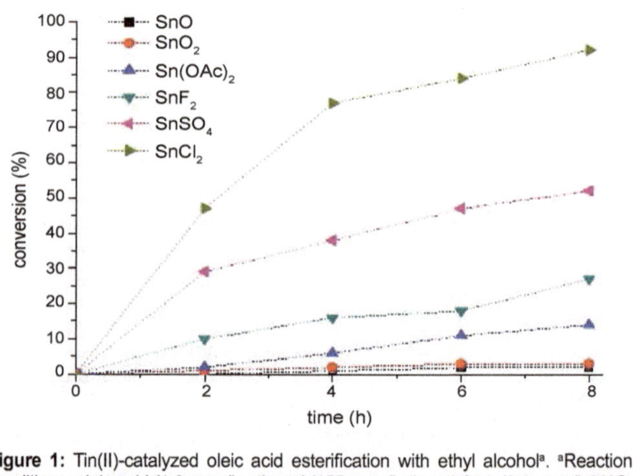

Figure 1: Tin(II)-catalyzed oleic acid esterification with ethyl alcohol[a]. [a]Reaction conditions: oleic acid (1.0 mmol); ethanol (155 mmol); tin catalyst (0.1 mmol) 60°C.

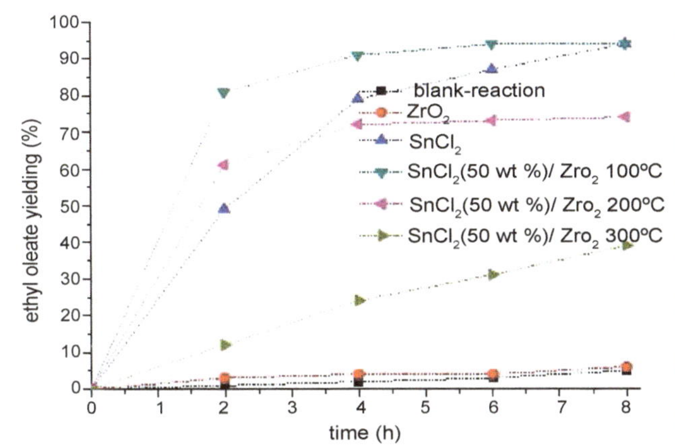

Figure 2: $SnCl_2/ZrO_2$-catalyzed oleic acid esterification with ethyl alcohol: effects of temperature of thermal treatment[a]. [a]Reaction conditions: oleic acid (1.0 mmol); ethyl alcohol (155 mmol); catalyst (50 mg of 50 % wt. $SnCl_2/ZrO_2$); temperature (60°C).

out with a large excess of ethyl alcohol, the reaction-blank achieved only a poor yielding (*ca.* 5%). In the presence of ZrO_2 support, the esterification of oleic acid reached <10% of conversion. It can be attributed to the presence of the weak Brønsted acid sites present in the zirconia [23]. The results obtained in $SnCl_2$ (50% wt.)/ ZrO_2- catalyzed reactions (Figure 2) showed that the thermal treatment temperature was an extremely important factor for the activity of tin composites. The catalyst treated to 100C temperatures gave best results followed by the one treatment at 200°C. Indeed, the $SnCl_2$ (50% wt.)/ZrO_2 100°C solid catalyst was more active than soluble $SnCl_2$. Conversely, for the composite treated to 300°C, the lowest conversion was achieved (*ca.* 43%). The literature describes that at temperatures higher than 80°C the $SnCl_2 \cdot 2H_2O$ loses its 2 water molecules [24]. Therefore, as water has negative contribution in esterification reaction (i.e., it is a by-product and can shift equilibrium toward reactants) the catalysts treated to 100°C was thus more active than hydrate tin chloride, even being this latter a homogeneous catalyst. The catalyst treated to 200 C displayed a high initial reaction rate, however, at reaction end, the conversion reached was lower than one obtained by the $SnCl_2$. Probably, at this temperature, the catalyst comprises of mixture of $SnCl_2$ and SnO, as may be inferred if we comparing the color of samples of these composites (Figure 3).

Figure 3 shows that the composite $SnCl_2$ (50 wt.%)/ ZrO_2-100°C has white coloration, while the composite treated to 200°C darkens, possibly suggesting that there is the formation of SnO, which has brown color. When the sample is heated to 300°C the composite to turn white, suggesting that prevails such as SnO_2 species (Figure 3). However, when highest treatment temperatures was used (i.e., 300°C), a drastic drop on yielding happened; only 44% was reached. This fact could be triggered by the transformation of tin(II) chloride to tin(IV) oxide, which has low catalytic activity (see Figure 3). Therefore, it can be seen that the performance of zirconium-supported $SnCl_2$ undergoes a strong influence of the thermal treatment, and the best and worst results were observed for the catalysts treated to 100 and 300°C, respectively.

Esterification of oleic acid over $SnCl_2/Nb_2O_5$: Effect of the thermal treatment

When we analyzing the reactions carried over tin/ niobia catalysts, we can realize that thermal treatment temperature plays a crucial role on activity of these composites. The following increasing trend of activity was observed: $SnCl_2$/ Nb_2O_5 (100°C)>$SnCl_2/Nb_2O_5$ (200°C)>$SnCl_2$>$SnCl_2/Nb_2O_5$ (300°C) (Figure 4). The tendency observed was similar to that verified in reactions with zirconia supported catalysts. The alone support had negligible activity and once more, the treatment temperature drastically affected the activity of supported catalysts.

Esterification of oleic acid over $SnCl_2/SiO_2$: Effect of the thermal treatment

The tendency to low conversions for the composites treated to 300°C temperature continues, therefore, low conversions were obtained as confirming the results shown in Figure 5. It is suggestive that part of the tin chloride is truly becoming tin(IV) oxide. As observed for the other supports, the reaction catalyzed by the composite treated to 100°C achieved the highest ethyl oleate yielding (i.e., $SnCl_2$ (50 wt.%)/ SiO_2 100°C).

Assessment the stability of solid supported tin(II) catalysts

The catalysts treated to 100 and 300 degrees were selected to investigate the leaching. Figures 6-8 displays the main results. To do it, the reactions were carried by 30 minutes and afterward, the catalysts were filtered and the reaction monitored during 8 hours without the presence of catalyst. The Figure 6a and 6b shows that Sn(II) cations are leaching to the reaction solution, because even after the catalyst be removed, the reaction had its progress continued, mainly for the $SnCl_2$/ Nb_2O_5 catalyst treated to 100°C. The stability of tin/zirconia supported

Figure 3: Samples of commercial and synthesized $SnCl_2/ZrO_2$ supported catalysts[a]. [a]From left to right: tin(II) chloride dihydrate, tin(II) oxide, tn(IV) oxide, zirconium oxide, and composites $SnCl_2$ (50 wt. %)/ZrO_2-100°C, $SnCl_2$ (50 wt. %)/ZrO2-200°C and $SnCl_2$ (50 wt. %)/ZrO_2-300°C, respectivamente.

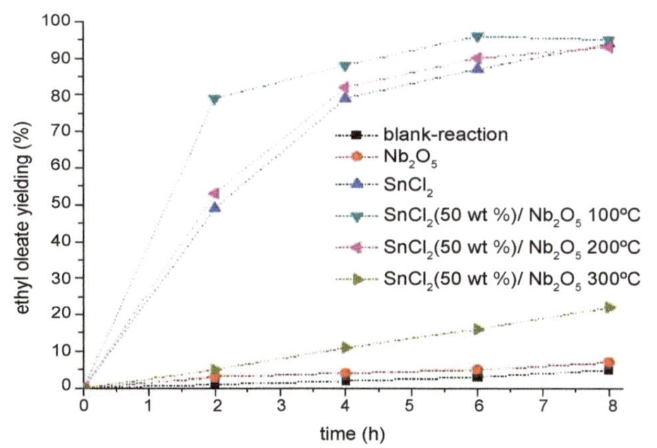

Figure 4: $SnCl_2/Nb_2O_5$-catalyzed oleic acid esterification with ethyl alcohol: effects of temperature of thermal treatment[a]. [a]Reaction conditions: oleic acid (1.0 mmol); ethyl alcohol (155 mmol); catalyst (50 mg of 50% wt. $SnCl_2/Nb_2O_5$); temperature (60°C).

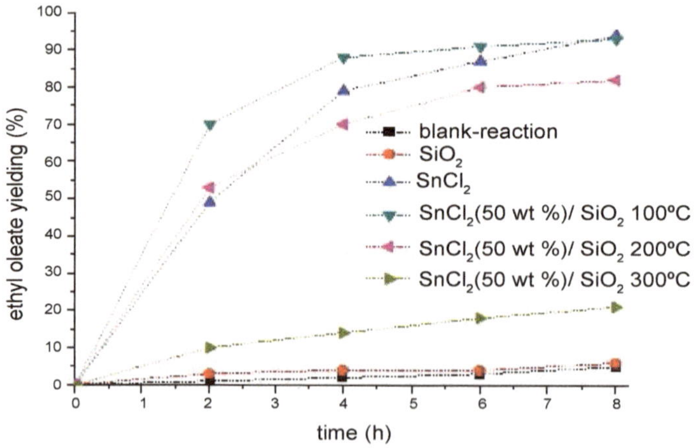

Figure 5: $SnCl_2/SiO_2$-catalyzed oleic acid esterification with ethyl alcohol: effects of temperature of thermal treatment[a]. [a]Reaction conditions: oleic acid (1.0 mmol); ethyl alcohol (155 mmol); catalyst (50 mg of 50% wt. $SnCl_2/SiO_2$); temperature (60°C).

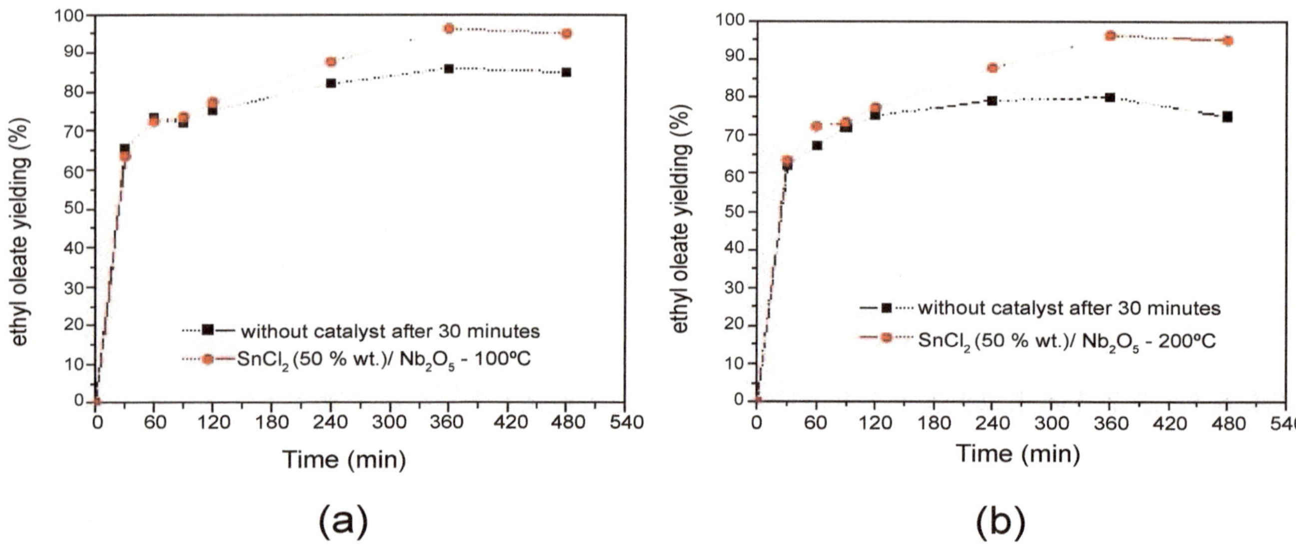

Figure 6: $SnCl_2/Nb_2O_5$-catalyzed oleic acid esterification with ethyl alcohol: assessment of catalyst leaching[a]. [a]Reaction conditions: oleic acid (1.0 mmol); ethyl alcohol (155 mmol); catalyst (50 mg of 50% wt. $SnCl_2$/ Nb_2O_5); temperature (60°C).

solid catalysts in the reaction medium was also low as to that verified for the tin/ niobium catalysts (Figure 7). Conversely, an auspicious result was obtained for the silica supported catalyst treated to 200°C; almost no conversion was observed in the reaction after the removing of solid catalyst (Figure 8). Therefore, this catalyst was selected to be recovered and reused in consecutive catalytic tests (Figure 9).

Heterogeneous catalysts have the main advantage over homogeneous catalysts the possibility of recovery and reuse. However, the leaching of the active component of the catalyst can cause its deactivation and contaminate the products. So, recovery and reuse tests should be performed only for catalysts that do not leach easily. The results show that the catalytic activity of $SnCl_2$ (50% wt.)/ SiO_2 200°C) suffered a decrease after three cycles of recovery/reuse (Figure 9).

The catalyst recovery rate varied between 86 and 82%, but, as we have seen, the $SnCl_2$ (50% wt.)/ SiO_2-200°C does not suffer a significant leaching process (Figure 8b). Thus is possible that some catalyst have been lost in steps of centrifugation, filtration or washing of catalyst.

FTIR spectra of the solid-supported tin(II) catalysts: $SnCl_2$/ ZrO_2, $SnCl_2/Nb_2O_5$ and $SnCl_2/SiO_2$: Effect of thermal treatment

The supported tin(II) composites were analyzed by FT-IR aims to assess if the presence of tin(II) when the support were thermally treated had provoke any significant change. FT-IR spectra were obtained **from** all samples with different content of Sn(II) (*ca.* 30 and 50% w/w). The FT-IR spectra of solid supported catalysts were measured in the wavenumber range between 500- 4000 cm^{-1} (Figure 10). In general, all bands were overlapping by support bands. No significant changes could be observed after thermal treatment to 300 degrees temperature. At the same time, the characteristic bands of Sn-O or Sn-Cl were absent. Indeed, literature describes that stretching frequencies of Sn-

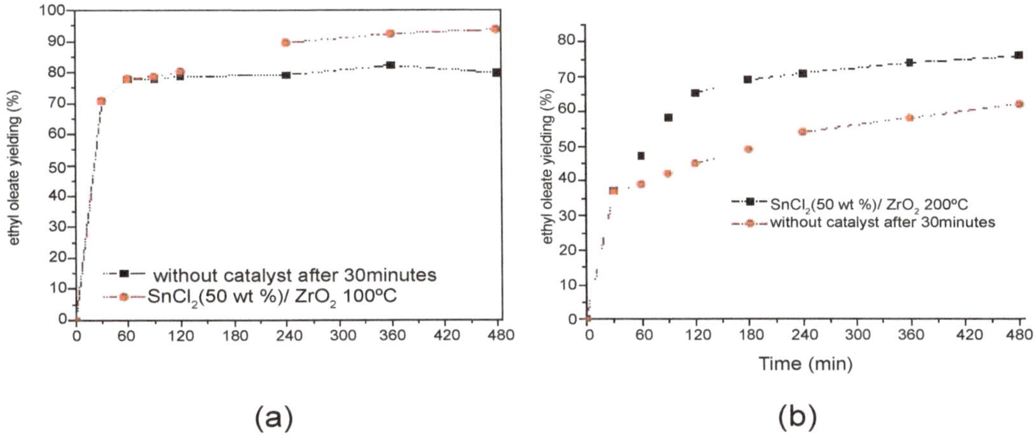

Figure 7: $SnCl_2/ZrO_2$-catalyzed oleic acid esterification with ethyl alcohol: assessment of catalyst leaching[a]. [a]Reaction conditions: oleic acid (1.0 mmol); ethyl alcohol (155 mmol); catalyst (50 mg of $SnCl_2$ (50% wt.)/ZrO_2); temperature (60°C).

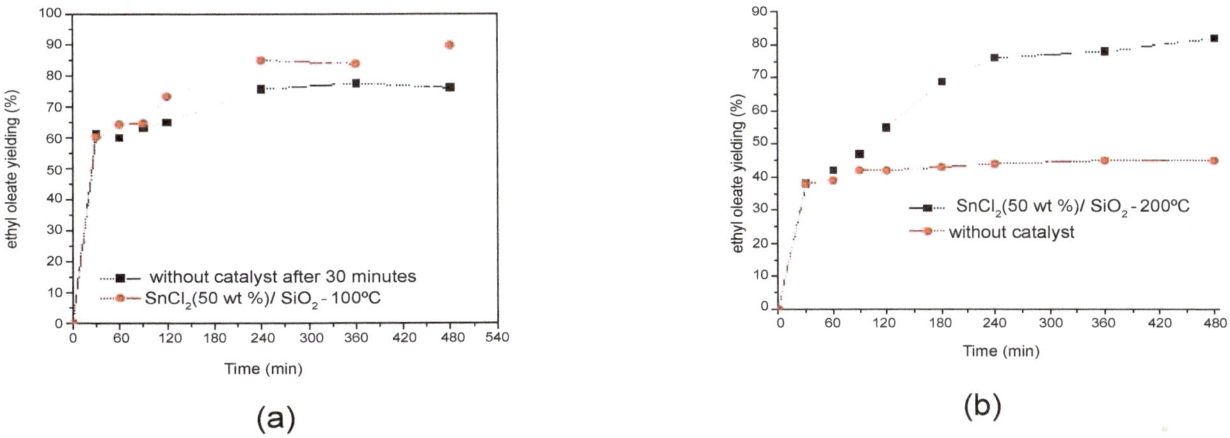

Figure 8: $SnCl_2/SiO_2$-catalyzed oleic acid esterification with ethyl alcohol: assessment of catalyst leaching[a]. [a]Reaction conditions: oleic acid (1.0 mmol); ethyl alcohol (155 mmol); catalyst (50 mg of $SnCl_2$ (50% wt.)/SiO_2); temperature (60°C).

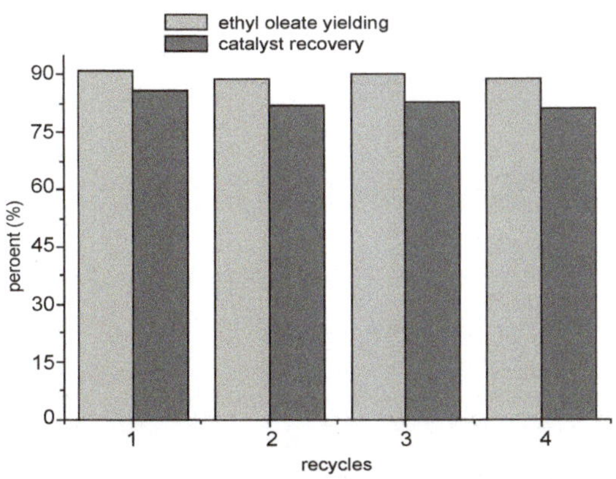

Figure 9: Recovery rates and ethyl oleate yielding of $SnCl_2$ (50% wt.)/ SiO_2 200°C catalyzed oleic acid esterification reactions[a]. [a]Reaction conditions: oleic acid (1.0 mmol); ethyl alcohol (155 mmol); catalyst (50 mg of $SnCl_2$ (50% wt.)/ SiO_2); temperature (60°C).

Cl bond are commonly observed between 325 and 384 cm^{-1} [25]. The characteristic bands of vibration of Sn-O are typically found at 500-523 cm^{-1} or 509 to 581 cm^{-1} wavenumbers [26,27]. However, these regions were overlapped by the support bands.

Determination of tin load in the solid supported catalysts

AAS analyses confirmed that the tin content in all samples synthesized, regardless support employed (Table 1).

Composite	Theoretical amount (wt. %)	Measured amount (wt. %)
$SnCl_2/SiO_2$	30.0	29.3
$SnCl_2/ZrO_2$	30.0	30.2
$SnCl_2/Nb_2O_5$	30.0	29.4

[a]Analysis conditions: See Materials and Methods (Determination of the Sn content)

Table 1: AAS analyses of samples of solid supported tin catalysts[a].

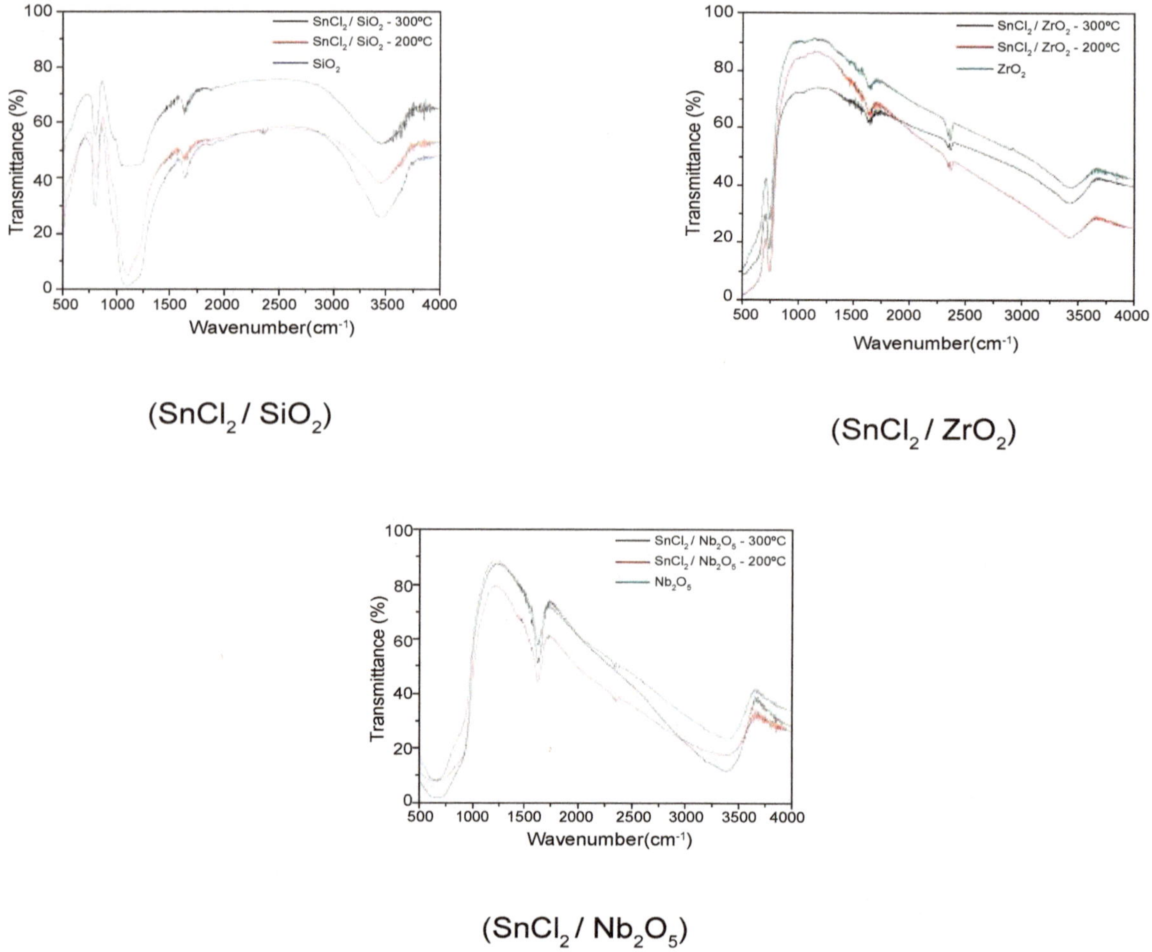

Figure 10: FT-IR spectra of solid supported tin catalysts.

Conclusion

The method used was efficient for the impregnation of $SnCl_2.2H_2O$ in different support. The thermal treatment had a crucial role on catalytic activity of supported tin composites. In general, high yielding were achieved when the catalyst was treated to 100°C or 200°C. However, in these cases, a leaching of active phase was responsible by the activity, regardless support employed. When the temperature of treatment was 300°C, the change on the coloring of support indicated that the tin was converted to SnO_2 or SnO. Previous tests had supported that the SnO and SnO_2 pure solids were less active catalysts in these reactions. Nonetheless, the results obtained in esterification reactions in heterogeneous phase were highly promising; in the case of $SnCl_2/SiO_2$-200°C; this was the most efficient and stable catalyst, and was recovered and reused without loss activity.

Acknowledgements

The authors thank to the CAPES, CNPq and FAPEMIG, Brazilian founding agency.

References

1. Santacesaria E, Martinez-Vicente G, Di Serio M, Tessera R (2012) Main technologies in biodiesel production: State of the art and future challenges. Catal Today 195: 2-13.
2. Leung DYC, Wu X, Leung MKH (2010) A review on biodiesel production using catalyzed transesterification. Appl Energy 87: 1083-1095.
3. Singh AP, Chouhan AKS (2011) Modern heterogeneous catalysts for biodiesel production: A comprehensive review. Renew Sust Energ Rev 15: 4378-4399.
4. Lopez DE, Goodwin JG, Bruce DA, Lotero E (2005) Transesterification of triacetin with methanol on solid acid and base catalysts. Appl Catal A 295: 97-105.
5. Semwal S, Arora AK, Badoni RP, Tuli DK (2011) Biodiesel production using heterogeneous catalysts. Bioresour Technol 102: 2151-2161.
6. Su F, Guo Y (2014) Advancements in solid acid catalysts for biodiesel production. Green Chem 16: 2934-2957.
7. Zheng S, Kates M, Dub MA, McLean DD (2006) Acid-catalyzed production of biodiesel from waste frying oil. Biomass Bioenergy 30: 267-272.
8. Di Serio M, Tesser R, Dimiccoli M, Cammarota F, Nastasi M, et al. (2005) Synthesis of biodiesel via homogeneous Lewis acid catalyst. J Mol Catal A 239: 111-115.
9. Yunuskhan TM, Atabani AE, Badruddin IA, Badarudin MSA, Khayoon M, et al. (2014) Recent scenario and technologies to utilize non-edible oils for biodiesel production. Renew Sust Energ Rev 37: 840-851.
10. Endalew AK, Kiros Y, Zanzi R (2011) Inorganic heterogeneous catalysts for biodiesel production from vegetable oils. Biomass Bioenergy 35: 3787-3809.
11. Yan S, Di Maggio C, Mohan S, Kim M, Salley SO, et al. (2010) Advancements in Heterogeneous Catalysis for Biodiesel Synthesis. Topic Catal 53: 721-736.
12. Shahid EM, Jamal Y (2011) Production of biodiesel: a technical review. Renew Sustain Energy Rev 15: 4732-4745.
13. Di Serio M, Tesser R, Casale L, D'Angelo A, Trifuoggi M, et al. (2010) Top Cata 53: 811.

14. Lee AF, Wilson K (2015) Recent developments in heterogeneous catalysis for the sustainable production of biodiesel. Catal Today 242: 3-18.

15. Silva MJ, Cardoso AL, Neves SCG (2009) Kinetic Study of Alcoholysis of the Fatty Acids Catalyzed by Tin Chloride(II): An Alternative Catalyst for Biodiesel Production. Energ Fuels 23: 1718-1722.

16. Da Silva ML, Figueiredo AP, Cardoso AL, Natalino R, da Silva MJ (2011) Effect of Water on the Ethanolysis of Waste Cooking Soybean Oil Using a Tin(II) Chloride Catalyst. J Am Oil Chem Soc 88: 1431-1437.

17. Silva MJ, Cardoso AL, Neves SCG (2008) Esterification of Oleic Acid for Biodiesel Production Catalyzed by SnCl2: A Kinetic Investigation. Energies 1: 79-92.

18. Silva MJ, Cardoso AL, Natalino R (2010) I J Chem React Eng 8: 1.

19. Goncalves CE, Laier LO, da Silva MJ (2011) Novel Esterification of Glycerol Catalysed by Tin Chloride (II): A Recyclable and Less Corrosive Process for Production of Bio-Additives. Catal Lett 141: 1111-1117.

20. Xie W, Wang H, Li H (2012) Silica-Supported Tin Oxides as Heterogeneous Acid Catalysts for Transesterification of Soybean Oil with Methanol. Ind Eng Chem Res 51: 225-231.

21. Pereira CO, Portilho MF, Henriques CA, Zotin FMZ (2014) J Braz Chem Soc 25: 2123.

22. Cardoso AL, da Silva MJ (2013) Novel and Highly Efficient SnBr2-Catalyzed Esterification Reactions of Fatty Acids: The Notable Anion Ligand Effect. Catal Lett 143: 1240-1246.

23. Peng G, Wang X, Chen X, Jiang Y, Mu X (2014) Zirconia-supported niobia catalyzed formation of propanol from 1, 2-propanediol via dehydration and consecutive hydrogen transfer. J Ind Eng Chem 20: 2641-2645.

24. Al-Gaashani R, Radimana S, Tabet N, Dauda AR (2012) Mat Science Engin B 177: 462.

25. Taimsalu P, Wood JL (1984) Spectrochmiea Acta 20: 1043.

26. Abu-Samn RH (1983) J Chem Soc Pak 5: 23.

27. Nath M, Pokharia S, Song X, Eng G, Gielen M, et al. (2003) Appl Organomet Chem 17: 305.

Microbial Fuel Cells: Influence of External Resistors on Power, Current and Power Density

Kamau JM*, Mbui DN, Mwaniki JM, Mwaura FB and Kamau GN

Department of Chemistry, University of Nairobi, PO Box 30197-00100, Nairobi, Kenya

Abstract

The effect of external resistance on voltage, current, power, power density and current density is investigated in a microbial fuel cells using cowdung is investigated. This involves use of varying resistance resistors. The fuel cells were operated under anaerobic condition for 9 days. PVC pipe was used to make a salt bridge using lamp wicks and potassium chloride. The obtained results indicated that the maximum voltage was on day 7 with 0.153 V across 33 kΩ resistor. The power was in the range of 0.000001 to 0.01 mW, current density was in the 0.1 to 23.29 mA/m^2 range while the power density was in 7.5 × 10^{-7} to 3.1036 mW/m^2 range.

Keywords: Microbial fuel cells; External resistors; Power; Power density

Introduction

Moon et al. [1] defines Microbial fuel cells (MFCs) as electrochemical bioreactors which converts chemical energy to electricity through biocatalysis of microbes. In a MFC, organic matter is oxidized at an anode by biological process where microorganisms deliver the electrons to the anodic electrode. These electrons flow through an external load and are released at the cathode where they are consumed to reduce an oxidant agent, usually oxygen. This biological system has the potential to oxidize a large variety of organic compounds producing at the same time electricity [2,3]. In literature the electricity generation using MFCs has been studied using pure organic compounds, such as acetate, butyrate, and glucose, sewage sludge and petroleum hydrocarbon [4-9]. The main advantage of electrogenic microorganisms is their ability to oxidize wastes with simultaneous energy generation [7].

Resistance in MFC

The electrical resistance of an object is a measure of its opposition to the passage of a steady electric current. For a uniform material of electrical resistivity ρ (Ωm) surface S (m^2) and distance L (m) it is given by the following equation:

$$R = \rho \frac{L}{S} \tag{1}$$

Typical values of the electrical resistivity ρ for common materials at 20°C range from 1.59 × 10^{-8} Ω m for silver to 7.5 × 10^{17} Ω m for quartz and even more for engineered materials like polytetrafluoroethylene (PTFE). A Fuel Cell is not meant to have an electric current passing through it but to produce one. Its electrical resistance is not defined. Nevertheless internal resistance is a concept that helps to model the electrical consequences of the processes happening inside it.

Internal resistance of an MFC

When a cell delivers a current, the measured voltage output is lower than when there is no current delivered. This is because when electrons flow, they have to face the resistivity of the materials composing the fuel cell. The internal resistance of an MFC can be distributed into anodic resistance, cathodic resistance, and electrolyte (including the membrane if present) resistance [10,11]. In an MFC system, where electrochemical reactions are under proton diffusion control we will see that the electrolyte resistance can be assimilated to the Warburg impedance [12]. since it requires a current to be observed, the internal resistance of a battery cannot be measured using a conventional ohmmeter. Other ways have to be used to determine it. According to

$$E_{cell} = E_{emf} - R_{int} I_{cell} \tag{2}$$

The slope of the linear section of the polarization curve represents the internal resistance of an MFC. MFC generates its maximum power (P_{max}, W) when R_{int}=R_{ext} [13], where R_{int} can be determined as:

$$R_{int} = (E_{emf} - E_{max}) / I_{max} \tag{3}$$

Where E_{max} (V) and I_{max} (A) are the cell voltage and current that give the maximum power.

At the same time, following Ohm's law

$$R_{ext} = \frac{E_{max}}{I_{max}} \tag{4}$$

Hence, when R_{int}=R_{ext},

$$E_{max} = \frac{E_{emf}}{2} \tag{5}$$

A schematic representation of MFC is shown in Figure 1.

Materials and Methods

A double chamber MFC was fabricated using inexpensive materials. They include 16.3 cm to 15.3 cm diameter and 7.4 cm to 9.4 cm long plastic containers, PVC pipes, adhesive glue, scissors, wicks, driller, masking tape and pipes joiners. The following materials were also used copper wires, graphite rod electrodes, 1 Ω-33 kΩ resistors, and digital multi-meters. The substrate used in this work is fresh cow dung with sucrose as bacteria food.

***Corresponding author:** Kamau JM, Department of Chemistry, University of Nairobi, PO Box 30197-00100, Nairobi, Kenya
E-mail: djames085@gmail.com

MFC construction

Two 1.2 L containers were prepared as an anode chamber and a cathode chamber. Two small holes were made in the caps of the bottles as to insert wire through. One end of the copper wire was attached to 5.7 cm long and 0.7 cm diameter graphite rod electrodes. The anode chamber was filled with the substrate (0.1 kg of cow dung mixed with 1 L of water) while the cathode chamber was filled with plain water of 1 L. Other two ends of wires were attached to digital multimeter for open circuit voltages and different 1 to 33 kΩ resistors.

Salt bridge

Salt bridge was prepared using 1.5 liter of 1M KCl solution and lamp wicks. The wicks were boiled in the KCl solution for 10 minutes after which it was kept in the freezer at -4°C for solidification. The solidified salt bridge was passed through PVC pipes and attached to the chambers using araldite adhesive which makes them leak proof.

Circuit assembly

The assembly of the H-shaped MFC was done as shown in Figure 2.

Experimental variations

To characterize voltage, current, power, current and power density across different resistors, 600 ml of cow-dung was added to the anodic chamber and 25 ml of sugar solution added via a pipette. 1 Ω, 1 kΩ, 2 kΩ, 3.3 kΩ, 5.1 kΩ, 10 kΩ, 15 kΩ, 20 kΩ, 33 kΩ resistors were attached to the terminals from cathodic and anodic chamber. The amount of voltage and current from the cells were recorded daily for 9 days across the attached resistors.

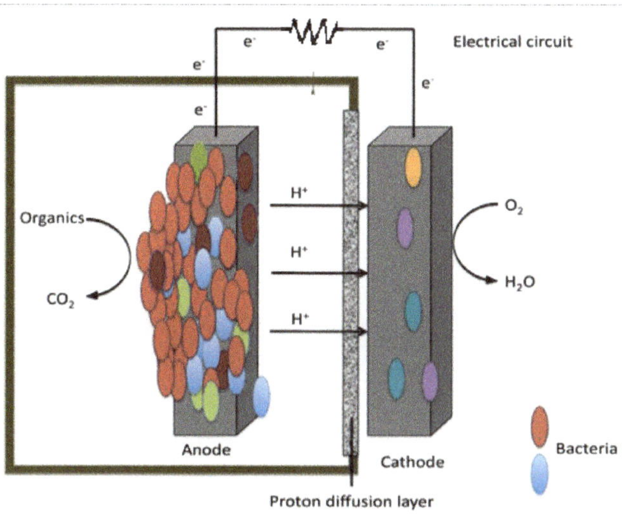

Figure 1: Microbial fuel cell [10].

Figure 2: Set-up of H-shaped microbial fuel cells.

Data collection and observation

The generated voltage and current was recorded from the digital multimeter at an interval of 24 hour for 9 days. The corresponding power, current density and power density was calculated by using formula

$$P = VI \tag{6}$$

$$Power\ Density = \frac{Power}{A_{area}} \tag{7}$$

$$Power\ Density = \frac{Power}{A_{area}} \tag{8}$$

Results and Discussion

Some resistance must be overcome if MFC technology is to be used in highly powered devices. In the current study resistors of varying resistances were used. The current and voltage across these resistors were monitored using a digital voltmeter. The obtained results are shown in Figure 3. From the graph, the maximum voltage was on day 7 with 0.153 V across 33 kΩ resistor. The power was in the range of 0.000001 to 0.01 mW, current density was in the 0.1 to 23.29 mA/m^2 range while the power density was in 0.1 to 3.26 mW/m^2 range. They were all dependent on the resistors resistance as shown in Figure 4. The high values across 33 kΩ are attributed to the high amount of work to be done to overcome the high resistance.

Previous studies by Menicucci et al. [14] using 6 to 0.125 KΩ revealed that, the cell voltage decreases when external resistance decreases. This is explained by the limitations imposed on the electrode reaction kinetics, on mass transfer, and on charge-transfer processes at the current-limiting electrode. In another study, Ghangrekar and Shinde [15] observed that cell voltage increased with an increase in external resistance from 0 to 4,000 Ω; the maximum voltage of 358 mV was observed at an external resistance of 4,000 Ω. Later on, Rismani-Yazdi et al. [16] obtained similar cathode potentials at different external resistances. However, anode potential varied under different external resistance employed. MFCs with lower external resistances resulted in higher anode potentials. This was also observed in the study of Song et al. [17] carried out using a sediment microbial fuel cell (SMFC).

High electron transfer from the anodic to the cathodic chamber leads to high current. In this study, when the external resistance was increased from 1 kΩ to 15 kΩ, the current decreased from 0.0024 mA to 0.0020 mA in day 6. Similar trends were observed for other days as shown in Figure 3. This is in agreement with previous studies [16-20].

A stepwise descent in external resistance have been observed to improve current generation over a given time. For instance, Aelterman et al. [21] observed that decreasing the resistance from 50 to 10.5 Ω resulted in significant upward trend of current generation. They observed that operating a MFC at lower resistances, the mass transfer or kinetic limitations during polarization lowered resulted in less steep descent of current density. Similar to previous studies, Zhang et al. [22] observed that a conventional MFC with a lower external resistance showed a higher current generation after start-up. The study further observed that when operating at high external resistance that almost

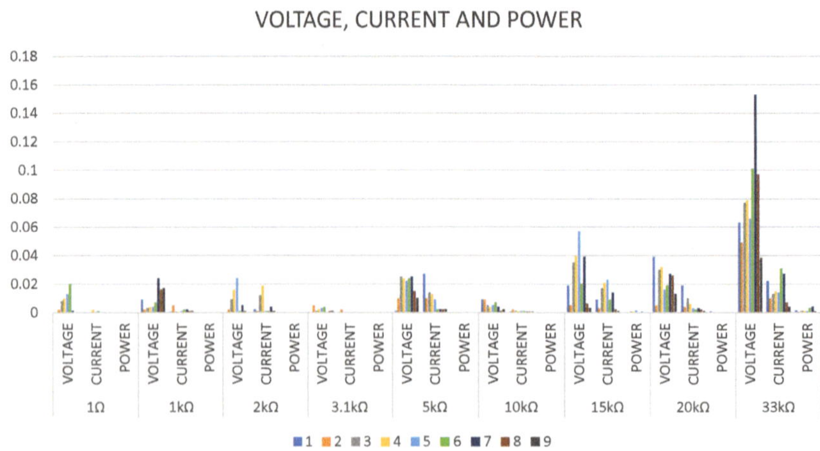

Figure 3: Graphs of voltage, current and power generated against days for 500 g cowdung across different resistors.

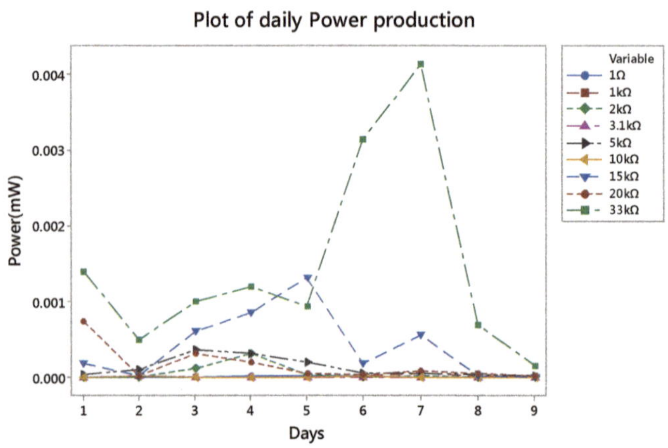

Figure 4: Plot of power against days.

mimics an open circuit, microorganisms were unable to transfer their electrons to such an unfavorable electron acceptor. In this situation, the high external resistance restricts the current that is able to flow from anode to cathode, and this may affect which microorganisms are able to colonize the anode [23,24]. In the study of Ghangrekar and Shinde [15], it was demonstrated that when external resistance is very high, the current was brought to minimum and nearly constant value. Moreover, in this case, the limiting factor is the external resistance, and the current production is almost independent of another factor, such as the distance between the electrodes and surface of the anode.

The power resulting from use of 1 Ω for day 1 to 9 was in the range of 1.02×10^{-9} mW to 2.0×10^{-5} mW. Increasing the external resistance from 1 kΩ to 33 kΩ translated to power increment of 4.5×10^{-6} mW to 1.386×10^{-3} mW in day 1 as shown in Figures 3 and 5. Similar trend was witnessed in the other days. Earlier studies on influence of external resistance by Gonza´lez et al. [25] indicates that decreasing the external resistance from 2.7 kΩ to 2.2 kΩ result in power decrease from 1.69×10^{-3} mW to 1.27×10^{-3} mW. It was observed in this study that the power generated increased from 0.00002 mW to 0.003131 mW when the external resistance was varied from 1 Ω to 33 k Ω on day 6. A clear illustration of the observation is shown in Figure 5.

The potential of the cell adjusts to the external resistance. The maximum power transfer theorem states that maximum power is drawn when the external resistance (R_{ext}) of electric power source equals internal resistance of power sources. This means that the power generated by MFC is highly dependent on the current.

$$I = \frac{V}{R_{int} + R_{ext}} \tag{9}$$

And hence,

$$P = I^2.R_{ext} = \frac{V^2}{\left(R_{int} + R_{ext}\right)^2}.R_{ext} = \frac{V^2}{\frac{R_{int}^2}{R_{ext}} + 2R_{int} + R_{ext}}.R_{ext} \tag{10}$$

P is maximum when the denominator is minimum. Differentiating the denominator with respect to R_{ext},

$$\frac{d\left(\frac{R_{int}^2}{R_{ext}} + 2R_{int} + R_{ext}\right)}{dR_{ext}} = -\frac{R_{int}^2}{R_{ext}} + 1 \tag{11}$$

For maximum or minimum, the first derivative is zero, so

$$-\frac{R_{int}^2}{R_{ext}^2} = 1$$

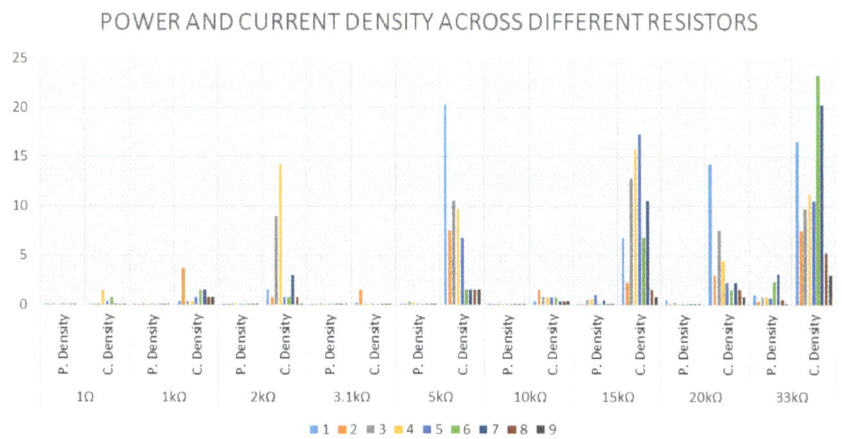

Figure 5: Graph of power and current density against days for 500 g cowdung across different resistors.

Therefore, the power is maximum when $R_{ext}=R_{int}$.

If external resistance is higher or lower than internal resistance, generated power will decrease. Many research have be carried out to investigate the applicability of maximum power transfer theorem [26-32]. Figure 4 shows the power density and the current density obtained in this work. It is clear from the figure that increasing the external resistance from 1 Ω to 33 kΩ resulted in power density increase of 23.2757 mW m^{-2} in day 6 with similar trend observed for the other days. The power density increases from day 1 through to 6 and then decreases. Similar observations were made by Lyon et al. [28], who observed that at an internal resistance of 300 Ω, high power production was obtained with an external resistance of 470 Ω, followed by 1,000 Ω, 100 Ω, 10 KΩ, and finally 10 Ω. External resistance of 10 Ω produced the weakest power production as the resistance was too low compared to the internal resistance (300 Ω). In other studies, Ren et al. [28], showed that higher power density values were observed when external resistances were changed from high to low (from 5,000 to 10 Ω) compared to low to high (from 10 to 5,000 Ω). In the study of Zhang et al. [22], the power obtained after start-up increased from 1.96 to 6.05 mW when external resistance increased from 10 to 50 Ω. Then, when external resistance increased from 50 to 1,000 Ω, the power decreased to 0.64 mW.

The theorem is not always true as observed by Ref. [17,33]. Song et al. [17] observed that power density increased with the increase in external resistance and the highest power density of 3.15 mW/m^{-2} was obtained at an external resistance of 1,000 Ω. However, the internal resistance varied between 132 and 214 Ω. This fluctuation could be explained because internal resistance is not a system constant and depends on the external resistance applied to the MFC. In the study of Lee et al. [33], MFC with highest external resistance had better power densities. In the current study, the same trend was also observed in some cases. For example, the highest power densities were recorded for 33 kΩ external resistor which ranged from 0.1142 to 3.103 mW/m^2. Figure 6 shows the relationship between the power density and the current density across different resistors. As shown Figure 6, the power density increases with increase in current density for the first 6 days, reaching maximum at day 6 or 7 then decrease to day 9. Similar trend was observed by Zhang [33] in 2011.

Conclusion

The external resistance directly influences on anode potential and current, and they influence on other variables in a MFC, such as microbial diversity, biofilm morphology, power generated, coulombic efficiency, and MFC stability among other variables. In this way, the selection of the optimal external resistance in order to get the best performance in the MFC is very important.

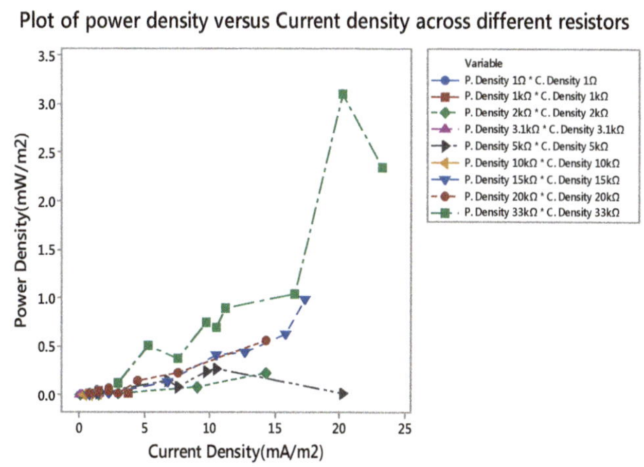

Figure 6: Graph of power density versus current density across different resistors.

References

1. Moon H, Chang IS, Kim BH (2006) Continuous electricity production from artificial wastewater using a mediator-less microbial fuel cell. Bioresource Technol 97: 621-627.
2. Logan BE (2008) Microbial fuel cells. Wiley, New Jersey, USA.
3. Debabov VG (2008) Electricity from microorganisms. Mikrobiologiia 77: 129-157.
4. Chaudhuri SK, Lovley DR (2003) Electricity generation by direct oxidation of glucose in mediatorless microbial fuel cells. Nat Biotechnol 21: 1229-1232.
5. Min B, Logan BE (2004) Continuous electricity generation from domestic wastewater and organic substrates in a flat plate microbial fuel cell. Environ Sci Technol 38: 5809-5814.
6. Liu H, Cheng S, Logan BE (2005) Production of electricity from acetate or butyrate using a single-chamber microbial fuel cell. Environ Sci Technol 39: 658-662.
7. Park DH, Zeikus JG (2003) Improved fuel cell and electrode designs for producing electricity from microbial degradation. Biotechnol Bioeng 81: 348-355.

8. Rabaey K, Lissens G, Siciliano SD, Verstraete W (2003) A microbial fuel cell capable of converting glucose to electricity at high rate and efficiency. Biotechnol Lett 25: 1531-1535.

9. Cheng S, Liu H (2006) Power densities using different cathode catalysts (Pt and CoTMPP) and polymer binders (Nafion and PTFE) in single chamber microbial fuel cells. Environmental Science & Technology 40: 364-369.

10. Lovley DR (2006) Bug juice: harvesting electricity with microorganisms. Nat Rev Microbiol 4: 497-508.

11. Fan YZ, Sharbrough E (2008) Quantification of the Internal Resistance Distribution of Microbial Fuel Cells. Environmental Science & Technology 42: 8101-8107.

12. Muralidharan VS (1997) Warburg impedance - Basics revisited. Anti-Corrosion Methods and Materials 44: 26-46.

13. Hoboken NJ (2005) Impedance spectroscopy: Theory, experiment, and applications. New York, USA.

14. Menicucci J, Beyenal H, Marsili E, Veluchamy RA, Demir G, et al. (2006) Procedure for determining maximum sustainable power generated by microbial fuel cells. Environ Sci Technol 40: 1062-1068.

15. Ghangrekar MM, Shinde VB (2007) Performance of membrane-less microbial fuel cell treating wastewater and effect of electrode distance and area on electricity production. Bioresour Technol 98: 2879-2885.

16. Rismani-Yazdi H, Christy AD, Carver SM, Yu Z, Dehority BA, et al. (2011) Effect of external resistance on bacterial diversity and metabolism in cellulose-fed microbial fuel cells. Bioresour Technol 102: 278-283.

17. Song TS, Yan ZS, Zhao ZW, Jiang HL (2010) Removal of organic matter in freshwater sediment by microbial fuel cells at various external resistances. J Chem Technol Biotechnol 85: 1489-1493.

18. Chae KJ, Choi MJ, Kim KY, Ajayi FF, Park W, et al. (2010) Methanogenesis control by employing various environmental stress conditions in two-chambered microbial fuel cells. Bioresour Technol 101: 5350-5357.

19. Jadhav GS, Ghangrekar MM (2009) Performance of microbial fuel cell subjected to variation in pH, temperature, external load and substrate concentration. Bioresour Technol 100: 717-723.

20. Jang JK, Pham TH, Chang IS, Kang KH, Moon H, et al. (2004) Construction and operation of a novel mediator and membrane-less microbial fuel cell. Process Biochem 39: 1007-1012.

21. Aelterman P, Versichele M, Marzorati M, Boon N, Verstraete W (2008) Loading rate and external resistance control the electricity generation of microbial fuel cells with different threedimensional anodes. Bioresour Technol 99: 8895-8902.

22. Zhang L, Zhu X, Li J, Qiang L, Ye D (2011) Biofilm formation and electricity generation of a microbial fuel cell started up under different external resistances. J Power Sources 196: 6029-6035.

23. Lyon DY, Buret F, Vogel TM, Monier JM (2010) Is resistance futile? Changing external resistance does not improve microbial fuel cell performance. Bioelectrochemistry 78: 2-7.

24. Ren Z, Yan H, Wang W, Mench MM, Regan JM (2011) Characterization of microbial fuel cells at microbially and electrochemically meaningful time scales. Environ Sci Technol 45: 2435-2441.

25. del Campo AG, Cañizares P, Lobato J, Rodrigo M, Morales FF (2014) Effects of External Resistance on Microbial Fuel Cell's Performance. In: Environment, Energy and Climate Change, pp: 175-197.

26. Aelterman P, Versichele M, Marzorati M, Boon N, Verstraete W (2008) Loading rate and external resistance control the electricity generation of microbial fuel cells with different three dimensional anodes. Bioresour Technol 99: 8895-8902.

27. Ren Z, Yan H, Wang W, Mench MM, Regan JM (2011) Characterization of microbial fuel cells at microbially and electrochemically meaningful time scales. Environ Sci Technol 45: 2435-2441.

28. Katuri KP, Scott K, Head IM, Picioreanu C, Curtis TP (2011) Microbial fuel cells meet with external resistance. Bioresour Technol 102: 2758-2766.

29. Logan BE, Regan JM (2006) Electricity-producing bacterial communities in microbial fuel cells. Trends Microbiol 14: 512-518.

30. Manohar AK, Bretschger O, Nealson KH, Mansfeld F (2008) The use of electrochemical impedance spectroscopy (EIS) in the evaluation of the electrochemical properties of a microbial fuel cell. Bioelectrochemistry 72: 149-154.

31. Jung S, Regan JM (2011) Influence of external resistance on electrogenesis, methanogenesis, and anode prokaryotic communities in microbial fuel cells. Appl Environ Microbiol 77: 564-571.

32. Lee CY, Chen JH, Cai YY (2010) Bioelectricity generation and organic removal in microbial fuel cells used for treatment of wastewater from fish-market. J Environ Eng Manage 20: 173-180.

33. Zhang Y, Mo G, Li X, Zhang W, Zhang J, et al. (2011) A graphene modified anode to improve the performance of microbial fuel cells. J Power Sources 196: 5402-5407.

Deactivation of Mesoporous Titanosilicate-supported WO_3 Catalyst for Metathesis of Butene to Propene by Coke

Derun Hua[1]*, Zheng Zhou[2], Jian Li[1], Yongrong Xie[1] and Xinning Lu[1]

[1]*Chemical Institute of Chemical Industry, Gannan Normal University, Jingxi, 341000, China*
[2]*Natural Gas Processing Plant of Sinopec Zhongyuan Oilfield Branch, Henan, 457000, China*

Abstract

Catalyst deactivation by coke was studied during metathesis of butene to propene on the WO_3/MTS-9 catalyst. Characterization results of the coked catalyst showed that the pore volume, pore size, and specific surface area of the catalyst decreased significantly for coke; the coke amount in larger pore of catalyst was higher than that in smaller pores of the catalyst. Coke resulted in the decrease of the acidic amount and the activity of catalyst. The component of carbon deposits was aromatic-olefin compounds.

Keywords: Deactivation; Metathesis; Coke; MTS-9

Introduction

Olefin metathesis reported in the 1960s by Banks is an organic reaction that entails the redistribution of fragments of alkenes (olefins) by the scission and regeneration of carbon-carbon double bonds [1]. Olefin metathesis has been used as an important way for olefin conversion and offers a path to convert low value olefins into high-value olefins, such as propene, ethene [2-5], short-chain olefin to long-chain olefin [6,7], and butene to propene [8,9]. Supported WO_3 catalysts are known to catalyze the reaction [8-11] and are currently applied industrially for heterogeneous catalysis metathesis reactions. Now, many studies have been carried out to improve performance of supported WO_3 catalysts [12,13]. Many factors, such as the method of catalyst preparation, reaction conditions, structure of the support and so on, affect catalytic activity. Debecker et al. [14] and Debecker et al. [15] prepared WO_3-based metathesis catalyst with high catalytic performance and shorter induction period by the one-step Sol-gel process. One-pot sol-gel catalysts perform better than corresponding impregnated catalysts, which were attributed to fact that the sol-gel method can control well on the final composition and textures of catalyst. Piyasan et al. [16] found that WO_3/SiO_2 catalysts treated by NaOH held a shorter induction period and low coke at the same metathesis activity. Piyasan et al. [17] suggested that silica-supported tungsten catalysts mixed with micro- and nano-sized SiO_2 excess support favored to form isolated surface tetrahedral tungsten oxide species. Phatanasri et al. [18] found that the calcination temperature markedly affected physico-chemical properties of the catalysts, particularly the active species of WO_3/SiO_2 catalysts. Khattaf et al. [19] investigated MCM-41- and SBA-15-supported tungsten catalyst. Supported oxide catalysts and mesoporous catalysts have been reviewed recently [20,21].

As heterogeneous catalysis, catalysts deactivation often occurs during olefin-metathesis reactions due to carbonaceous species formed on catalyst surface. The carbonaceous species deposit on the surface of catalyst, and results in deactivation of the catalyst. The mechanism for catalyst deactivation is complex. The causes of deactivation are various, such as poisoning, valence change of active sites, and active-site coverage [22]. For different types of catalysts employed to olefin-metathesis reactions, their deactivation reasons are different. For Re-supported catalysts, deactivation results from poisoning, and poison are adsorbed strongly on the surface of catalyst and hinder reactant to access the active sites [23]. For Mo-supported catalyst, the deep-reduction of Mo species may be the main reason for its deactivation during metathesis [24]. For the tungsten catalytic system, high reaction temperatures (300–500°C) have to be used in order to achieve high metathesis activity and avoid poisoning [25], but coke is easy to form at a high temperature and high acid, and deposits on the surface of catalyst [16,26]. In the last few years, a large amount of investigations have been undertaken to study coke deposits (location and total amount) and to obtain information regarding the deactivation mechanism and regeneration. Moodley [27] investigated the effect of the coke amount on catalytic activity and the coking location, and found that 2-Pentanone was the most effective inhibitor for coking, and the location of coke deposition is a more important factor than the coke amount. In the case of alkene metathesis, deep-reduction may result in deactivation [28]. Besides, it was found that water and polar solvents are useful to coke retardation [25,29] at a high temperature. Unfortunately, the coking location and the structure of carbon species are not clear.

In this study, the optimized 8.0 wt.% W supported on MTS-9 was used as the catalyst for metathesis of butene to propene. The deactivated catalyst was characterized, and regenerated by treating at 500°C in atmosphere. N_2 adsorption-desorption was used to detect pore size and pore volume of the catalyst before and after reaction. ^{13}C-NMR and UV-Raman are used to investigate the nature of the carbonaceous species. STEM (scanning transmission electron microscopy) and EFTEM (energy filtered transmission electron microscopy) are used to monitor the coked location.

Experimental

Preparation of catalyst

MTS-9 was hydrothermally synthesized in a two-step procedure as given in the literature. First, precursors containing titanosilicate (TS-1) nanoclusters were prepared. Second, the preformed precursors were assembled with triblock copolymers in a strong acid medium

***Corresponding author:** Derun Hua, Chemical Institute of Chemical Industry, Gannan Normal University, Jingxi, 341000, China
E-mail: dr_hua2011@sina.com

(pH<1) [30,31]. WO_3/MTS-9 catalysts with tungsten loading (8.0 wt.%) was prepared by a wet impregnation method with water solutions of ammonium metatungstate. The impregnated product was dried at 80°C for 12 h; then the catalyst was thermally treated at 550°C for 4 h in a temperature programmed furnace (1 K/min) [32]. Texture of support and catalyst was listed in Table 1.

Characterization of catalyst

N_2 adsorption–desorption isotherms at -196°C were recorded with a Micromeritics ASAP 2010 automatic sorption analyzer. The BET surface areas were calculated from the desorption isotherms in the relative pressure range of P/P_0=0.01~1.0. UV-Raman spectra were obtained on a HR800 UV-Raman Spectrograph (Horiba Jobin Yvon Company. France). The 44.0 nm line from a He–Cd was used as the excitation source, and the spectrum resolution was estimated to be 4.0 cm^{-1}. STEM and EFTEM images were obtained with a Tecnai G2 F20 Super-twin (FEI Company). A ^{13}C-NMR study was conducted on a Bruker Avance DSX400 NMR spectrometer at a frequency of 100.6 MHz. The peaks were referenced to tetramethylsilane (TMS) as an external standard. The spinning rate was 3-4 kHz. Temperature programmed desorption (TPD) was conducted by ramping to 500°C at 10°C/min and NH_3 (m/e=16) in the effluent was detected and recorded as a function of temperature by a thermal conductivity detector (TCD) (BJGC, GSD301, China). All products were analyzed using a gas chromatograph equipped with a flame ionization detector (FID) and a 50 m PONA capillary column. The acid sites amount and strength distribution of samples were determined by a temperature-programmed desorption of ammonia (NH_3-TPD). The ammonia in the effluent gas was detected by a TCD detector, and then adsorbed with 0.02 mol/L HCl solution. The total acidic amount was determined by back-titrating the HCl solution using 0.01 mol/L NaOH solution, and bromocresol green/methyl red as indicator.

Reaction

The metathesis reaction of butene (1-butene and 2-butene) was carried out in a downflow fixed bed stainless steel reactor (i.d., 10 mm) with 0.375 g catalyst. The C_4 stream, which are from MTBE chemistry in Shanghai and compose of 2-butene (ca. 78%, wt/wt) and butane (18%), was treated through BASF adsorbents Selexsorb CD and COS to remove oxygenated organic compounds (methanol, MTBE) and other trace contaminants, which can cause catalyst deactivation. Before the introduction of C_4 stream into the reactor, the catalyst was pretreated in situ with a N_2 and H_2 (2%, vol/vol) mixture flow for 30 min at 420°C and atmospheric pressure. The weight hour space velocity (WHSV) of the metathesis reaction was 3.2 h^{-1}.

Results and Discussion

After 320 h, the conversion of butene decreases from ca. 50% to ca. 25% (see Figure 1). The catalyst was discharged, then characterized to study the mechanism of coking and the deactivation reason. The N_2 adsorption-desorption isotherm of the fresh and spent catalysts is compared (see Figure 2a). It is found that the spent catalyst had a smaller pore volume than the fresh catalyst, suggesting that pore blockage occurs for carbon deposition. The texture of the catalyst is listed in Table 1. The distribution of the pore size of the spent catalyst is presented in Figure 2b. The Figure 2b shows that carbonaceous deposit filled the pores of the catalyst. The carbon deposit amount in the larger pores is higher than that in the smaller pores. Coking depends mainly on feedstock, temperature, catalyst acidity and pore size [33]. At the same conditions, carbon deposit depends on the texture of the catalyst. The available space was larger. The coking amount was greater.

The scanning transmission electron microscopy (STEM) images of the spent and fresh catalyst are shown in Figure 3. The lightest regions (dashed circle) show the crystallites of tungsten oxide on the silica support. The lighter regions (dashed rectangle) show the surface tungsten interacting with support. The darkest regions (solid rectangle) show the support. The pore fingerprint of the fresh catalyst is clear in Figure 3A, but that of the spent catalyst is not found in Figure 3B.

Sample	Pore diameter (nm)	Tatol pore volume (cm^3/g)	Micropore volume (cm^3/g)	Mesopore surface area (m^2/g)
MTS-9	6.75	0.987	0.048	489.1
Catalyst	6.346	0.7125	0.0205	436.5
Spent catalyst	5.84	0.2944	0.002508	201.6

Table 1: Textural features of support and catalyst.

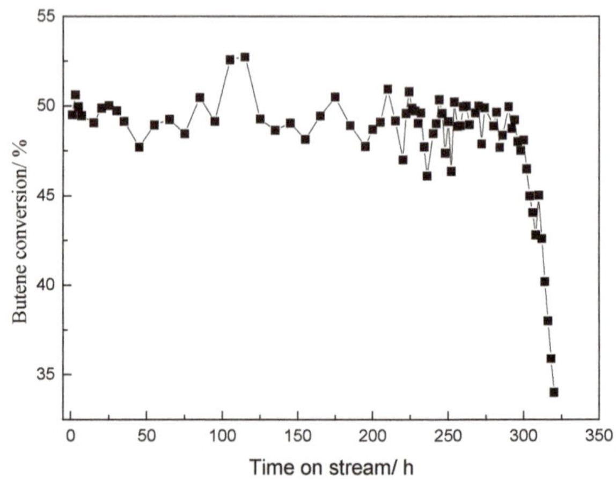

Figure 1: Function of butene conversion with time.

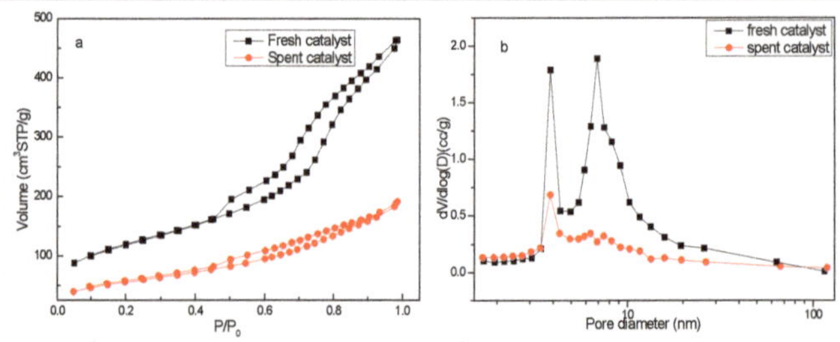

Figure 2: N_2-adsorption isotherm and the distribution of the pore size of spent and fresh catalyst.

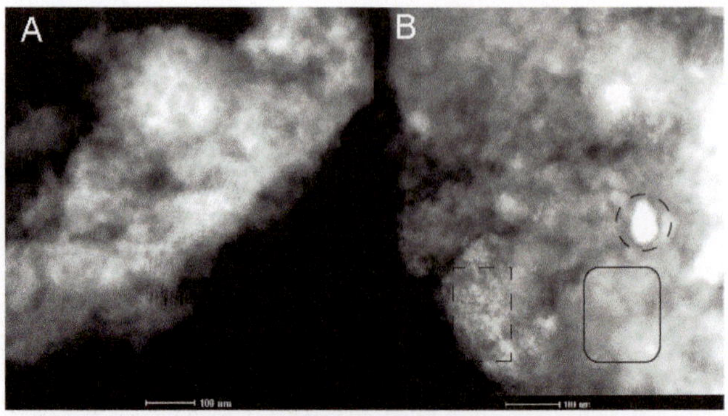

Figure 3: STEM micrograph of the fresh (A) and spent (B) catalysts.

The EFTEM technique is used to acquire the location of coke on the WO_3/MTS-9 catalyst. The carbon map of the catalyst is shown in Figure 4A; the tungsten map of the catalyst is shown in Figure 4B. In Figure 4C, red regions are carbon maps, and the blue regions are tungsten maps. The carbon maps in Figure 4C show that all the carbon (red region) surrounds the tungsten oxide (blue region in the centre), that is, carbon deposition occurs around the tungsten, and tungsten is not covered by carbon. Effect of coke on the acidic amount of catalysts is listed in Table 2. A comparison of the acidic amount of catalysts, it is found that the acidic amount of the fresh catalyst is more than that of support, which indicates that the formation of new acidic sites is resulted from the interaction of tungsten and support. However, the acidic amount of the spent catalyst is less than that of the fresh catalyst, even is less than that of support, which indicates the acidic sites are covered by coke. Namely, the coking location is at the acidic sites.

We attempt to study the composition of coke by ^{13}C-NMR (Figure 5). As shown in Figure 5, the resonance line of the spent catalyst is wide, indicating that condensed carbonaceous compounds are present [34]. Besides the above information, the ^{13}C-NMR spectrum of spent catalyst exhibits two distinct groups of resonance lines, which are assigned to aliphatic (0-50 ppm) and aromatic (100-150 ppm) carbons. The resonance peak at 25 ppm is attributed to methyl group attached to aromatic ring [35]. The resonance peak at 47 ppm is assigned to methyl-substituted cyclopentenyl carbocation, which was reported as intermediates in the formation of aromatic hydrocarbon [36]. The resonance peak at 127 ppm corresponds to aromatic C-H bonds or olefin [37,38]. Since the operation temperature is above 350°C, coke components are polyaromatic regardless of other conditions [33].

Figure 6 shows the ultraviolet Raman spectrum of the fresh and spent catalysts. For the fresh catalyst, Raman spectrum consisted mainly of bands at 995 and 1087 cm^{-1}. Raman bands of 1087 and 995 cm^{-1} are the characteristic of surface tungsten oxide species (active species). Bands at 1087 and 995 cm^{-1} corresponded to the stretching vibration mode of the terminal bond of isolated surface tungsten oxide species and surface poly tungsten, respectively [39,40]. On the spent catalyst, the strong bands of 1396 and 1610 cm^{-1} were observed. It was interesting that bands of 1087 and 995 cm^{-1} disappear or become weak. This is because the bands due to carbon are more intense and flatten Raman bands at 1087 and 995 cm^{-1}.

Raman bands assignments for vibrations in the region from 1000 to 1650 cm^{-1} are listed in Table 3 [41]. Compared Table 2 with the UV-Raman spectra of the spent catalyst in Figure 6, bands at 1396 and 1610 cm^{-1} correspond to the ring stretches of poly aromatic species. Based on the results of ^{13}C-NMR (Figure 5) and UV-Raman spectra (Figure 6) of the spent catalyst, it is definite that the component of coke is poly aromatic species.

Catalysts	WO_3	MTS-9	Fresh catalyst	Spent catalyst
Acidic amount (mmol/g)	0	0.188	0.261	0.157

Table 2: The acidic amount of catalysts.

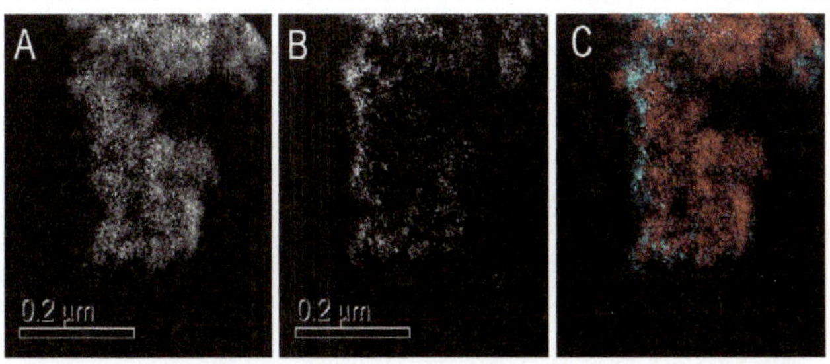

Figure 4: EFTEM of spent WO_3/MTS-9 (A: carbon map, B: tungsten map, C: Contrast situation of carbon and tungsten).

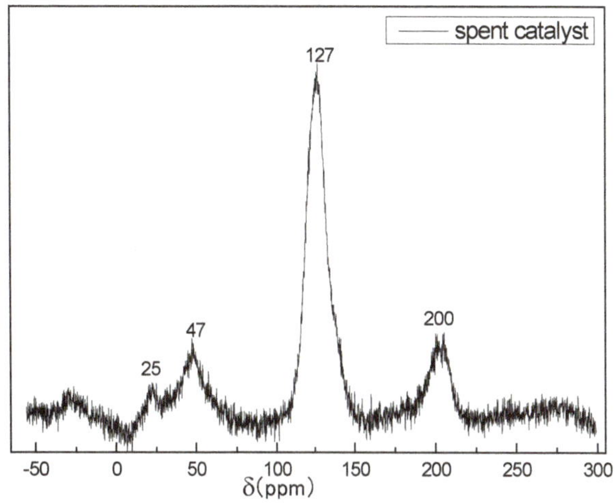

Figure 5: ^{13}C-NMR of the spent catalyst.

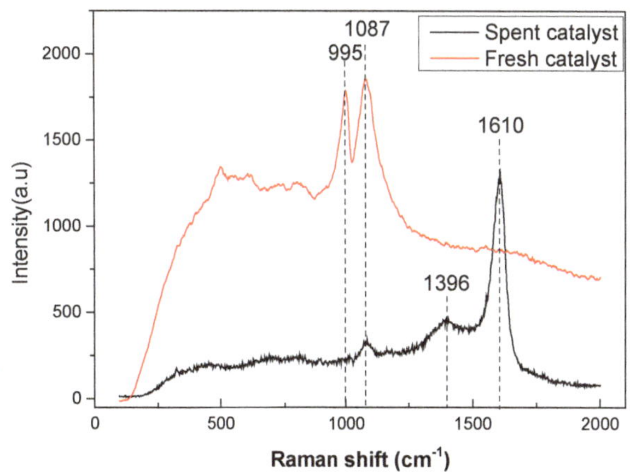

Figure 6: Raman spectra of the fresh and spent catalysts.

The UV-Raman spectrum of the spent catalyst is compared with the UV-Raman spectra of a series of polyaromatic compounds, such as naphthalene, phenanthrene, anthracene, fluorene, pyrene, pentacene, and coronene. UV-Raman measurements of these compounds have been reported previously by Chua and Stair [41,42], and Johnson [29]. By contrast, the spectrum of carbon deposits on WO_3/MTS-9 is similar to that of the chain-like polyaromatic compounds, such as naphthalene and pentacene.

The ^{13}C-NMR and UV-Raman spectrum provides some evidence for coke configuration. The formation of coke molecules includes condensation, hydrogen transfer and dehydrogenation processes. So the following reaction scheme (Figure 7) can be proposed to explain the formation of coke molecules. Benzenic compounds are formed through the scheme of aromatization on the WO_3/MTS-9 catalyst.

The deactivated catalyst was regenerated at 500°C under air flow, and the performance of the fresh and regenerated catalyst was compared, as shown in Figure 8a. It can be seen that the activity of the regenerated catalyst is comparable to that of the fresh one, suggesting that the regeneration of the deactivated catalyst is feasible. Figure 8b shows Raman spectra of the regenerated and fresh catalyst.

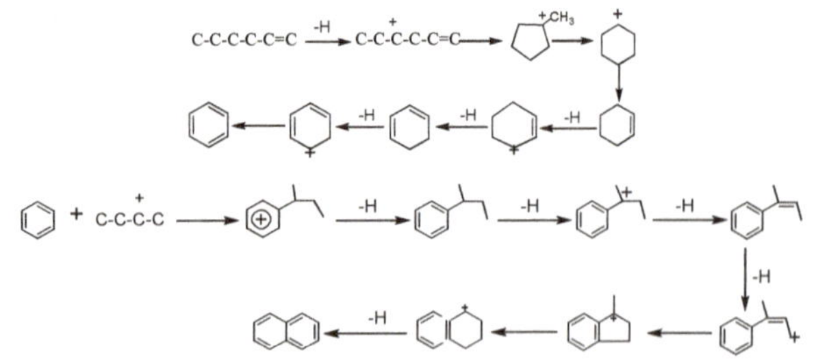

Figure 7: Formation of carbonaceous compounds on WO_3/MTS-9 catalyst.

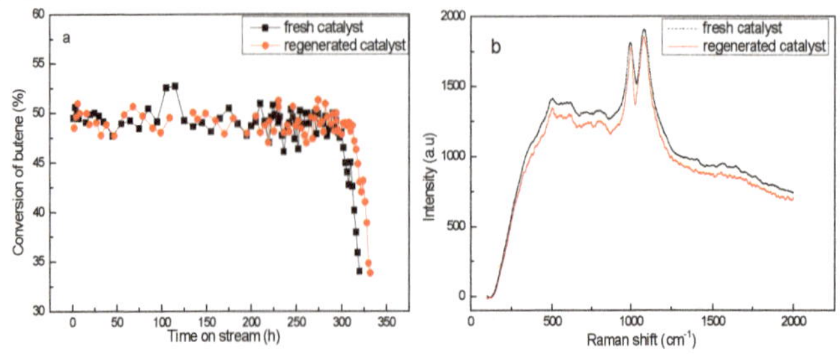

Figure 8: Catalytic activity and Raman spectra of fresh and regenerated catalysts.

Raman shift (cm^{-1})	Raman band assignments
1605-1615	Ring stretches of polyaromatic species
1360-1410	Ring stretches of polyaromatic species
1200-1210	C-C stretches of polyaromatic species
1545-1550	C=C stretches of conjugated olefins
1483	In-phase C=C stretch of cyclopentadienyl species

Table 3: Raman band assignments.

In comparison with Raman spectra of the fresh catalyst, there are no changes. So the fresh and regenerated catalyst holds the same activity.

Conclusions

According to results of BET, EFTEM and Raman, deactivation of WO_3/MTS-9 catalyst is attributed to the pore blockage. Coke results in the decrease of the acidic amount of the spent catalyst. According to our experiments, the coking location successfully determined is on the support and the active sites rather than the tungsten oxide crystallite and results indicate that the decrease of catalytic activity results from the coverage of the active sites. The deposited carbon is aromatic compounds, which the process of formation is speculated. Moreover, the deactivated catalyst may be regenerated at 500°C under air flow, and the regenerated catalyst holds as high activity as the fresh catalyst.

Acknowledgements

The authors are grateful for the financial supports from the National Natural Science Foundation of China (No. 21466001).

References

1. Banks R, Balley G (1964) Olefin Disproportionation. A New Catalytic Process. Ind Eng Chem Prod Res Dev 3: 170-173.
2. Basrur AG, Patwardhan SR, Vyas SN (1991) Propene metathesis over silica-supported tungsten oxide catalyst—Catalyst induction mechanism. J Catal 127: 86-95.
3. Andreini A, MOL JC (1985) Journal of the Chemical Society, Faraday Transactions 1: Physical Chemistry in Condensed Phases. J Chern Soc Faraday Trans I 81: 1705-1714.
4. Schrock RR (1999) Olefin metathesis by molybdenum imido alkylidene catalysts. Tetrahedron 55: 8141-8153.
5. Andreini A, Xu XD, Mol JC (1986) Activity of $Re_2O_7/SiO_2{\cdot}Al_2O_3$ catalysts for propene metathesis and the influence of alkyltin promotors. Appl Catal 27: 31-40.
6. Topka P, Balcar H, Rathouský J, Zilková N, Verpoort F, et al. (2006) Metathesis of 1-octene over MoO3 supported on mesoporous molecular sieves: The influence of the support architecture. Microporous and Mesoporous Mater 96: 44-54.
7. Aguado J, Escola J, Castro M, Paredes B (2005) Metathesis of 1-hexene over rhenium oxide supported on ordered mesoporous aluminas: comparison with Re2O7/γ-Al_2O_3. Appl Catal A: General, 284: 47-57.
8. Zhang XJ, Huang W, Yang S, Sun LD, Zhang FY, et al. (2009) Psoriasis genome-wide association study identifies susceptibility variants within LCE gene cluster at 1q21. Nat Genet 41: 205-210.
9. Aad G, Abbott B, Abdallah J, Abdinov O, Aben R, et al. (2015) Combined Measurement of the Higgs Boson Mass in pp Collisions at sqrt[s]=7 and 8 TeV with the ATLAS and CMS Experiments. Phys Rev Lett 114: 191803.
10. Debecker DP, Stoyanova M, Rodemerck U, Colbeau-Justin F, Boissère C, et al. (2014) Aerosol route to nanostructured WO_3-SiO_2-Al_2O_3 metathesis catalysts: Toward higher propene yield. Appl Catal A: General 470: 458-466.
11. Cheng Z, Lo CS (2012) Formation of Active Sites on WO3 Catalysts: A Density Functional Theory Study of Olefin Metathesis. ACS Catal 2: 341-349.
12. Huang S, Liu S, Zhu Q, Zhu X, Xin W, et al. (2007) The effect of calcination time on the activity of WO_3/Al_2O_3/HY catalysts for the metathesis reaction between ethene and 2-butene. Appl Catal A: General 323: 94-103.
13. Huang S, Liu S, Xin W, Bai J, Xie S, et al. (2005) Metathesis of ethene and 2-butene to propene on W/Al2O3–HY catalysts with different HY contents. Mol Catal A: Chemical 226: 61-68.
14. Maksasithorn S, Praserthdam P, Suriye K, Debecker DP (2015) Preparation of super-microporous WO_3–SiO_2 olefin metathesis catalysts by the aerosol-assisted sol–gel process. Microporous and Mesoporous Mater 213: 125-133.
15. Maksasithorn S, Praserthdam P, Suriye K, Devillers M, Debecker DP, et al. (2014) WO_3-based catalysts prepared by non-hydrolytic sol-gel for the production of propene by cross-metathesis of ethene and 2-butene. Appl Catal A: General 488: 200-207.
16. Maksasithorn S, Debecker DP, Praserthdam P, Panpranot J, Suriye K, et al. (2014) NaOH modified WO_3/SiO_2 catalysts for propylene production from 2-butene and ethylene metathesis. Chin J Catal 35: 232-241.
17. Limsangkass W, Praserthdam P, Phatanasri S, Panpranot J, Chaemchuen S, et al. (2014) Influence of micro- and nano-sized SiO_2 excess support on the metathesis of ethylene and trans-2-butene to propylene over silica-supported tungsten catalysts. React Kinet Mech Catal 113: 225-240.
18. Chaemchuen S, Phatanasri S, Verpoort F, Sae-ma N, Suriye K, et al. (2012) The structure-reactivity relationship for metathesis reaction between ethylene and 2-butene on WO_3/SiO_2 catalysts calcinated at different temperatures. Kinet Catal 53: 247-252.
19. Bhuiyan TI, Arudra P, Akhtar MN, Aitani AM, Abudawoud RH, et al. (2013) Metathesis of 2-butene to propylene over W-mesoporous molecular sieves: A comparative study between tungsten containing MCM-41 and SBA-15. Appl Catal A: General, 467: 224-234.
20. Balcar H, Cejka J (2013) Mesoporous molecular sieves as advanced supports for olefin metathesis catalysts. Coord Chem Rev 257: 3107-3124.
21. Lwin S, Wachs IE (2014) Olefin Metathesis by Supported Metal Oxide Catalysts. ACS Catal 4: 2505-2520.
22. Guisnet M, Magnoux P Catal (1997) Deactivation by coking of zeolite catalysts. Prevention of deactivation. Optimal conditions for regeneration. Today 36: 477-483.
23. Behr A, Schüller U, Bauer K, Maschmeyer D, Wiese KD, et al. (2009) Investigations of reasons for the deactivation of rhenium oxide alumina catalyst in the metathesis of pentene-1. Appl Catal A: General 357: 34-41.
24. Li X, Zhang W, Li X, Liu S, Huang H, et al. (2009) EPR study of the superoxide ion in molybdenum zeolites. J Phys Chem C 113: 8228-8233.
25. Schalkwyk CV, Spamer A, Moodley DJ, Dube T, Reynhardt J, et al. (2003) Factors that could influence the activity of a WO_3/SiO_2 catalyst: Part III. Appl Catal A: General 255: 143-152.

26. Gangwal SK, Fathi-kalajahi J, Wills GB (1977) Break-in Behavior of a Tungsten Oxide on Silica Catalyst in Propylene Disproportionation. Ind Eng Chem Prod Res Dev 16: 237-241.

27. Moodley DJ, Schalkwyk CV, Spamer A, Botha JM, Datye AK (2007) Coke formation on WO_3/SiO_2 metathesis catalysts. Appl Catal A: General 318: 155-159.

28. Mol JC (1997) Handbook of Heterogeneous Catalysis. 2nd edition.

29. Siddiqui MRH, Holmes S, He H, Smith W, Coker EN, et al. (2000) Coking and regeneration of palladium-doped $H_3PW_{12}O_{40}/SiO_2$ catalysts. Catal Lett 66: 53-57.

30. Khomane RB, Kulkarni BD, Paraskar A, Sainkar SR (2002) Synthesis, characterization and catalytic performance of titanium silicalite-1 prepared in micellar media. Mater Chem Phys 76: 99-103.

31. Meng X, Li D, Yang X, Yu Y, Wu S, et al. (2003) Synthesis, Characterization, and Catalytic Activity of Mesostructured Titanosilicates Assembled from Polymer Surfactants with Preformed Titanosilicate Precursors in Strongly Acidic Media. J Phys Chem B 107: 8972-8980.

32. Hua D, Chen S-L, Yuan G, Wang Y, Zhao Q, et al. (2011) Metathesis of butene to propene and pentene over WO_3/MTS-9. Microporous and Mesoporous Mater 143: 320-325.

33. Guisnet M, Magnoux P (2001) Organic chemistry of coke formation. Appl Catal A: General 212: 83-96.

34. Bibby DM, Howe RF, McLellan GD (1992) Coke formation in high-silica zeolites. Appl Catal A: General 93: 1-34.

35. Mellouki A, Liévin J, Herman M (2001) The vibrational spectrum of pyrrole (C_4H_5N) and furan (C_4H_4O) in the gas phase. Chem Phys 271: 239-266.

36. Stepanov AG, Luzgin MV, Arzumanov SS, Ernst H, et al. (2002) *n*-Butene Conversion on H-Ferrierite Studied by ^{13}C MAS NMR. J Catal 211: 165-172.

37. Bigey C, Hilaire L, Maire G (1999) Catalysis on Pd/WO_3 and Pd/WO_2: Effect of the Modifications of the Surface States Due to Redox Treatments on the Skeletal Rearrangement of Hydrocarbons: Part I. Physical and Chemical Characterizations of Catalysts by BET, TPR, XRD, XAS, and XPS. J Catal 184: 406-420.

38. Bartle KD, Ladner WR, Martin TG, Snape CE, et al. (1979) Structural analysis of supercritical-gas extracts of coals. Fuel 58: 413-422.

39. Kim DS, Ostromecki M, Wachs IE (1996) Surface structures of supported tungsten oxide catalysts under dehydrated conditions. J Mol Catal A: Chemical 106: 93-102.

40. Ramírez J, Gutiérrez-Alejandre A (1997) Characterization and Hydrodesulfurization Activity of W-Based Catalysts Supported on Al_2O_3–TiO_2Mixed Oxides. J Catal 170: 108-122.

41. Chua YT, Stair PC (2003) An ultraviolet Raman spectroscopic study of coke formation in methanol to hydrocarbons conversion over zeolite H-MFI. J Catal 213: 39-46.

42. Asher SA, Johnson CR (1984) Raman spectroscopy of a coal liquid shows that fluorescence interference is minimized with ultraviolet excitation. Science 225: 311-313.

Exploration of Solution Behaviour of Potassium Halides in Mixtures of L-Proline and Water at 298.15, 308.15 and 318.15 K

Mahendra Nath Roy*, Partha Sarathi Sikdar and Pritam De

Department of Chemistry, University of North Bengal, Darjeeling 734013, India

Abstract

Apparent molar volume (ϕ_V) and viscosity B-coefficients were estimated for potassium chloride, potassium bromide and potassium iodide in aqueous mixture of L-Proline from measured solution density (ρ) and viscosity (η) at 298.15, 308.15 and 318.15 K at various electrolyte concentrations. The experimental density data were evaluated by Masson equation and the derived data were interpreted in terms of ion-solvent and ion-ion interactions. The viscosity data has been analyzed using Jones-Dole equation and the derived parameters, B and A, have also been interpreted in terms of ion-solvent and ion-ion interactions respectively. The structure-making or breaking capacity of the electrolyte under investigation has been discussed in terms of sign$(\partial\phi_V^0/\partial T)_P$.

Keywords: Potassium chloride; Potassium bromide; Potassium iodide; L-Proline; Ion-solvent interaction

Introduction

Studies on densities (ρ) and viscosities (η) of electrolyte solutions are of great importance in characterizing the properties and structural aspects of solutions. The addition of an electrolyte to an aqueous organic solution alters the pattern of ion solvation and causes phenomenal changes in the behaviour of the dissolved electrolyte. Hence studies on the limiting apparent molar volume and viscosity B-coefficients of electrolyte provide us valuable information regarding ion-ion, ion-solvent and solvent-solvent interactions [1-3]. It has been found by a number of workers [4-6] that the addition of an electrolyte could either make or break the structure of a liquid. As the viscosity of a liquid depends on the intermolecular forces, the structural aspects of the liquid can be inferred from the viscosity of solutions at various electrolyte concentrations and temperature.

In this paper we have attempted to report the limiting apparent molar volume (ϕ_V^0), experimental slopes (S_V^*) and viscosity B-coefficients for potassium chloride, potassium bromide and potassium iodide in aqueous mixture of L-Proline at 298.15, 308.15 and 318.15 K. Since potassium ion being a common cation for all of the electrolytes under investigation, the present work enables us to have a qualitative comparison of the role of anion in aqueous L-Proline in terms of various derived parameters obtained from viscosity (η) and density (ρ) measurements.

Experimental Methods

Materials

L-Proline (SD. Fine Chemicals) was purified by standard methods [7]. The purity of the solvent was checked by measuring the viscosity (η) and density (ρ) at 298.15 K which was in good agreement with the literature values. Doubly distilled, degassed and deionised water with a specific conductance of 1×10^{-6}/Ω/cm was used. Potassium chloride, potassium bromide and potassium iodide (Sigma-Aldrich, Germany) were purified by re-crystallizing twice from conductivity water and then dried in a vacuum desiccator over P_2O_5 for 24 h before use. The experimental values of viscosity (η) and density (ρ) of aqueous mixtures of 0.01 M, 0.03 M and 0.05 M L-Proline at different temperatures are listed in Table 1.

Apparatus and procedure

Densities (ρ) were measured with an Anton Paar density-meter (DMA 4500M) with a precision of 0.0005 g/cm³. The calibration was done by double-distilled water and dry air and uncertainty in density was ± 0.00005 g/cm.

The measurements were done in a thermostat bath controlled to ± 0.01 K. Viscosity (η) was measured by means of Brookfield DV-III Ultra Programmable Rheometer with spindle size-42 having an accuracy of 1.0% and fitted to a Brookfield Digital Bath TC-500 at 298 K using density and viscosity values from the literature [8-10]. The uncertainty in viscosity measurements is within ± 0.003 mPa s. The mixtures were prepared by mixing known volume of pure liquids in airtight-stopper bottles and each solution thus prepared was distributed into three recipients to perform all the measurements in triplicate, with the aim of determining possible dispersion of the results obtained. Adequate precautions were taken to minimize evaporation loses during the actual measurements.

The electrolyte solutions studied here were prepared by mass and the conversion of molality into molarity was accomplished [3] using experimental density values. The experimental values of concentrations (c), densities (ρ), viscosities (η), and derived parameters at various temperatures are reported in Table 2.

Results and Discussion

Density calculation

The apparent molar volumes (ϕ_V) were determined from the solution densities using the following Equation [3]:

***Corresponding author:** Mahendra Nath Roy, Department of Chemistry, University of North Bengal, Darjeeling-734013, India
E-mail: mahendraroy2002@yahoo.co.in

$$\phi_V = M/\rho - (\rho - \rho_o)/c\rho\rho_o \qquad (1)$$

Where M is the molar mass of the solute, c is the molarity of the solution; ρ_0 and ρ are the densities of the solvent and the solution respectively. The limiting apparent molar volumes ϕ_V^0 was calculated using a least-squares treatment to the plots of ϕ_V versus $\sqrt{c}$ using the following Masson equation [11]:

$$\phi_V = \phi_V^0 + S_V^* \cdot \sqrt{c} \qquad (2)$$

Where (ϕ_V^0) is the apparent molar volume at infinite dil ution and (S_V^*) is the experimental slope. The plots of (ϕ_V^0) against square root of molar concentration ($\sqrt{c}$) were found to be linear as depicted graphically in Figures 1-9 with negative slopes. Values of ϕ_V^0 and S_V^* are reported in Table 3.

In these systems the ion-solvent and ion-ion interactions can be interpreted in terms of structural changes between various components of the solvent and solution systems. ϕ_V^0 can be used to interpret ion-solvent interactions. A perusal of Table 3 reveals that the ϕ_V^0 values are positive and increase with a rise in both the temperature and amount of L-Proline in the mixtures. This indicates the presence of strong ion-solvent interactions and these interactions are further strengthened at higher temperatures and higher molar mass of L-Proline in the mixtures, suggesting larger electrostriction at higher temperatures and in enhanced amount of L-Proline.

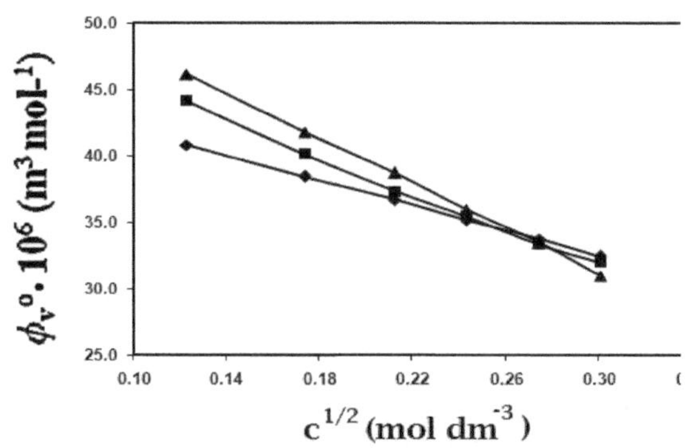

Figure 1: The apparent molar volume (ϕ_V) and the square root of concentrations ($\sqrt{c}$) for potassium chloride in different aqueous mixtures 0.01 M L-Proline at 298.15 K (—♦—), 308.15 K (—■—) and 318.15 K (—▲—).

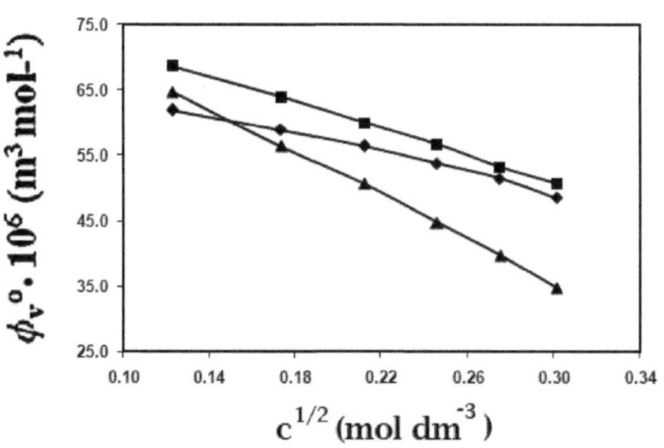

Figure 2: The apparent molar volume (ϕ_V) and the square root of concentrations ($\sqrt{c}$) for potassium chloride in different aqueous mixtures 0.03 M L-Proline at 298.15 K (—♦—), 308.15 K (—■—) and 318.15 K (—▲—).

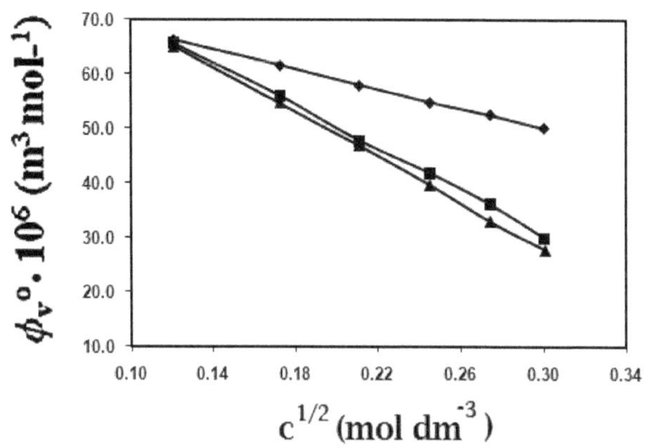

Figure 3: The apparent molar volume (ϕ_V) and the square root of concentrations ($\sqrt{c}$) for potassium chloride in different aqueous mixtures 0.05 M L-Proline at 298.15 K (—♦—), 308.15 K (—■—) and 318.15 K (—▲—).

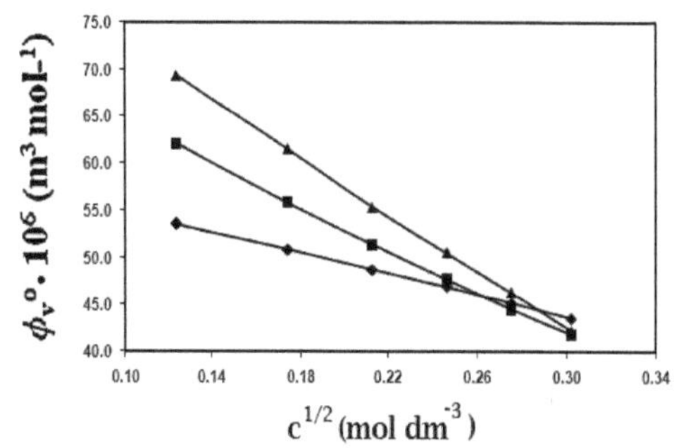

Figure 4: The apparent molar volume (ϕ_V) and the square root of concentrations ($\sqrt{c}$) for potassium bromide in different aqueous mixtures 0.01 M L-Proline at 298.15 K (—♦—), 308.15 K (—■—) and 318.15 K (—▲—).

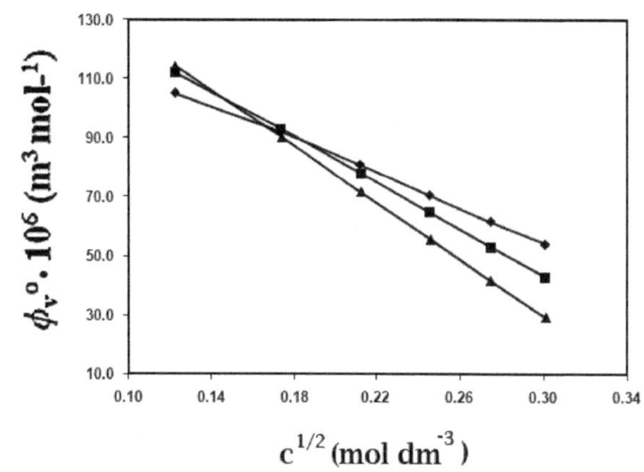

Figure 5: The apparent molar volume (ϕ_V) and the square root of concentrations ($\sqrt{c}$) for potassium bromide in different aqueous mixtures 0.03 M L-Proline at 298.15 K (—♦—), 308.15 K (—■—) and 318.15 K (—▲—).

A perusal of Table 3 also reveals that the S_V^* values are negative for all the solutions at all the experimental temperatures and S_V^* values decrease as the experimental temperature and amount of L-Proline in the mixtures increases. Since S_V^* is a measure of ion-ion interactions, the results indicate the presence of weak ion-ion interactions in the

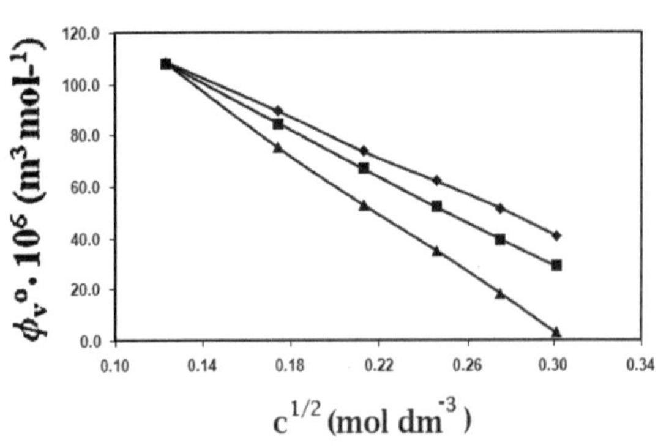

Figure 6: The apparent molar volume (ϕ_V) and the square root of concentrations (√c) for potassium bromide in different aqueous mixtures 0.05 M L-Proline at 298.15 K (—♦—), 308.15 K (—■—) and 318.15 K (—▲—).

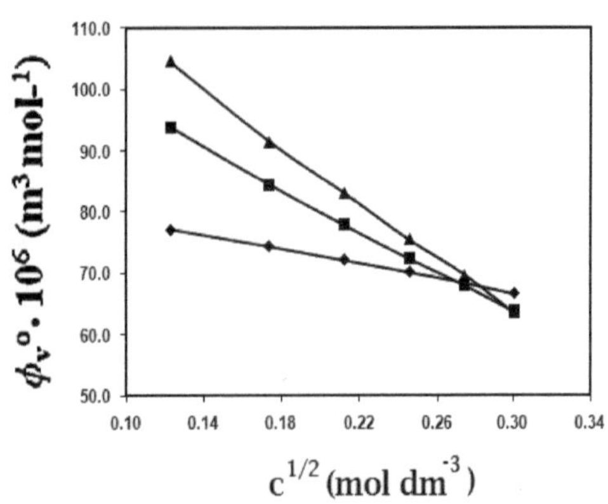

Figure 7: The apparent molar volume (ϕ_V) and the square root of concentrations (√c) for potassium iodide in different aqueous mixtures 0.01 M L-Proline at 298.15 K (—♦—), 308.15 K (—■—) and 318.15 K (—▲—).

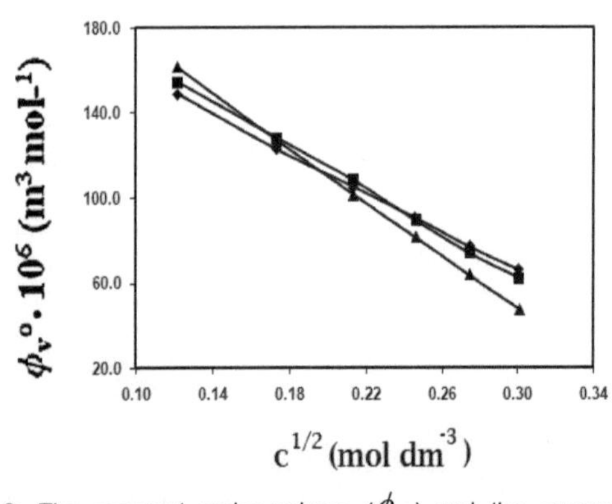

Figure 8: The apparent molar volume (ϕ_V) and the square root of concentrations (√c) for potassium iodide in different aqueous mixtures 0.03 M L-Proline at 298.15 K (—♦—), 308.15 K (—■—) and 318.15 K (—▲—).

solutions at all the experimental temperatures and these interactions further decrease with a rise in temperature and increase in molar mass L-Proline in the mixtures. In other words, it may be said that the solvation of electrolyte/ions increases with the increase of L-Proline content in water. This is probably due to more violent thermal agitation at higher temperatures, resulting in diminishing the force of ion-ion interactions (ionic-dissociation) [12]. This suggests that ion-solvent interactions dominate over ion-ion interactions in all the solutions and at all experimental temperatures.

The variation of ϕ_V^0 with temperature of potassium chloride, potassium bromide and potassium iodide in aqueous mixture of L-Proline follows the polynomial,

$$\phi_V^0 = a_0 + a_1 T + a_2 T^2 \tag{3}$$

Over the temperature range under study where T is the temperature in K. Values of coefficients of the above equation for potassium chloride, potassium bromide and potassium iodide in aqueous mixture of L- Proline are reported in Table 4.

The apparent molar expansibilities (ϕ_E^0) can be obtained by the following equation:

$$\phi_E^0 = (\partial \phi_V^0 / \partial T)_P = a_1 + 2a_2 T \tag{4}$$

The values of ϕ_E^0 for different solutions of the studied electrolytes at 298.15, 308.15 and 318.15 K are reported in Table 5. From the table it is evident that the values of ϕ_E^0 for potassium chloride increases with the increase in the amount of L- Proline in the mixture. However, for potassium bromide and potassium iodide the ϕ_E^0 values were found to be rather complicated to explain. During the past few years it has been emphasized by a number of workers that S_V^* is not the sole criterion for

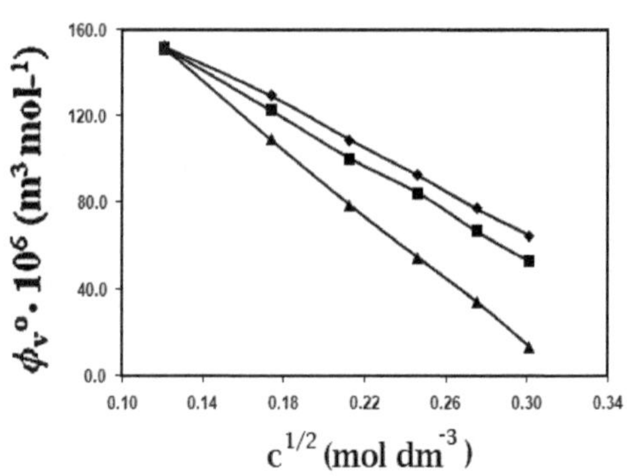

Figure 9: The apparent molar volume (ϕ_V) and the square root of concentrations (√c) for potassium iodide in different aqueous mixtures 0.05 M L-Proline at 298.15 K (—♦—), 308.15 K (—■—) and 318.15 K (—▲—).

Temperature (K)	$\rho \times 10^{-3}$(kg/m)	η (mPa.s)
0.01 M L-proline		
298.15 K	0.99753	0.909
308.15 K	0.99605	0.771
318.15 K	0.99466	0.639
0.03 M L-proline		
298.15 K	0.99902	0.916
308.15 K	0.99763	0.778
318.15 K	0.99604	0.645
0.05 M L-proline		
298.15 K	1.00045	0.923
308.15 K	0.99941	0.786
318.15 K	0.99772	0.652

Table 1: Density (ρ, kg/m) and viscosity (η, mPa.s) of aqueous mixtures of 0.01 M L-proline, 0.03 M L-proline, 0.05 M L-proline at different temperatures.

c (mol/dm³)	ρ ×10⁻³ (kg/m)	η (mPa.s)	ϕ_V ×10⁶ (m³/mol)	$(\eta/\eta_0-1)/c^{1/2}$ (kg$^{1/2}$ mol$^{-1/2}$)
Potassium chloride in aqueous mixture 0.01M L-proline				
298.15 K				
0.0151	0.99804	0.918	40.7724	0.081
0.0302	0.99862	0.923	38.4219	0.089
0.0452	0.99924	0.927	36.6533	0.093
0.0590	0.99985	0.931	35.1369	0.100
0.0752	1.00059	0.935	33.7388	0.104
0.0903	1.00132	0.938	32.4331	0.106
Potassium chloride in aqueous mixture 0.01M L-proline				
308.15 K				
0.0151	0.99651	0.779	44.1206	0.084
0.0302	0.99709	0.784	40.0940	0.097
0.0452	0.99773	0.789	37.3202	0.110
0.0590	0.99835	0.793	35.4719	0.117
0.0752	0.99914	0.798	33.3263	0.128
0.0903	0.99988	0.802	31.9725	0.134
Potassium chloride in aqueous mixture 0.01M L-proline				
318.15 K				
0.0151	0.99509	0.647	46.1479	0.102
0.0302	0.99565	0.653	41.7753	0.126
0.0452	0.99628	0.659	38.6617	0.147
0.0590	0.99693	0.664	35.9803	0.161
0.0752	0.99774	0.670	33.4493	0.177
0.0903	0.99858	0.675	30.9510	0.187
Potassium bromide in aqueous mixture 0.01M L-proline				
298.15 K				
0.0153	0.99853	0.922	53.5574	0.116
0.0303	0.99959	0.931	50.8657	0.139
0.0454	1.00071	0.940	48.7479	0.160
0.0607	1.00189	0.948	46.9048	0.174
0.0759	1.00311	0.957	45.1597	0.192
0.0912	1.00438	0.966	43.5138	0.208
Potassium bromide in aqueous mixture 0.01M L-proline				
308.15 K				
0.0153	0.99692	0.783	62.1030	0.126
0.0303	0.99796	0.792	55.8276	0.156
0.0454	0.99911	0.801	51.3776	0.183
0.0607	1.00036	0.810	47.6963	0.205
0.0759	1.00168	0.819	44.4546	0.226
0.0912	1.00305	0.828	41.8137	0.245
Potassium bromide in aqueous mixture 0.01M L-proline				
318.15 K				
0.0153	0.99542	0.651	69.3779	0.152
0.0303	0.9964	0.661	61.4873	0.198
0.0454	0.99754	0.671	55.3595	0.235
0.0607	0.9988	0.681	50.4901	0.267
0.0759	1.00015	0.690	46.2727	0.290
0.0912	1.00162	0.701	42.2061	0.321
Potassium iodide in aqueous mixture 0.01M L-proline				
298.15 K				
0.0151	0.99887	0.926	77.1285	0.152
0.0302	1.00029	0.941	74.3642	0.203
0.0451	1.00175	0.955	72.0751	0.238
0.0604	1.00329	0.968	70.1716	0.264
0.0755	1.00486	0.982	68.3442	0.292
0.0906	1.00647	0.995	66.6517	0.314
Potassium iodide in aqueous mixture 0.01M L-proline				

308.15 K				
0.0151	0.99714	0.788	93.7994	0.179
0.0302	0.99851	0.803	84.3485	0.239
0.0451	1.00001	0.818	77.8489	0.287
0.0604	1.00168	0.833	72.2998	0.327
0.0755	1.00341	0.847	67.9012	0.359
0.0906	1.00525	0.861	63.7203	0.388
Potassium iodide in aqueous mixture 0.01M L-proline				
318.15 K				
0.0151	0.99559	0.656	104.5438	0.217
0.0302	0.99691	0.672	91.3819	0.297
0.0451	0.99839	0.688	82.9873	0.361
0.0604	1.00010	0.703	75.4455	0.408
0.0755	1.00189	0.718	69.5956	0.450
0.0906	1.00387	0.733	63.5555	0.489
Potassium chloride in aqueous mixture 0.03M L-proline				
298.15 K				
0.0150	0.99921	0.923	61.9450	0.062
0.0301	0.99949	0.927	58.9421	0.069
0.0451	0.99984	0.931	56.3604	0.077
0.0603	1.00027	0.935	53.7704	0.085
0.0754	1.00075	0.938	51.5348	0.087
0.0907	1.00137	0.942	48.5495	0.094
Potassium chloride in aqueous mixture 0.03M L-proline				
308.15 K				
0.0151	0.99772	0.785	68.7139	0.073
0.0301	0.99795	0.790	64.0361	0.089
0.0451	0.99829	0.795	59.9847	0.103
0.0603	0.99871	0.799	56.6853	0.110
0.0754	0.99924	0.804	53.2104	0.122
0.0907	0.99979	0.808	50.6903	0.128
Potassium chloride in aqueous mixture 0.03M L-proline				
318.15 K				
0.0151	0.99619	0.652	64.8067	0.088
0.0302	0.99659	0.658	56.4412	0.116
0.0451	0.99712	0.664	50.6549	0.139
0.0604	0.99784	0.669	44.7281	0.151
0.0756	0.99866	0.675	39.7752	0.169
0.0907	0.99964	0.679	34.7144	0.175
Potassium bromide in aqueous mixture 0.03M L-proline				
298.15 K				
0.0150	0.99923	0.928	105.1030	0.107
0.0301	0.99983	0.938	92.0902	0.138
0.0452	1.00074	0.947	80.8570	0.159
0.0603	1.00193	0.956	70.5692	0.178
0.0754	1.00332	0.965	61.7272	0.195
0.0905	1.00486	0.975	54.1642	0.214
Potassium bromide in aqueous mixture 0.03M L-proline				
308.15 K				
0.0151	0.99774	0.790	111.9319	0.126
0.0302	0.99842	0.802	92.8868	0.178
0.0453	0.99948	0.812	78.0739	0.205
0.0604	1.00089	0.822	64.8203	0.230
0.0755	1.00258	0.832	53.1259	0.253
0.0906	1.00448	0.842	42.9908	0.273
Potassium bromide in aqueous mixture 0.03M L-proline				
318.15 K				
0.0151	0.99612	0.657	114.1186	0.152
0.0302	0.99691	0.669	90.3578	0.214
0.0453	0.99818	0.680	71.7285	0.255
0.0604	0.99985	0.691	55.7207	0.290

0.0755	1.00185	0.702	41.6985	0.322
0.0906	1.00412	0.713	29.3384	0.350
Potassium iodide in aqueous mixture 0.03M L-proline				
298.15 K				
0.0149	0.99928	0.934	148.6430	0.161
0.0302	1.00032	0.950	122.8749	0.214
0.0456	1.00180	0.965	104.7895	0.251
0.0604	1.00356	0.980	90.4247	0.284
0.0755	1.00571	0.996	76.8781	0.318
0.0906	1.00802	1.012	66.0675	0.348
Potassium iodide in aqueous mixture 0.03M L-proline				
308.15 K				
0.0149	0.99781	0.796	154.2313	0.190
0.0302	0.99878	0.813	127.9890	0.259
0.0456	1.00027	0.829	107.9414	0.307
0.0605	1.00225	0.845	89.2142	0.350
0.0756	1.00455	0.861	73.9113	0.388
0.0907	1.00701	0.879	61.9273	0.431
Potassium iodide in aqueous mixture 0.03M L-proline				
318.15 K				
0.0149	0.99612	0.663	161.2379	0.229
0.0302	0.99723	0.680	126.7934	0.312
0.0456	0.99899	0.697	101.1548	0.378
0.0605	1.00115	0.714	81.1575	0.435
0.0756	1.00374	0.731	63.5879	0.485
0.0907	1.00674	0.749	47.3012	0.535
Potassium chloride in aqueous mixture 0.05M L-proline				
298.15 K				
0.0147	1.00058	0.931	66.1920	0.071
0.0298	1.00085	0.937	61.6009	0.088
0.0449	1.00122	0.942	57.8579	0.097
0.0602	1.00167	0.947	54.7120	0.106
0.0754	1.00215	0.952	52.4210	0.114
0.0904	1.00271	0.957	49.9364	0.123
Potassium chloride in aqueous mixture 0.05M L-proline				
308.15 K				
0.0147	0.99955	0.794	65.5701	0.084
0.0298	0.99998	0.800	55.9323	0.103
0.0449	1.00064	0.806	47.6291	0.120
0.0602	1.00141	0.812	41.7613	0.135
0.0754	1.00234	0.817	36.1032	0.144
0.0903	1.00348	0.823	29.8654	0.157
Potassium chloride in aqueous mixture 0.05M L-proline				
318.15 K				
0.0147	0.99787	0.660	64.9810	0.101
0.0298	0.99833	0.667	54.6447	0.133
0.0449	0.99899	0.674	46.7675	0.159
0.0603	0.99985	0.681	39.6604	0.181
0.0754	1.00089	0.688	32.9022	0.201
0.0904	1.00199	0.695	27.6887	0.219
Potassium bromide in aqueous mixture 0.05M L-proline				
298.15 K				
0.0152	1.00061	0.935	108.4123	0.105
0.0304	1.00135	0.945	89.2875	0.137
0.0454	1.00251	0.954	73.4616	0.158
0.0605	1.00389	0.964	61.9252	0.181
0.0756	1.00556	0.974	51.1534	0.201
0.0907	1.00753	0.983	40.6693	0.216
Potassium bromide in aqueous mixture 0.05M L-proline				
308.15 K				
0.0152	0.99958	0.798	107.8545	0.124

0.0304	1.00046	0.809	84.4013	0.168
0.0454	1.00177	0.820	66.8686	0.203
0.0605	1.00345	0.830	52.0043	0.228
0.0756	1.00542	0.841	39.2430	0.254
0.0907	1.00756	0.851	28.8720	0.275
Potassium bromide in aqueous mixture 0.05M L-proline				
318.15 K				
0.0152	0.99789	0.664	108.0181	0.149
0.0304	0.99906	0.676	74.8907	0.211
0.0454	1.00072	0.687	52.7317	0.252
0.0605	1.00279	0.698	34.9094	0.287
0.0756	1.00533	0.709	18.0126	0.318
0.0907	1.00822	0.721	2.9450	0.351
Potassium iodide in aqueous mixture 0.05M L-proline				
298.15 K				
0.0146	1.00065	0.941	152.2114	0.161
0.0303	1.00155	0.959	129.5148	0.224
0.0452	1.00301	0.976	109.0648	0.270
0.0607	1.00488	0.993	92.6019	0.308
0.0759	1.00713	1.009	77.4794	0.338
0.0908	1.00958	1.025	64.8759	0.367
Potassium iodide in aqueous mixture 0.05M L-proline				
308.15 K				
0.0146	0.99962	0.804	151.6684	0.190
0.0303	1.00072	0.823	122.6546	0.270
0.0452	1.00236	0.841	100.5065	0.329
0.0607	1.00433	0.859	84.5345	0.377
0.0759	1.00688	0.876	67.0643	0.416
0.0908	1.00963	0.893	52.8721	0.452
Potassium iodide in aqueous mixture 0.05M L-proline				
318.15 K				
0.0146	0.99794	0.670	151.2114	0.228
0.0303	0.99944	0.690	109.1685	0.335
0.0453	1.00165	0.709	78.8492	0.411
0.0607	1.00446	0.728	54.4682	0.473
0.0759	1.00769	0.746	34.0835	0.523
0.0908	1.01154	0.764	13.2989	0.570

Table 2: The concentration (*c*), density (ρ), viscosity (η), apparent molar volume (ϕ_V), and $(\eta/\eta_0-1)/c^{1/2}$ of potassium chloride, potassium bromide and potassium iodide in different aqueous mixtures 0.01M L-proline , 0.03M L-proline, 0.05M L-proline at different temperatures.

Molarity of L-proline	**a_0 (m^3/mol)**	**a_1 (m^3/mol/K)**	**a_2 (m^3/mol/K)**
Potassium chloride			
0.01	-686.321	4.2917	-0.0062
0.03	-2922.861	18.7939	-0.0293
0.05	-5619.977	36.3861	-0.0579
Potassium bromide			
0.01	-1817.842	10.9216	-0.0155
0.03	-2976.151	18.7747	-0.0279
0.05	4075.112	-26.5645	0.0451
Potassium iodide			
0.01	-6181.141	38.5014	-0.0586
0.03	2016.222	-13.3537	0.0244
0.05	9044.573	-58.7362	0.0976

Table 3: Limiting apparent molar volumes (ϕ_V^0) and experimental slopes (S_V^*) of potassium chloride, potassium bromide and potassium iodide in different aqueous mixtures 0.01M L-proline , 0.03M L-proline, 0.05M L-proline at different temperatures.

Molarity of L-proline	a_0 (m³/mol)	a_1 (m3/mol/K)	a_2 (m³/mol/K)
	Potassium chloride		
0.01	-686.321	4.2917	
0.03	-2922.861	18.7939	
0.05	-5619.977	36.3861	
	Potassium bromide		
0.01	-1817.842	10.9216	-0.0155
0.03	-2976.151	18.7747	-0.0279
0.05	4075.112	-26.5645	0.0451
	Potassium iodide		
0.01	-6181.141	38.5014	-0.0586
0.03	2016.222	-13.3537	0.0244
0.05	9044.573	-58.7362	0.0976

Table-4: Values of the coefficients of Eq. (4) for potassium chloride, potassium bromide and potassium iodide in different aqueous mixtures 0.01 M L-proline, 0.03 M L-proline, 0.05M L-proline at different temperatures.

Molarity of L-proline	ϕ_E^0 (m³/mol)			$(\partial\phi_E^0/\partial T)_P$ (m³/mol/K)
	298.15 K	308.15 K	318.15 K	
		Potassium chloride		
0.01	0.0595	0.47064	0.34664	-0.0124
0.03	1.3223	1.7363	0.1505	-0.0586
0.05	1.8603	0.7023	-0.4556	-0.1158
		Potassium bromide		
0.01	1.6789	1.3690	1.0589	-0.0310
0.03	2.1379	1.5799	1.0220	-0.0558
0.05	0.3286	1.2306	2.1326	0.0902
		Potassium iodide		
0.01	3.5582	2.3862	1.2142	-0.1172
0.03	1.1960	1.6840	2.1720	0.0488
0.05	-0.5373	1.4146	3.3666	0.1952

Table-5: Limiting partial molar expansibilities for potassium chloride, potassium bromide and potassium iodide in different aqueous mixtures 0.01 M L-proline , 0.03 M L-proline, 0.05 M L-proline at different temperatures.

Molarity of L-proline	*B*-coefficient (dm$^{3/2}$/mol$^{1/2}$)			*A*-coefficient (dm³/mol)		
	298.15 K	308.15 K	318.15 K	298.15 K	308.15 K	318.15 K
			Potassium chloride			
0.01	0.149	0.283	0.488	0.062	0.048	0.042
0.03	0.179	0.308	0.496	0.039	0.035	0.029
0.05	0.278	0.403	0.658	0.038	0.034	0.020
			Potassium bromide			
0.01	0.511	0.669	0.938	0.050	0.041	0.035
0.03	0.588	0.810	1.103	0.034	0.031	0.019
0.05	0.625	0.849	1.116	0.027	0.019	0.013
	Potassium iodide					
0.01	0.903	1.177	1.525	0.043	0.035	0.031
0.03	1.037	1.325	1.706	0.032	0.026	0.016
0.05	1.137	1.451	1.889	0.025	0.016	0.004

Table-6: Values of A and B coefficients for potassium chloride, potassium bromide and potassium iodide in different aqueous mixtures 0.01M L-proline , 0.03M L-proline, 0.05M L-proline at different temperatures.

determining the structure making or breaking tendency of any solute. Hepler [13] developed a technique of examining the sign of $(\partial\phi_E^0/\partial T)_P$ for the solute in terms of long-range structure-making and breaking capacity of the electrolytes in the mixed solvent systems. The general thermodynamic expression used is as follows

$$(\partial\phi_E^0/\partial T)_P = 2a_2 \quad (5)$$

If the sign of $(\partial\phi_E^0/\partial T)_P$ is positive or small negative [14,15] the electrolyte is a structure maker and when the sign of $(\partial\phi_E^0/\partial T)_P$ is negative, it is a structure breaker. As is evident from Table 5, the electrolyte under investigation generally acts as a structure breaker.

The viscosity data of solutions for the electrolytes in 0.01 M, 0.03M, 0.05M L-Proline have been analyzed using Jones-Dole [16] equation:

Where η_0 and η are the viscosities of the solvent and solution respectively. A and B are the constants estimated by least square method and are reported in Table 6. From the table it is evident that the values of the A-coefficient are very small for all the solutions under investigation at all experimental temperatures. These results indicate the presence of weak ion-ion interactions, and these interactions further decrease with both rise of experimental temperatures and amount of L-Proline suggesting an increase in ion-solvation. Interestingly, values are found to be smallest for Potassium iodide and hence it may be concluded that solubility in aqueous L-Proline solutions is maximum for Potassium iodide and minimum for Potassium chloride.

The effects of ion-solvent interactions on the solution viscosity can be inferred from the B-coefficient [17,18]. The viscosity B-coefficient is a valuable tool to provide information concerning the solvation of the solutes and their effects on the structure of the solvent. From Table 6 it is evident that the values of the B-coefficient of potassium chloride, potassium bromide and potassium iodide in the studied solvent systems are more positive than A-coefficients, thereby suggesting the presence of strong ion-solvent interactions, and these types of interactions are strengthened with a rise in both temperature and amount of L-Proline in solutions. These conclusions are in excellent agreement with those drawn from ϕ_V^0 values discussed earlier. It has been reported in a number of studies [19,20] that dB/dT is a better criterion for determining the structure-making/breaking nature of any solute rather than simply the value of the B-coefficient. It is found from Table 6 that the values of the B-coefficient increase with a rise in temperature (positive dB/dT) suggesting the structure-breaking tendency of potassium chloride, potassium bromide and potassium iodide in the solvent systems.

Conclusion

Extensive study of potassium chloride, potassium bromide and potassium iodide in aqueous mixture of L- Proline reveals that potassium iodide is more associated in L-Proline than the other two halides. The ion-association is found minimum in the case of potassium chloride in L- Proline. The said interaction of potassium bromide arises in the intermediacy of potassium iodide and potassium chloride. The present study reveals the predominance of ion-solvent interaction over the ion-ion interaction in all the solution under investigation.

Acknowledgements

The authors are grateful to the UGC supported Major research project, Ref. No. RP/5032/FCS/2011 New Delhi for financial support in order to continue this research work.

One of the authors, Prof. M. N. Roy is thankful to University Grant Commission, New Delhi, Government of India for being awarded one time grant under Basic Scientific Research via the grant-in-Aid No. F.4-10/2010 (BSR) regarding his active service for augmenting of research facilities to facilitate further research work.

References

1. Mc Dowali JM, Vincent CA (1974) Viscosity behaviour of some simple electrolytes in formamide solution. J Chem Soc Faraday Trans 1: 1862-1868.
2. Dack MRJ, Bird KJ, Parker AJ (1975) Solvation of ions XXV Partial molal volumes of single ions in protic and dipolar aprotic solvents. Aust J Chem 28: 955-963.
3. Roy MN, Sinha B, Dey R, Sinha A (2005) Solute-Solvent and Solute-Solute Interactions of Resorcinol in Mixed 1,4-Dioxane-Water Systems at Different Temperatures. Int J Thermophys 26: 1549-1563.
4. Stokes RH, Mills R (1965) Int Encyclopedia of Physical Chemistry and Chemical Physics.
5. Nikam PS, Mehdi H (1988) Ultrasonic velocity and adiabatic compressibility of monochloroacetic acid in aqueous ethanol at various temperatures. J Chem Eng Data 33: 165-169.
6. Nandi D, Roy MN, Hazra DK (1993) J Indian Chem Soc 70: 305-310.
7. Perrin DD, Amarego WLF (1966) Purification of Laboratory Chemicals (4thedn.) Reed Educational and Professional Publishing Ltd.
8. Sinha B, Dakua VK, Roy MN (2007) Apparent Molar Volumes and Viscosity *B*-Coefficients of Some Amino Acids in Aqueous Tetramethylammonium Iodide Solutions at 29815 K. J Chem Eng Data 52: 1768-1772.
9. Dean JA (1973) Lange's Handbook of Chemistry (11thedn.) McGraw-Hill Book Company, New York.
10. Chatterjee BD (2006) Electrical Conductances of Tetrabutylammonium Bromide, Sodium Tetraphenylborate, and Sodium Bromide in Methanol (1) + Water (2) Mixtures at (29815, 30815, and 31815) K. J Chem Eng Data 51: 1352-1355.
11. Masson DO (1929) Phila Mag 8: 218-226.
12. Millero FJ (1972) Structure and Transport Process in Water and Aqueous Solutions, RAHorne, New York
13. Hepler LG (1969) Thermal expansion and structure in water and aqueous solutions Can. J Chem 47: 4613-4617.
14. Sarkar BK, Sinha B, Roy MN (2008) Ion-solvent and ion-ion interactions of sodium molybdate and sodium tungstate in mixtures of acetonitrile and water at 29815, 30815, and 31815 K. Russ J Phys Chem 82: 960-966.
15. Parmar ML, Banyal DS (2005) Effect of temperature on the partial molar volumes of some bivalent transition metal nitrates and magnesium nitrate in DMF + water mixtures Indian. J Chem 44A: 1582-1588.
16. Jones G, Dole M (1929) The viscosity of aqueous solutions of strong electrolytes with special reference to barium chloride. J Am Chem Soc 51: 2950-2964.
17. Millero FJ (1971) Molal volumes of electrolytes. Chem Rev 71: 147-176.
18. Millero FJ, Lo Surdo A, Shin C (1978) The apparent molal volumes and adiabatic compressibilities of aqueous amino acids at 25°C. J Phys Chem 82: 784-792.
19. Gopal R, Siddiqi MA (1969) Variation of partial molar volume of some tetra-alkylammonium iodides with temperature in aqueous solutions. J Phys Chem 72: 1814-1817.
20. Saha N, Das B (1997) Apparent Molar Volumes of Some Symmetrical Tetraalkylammonium Bromides in Acetonitrile at (29815, 30815, and 31815) K. J Chem Eng Data 42: 227-229.

Low-Temperature Selective Catalytic Reduction of NO with NH_3 over Manganese Oxides Supported on Fly Ash-Palygorskite

Xianlong Zhang*, Shuangshuang lv, Xiaobin Jia, Xueping Wu and Fanyue Meng

School of Chemistry and Chemical Engineering, Hefei University of Technology, Hefei, 230009, PR China

Abstract

A catalyst of manganese oxides (MnOx) supported on fly ash-palygorskite (MnOx/FA-PG) was prepared for low-temperature selective catalytic reduction (SCR) of nitric oxides (NOx) by ammonia (NH_3). The influences of the preparation method, active species precursors, calcination temperature, calcination time and particle size on the SCR performance of the MnOx/FA-PG catalyst were studied. Catalysts were characterized by Brunauer-Emmett-Teller (BET) analysis, scanning electron microscopy (SEM), X-ray diffraction (XRD), X-ray fluorescent spectrometry (XRF), and NH_3 adsorption and temperature programmed desorption (TPD) to illustrate the physic-chemical properties of catalysts. Results showed that a negligible difference in SCR activity is observed on the MnOx/FA-PG catalyst prepared by heterogeneous deposition and wetness impregnation. Activity of the MnOx/FA-PG catalyst prepared by wetness impregnation with $Mn(NO_3)_2$ as precursor was significantly superior to that of prepared with $Mn(CH_3COO)_2$ as precursor. Moreover, calcination temperature significantly influenced catalytic activity. Over 70% NO conversion could be achieved at 100°C for Mn_8/FA-PG catalysts calcined at 400°C for 3 h. The active component of the catalysts mainly existed in Mn^{4+}. According to the results of NH_3 adsorption and temperature programmed desorption (TPD), adsorption capacity of NH_3 was not the decisive factor of catalytic activity and the key was the activation of adsorbed NH_3.

Keywords: Fly ash-palygorskite; Manganese oxides; Low-temperature SCR

Introduction

Over the past decades, nitrogen oxides from power plants and motor vehicles have been remained as major air pollutants. They can result in acid rain, photochemical smog, and ozone depletion [1,2]. Selective catalytic reduction (SCR) of NOx with NH_3 in the presence of oxygen is a proven technique for the removal of NOx from flue gases [3]. A large number of SCR catalysts have been explored, including noble metals [4,5], transition metal oxides [6,7] and zeolites [8,9]. Especially, V_2O_5-WO_3/TiO_2 based catalysts have been extensively used in the industry [10,11]. However, the operating temperature for these catalysts is usually above 300°C, and the exhaust gases usually contain a large number of fly ash and SO_2, which can easily deactivate the catalysts. Thus, the development of active SCR catalysts that can be operated at relatively low temperatures is needed.

At present, many transition metal oxides are extensively studied and proved to be highly active for low temperature SCR of NO with NH_3, among which MnO_x-supported catalysts attract the most concern [10,12-14]. Zhao et al. found that $Mn_{0.06}/Al_2O_3$ yields the highest NO conversion and 80% of NO conversion is obtained at 250°C and 15,000 h^{-1} [15]. Zhang et al. reported that Mn/TiO_2 catalysts exhibit a high NO conversion of 90% at 100°C [16]. The high SCR activity for MnO_x-supported catalysts implies that there may be an intimate connection between manganese oxides and carrier. Smirniotis et al. compared TiO_2-, Al_2O_3- and SiO_2-supported MnOx catalysts and found that the SCR performance of the supported Mn catalysts decreased in the following order: TiO_2 (anatase, high surface area), TiO_2 (rutile), TiO_2 (anatase, rutile), γ-Al_2O_3, SiO_2, TiO_2 (anatase, low surface area) [17]. The material of carrier is directly related to the SCR activity for manganese-based catalysts.

In this paper, a solid waste material, fly ash (FA), was used as support, which can cause many problems and has become an important environmental issue [18]. Due to its moderate pore volume and pore structure, as well as its high silica and alumina contents, FA is used to prepare zeolite materials. Besides, a clay mineral, palygorskite (PG) [19,20], was added as a binder to FA. The materials were mixed well after machine extrusion to provide an inexpensive catalyst support [21]. Manganese oxide (MnOx) was loaded as the active ingredient to produce a MnOx -supported catalyst (MnOx/FA-PG). This catalyst was then characterized through experiments in a fixed bed reactor to investigate the feasibility of using FA-PG as a catalyst support. Specific surface area [Brunauer-Emmett-Teller (BET)] analysis, scanning electron microscopy (SEM), X-ray diffraction (XRD) and X-ray fluorescent spectrometry (XRF) were used to characterise the surface properties and crystalline structures of the catalyst. NH_3 adsorption and temperature programmed desorption (TPD) were performed to evaluate the effect of manganese species on the surface acidity of the catalyst.

Experimental

Catalyst preparation

FA was obtained from the Anqing Power Plant in Anhui Province, and natural PG was acquired from Jiashan County. Main chemical composition of the molded fly ash and palygorskite are listed in Table 1. Four particle size ranges, i.e., 5-8 mm, 2-5 mm, 0.38-0.83 mm and 0.25-0.38 mm, were chosen as the experimental samples. FA-PG was pretreated at 300°C for 3 h to remove surface water. The Mnx/FA-PG catalysts were prepared firstly by heterogeneous precipitation and wetness impregnation of heat-pretreated FA-PG with Mn nitrate aqueous solution. The concentration of the solution was varied to yield

***Corresponding author:** Xianlong Zhang, School of Chemical Engineering, Hefei University of Technology, Hefei, 230009, PR China
E-mail: zhangxianlong78@126.com

Sample	SiO_2	AlO_3	Fe_2O_3	CaO	MgO	TiO_2	K_2O	MnO	P_2O_5	SO_3	FeO
FA	48.29	21.64	3.05	20.33	1.74	0.98	1.66	0.04	0.26	1.87	/
PG	69.44	11.84	5.139	0.211	12.16	0.43	0.5	0.046	0.03	/	0.08

Table 1: Main chemical composition of FA and PG (%).

different Mn loading contents. The samples were dried overnight at 110°C. Then, the specific particle sizes of catalysts were calcined in O_2 stream at different temperatures and times. The catalyst samples were labeled Mnx/FA-PG, MA-Mnx/FA-PG and Mnx(P)/FA-PG based on loading of the active component with different precursors and preparation methods ("x", "MA" and "P" refer to the weight percentage of Mn loading, the $Mn(CH_3COO)_2$ solution and the heterogeneous precipitation method, respectively).

Catalyst characterization

X-ray fluorescent spectrometry (XRF) was performed on a XRF-1800 analyses (Shimadzu, Japan) to measure the chemical composition of the catalyst. X-ray Diffraction (XRD) was performed on a D/MAx2500V diffractometer (Rigaku, Japan) equipped with a monochromated Cu Kα radiation source. The catalysts were scanned at 2θ ranging from 10°-70°. Scanning electron microscopy (SEM) images of the samples were obtained by a JSM-6490LV (Hitachi, Japan) analyser.

Brunauer-Emmett-Teller (BET) surface area was measured from the N_2 adsorption and desorption isotherms at -196°C using the NOVA 2200e (Quantachrome, American) adsorption analyser. Prior to analysis, 0.08-0.1 g of catalyst was preheated at 100°C for 2 h under a N_2 atmosphere. NH_3 adsorption and temperature programmed desorption (TPD) were conducted in a quartz microreactor. Then, 100 mg of the sample was preheated in Ar at 473 K for 1 h to remove any adsorbed species. Afterward, the sample was cooled to 323 K in Ar (100 mL·min^{-1}) and 1500 ppm NH_3 balanced by Ar was adsorbed at this temperature with a total flow rate of 100 mL·min^{-1} until no signal variation of NH_3 was detected. Subsequently, the sample was purged with 100 mL·min^{-1} Ar until no NH_3 was detected in the outlet, and then ramped to 1073 K at a linear heating rate of 10 K·min^{-1} in Ar (100 mL·min^{-1}). The analysis of gases was performed using an online mass spectrometer QIC2000 (Hidden, Germany).

Activity measurement

Catalyst activity was evaluated in a quartz fixed bed continuous flow reactor that operated at a pressure of 0.1 MP, and 3 g of sample was used in each test. Gas composition was 600 ppm NO, 600 ppm NH_3 and 3 vol% O_2 balanced by Ar. Total flow rate was 350 mL·min^{-1}, which corresponds to a gas hour space velocity of 4000 h^{-1}. NO concentrations in the inlet and outlet gases were measured online with a flue gas analyser (testo350-XL). Catalyst performance was estimated through NO conversion, which was calculated by the following equation:

$$X_{NO} = \frac{[NO]_{in} - [NO]_{out}}{[NO]_{in}} \times 100\% , \quad (1)$$

Where $[NO]_{in}$ and $[NO]_{out}$ denote the inlet and outlet concentrations of NO, respectively.

Results and Discussion

Effect of different precursors and loadings on SCR activity over catalysts

NO conversion of Mnx/FA-PG catalyst prepared using $Mn(NO_3)_2$ as a precursor is shown in Figure 1. NO conversion increased with temperature increasing at the temperature range of 100-250°C with Mn-loading. Then, it started to decrease at temperature higher than 250°C. Besides, at the same temperature, NO removal efficiency increased with increasing Mn-loading. NO conversion for Mn_{24}/FA-PG catalyst could reach to 100% at 100°C. For FA-PG, the NO conversion was much lower in a temperature range of 100-200°C, indicating that the low temperature SCR performance of Mn-loaded catalyst was not contributed by the carrier but the MnOx itself. As shown in Figure 2, NO conversion for the MA-Mnx/FA-PG catalyst prepared using $Mn(CH_3COO)_2$ as precursor was higher than that of FA-PG over the entire reaction temperature, but significantly lower than that of the Mnx/FA-PG catalyst under the same Mn-loading. It is reasonable deduced that the difference of the precursors may form different active species after calcination. Researchers also investigated the characteristics of the catalysts prepared by different precursors. However, no consensus was reached. Li et al. [22] reported that the catalyst prepared using $Mn(NO_3)_2$ as precursor after calcination at 400°C mainly formed MnO_2 species. Meanwhile, Mn_2O_3 species was formed when $Mn(CH_3COO)_2$ was used as precursor. In addition, the catalytic activity of Mn_2O_3 was higher than that of MnO_2 in the unit area. However, Wu et al. [23] considered that the catalytic activity of MnOx could be sequenced as follows: $MnO_2 > Mn_5O_8 > Mn_2O_3 > Mn_3O_4 > MnO$. Further explanations about the characteristics of the catalysts prepared by different precursors will illustrate in the following.

Effect of preparation method on SCR activity

From above results, catalysts prepared by $Mn(NO_3)_2$ as precursor had high SCR activity. Thus, the NO conversion of the Mn_8/FA-PG/300°C and Mn_8(P)/FA-PG/300°C catalysts prepared by $Mn(NO_3)_2$ as precursor at the entire reaction temperature range is shown in Figure 3. As shown in Figure 3, Mn_8(P)/FA-PG catalyst had higher NO conversion than Mn_8/FA-PG catalyst in the temperature range of 100-150°C. However, at the temperature higher than 200°C, NO conversion of Mn_8(P)/FA-PG was lower than that of the Mn_8/FA-PG catalyst. Besides, the preparation process for impregnation method was complex and led to the loss of raw materials during catalyst preparation. Based on the aforementioned results, the impregnation method is optimal and relatively economic.

Effect of calcination temperature and calcination time on SCR activity

Based on the preceding catalyst characterization, calcination temperature and calcination time have a significant effect on the characteristics of the catalysts. In this work, we also study the effect of calcination temperature and calcination time on SCR activity. As shown in Figure 4a, the NO conversion of the Mnx/FA-PG catalyst first gradually increased, and then decreased slightly with increasing calcination temperature. When calcination temperature was 400°C, the highest SCR activity was obtained at a low temperature. In particular, over 70% NO conversion was got at 100°C and approximately 100% at 150°C. Hence, the Mnx/FA-PG catalyst that calcined at 400°C had higher SCR activity. A low calcination temperature may lead to the incomplete decomposition of $Mn(NO_3)_2$ which is accompanied with NOx production during the reaction process. However, at a high calcination temperature, the structure of gesso-water and structure-water in PG might fold, which can lead to the collapse of passageways in FA-PG carriers; this condition is disadvantageous for

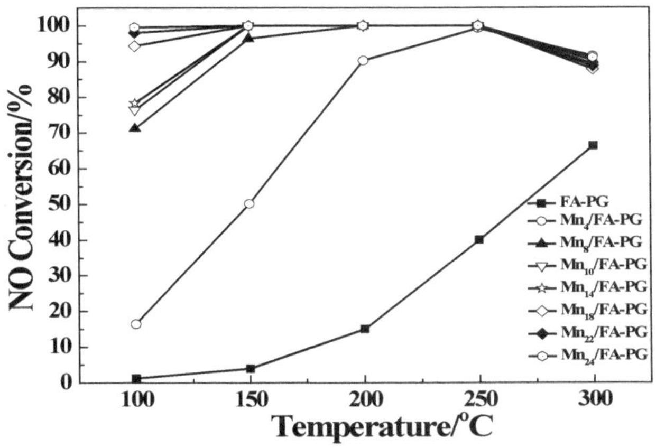

Figure 1: NO conversion of the Mnx/FA-PG catalyst with $Mn(NO_3)_2$ as precursor.

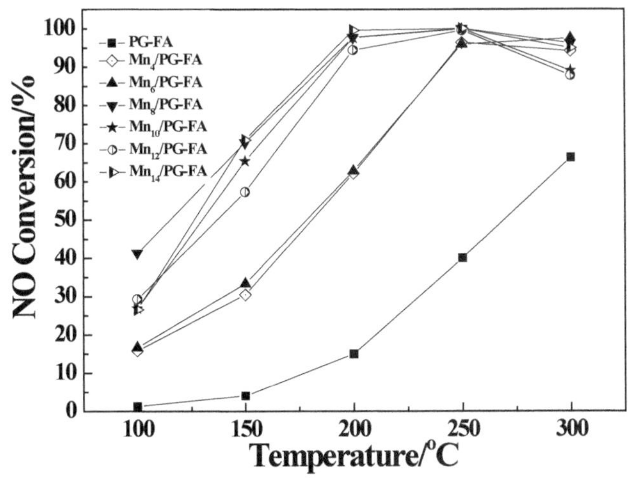

Figure 2: NO conversion of the Mnx/FA-PG catalyst with $Mn(CH_3COO)_2$ as precursor.

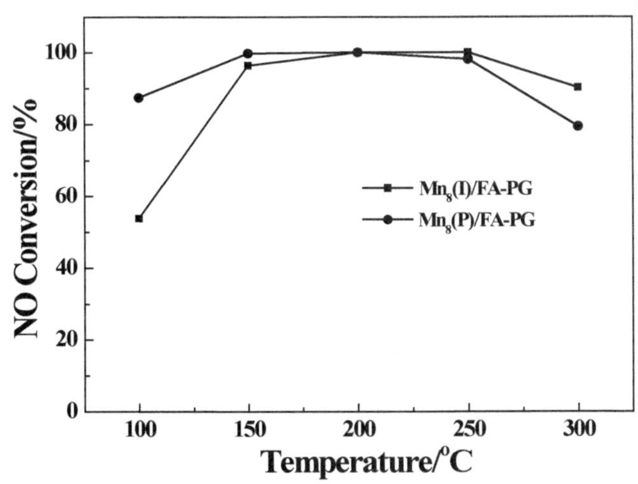

Figure 3: NO conversion of the Mnx/FA-PG catalyst prepared through different methods.

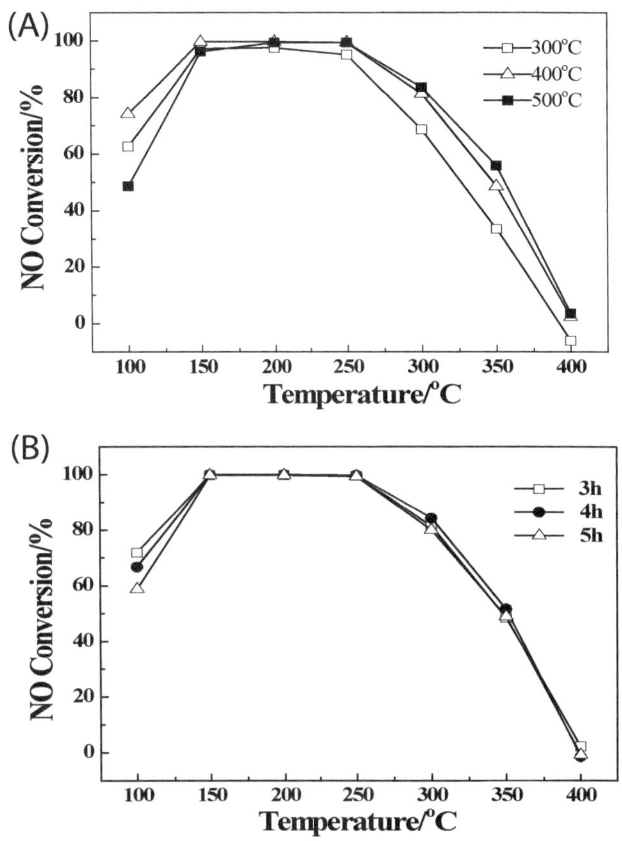

Figure 4: NO conversion of the Mn_8/FA-PG catalyst. (a) calcined at different temperatures, (b) calcined at 400°C for different times.

Sample	Parameter	S_{BET} [m²/g]
FA-PG	0.38-0.83mm	37.20
Mn_8/FA-PG [a]	Impregnation	68.70
Mn_8(P)/FA-PG [b]	Precipitation	72.10
Mn_8/FA-PG	$Mn(NO_3)_2$	68.70
MA-Mn_8/FA-PG [c]	$Mn(CH_3COO)_2$	44.04
Mn_8/FA-PG/300°C [d]	300°C	56.28
Mn_8/FA-PG/400°C [e]	400°C	49.05
Mn_8/FA-PG/500°C [f]	500°C	29.22
Mn_8/FA-PG	4.00-8.00 mm	63.32
Mn_8/FA-PG	2.00-4.00 mm	65.66
Mn_8/FA-PG	0.38-0.83 mm	68.70
Mn_8/FA-PG	0.25-0.38 mm	64.53

S_{BET}: Specific surface area

Table 2: BET surface area of catalysts. (a) Mn_8/FA-PG was prepared by impregnation method with $Mn(NO_3)$ as precursor. (b) Mn_8(P)/FA-PG was prepared via the heterogeneous precipitation method with $Mn(NO_3)_2$ as precursor. (c) MA-Mnx/FA-PG prepared using $Mn(CH_3COO)_2$ as precursor. (d) Mn_8/FA-PG/300°C calcined at 300°C. (e) Mn_8/FA-PG/400°C calcined at 400°C. (f) Mn_8/FA-PG/500°C calcined at 500°C.

NH_3 adsorption [24]. The surface areas of Mnx/FA-PG catalysts under different calcination temperatures are listed in Table 2. Compared with the carrier of FA-PG, the catalysts calcined at 300°C and 400°C had large BET surface areas, but the difference between them was minimal. Meanwhile, for the catalyst calcined at 500°C, the BET surface area decreased to a certain extent. A conclusion could be drawn that an extremely high calcination temperature leads to the destruction of the support structural. Besides, the NO conversion of the Mn_8/FA-PG catalyst calcined at 400°C for different times is shown in Figure 4b. As shown in Figure 4b, the NO conversion of the Mnx/FA-PG catalyst gradually decreased with increasing calcination times. Highest SCR activity was obtained when calcined at 400°C for 3 h. Thus, catalyst of Mn_8/FA-PG mentioned below was all calcined at 400°C for 3 h.

Effect of particle size on SCR activity over catalysts

Catalyst particle size has a significant effect on SCR activity. In the preparation method described above, the Mn_8/FA-PG (4.00-8.00 mm), Mn_8/FA-PG (2.00-4.00 mm), Mn_8/FA-PG (0.38-0.83 mm) and Mn_8/FA-PG (0.25-0.38 mm) catalysts were prepared to evaluate the influence of particle size on SCR activity for Mn-based catalysts. As illustrated in Figure 5, the NO conversion of the Mn_8/FA-PG catalyst first gradually increased, and then decreased slightly with decreasing particle size. The Mn_8/FA-PG catalyst with a particle size of 0.38-0.83 mm had a highest SCR activity, which is different from the results of Xuan [25,26]. The difference can be attributed to the material of the carrier and the preparation methods. The specific surface areas of Mn_8/FA-PG catalysts under different situations are listed in Table 2. Mn_8/FA-PG catalyst with a particle size of 0.38-0.83 mm had largest surface area. According to Eley-Ridal mechanism [27], the catalyst with large surface area is more conductive to the reaction gases (NH_3 and NOx) adsorbed on the carrier leading to the enhancement of SCR activity.

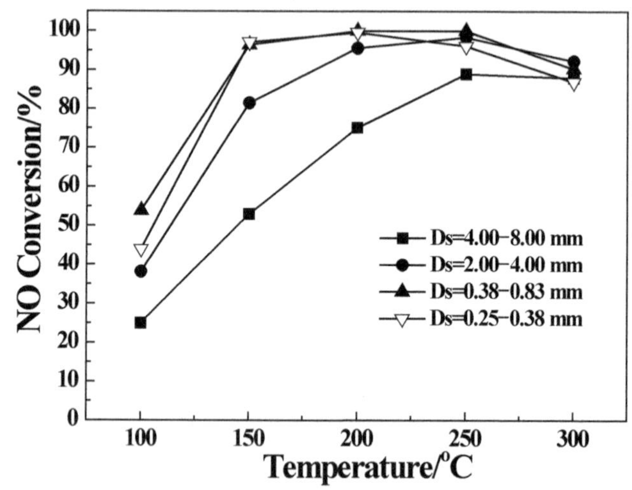

Figure 5: SCR activity of the Mn_8/FA-PG catalyst with different size of particle.

XRD analyses

The XRD patterns of the Mnx/FA-PG catalyst prepared by wetness impregnation with $Mn(NO_3)_2$ and $Mn(CH_3COO)_2$ as precursors are shown in Figure 6. As shown in Figure 6a, all these patterns presented intense peaks of PG, kaolinite and dolomite due to the complex composition of FA-PG. However, there were no diffraction peaks of manganese oxides for Mn_8/FA-PG and MA-Mn_8/FA-PG catalysts calcined at 400°C indicating that the complex composition of FA-PG may influence the diffraction of MnOx or MnOx is uniformly dispersed on the surface of the carrier with a low percentage of active components. Thus, in this study, we particularly selected the Mn_{12}/FA-PG and MA-Mn_{12}/FA-PG catalysts for XRD analyses to illustrate the presence of manganese oxides. As shown in Figures 6d and 6e, the diffraction peaks of crystalline MnO_2 and Mn_2O_3 phases were observed in the Mn_{12}/FA-PG and MA-Mn_{12}/FA-PG catalysts respectively, suggesting that MnOx in the amorphous phase is uniformly dispersed on the surface of the carrier, which is consistent with the founding by Li [22]. This finding indicated that MnOx was mainly in the form of MnO_2 when the Mnx/FA-PG catalyst was calcined at 400°C with $Mn(NO_3)_2$ as precursor. Meanwhile, the $Mn(CH_3COO)_2$ precursor mainly formed Mn_2O_3 species. Based on the above results in Figures 1 and 2, it is reasonable deduced that catalytic activity of MnO_2>Mn_2O_3 in the unit area. The XRD patterns of the Mn_{12}/FA-PG catalyst calcined at different temperatures are shown in Figure 7. The intensity of the diffraction peak of PG, kaolinite and dolomite for Mn-based catalyst decreased with increasing calcination temperature indicating that a high calcination temperature may destroy the primary structure of the carrier. Besides, the peaks corresponding to crystalline MnO_2 phase

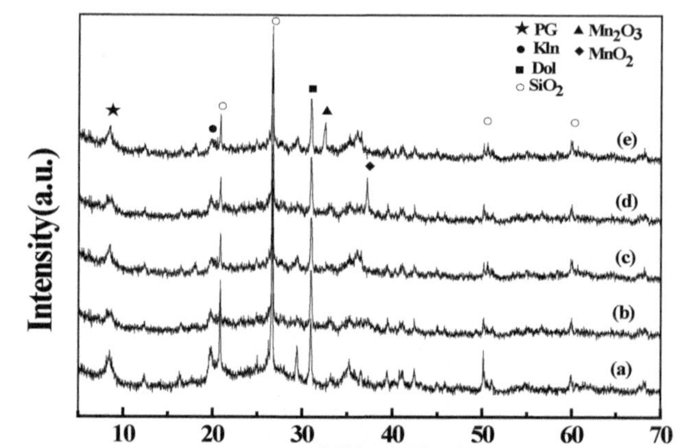

Figure 6: XRD patterns of different samples: (a) FA-PG, (b) Mn_8/FA-PG, (c) MA-Mn_8/FA-PG, (d) Mn_{12}/FA-PG and (e) MA-Mn_{12}/FA-PG.

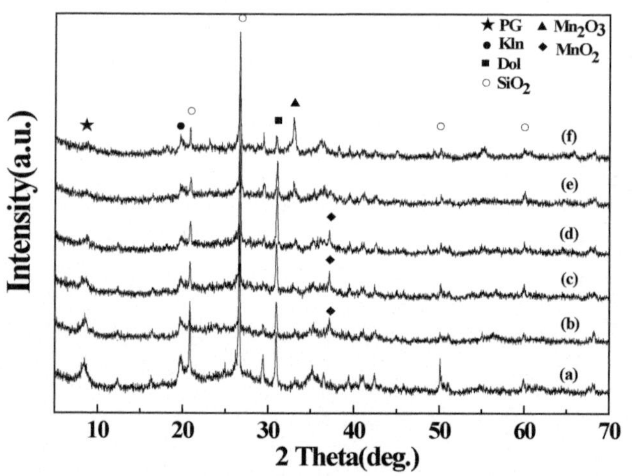

Figure 7: XRD patterns of the catalysts calcined at different temperatures: (a) Mn_{12}/FA-PG, (b) Mn_{12}/FA-PG/200°C, (c) Mn_{12}/FA-PG/300°C, (d) Mn_{12}/FA-PG/400°C, (e) Mn_{12}/FA-PG/500°C and (f) Mn_{12}/FA-PG/600°C.

diminished remarkably along with increasing calcination temperature and disappeared completely when calcination temperature reached over 400°C. When calcination temperature was 500°C, the diffraction peaks of Mn_2O_3 appeared. Moreover, the diffraction peaks of Mn_2O_3 were most prominent when calcination temperature reached to 600°C. This phenomenon is accordance with the founding by Kapteijn [28] that different calcination temperatures result in various oxidation states of manganese oxides and a high calcination temperature leads to a low valence of Mn.

SEM of the catalysts

The electron microscopy images of the Mnx/FA-PG catalyst prepared via different methods are shown in Figure 8. These images (b-d) pertaining to the Mn_8/FA-PG catalyst prepared by wetness impregnation with $Mn(NO_3)_2$, $Mn(CH_3COO)_2$ as precursors and heterogeneous precipitation method, respectively. Compared to the carrier of FA-PG, there was a certain amount of white particles existed for all of the Mn-loaded catalysts, revealing that MnOx has dispersed on the surface of the catalyst after calcination. However, for Mn_8/FA-PG catalyst prepared with $Mn(NO_3)_2$ as precursor, the particles on the surface of the catalyst were obviously more and higher uniformly dispersed than that of MA-Mn_8/FA-PG catalyst prepared with $Mn(CH_3COO)_2$, indicating that the precursor remarkably influences the morphologies of catalysts. Moreover, for MA-Mn_8/FA-PG catalyst, there was a slight sintering phenomenon after being calcined, which is not conducive to improving catalyst activity. The SEM images of Mn_8/FA-PG and MA-Mn_8/FA-PG were agreed with the activity test in Figures 1 and 2. Besides, as shown in Figure 8d, most white globular substances (MnOx) appeared on the surface of Mn_8/FA-PG catalyst prepared by heterogeneous deposition method illustrated that the catalyst exhibited the best SCR activity compared to the others [29].

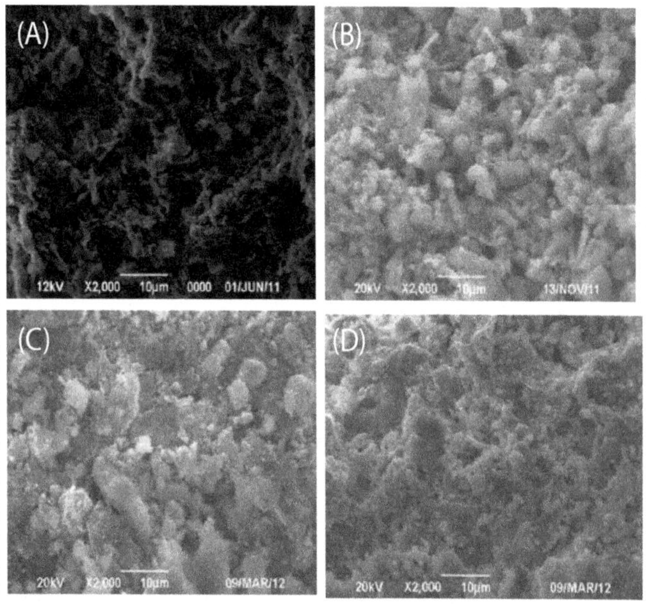

Figure 8: SEM images of the catalysts: (a) FA-PG, (b) Mn_8/FA-PG, (c) MA-Mn_8/FA-PG and (d) Mn_8(P)/FA-PG.

BET analyses

The physical properties of catalysts are important in determining the adsorption-desorption phenomena of gases onto its surface and influencing the activity of catalysts. Catalyst with large surface area is more conductive to the reaction gases (NH_3 and NOx) adsorbed on the carrier leading to the enhancement of SCR activity. Thus, the BET analyses are given and the textural properties are summarized in Table 2. Compared to the carrier of FA-PG, the surface area of catalysts increased after Mn-loaded. Mn_8(P)/FA-PG prepared by heterogeneous precipitation had larger surface area than Mn_8/FA-PG with impregnation. Combined with activity tests in Figure 3, it can be concluded that the larger surface area accounts for the improvement of SCR activity for Mn_8(P)/FA-PG. Besides, for Mn_8/FA-PG prepared with $Mn(NO_3)_2$ as precursor, the BET surface area was higher than that of MA-Mn_8/FA-PG prepared with $Mn(CH_3COO)_2$ as precursor. Similar result was obtained for Mn_8/FA-PG prepared with $Mn(NO_3)_2$ as precursor. For Mnx/FA-PG catalyst with different size of particles, the catalyst with a particle size of 0.38 mm to 0.83 mm has largest surface area corresponding to the SCR activity tests in Figure 5. The specific surface areas of the Mnx/FA-PG catalyst calcined at 300°C and 400°C both exhibited a significant increase compare to the FA-PG carrier. However, the NO conversion of the Mnx/FA-PG decreased when calcined at 500°C because the carrier structure was destroyed when calcination temperature was too high.

NH_3 adsorption analyses

A comparative study of NH_3 adsorption for the Mn(x)/FA-PG catalyst with different Mn loadings is presented in Figure 9. The penetration time of NH_3 was observed to decrease as Mn loading increased. This finding showed that the FA-PG carrier has a higher adsorption capacity than others. However, according to the results of BET analyses and activity test in Figure 1, the surface area of catalysts increased and the activity was improved after Mn-loaded. Thus, it can be deduced that the surface area of catalysts is not the decisive factor for NH_3 adsorption and the adsorption capacity of NH_3 not directly influence the SCR activity. Busca et al. [11] determined that the adsorption capacity of NH_3 influences catalytic activity. However, it was not the decisive factor and the key was the activation of adsorbed NH_3.

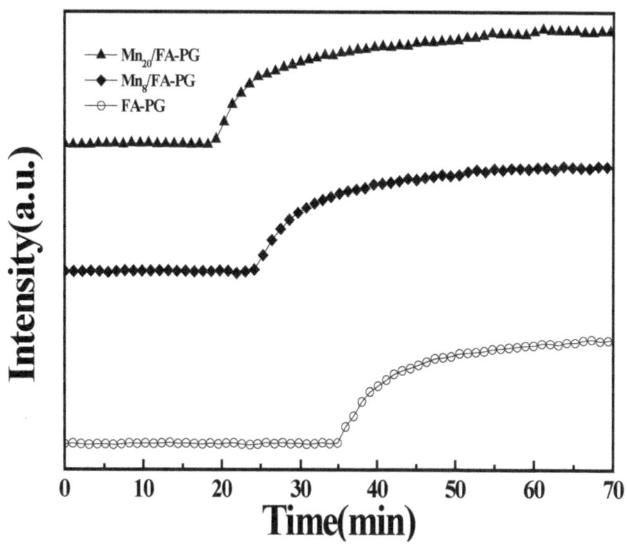

Figure 9: NH_3 adsorption profiles of catalysts at 50°C.

NH_3-TPD analyses of catalysts

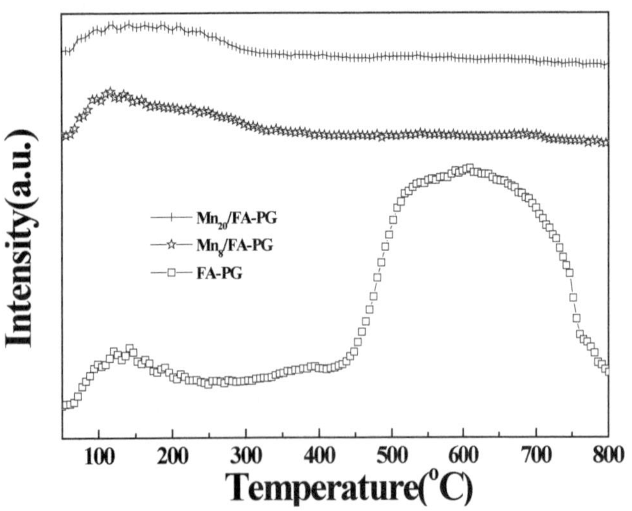

Figure 10: NH_3-TPD profiles of catalysts.

NH_3-TPD profiles of different Mnx/FA-PG catalysts are shown in Figure 10. Three main desorption peaks with different strength were observed at 132°C, 390°C and 603°C for FA-PG in the entire temperature range indicating that the carrier of FA-PG has three obvious acid sites. However, after manganese oxides loaded on the FA-PG, only one desorption peak at 132°C was found, and the desorption peak at 390°C and 603°C were disappeared. The peak at 132°C for Mnx/FA-PG could be assigned to the physically adsorbed of NH_3. According to the founding by Greenhalgh [30], the strength of B-acid site gradually decreased with the exchange of metal ion due to the reaction between metal ions of the precursors and OH on the support surface. Therefore, the peak at 390 °C and 603 °C may be caused by desorption of NH_3 that was adsorbed on the B-acid site of the carrier itself. Besides, Busca et al. [11] indicated that, the SCR performance of Mn-based catalyst is not related to the adsorption and activation of the B-acid site, but the L-acid site.

Conclusion

Mnx/FA-PG catalyst was highly active for SCR of NO with NH_3 at a temperature range of 150-300°C. Catalysts prepared by impregnation with $Mn(NO_3)_2$ as precursor not only efficient but also relatively economic owning to the uniformly dispersed and high catalytic activity of MnOx on the surface of the catalyst. NO conversion closed to 100% at 100°C under Mn-loading of 24% with particle size of 0.38 mm to 0.83 mm. Moreover, catalysts calcined at 400°C for 3 h exhibited the highest activity. Most important, the SCR performance of Mn-based catalyst was not related to the adsorption and activation of the B-acid site, but the L-acid site.

Acknowledgements

The authors gratefully acknowledge the financial support received from the National Natural Science Foundation of China (Grants Nos. 40902020, 51002042) and the Ph.D. Programmes Foundation of the Ministry of Education of China (Grant No. 20090111120019).

References

1. Zhao QS, Xiang J, Sun LS, Su S, Hu S (2009) Adsorption and Oxidation of NH3 and NO over Sol-Gel-Derived CuO-CeO2-MnOx/?-Al2O3 Catalysts. Energy Fuel 23: 1539-1544.
2. Blanco J, Avila P, Suárez S, Yates M, Martin JA, et al. (2004) CuO/NiO monolithic catalysts for NOx removal from nitric acid plant flue gas. Chem Eng J 97: 1-9.
3. Liu ZM, Woo SI (2006) Recent Advances in Catalytic DeNOx Science and Technology. Catal Rev 48: 43-89.
4. Lee JH, Schmieg SJ, Oh SH (2008) Improved NOx reduction over the staged Ag/Al2O3 catalyst system. Appl Catal A: Gen 342: 78-86.
5. Itoh M, Saito M, Takehara M, Motoki K, Iwamoto J, et al. (2009) Influence of supported-metal characteristics on deNOx catalytic activity over Pt/CeO2. J Mol Catal A: Chem 304: 159-165.
6. Shi YN, Chen S, Sun H, Shu Y, Quan X (2013) Low-temperature selective catalytic reduction of NOx with NH3 over hierarchically macro-mesoporous Mn/TiO2. Catal Commun 42: 10-13.
7. Schneider H, Maciejewski M, Kohler K, Wokaun A, Baiker A (1995) Chromia Supported on Titania: VI. Properties of Different Chromium Oxide Phases in the Catalytic Reduction of NO by NH3 Studied by in Situ Diffuse Reflectance FTIR Spectroscopy. J Catal 157: 312-320.
8. Long RQ, Yang RT (2000) Characterization of Fe-ZSM-5 Catalyst for Selective Catalytic Reduction of Nitric Oxide by Ammonia. J Catal 194: 80-90.
9. Li J, Zhu R, Cheng Y, Lambert CK, Yang RT (2010) Mechanism of propene poisoning on Fe-ZSM-5 for selective catalytic reduction of NO(x) with ammonia. Environ Sci Technol 44: 1799-1805.
10. Qi GS, Yang RT (2003) Low-temperature selective catalytic reduction of NO with NH3 over iron and manganese oxides supported on titania. Applied Catalysis B: Environmental 44: 217-225.
11. Busca G, Lietti L, Ramis G, Berti F (1998) Chemical and mechanistic aspects of the selective catalytic reduction of NOx by ammonia over oxide catalysts: A review. Applied Catalysis B: Environmental 18: 1-36.
12. Kim YJ, Kwon HJ, Nam IS, Choung JW, Kil JK et al. (2010) Deactivation of Pt catalysts during wet oxidation of phenol. Catal Today 15: 244-250.
13. Peña DA, Uphade BS, Smirniotis PG (2004) TiO2-supported metal oxide catalysts for low-temperature selective catalytic reduction of NO with NH3: I. Evaluation and characterization of first row transition metals. J Catal 221: 421-431.
14. Zhang LF, Zhang XL, Lv SS, Wu XP, Wang PM (2015) Promoted performance of MnOx/PG catalyst for low-temperature SCR against SO2 poisoning by addition of cerium oxide. RSC Advances 101: 82952-82959.
15. Zhao W, Li C, Lu P, Wen Q, Zhao Y, et al. (2013) Iron, lanthanum and manganese oxides loaded on gamma-Al2O3 for selective catalytic reduction of NO with NH3 at low temperature. Environ Technol 34: 81-90.
16. Zhang Y, Zhao X, Xu H, Shen K, Zhou C, et al. (2011) Novel ultrasonic-modified MnOx/TiO2 for low-temperature selective catalytic reduction (SCR) of NO with ammonia. J Colloid Interface Sci 361: 212-218.
17. Smirniotis PG, Sreekanth PM, Peña DA, Jenkins RG (2006) Manganese oxide catalysts supported on TiO2, Al2O3, and SiO2: A comparison for low-temperature SCR of NO with NH3. Industrial & Engineering Chemistry Research 45: 6436-6443.
18. Rubel A, Andrew R, Gonzalez R, Groppo J, Robl T et al. (2005) Adsorption of Hg and NOx on coal by-products. Fuel 84: 911-916.
19. Chen TH, Xu XC (2004) Nanometer Mineralogy and Geochemistry of palygorskite clays in the Border of Tiangsu and Anhui provinces. Beijing: Science press.
20. Zhang XL, Lv SS, Zhang CP, Wu XP, Zhang LF, et al. (2015) Effect of SO2 on SCR activity of MnOx/PG catalysts at low temperature. Chemical Papers 69: 1548-1555.
21. Sun XG, Yun D, Zhang P, Guo LY, Yao Q (2010) De-NOx Performance of Modified Fly Ash Supporting SCR Catalyst Under Various Molding Techniques. Journal of Combustion Science and Technology 2: 133-136.
22. Li JH, Chen JJ, Ke R, Luo CK, Hao JM (2007) Effects of precursors on the surface Mn species and the activities for NO reduction over MnOx/TiO2 catalysts. Catal Commun 8: 1896-1900.
23. Wu BJ, Liu XQ, Wang SG, Zhu L, Cao LY (2008) Investigation and characterization on MnOx/TiO2 for low-temperature selected catalytic reduction of NOx with NH3. Journal of Combustion Science and Technology 14: 221-226.
24. Chen TH, Wang J, Qing CS, Peng SC, Song YX (2006) Effect of heat treatment on the Palygorskite structure morphology, and surface properties, Ceramic Society 34: 1406-1410.

25. Xuan XP, Yue CT, Yao Q, Li SY (2003) Selective catalytic reduction of NO by ammonia with fly ash catalyst. Energy & Fuel 82: 575-579.

26. Xuan XP, Yue CT (2003) The selective catalytic reduction of NO by ammonia with fly ash catalyst. Journal of Environmental Science 23: 33-38.

27. Marban G, Valdes-Solis T (2004) Mechanism of low-temperature selective catalytic reduction of NO with NH3 over carbon-supported Mn3O4 Role of surface NH3 species: SCR mechanism. Journal of catalysis 226: 138-155.

28. Kapteijn F, Smgoredjo L, Andreml A (1994) Activity and selectivity of pure manganese oxides in the selective catalytic reduction of nitric oxide with ammonia [J]. Applied Catalysis B: Environmental 3: 173-189.

29. Jiang B, Liu Y, Wu Z (2009) Low-temperature selective catalytic reduction of NO on MnO(x)/TiO(2) prepared by different methods. J Hazard Mater 162: 1249-1254.

30. Greehalgh B, Fee M, Dobri A, Moir J, Burich R (2010) DeNOx activity-TPD correlations of NH3-SCR catalysts. Journal of Molecular Catalysis A: Chemical 333: 121-127.

Biofield Energy Treatment: A Potential Strategy for Modulating Physical, Thermal and Spectral Properties of 3-Chloro-4-fluoroaniline

Mahendra Kumar Trivedi[1], Rama Mohan Tallapragada[1], Alice Branton[1], Dahryn Trivedi[1], Gopal Nayak[1], Rakesh Kumar Mishra[2] and Snehasis Jana[2*]

[1]*Trivedi Global Inc., 10624 S Eastern Avenue Suite A-969, Henderson, NV 89052, USA*
[2]*Trivedi Science Research Laboratory Pvt. Ltd., Hall-A, Chinar Mega Mall, Chinar Fortune City, Hoshangabad Rd., Bhopal, Madhya Pradesh, India*

Abstract

3-Chloro-4-fluoroaniline (CFA) is used as an intermediate for the synthesis of pharmaceutical compounds. The objective of this study was to investigate the influence of biofield energy treatment on the physical, thermal and spectral properties of CFA. The study was performed in two groups (control and treated). The control group remained as untreated, and the treated group received Mr. Trivedi's biofield energy treatment. The control and treated CFA samples were further characterized by x-ray diffraction (XRD), differential scanning calorimetry (DSC), thermogravimetric analysis (TGA), fourier transform infrared (FT-IR) spectroscopy, and ultra violet-visible spectroscopy (UV-vis) analysis. The XRD analysis of treated CFA showed significant changes in the intensity of peaks as compared to the control. However, the average crystallite size (G) was significantly decreased by 22.08% in the treated CFA with respect to the control. The DSC analysis showed slight decrease in the melting temperature of treated CFA (47.56°C) as compared to the control (48.05°C). However, the latent heat of fusion in the treated sample was considerably changed by 4.28% with respect to the control. TGA analysis showed increase in maximum thermal decomposition temperature (Tmax) of the treated sample (163.34°C) as compared to the control sample (159.97°C). Moreover the onset temperature of treated CFA (148 °C) was also increased as compared to the control sample (140°C). Additionally, the weight loss of the treated sample was reduced (42.22%) with respect to the control (56.04%) that may be associated with increase in thermal stability. The FT-IR spectroscopic evaluation showed emergence of one new peak at 3639 cm-1 and alteration of the N-H (stretching and bending) peak in the treated sample as compared to the control. Overall, the result demonstrated that Mr. Trivedi's biofield energy treatment has paramount influence on the physical, thermal and spectral properties of CFA.

Keywords: 3-Chloro-4-Fluoroaniline; Biofield energy treatment; Thermal analysis; X-ray diffraction; Fourier transform infrared analysis; Ultra violet-visible spectroscopy

Abbreviations: CFA: 3-Chloro-4-fluoroaniline; XRD: X-Ray Diffraction; DSC: Differential Scanning Calorimetry; TGA: Thermo Gravimetric Analysis; FT-IR: Fourier Transform Infrared Spectroscopy; UV-vis: Ultra Violet-Visible Spectroscopy; CAM: Complementary and Alternative Medicine

Introduction

The amine derivatives are used as the building units in the construction of channel type supramolecular structure revealing its catalytic and separation properties due to their ability to generate intermolecular interactions [1,2]. 3-Chloro-4-fluoroaniline (CFA) is an amine derivative which is used as intermediate compound for the synthesis of herbicides [3]. Moreover, CFA is used as an intermediate for the synthesis of ciprofloxacin hydrochloride which is approved for the treatment of bone, joint infections, diarrhea, lower respiratory tract infections, and urinary tract infections [4]. Further, CFA is also used for the synthesis of norfloxacin that is the first representative of fluorinated quinolone derivatives. Norfloxacin drug has wide antimicrobial actions, and it was approved for the treatment of respiratory tract, ear, throat, nose and other infective and inflammatory diseases [5]. The chemical and physical stability of the pharmaceutical compounds are more desired quality attributes that directly affect its safety, efficacy, and shelf life [6]. Hence, it is required to explore some new alternate approach that could alter the physical and thermal properties of the compounds. Recently, biofield energy treatment has substantially changed the physical and thermal properties of metals [7,8], ceramics [9], organic product [10] and spectral properties of various pharmaceutical drugs [11]. After considering the pharmaceutical applications of CFA as intermediate, authors wish to investigate the impact of biofield energy treatment on CFA and analyzed its physical, thermal and spectral properties. The National Center for Complementary and Alternative Medicine (NCCAM), a part of the National Institute of Health (NIH), recommends the use of Complementary and Alternative Medicine (CAM) therapies as an alternative to the healthcare sector and about 36% of Americans regularly uses some form of CAM [12]. CAM includes numerous energy-healing therapies; biofield therapy is one of the energy medicine used worldwide to improve the health. The biofield treatment is being used in healing process to reduce pain, anxiety and to promote the overall health of human being [13,14]. Recently it was discovered that electrical process occurring in the human body has a relation with the magnetic field. According to Ampere's law, the moving charge produces the magnetic field in surrounding space. Likewise, human body emits the electromagnetic waves in the form of bio-photons, which surrounds the body, and it is commonly known as biofield. Therefore, the biofield consists of an electromagnetic field, being generated by moving electrically charged particles (ions,

***Corresponding author:** Snehasis Jana, Trivedi Science Research Laboratory Pvt. Ltd., Hall-A, Chinar Mega Mall, Chinar Fortune City, Hoshangabad Rd., Bhopal-462 026, Madhya Pradesh, India, E-mail: publication@trivedisrl.com

cell, molecule, etc.) inside the human body [15,16]. Rivera-Ruiz *et al.* reported that electrocardiography has been extensively used to measure the biofield energy of the human body [17]. Thus, human beings have the ability to harness the energy from the environment/Universe and can transmit into any object (living or non-living) around the Globe. The object(s) will always receive the energy and responding in a useful manner that is called biofield energy. Mr. Trivedi's unique biofield treatment is also known as The Trivedi Effect®. Mr. It is known to transform the characteristics of various living and nonliving things. Moreover, the biofield treatment has improved the growth and production of agriculture crops [18-20] and significantly altered the phenotypic characteristics of various pathogenic microbes [21,22]. The present work was focused to study the impact of biofield treatment on physical, thermal and spectral properties of CFA and characterized by XRD, DSC, TGA, FT-IR and UV-visible spectroscopy analysis.

Materials and Methods

3-Chloro-4-fluoroaniline (CFA) was procured from Sisco Research Laboratories (SRL), India.

Biofield treatment

CFA was divided into two parts; one was kept as a control sample, while the other was subjected to Mr. Trivedi's biofield treatment and coded as treated sample. The treatment group was in sealed pack and handed over to Mr. Trivedi for biofield treatment under laboratory condition. Mr. Trivedi provided the treatment through his energy transmission process to the treated group without touching the sample. After biofield treatment the control and treated group was subjected to physicochemical characterization under standard laboratory conditions. The control and treated samples were characterized by XRD, DSC, TGA, FT-IR, and UV-visible analysis.

Characterization

X-ray diffraction (XRD) study: XRD analysis of control and treated CFA was carried out on Phillips, Holland PW 1710 X-ray diffractometer system, which had a copper anode with nickel filter. The radiation of wavelength used by the XRD system was 1.54056 Å. The data obtained from this XRD were in the form of a chart of 2θ *vs.* intensity and a detailed table containing peak intensity counts, d value (Å), peak width (θ°), relative intensity (%) etc. The average crystallite size (G) was calculated by using formula:

$$G=k\lambda/(b\cos\theta)$$

Here, λ is the wavelength of radiation used, b is full width half-maximum (FWHM) of peaks and k is the equipment constant (=0.94).

Percentage change in average crystallite size was calculated using following formula:

$$\text{Percentage change in crystallite size}=[(G_t-G_c)/G_c]\times 100$$

Where, G_c and G_t are the crystallite size of control and treated powder samples respectively.

Differential scanning calorimetry (DSC)

DSC was used to investigate the melting temperature and latent heat of fusion (ΔH) of samples. The control and treated CFA samples were analyzed using a Pyris-6 Perkin Elmer DSC at a heating rate of 10°C/min under air atmosphere and the air was flushed at a flow rate of 5 mL/min. Predetermined amount of sample was kept in an aluminum pan and closed with a lid. A blank aluminum pan was used as a reference. The percentage change in latent heat of fusion was calculated using following equations:

$$\%\text{Change in latent heat of fusion}=\frac{[\Delta H_{\text{Treated}}-\Delta H_{\text{Control}}]}{\Delta H_{\text{Control}}}\times 100$$

Where, $\Delta H_{\text{Control}}$ and $\Delta H_{\text{Treated}}$ are the latent heat of fusion of control and treated samples, respectively.

Thermogravimetric analysis-differential thermal analysis (TGA-DTA)

The thermal stability of control and treated CFA were analyzed by using Mettler Toledo simultaneous TGA and Differential thermal analyzer (DTA). The samples were heated from room temperature to 400°C with a heating rate of 5°C/min under air atmosphere.

FT-IR spectroscopy

The FT-IR spectra were recorded on Shimadzu's Fourier transform infrared spectrometer (Japan) with the frequency range of 4000-500 cm^{-1}. The analysis was accomplished to evaluate the effect of biofield treatment at an atomic level like dipole moment, force constant and bond strength in chemical structure [23]. The treated sample was divided in two parts T1 and T2 for FT-IR analysis.

UV-Vis spectroscopic analysis

UV spectra of the control and treated CFA samples were recorded on Shimadzu UV-2400 PC series spectrophotometer with 1 cm quartz cell and a slit width of 2.0 nm. The spectroscopic analysis was carried out using wavelength in the range of 200-400 nm and methanol was used as a solvent. The UV spectra was analyzed to determine the effect of biofield treatment on the energy gap of highest occupied molecular orbital and lowest unoccupied molecular orbital (HOMO–LUMO gap) [23]. The treated sample was divided in two parts T1 and T2 for the UV-Vis spectroscopic analysis.

Results and Discussion

XRD analysis

The XRD diffractogram of control and treated CFA are shown in Figure 1. XRD diffractogram of the control CFA sample showed intense crystalline peaks that can be correlated to its crystalline nature. The XRD diffractogram showed peaks at 2θ equal to 13.10°, 20.58°, 22.16°, 22.29°, and 28.40°. However, the treated sample also showed intense crystalline peaks at 2θ equal to 20.73°, 22.12°, 23.08°, 29.97°, and 30.30°. The XRD peaks originally present at 2-theta equal to 20.58°, 28.40°, 29.77° and 30.33° were shifted to Bragg's angle 2-theta equal to 20.73°, 28.37°, 29.97° and 30.30°. The intensities of these XRD peaks were increased substantially as compared to the control. This may be inferred as increase in the crystallinity of the treated CFA as compared to the control. The crystallite size was calculated using Scherrer formula and the results are presented in Figure 2. The crystallite size of the control sample was 94.80 nm, and it was decreased to 73.86 nm in the treated sample. The crystallite size was decreased by 22.08% in treated CFA with respect to control. The decrease in crystallite size of treated sample could be ascribed to biofield treatment that may cause increase in compressive stress that leads to increase in dislocation density and decrease in crystallite size. Zhao *et al.* showed that increase in compressive stress in the material causes decrease in crystallite size [24]. Moreover, Schicker *et al.* reported that crystallite size is inversely proportional to the milling speed, thus increase in milling speed decreases the crystallite size [25,26]. Mahmoud *et al.* reported that

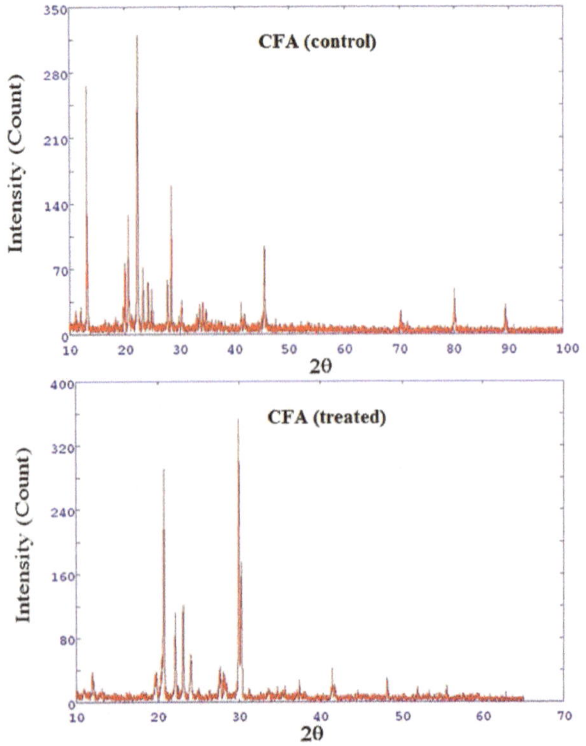

Figure 1: XRD diffractogram of control and treated 3-chloro-4-fluoroaniline.

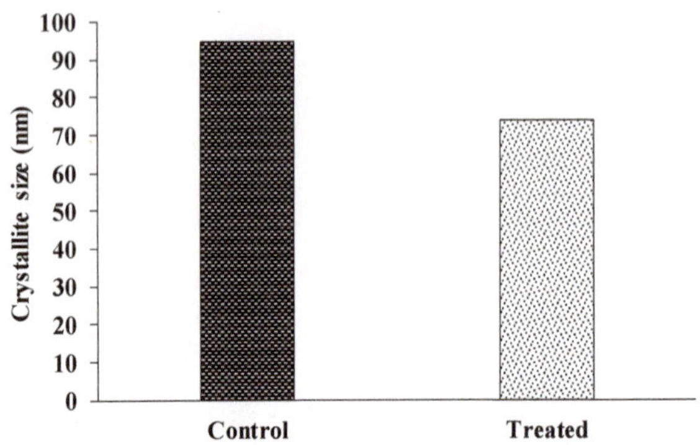

Figure 2: Crystallite size of control and treated 3-chloro-4-fluoroaniline.

lattice strain induced by mechanical milling may causes a significant reduction in crystallite size [27]. Previously our research group reported that biofield treatment had substantially reduced the crystallite size of vanadium pentoxide powders. It was proposed that internal strains made dislocations to move on the slip planes and intersecting slip planes built in stress concentration to such an extent causing the crystal to fracture at the sub boundaries [9]. Hence, it is assumed here that biofield energy treatment may provide the compressive stress through energy milling, which may lead to the reduction in the crystallite size of treated CFA as compared to the control.

It was previously suggested that nano scale particle size and small crystallite size can overcome slow diffusion rate by reducing overall diffusion distance and this enhances the net reaction rate [28,29]. Hence, it is assumed that the lower crystallite size of treated CFA may improve its reaction rate [30] and it could be utilized for the synthesis of pharmaceutical compounds.

DSC analysis

The DSC was used to investigate the melting temperature and latent heat of fusion of the control and treated CFA. The DSC thermograms of control and treated CFA are presented in Figure 3. DSC of control CFA showed a sharp endothermic peak at 48.05°C which corresponded to melting temperature of the sample. However, the treated CFA showed a slight change in the melting temperature (47.56°C) as compared to the control sample. It was reported that the thermal energy required for completing the phase change from solid to liquid of a substance is known as the latent heat of fusion (ΔH). The latent heat of fusion of control CFA was 159.98 J/g; however after biofield treatment it was changed to 153.14 J/g. The result showed that latent heat of fusion of the treated CFA was changed by 4.27% with respect to the control. It is assumed that biofield treatment might altered the internal energy stored in the treated CFA which may lead to change in latent heat of fusion with respect to the control.

TGA-DTA analysis

TGA was conducted to investigate the thermal stability of control and treated CFA. The TGA thermogram of control CFA showed one-step thermal degradation (Figure 4). The thermal degradation commenced at around 140°C, and it was terminated at around 189°C in control sample. During this process, the sample lost 56.04% of its weight. However, the TGA thermogram of treated CFA showed thermal degradation at around 148 °C, and it was terminated at 185°C. During this event, the sample lost around 42.22% of its weight. Hence, the onset temperature was increased in treated CFA as compared to control. Additionally, the result showed a reduction in weight loss of the treated CFA with respect to the control.

The DTA thermogram of control and treated CFA are presented in Figure 4. DTA thermogram of control sample showed an endothermic peak at 170.89°C, which corresponded to thermal degradation temperature of the sample. Nevertheless, the treated sample exhibited an endothermic peak at 172.64°C that attributed to thermal degradation of the sample. The DTG thermogram of control and treated samples are presented in Figure 4. The temperature where maximum thermal decomposition (T_{max}) occurred is recorded from DTG and data are tabulated in Table 1. DTG of control sample showed T_{max} at 159.97°C; however it was increased to 163.34 °C in treated compound. The increase in T_{max} of biofield treated compound showed an increase in the thermal resistance with respect to the control. Overall, the increase in onset temperature, T_{max} and reduction in weight loss of treated sample showed the superior thermal stability as compared to the control. The increase in thermal stability of treated CFA was may be due to conformational changes and crosslinking caused by biofield treatment [31].

FT-IR spectroscopy

The FT-IR spectra of control and treated samples are shown in

Parameter	Control	Treated
Latent heat of fusion ΔH (J/g)	159.98	153.14
Melting temperature (°C)	48.05	47.56
T_{max} (°C)	159.97	163.34
Weight loss (%)	56.04	42.22

Table 1: Thermal analysis data of control and treated 3-chloro-4-fluoroaniline. T_{max}: Maximum thermal decomposition temperature.

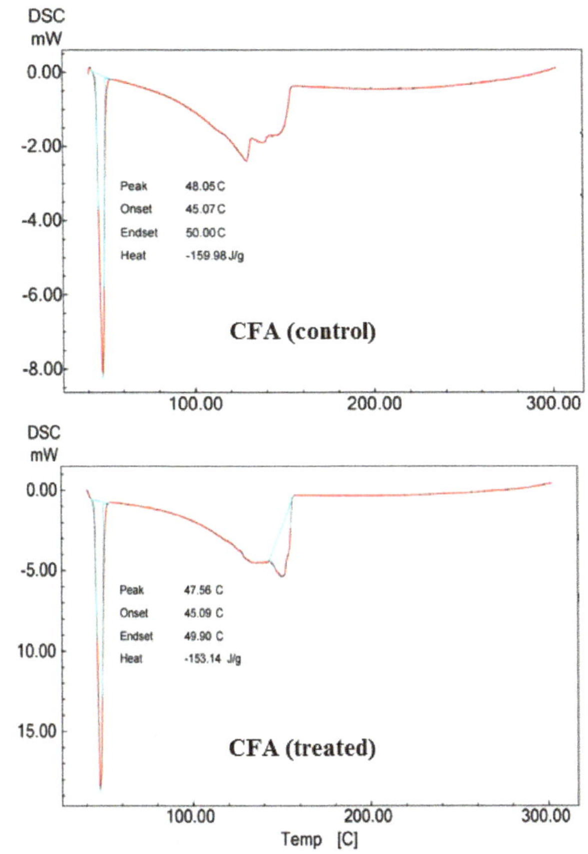

Figure 3: DSC thermogram of control and treated 3-chloro-4-fluoroaniline.

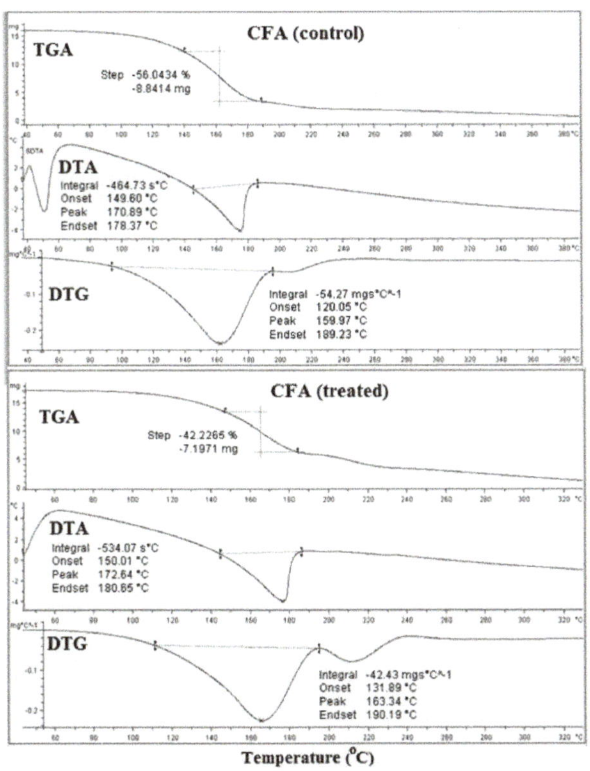

Figure 4: TGA thermogram of control and treated 3-chloro-4-fluoroaniline.

Figure 5. FT-IR spectrum of the control CFA showed characteristic peaks in the region of 3211-3439 cm^{-1} due to the N-H stretching vibration peaks. Vibration peak at 3049 cm^{-1} was due to the –CH aromatic stretching. Additionally, FT-IR peak at 1624 cm^{-1} was mainly due to the N-H bending vibration peak. The vibration peaks at 1500, 1599 cm^{-1} were corresponded to the C-C in ring stretching of aromatics. Absorption peaks at 1215 and 1259 cm^{-1} were due to the C-N stretching (aromatic) vibrations. The peak at 848 and 908 cm^{-1} were due to the –CH out of plane bending vibrations. Vibration peak at 769 cm^{-1} was attributed to chloro group attached to the phenyl ring [32]. The peak at 1045-1130 cm^{-1} were attributed to fluorine attached to the phenyl ring.

FT-IR spectrum of treated sample (T1) showed the N-H stretching vibration peaks in the region of 3211-3439 cm^{-1}. The –CH aromatic group stretching was observed at 3049 cm^{-1}. Vibrations peaks for the N-H bending and C-C in ring stretching were observed at 1620, 1599, and 1494 cm^{-1}. The C-N stretching vibration peaks were observed at 1215 and 1259 cm^{-1}. The C-H out of plane stretching vibrations were observed at 848 and 908 cm^{-1}. The C-Cl stretch was observed at 767 cm^{-1}. Additionally the absorption peaks in the region of 1045-1130 cm^{-1} were due to the fluorine group attached with phenyl ring.

The FT-IR spectrum of treated sample (T2) showed the N-H stretching vibration peaks in the region of 3215-3446 cm^{-1}. The –CH aromatic stretching was observed at 3049 cm^{-1}. Vibration peaks for the N-H bending and C-C in ring stretching were observed at 1637, 1602 and 1518 cm^{-1} respectively. The C-H out of plane bending vibrations were observed at 852 and 910 cm^{-1}. Vibration peak for a C-Cl stretch was observed at 775 cm^{-1}. The C-F stretching bond was observed in the range of 1057-1130 cm^{-1}. It is worthwhile to mention here that FT-IR spectrum of treated sample (T2) showed the emergence of new absorption peak at 3639 cm^{-1} that may be due to the intermolecular hydrogen bonding in the treated sample (T2). Moreover, the FT-IR spectrum of T2 sample showed an alteration in the region of 3439→3446 cm^{-1} (N-H stretch) and 1624-1637 cm^{-1} (N-H bending) which may be associated with an increase in hydrogen bonding after biofield treatment. Overall, the FT-IR result showed an increase in hydrogen bonding of treated CFA after biofield treatment as compared to the control.

UV-visible spectroscopy

UV spectra of control and treated CFA are shown in Figure 6. The UV spectrum of control CFA showed three absorption peaks *i.e.,* 205, 235 and 301 nm. The UV spectrum of CFA (T1) showed the occurrence of three absorption peaks *i.e.,* 210, 237 and 286 nm. However, the UV spectrum of CFA (T2) showed absorption peaks at 204, 235 and 301 nm. The results showed that as compared to the control alteration in absorption peaks was noticed in CFA (T1) sample. The absorption peak originally present at 205 nm in control, which was shifted to 210 nm in T1 sample. Additionally, the absorption peak present at 301 nm in control was shifted downward to 286 nm in T1 sample. According to Cinarli *et al.* the absorption peak appears in the region of 210-290 are mainly due to n→ π* and π - π* transition of the aromatic rings [33]. Hence it is assumed here that due to biofield treatment alterations occurred in electron *i.e.,* bonding (n→ π* and π→π* transition) from the ground state to excited state in the treated CFA as compared to the control.

Conclusion

In summary, the XRD analysis revealed a decrease in crystallite size by 22.08% in treated CFA as compared to the control. It is assumed

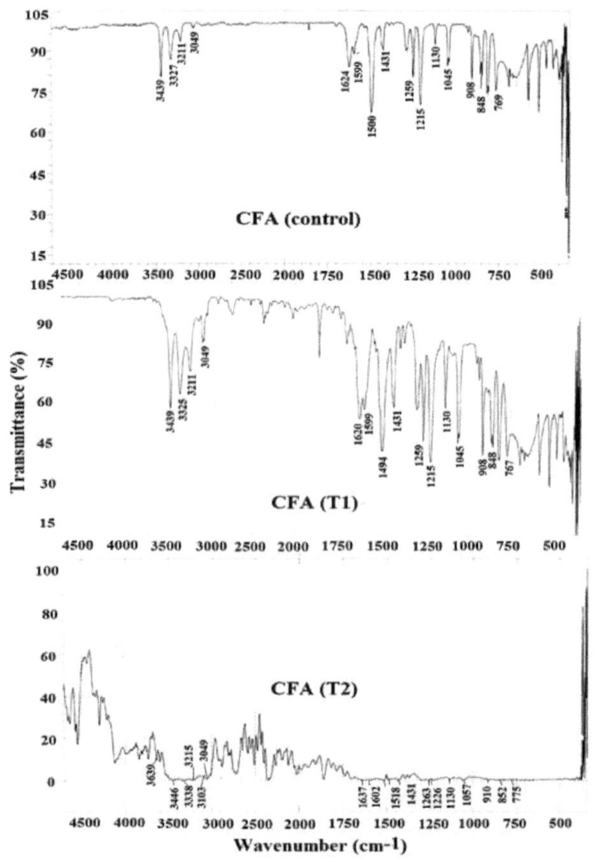

Figure 5: FTIR spectra of control and treated (T1 and T2) 3-chloro-4-fluoroaniline.

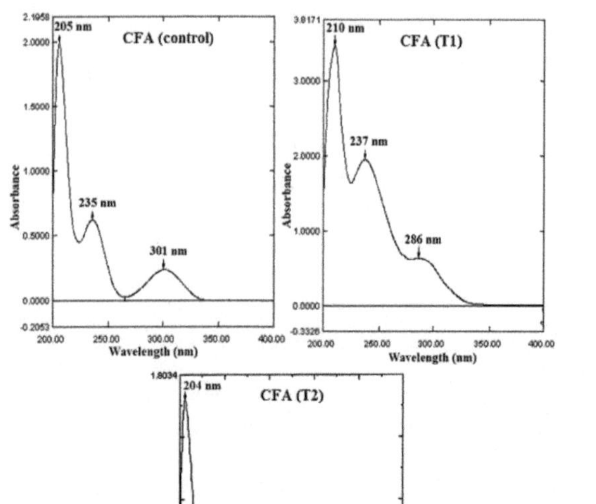

Figure 6: UV visible spectra of control and treated (T1 and T2) 3-chloro-4-fluoroaniline.

that occurrence of micro strain might cause a decrease in crystallite size in the treated CFA. However, the latent heat of fusion of treated sample was altered by 4.28% with respect to the control. The TGA analysis revealed an increase in T_{max}, and onset temperature as well as a reduction in the weight loss in treated sample as compared to the control. This indicated the increase in thermal stability of treated the CFA as compared to the control sample. The FT-IR spectroscopic analysis of treated CFA showed alterations in the N-H stretching and bending peaks as compared to the control. This may be due to biofield treatment that increased intermolecular hydrogen bonding in the sample as compared to the control. Additionally, UV-visible analysis showed alterations in absorption peaks of aromatic ring in the treated compound as compared to the control. The decrease in crystallite size, and increase in thermal stability of the treated sample showed that biofield energy treatment has significant impact on physical, thermal and spectral properties of CFA.

Acknowledgements

The authors would like to thank all the laboratory staff of MGV Pharmacy College, Nashik for their assistance during the various instrument characterizations. The authors would also like to thank Trivedi Science, Trivedi Master Wellness and Trivedi Testimonials for their support during the work.

References

1. Brycki B, Kowalczyk I, Werner J, Borowiak T, Wolska I, et al. (2006) Polyamines. I. Spectroscopic properties of N,N-bis-(phthalimidopropyl)-N-propylamine and supramolecular interactions in its crystals. J Mol Struct 791: 137-143.
2. Ranganathan A, Pedireddi VR, Rao CNR (1999) Hydrothermal synthesis of organic channel structures: 1:1 hydrogen-bonded adducts of melamine with cyanuric and trithiocyanuric acids. J Am Chem Soc 121: 1752-1753.
3. Boogaard PJ, Beulink GD, van Sittert NJ (1994) Biological monitoring of exposure to 3-chloro-4-fluoroaniline by determination of a urinary metabolite and a hemoglobin adduct. Environ Health Perspect 102 Suppl 6: 23-25.
4. Troy DB, Beringer P (2006) Remington: The Science and Practice of Pharmacy, Lippincott Williams and Wilkins, Baltimore, Maryland, USA.
5. Vardanyan R, Hruby V (2006) Synthesis of Essential Drugs. Elsevier, Amsterdam, Netherlands.
6. Blessy M, Patel RD, Prajapati PN, Agrawal YK (2014) Development of forced degradation and stability indicating studies of drugs- A review. J Pharm Anal 4: 159-165.
7. Trivedi MK, Patil S, Tallapragada RM (2013) Effect of biofield treatment on the physical and thermal characteristics of silicon, tin and lead powders. J Material Sci Eng 2: 125.
8. Trivedi MK, Patil S, Tallapragada RMR (2015) Effect of biofield treatment on the physical and thermal characteristics of aluminium powders. Ind Eng Manag 4: 151.
9. Trivedi MK, Patil S, Tallapragada RM (2013) Effect of biofield treatment on the physical and thermal characteristics of vanadium pentoxide powder. J Material Sci Eng S11: 001.
10. Trivedi MK, Nayak G, Patil S, Tallapragada RM, Jana S, et al. (2015) Bio-field treatment: An effective strategy to improve the quality of beef extract and meat infusion powder. J Nutr Food Sci 5: 389.
11. Trivedi MK, Patil S, Shettigar H, Bairwa K, Jana S (2015) Effect of biofield treatment on spectral properties of paracetamol and piroxicam. Chem Sci J 6: 98.
12. Barnes PM, Powell-Griner E, McFann K, Nahin RL (2004) Complementary and alternative medicine use among adults: United States, 2002. Adv Data: 1-19.
13. Aldridge D (1991) Spirituality, healing and medicine. Br J Gen Pract 41: 425-427.
14. Springhouse (1998) Nurses handbook of complementary and alternative therapies. (2nd edn), Lippincott Williams & Wilkins.
15. Rivera-Ruiz M, Cajavilca C, Varon J (2008) Einthoven's string galvanometer: the first electrocardiograph. Tex Heart Inst J 35: 174-178.

16. Movaffaghi Z, Farsi M (2009) Biofield therapies: biophysical basis and biological regulations? Complement Ther Clin Pract 15: 35-37.

17. Neuman MR (2000) Biopotential electrodes. The biomedical engg handbook: (2nd edn), Boca Raton: CRC Press LLC.

18. Shinde V, Sances F, Patil S, Spence A (2012) Impact of biofield treatment on growth and yield of lettuce and tomato. Aust J Basic Appl Sci 6: 100-105.

19. Sances F, Flora E, Patil S, Spence A, Shinde V, et al. (2013) Impact of biofield treatment on ginseng and organic blueberry yield. Agrivita J Agric Sci 35: 22-29.

20. Patil SA, Nayak GB, Barve SS, Tembe RP, Khan RR, et al. (2012) Impact of biofield treatment on growth and anatomical characteristics of Pogostemon cablin (Benth.). Biotechnology 11: 154-162.

21. Trivedi MK, Patil S, Shettigar H, Bairwa K, Jana S, et al. (2015) Phenotypic and biotypic characterization of Klebsiella oxytoca: An impact of biofield treatment. J Microb Biochem Technol 7: 203-206.

22. Trivedi MK, Patil S, Shettigar H, Gangwar M, Jana S, et al. (2015) An effect of biofield treatment on multidrug-resistant Burkholderia cepacia: A multihost pathogen. J Trop Dis 3: 167.

23. Pavia DL, Lampman GM, Kriz GS (2001) Introduction to spectroscopy. (3rd edn), Thomson Learning, Singapore.

24. Zhao Z, Du L, Tan Z (2014) Influence of electrodeposited crystallite size on interfacial adhesion strength of electroformed layers. Micro Nano Lett 9: 73-76.

25. Schicker S, Garcia DE, Gorlov I, Janssen R, Claussen N, et al. (1999) Wet milling of Fe/Al/Al2O3 and Fe2O3/Al/Al2O3 powder mixtures. J Am Ceram Soc 82: 2607-2612.

26. Muslimin M, Meor Yusoff MS (2009) The effect of high-energy milling on the crystallite size of alumina. JNRT 6: 95-102.

27. Mahmoud AE, Wasly HS, Doheim MA (2014) Studies of crystallite size and lattice strain in Al-Al2O3 powders produced by high-energy mechanical milling. J Eng Sci 42: 1430-1439.

28. Chaudhary AL, Sheppard DA, Paskevicius M, Webb CJ, Gray EM, et al. (2014) Mg2Si nanoparticle synthesis for high pressure hydrogenation. J Phys Chem C 118: 1240-1247.

29. Chaudhary AL, Sheppard DA, Paskevicius M, Saunders M, Buckley C (2014) Mechanochemical synthesis of amorphous silicon nanoparticles. R Soc Chem Adv 42: 21979-21983.

30. Chaudhary AL, Sheppard DA, Paskevicius M, Pistidda, C, Dornheim M, et al. (2015) Reaction kinetic behaviour with relation to crystallite/grain size dependency in the Mg–Si–H system. Acta Mater 95: 244-253.

31. Szabo L, Cik G, Lensy J (1996) Thermal stability increase of doped poly (3-hexadecylthiophene) by γ-radiation. Synt Met 78: 149-153.

32. http://orgchem.colorado.edu/Spectroscopy/specttutor/irchart.html.

33. Cinarli A, Gurbuz D, Tavman A, Birteksoz AS (2011) Synthesis, spectral characterizations and antimicrobial activity of some schiff bases of 4-chloro-2-aminophenol. Bull Chem Soc Ethiop 25: 407-417.

7

Modification of Activated Sludge Cohesion by Enzymatic Catalysis Process and Perspectives

Jacques Thierie*

Aquaphotomix Brussels Study Center (ABSC), 248 Beyseghem Street / 1120 NOH Brussels, Belgium

Abstract

It is possible to transform the structure of organic compounds so as to modify some of their properties to make them more "elastic", for example. Many solid-liquid separation processes use the cohesion property of the solid in suspension to increase the performance of the separation system. In this work, we examine some beneficial aspects that enzyme catalysis may reveal when acting on the strength of cohesion of a suspension. The topic concerns the activated sludge (AS) produced by wastewater of urban (or industrial) treatment plants. We used urban sludge and industrial liquid enzymes produced by the firm NOVO. To accelerate the research process, we employed enzyme mixtures (we called "enzyme cocktails") that highlighted the active mixtures, but until now we have not characterized specifically active enzymes. Despite a significant number of experiences, it is clear that given the number of different enzymes available and the almost infinite possible ratios of cocktails, only a small number of conceivable methods have been tested (and even less optimized). Nevertheless, a few cocktails were active and promising enough to improve solids contact processes in various industrial processes. We have found a range of combinations that increases the sludge cohesion coefficient and considered this enzyme treatment to strengthen the separation by solid contact structures mainly to tackle important hydraulic shock.

Keywords: Cohesion coefficient; Activated sludge; Wastewater treatment plant; Settler; Clarifier; Enzymatic catalysis

Introduction

The need to separate the non-biodegradable suspended solids from a liquid phase (water, for example) will become increasingly urgent over time. Not only because of solid waste, but also because of increased quantities of activated sludge from wastewater treatment plants. Among several methods, the "up flow" process involves passing a fluid stream containing suspended solids from the bottom to the top. Under good conditions, the solid tends to settle and the clarified water leaves the device, to be recovered (in a purification process) or to be released into the environment. Numerous methods based on the principle of an upstream liquid flow have emerged and are applied in various applications. In itself, the idea is simple (although the practical realization is not necessarily easy...). The basic principle of separation is shown in (Figure 1).

The "solid" part may be formed by a highly settling substrate (such as sand or slag) which helps in the suspended solids retention. In other cases, the retaining solid is absent and the "solid" (condensed) phase is only formed by the suspended particles (such as activated sludge of a wastewater treatment plant, for example). In a system like this, there are two flows in the opposite direction: an up flux liquid, flowing to the top and a downward solids settling flux. It is obvious that the process is applicable only if the considered compounds have particular characteristics, the most evident being the decant ability of the solid. But, that is not enough. The upstream liquid flow is characterized by an ascending velocity, v_{asc}, which is simply the ratio of the area of the container (the "cylinder" where clarification happens) to its section.

$$v_{asc} = Q(liquor\ volumic\ flux) / S(area\ of\ the\ clarifier)$$

The ascending rate has thus the dimensions of an area per unit time (m/h, for example). This up flow rate has a driving force on the suspended particles, from bottom to top. The less they are settleable, the faster they are driven upward. However, settleability is not the sole factor and they will also move upward more rapidly as they have less "solidarity". The ability of particles to "stick" to each other is called cohesion. This cohesion is quantified by the coefficient of cohesion, generally designated by K (in m/h, for example - ascending velocity units). The greater the interparticle cohesion, the greater the ratio of the lifting force that may be applied. Concurrently, the up flow clarifier performance increases as the ascending rate is higher (since the amount of clarified water is high).

The K factor of cohesion is therefore a crucial factor for clarifier performance since it increases the rate of climb (or overflow). For example, we give some useful overflow values (Table 1). Except for Densadeg, our device (see below) gives values close to systems used in the market. However, we still need to improve these values to improve the process and this is why we rely on an efficient enzymatic treatment. For reasons that will appear later, we focus here on activated sludge from urban wastewater treatment plants. The term "activated sludge" has an historical origin, and which is not really appropriate. This "mud" is in effect a rather compact biomass (several grams of dry matter per

TYPE	MANUFACTURER	OVERFLOW RATE (m/h)
PULSATOR	DEGRÉMONT	3-7
AQUACYCLE	S.A.U.R	3
CIRCULATOR	DEGRÉMONT	2
DENSADEG	DEGRÉMONT	10-20

Table 1: Sludge contact settlers in the market.

***Corresponding author:** Aquaphotomix Brussels Study Center (ABSC), 248 Beyseghem Street / 1120 NOH Brussels, Belgium, mail: thieriej@gmail.com

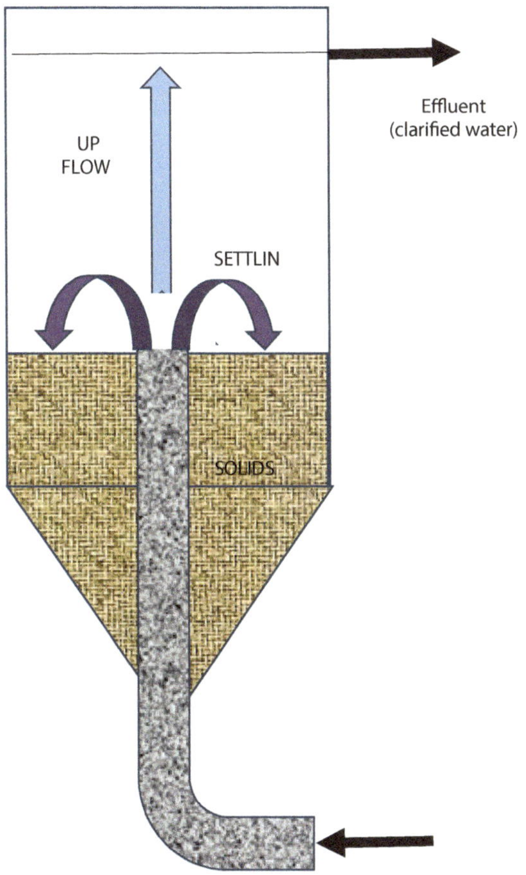

Figure 1: Schematic diagram of an up flow process.
The upwardly driving force of the liquor creates an ascendant velocity of the fluid; in the same time the solid particles settle. The volumic mass of the settling particles is generally very close to 1 g/cm^3, making the separation delicate.

liter), mainly composed of bacteria. These bacteria are embedded in flocs, a kind of "slime" structure of very complex composition which form a suspension in the mixed liquor. Just as for biofilms (which are fixed), the floc consists of microbial extracellular polymeric substances [2,3]. The chemical compositions thereof are very diverse, ranging from simple proteins, carbohydrates, polysaccharides, various nucleic acids, etc. The EPS turn out to be important, but play an only poorly understood role in floc cohesion, sludge grain and biofilms. The extraordinary diversity of chemical compounds and the relative misunderstanding of the phenomena dictated our strategy of using enzyme mixtures ("cocktails") to select more quickly effective mixtures although we will have to make it simpler later. Here we present our methodology, and some encouraging results. We are mindful of the immense work that continues to be answered.

Materials and Methods

Assembly

The apparatus (Figure 2) includes a large stand (2 m) in the top of which a graduated 2 L Squibb funnel is fixed (index: *amp*). A PTFE stopcock allows a continuous regulation of the flow from the funnel. A long silicone (or rubber) tube joined an ordinary glass tube that almost reached the bottom of a graduated cylinder (index: *GC*) 250 mL. The cylinder is placed in a tank which collects the excess effluent. On the course of the rubber tube a simple device (Figure 2a and 2b) allows to generate a discontinuous flow [1]. The dimensions of the graduated cylinder are: BRAND Din B (Silber BRAND 250 mL graduated 2 in 2 ml (error at 20°C: 1.5 mL) DURAN Germany External diameter: 3.86 cm internal diameter. 3.5 cm; Height: 31.5 cm; height to 250 mL: 25 cm. ΔH per 100 mL: 10 cm. (These dimensions are important for comparing data obtained elsewhere or to compare with the Degrémont method sometimes used for patents; Boutin, Combet, Communal) [4].

Filling

The filling liquid of the device is the effluent from the wastewater treatment plant where the sludge is coming from (to avoid differences in pH, osmotic strength, etc.). It is imperative to eliminate all bubbles in the unit (a small volume of effluent at the bottom of the graduated cylinder was often been useful). After completing the circuit, the Squibb stopcock is closed and the bulb carefully zeroed (always with the specific WWTP effluent of the studied sludge).

Sludge re-concentration

It is sometimes more effective to work with re-concentrated sludge. A sample of stirred AS V_{in}=100 ml (for example) is placed in a graduated vessel. After settling, a fixed volume of supernatant (V_{out}) is carefully removed (by pumping, for instance) without disturbing the settled sludge. The final concentration of the sludge is then $M_{AS}/(V_{in}-V_{out})$ (instead of M_{AS}/V_{in}).

Enzyme Cocktails

We used several enzyme mixtures of various formulas ("cocktails"). *NOVO* brand, they were used in liquid form and stored at less than 6°C. Proteases were added to the cocktail at the last moment. Unfortunately, the compositions of these cocktails are subject to a confidentiality agreement and cannot be disclosed here. (To quote the main: *NEUTRASE*: protease used to more or less deeply broken proteinaceous matter; *LIPOLASE*: the first fat-splitting enzyme; *VISCOZYME L*: cellulolytic enzyme mixture; *FLAVOURZYME*: an exo-protease; *AQUAZYM*: is used to remove starch, etc). The processed AS (index: *TR*) were mixed with the enzymes and agitated via magnetic stirrer for variable periods (from a few minutes to several hours), usually at room temperature (20-25°C).

Procedure

We tested the action of enzyme catalysis using two methods, starting in the same manner. An initial volume of AS (reconcentrated or not), V0, is introduced into the bottom of the graduated cylinder. During a very short time, the Squibb funnel stopcock is opened to adjust a "reasonable" flow of the effluent (neither too fast nor too slow).

Method 1 (M1)

This procedure consists in carrying out the following steps batch wise. The activated sludge being at its original volume (V0), discontinuous liquid volumes are added (thanks to device of Figure 3) to a desired volume AS (V_{SA}=75 mL, for example) in the graduated cylinder; the effluent is constantly discontinuously added to maintain the wanted volume for a fixed (long) time (two to three minutes). Stop the flow and reduce the AS back to the initial volume, V0; proceed as above for another value of V_{AS} (100 mL, then 150, and so on starting each time from V0 again). This method is relevant, if we admit that the AS swelling/deflation process is reproducible (or sufficiently reproducible). This method can be called "maintenance of swelling" at

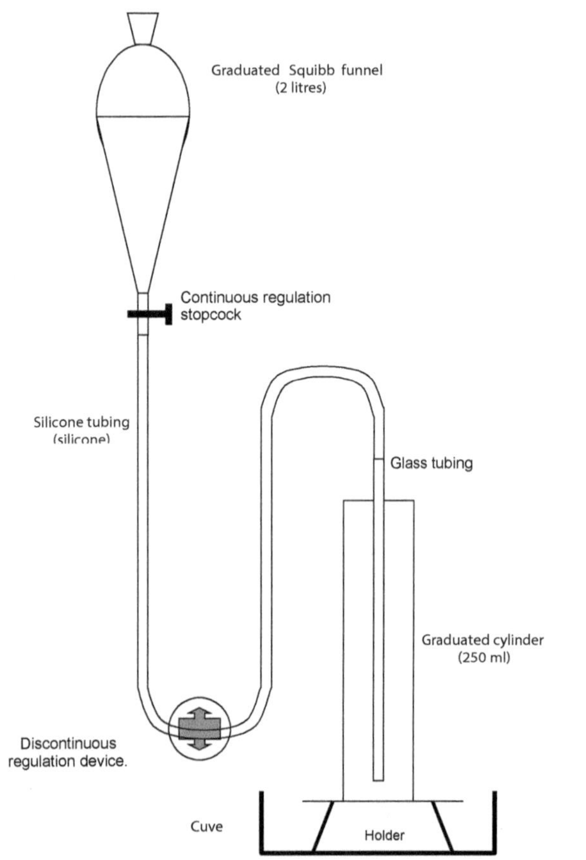

Figure 2: Scheme of the general assembly for determining the cohesion coefficient K.

a given value.

Method 2 (M2)

The previous procedure (M1) includes two factors: first, the activated sludge is brought to a desired volume; secondly, it is maintained in this volume. The flow time was short compared to the total time. In the second method (M2) conversely, the aim was to bring the sludge from V0 to V_{AS} as fast as possible, the principle being to keep a front sludge (sludge blanket) as demarcated as possible. The procedure was the same as in M1, but the measurement is interrupted when V_{AS} is reached. Here we speak about a "propelling" flow of AS rather than a maintenance flux.

In both cases, it is necessary to simultaneously measure:

• The flow volume in the Squibb funnel (V_{AMP});

• The corresponding time (t_{AMP}). This time corresponds to the duration of the flow admitted through the discontinuous device of Figure 3.

• The sludge volume in the graduated cylinder (V_{AS}).

Both methods give results which obviously depend on the cohesion of the sludge that is to say their ability to "stretch" and to maintain their consistency.

Results

We applied successively both methods (M1 and M2) on the same enzyme-treated (T) and untreated (NT) sludge. Treated or not, the AS have always been mechanically stirred in the same manner, in order to have identical manipulations in both cases (with and without treatment). We showed that this could be important for sticking phenomena and flocculation sludge [5] we performed a dozen experiments using different enzyme cocktails (differing both qualitatively and quantitatively). Each experiment included 4 results as well, according to the diagram below (Table 2).

Here are the synthetic results of a representative experiment (enzymatic cocktail: CKTL01). The sludge was taken at the wastewater treatment plant of Nivelles (Intercommunale du Brabant Wallon - IBW). The dry matter was 4.29 ± 0.13 gDW/L. The **T** sludge was treated by the CKTL01 enzyme mixture for 1 hour at 17.7°C, with magnetic stirring at 250 RPM. In all four cases, 50 mL of sludge were reconcentrated to 30 to 40 mL (V0) by careful aspiration of the supernatant. The dry matter was of course modified (Table 3a-3d).

The values of the sludge cohesion coefficient were obtained by linear regression using the graphs

$$v_{asc} = K\left(\frac{V}{V_0} - 1\right)$$

that is to say, straight lines [1] relating K and the ascending rate in the graduated cylinder. K has thus the units of an ascending velocity. The cohesive "force" of K is inversely proportional to its value as shown in Figure 4. (It is easily seen that the sludge volume in the graduated cylinder (V_{AS}) decreases rapidly with K. In our specific example, it is calculated that the sludge comes out of the cylinder for K ≈0.5 m/h). Figure 5a shows tests of treated sludge using method 1, keeping sludge level for 2-3 minutes (and EXP1 EXP3). The K values are very similar (2.5 and 2.62 m/h respectively). The enzymatic treatment doesn't seem to be efficient for the maintenance of sludge in the cylinder. On the contrary, Figure 5b shows a significant difference, with cohesion coefficients of 2.5 for treated sludge and 3.53 m/h for untreated sludge, so they are less cohesive.

Discussion

We have described a promising method for the modification of the cohesion coefficient of activated sludge. Earlier results specifically apply to resistance to hydrodynamic shocks (sludge displacement, not constant level change). Without wishing to insist any more on the issue, resistance to hydraulic shock is also an important advantage, particularly for more common "gravity" settlers, which likewise present an ascending rate, unavoidable for the flood. The problem is broader than our previous work suggests. (We have concentrated on the cohesion measurement methodology rather than on the usable practical results). In a critical discussion on the enzymes uses for improved treatment processes, it is not possible to ignore the high cost of enzymes usages. First, we tested at this point some industrial extracted enzymes, sometimes roughly purified, generally in a liquid form (sometimes solid, making them more stable). Once the studies performed with known products, it is obviously possible to use the whole microorganisms rather than expensive extracts. The downstream processing could obviously greatly be reduced. It is even conceivable to grow these producing microorganisms directly on the site of the plants, directly on unchanged, or only slightly amended wastewater, in parallel with the purification processes. (One can expect to get good results with fungi, robust and very rich in various enzymes). We believe then that the process we proposed is a preliminary step to an integrated work on site deployment. Our way of proceeding as described above, is expected to greatly facilitate the screening of enzymes and producing

Method 1 Maintenance flow	Without enzymatic treatment NT-EXP1	With enzymatic treatment T-EXP3
Method 2 Propelling flow	Without enzymatic treatment NT-EXP2	With enzymatic treatment T-EXP4

Table 2: The EXPz label designates method and the presence or absence of enzymes.

A

EXP 1	V_{AS} (mL)	t (s)	V_{AMP} (mL)	V_{asc} (m/h)	F(V) (nu)
	75	153	100	2.48	0.94
V0=39.6 mL	100	193	215	4.22	1.59
K (m/h)=2.50 ± 0.15	125	185	250	5.12	2.24
	150	182	350	7.29	2.89
r=0.99	175	124	275	8.40	3.53
	200	96	275	10.86	4.18

B

EXP 2	V_{AS} (mL)	t (s)	V_{AMP} (mL)	V_{asc} (m/h)	F(V) (nu)
	75	19	30	5.98	1.63
V0=28.5 mL	100	32	75	9.02	2.51
K (m/h)=3.53±0.45	125	35	100	10.92	3.39
	150	36	150	15.66	2.89
r=0.97	175	-	-		-
	200	77	240	11.89	7.18

C

EXP 3	V_{AS} (mL)	t (s)	V_{AMP} (mL)	V_{asc} (m/h)	F(V) (nu)
	75	183	100	2.07	0.94
V0=38.6 mL	100	177	210	4.50	1.59
3.53 ± 0.45	125	141	225	6.05	2.24
	150	137	275	7.60	2.89
r=0.99	175	-	-	-	-
	200	123	350	10.78	4.18

D

EXP 4	V_{AS} (mL)	t (s)	V_{AMP} (mL)	V_{asc} (m/h)	F(V) (nu)
	75	18	50	10.58	1.50
V0=30 mL	100	20	70	13.07	2.33
K (m/h)=2.5 ± 0.28	125	26	90	13.22	3.17
	150	28	125	16.90	4.00
r=0.98	175	-	-	-	-
	200	37	190	19.25	5.67

# EXP=Experiment number
V0=AS initial volume in the graduated cylinder (t=0 sec) in mL
K=AS cohesion coefficient (m/h)
r=Fitting correlation coefficient
V_{AS}=Sludge volume to obtain (or to maintain) in the graduated cylinder (mL)
t=Time (in sec) necessary to attain (or to maintain) V_{AS}
V_{AMP}=Elapsed volume from the Squibb funnel during the above time
V_{ASC}=Ascending velocity in the graduated cylinder $F(V)=K\left(\frac{V}{V_0}-1\right)$ where A is the section of the graduated cylinder minus the section of the glass tubing supplying the effluent; m/h)
F(V)=$F(V)=K\left(\frac{V}{V_0}-1\right)$ to keep the Degrémont [1] representation (nu=no units)

Table 3: Typical results of an experiment.

microorganisms. There is yet another aspect to consider about the cost price of the process. Following an estimate of Leonard [6] from the University of Liège, there may be an annual sludge production of around 13 10^6 tons (dry weight, DW) in 2020 in Europe (via Eurostat). According to an ADEME and CEMAGREF [7] study, the costs per ton of DW activated sludge could spread from 300 to 555 €; based on these partial data, we would spend around 4 to 6,500,000,000 euros merely to treat sewage sludge. Even approximately, these amounts are huge! Two approaches are possible to mitigate such costs: reducing the production and recovery of such waste. Clearly these two ways of remediation are

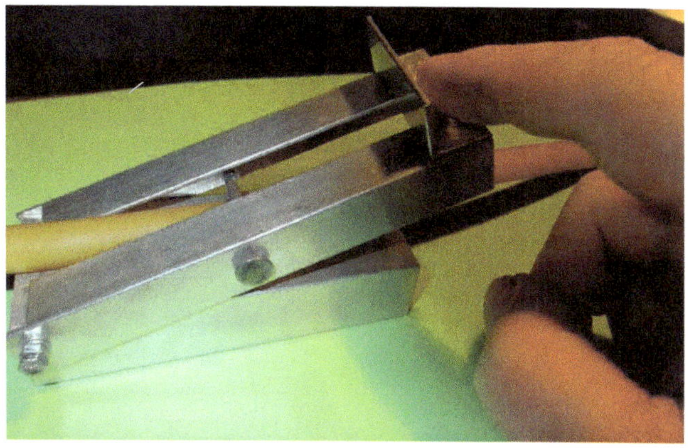

Figure 3a: The elasticity of the rubber tube maintains a high position of the lever. This lever comprises a clip that reposes on the tube but does not crush it: the fluid flows.

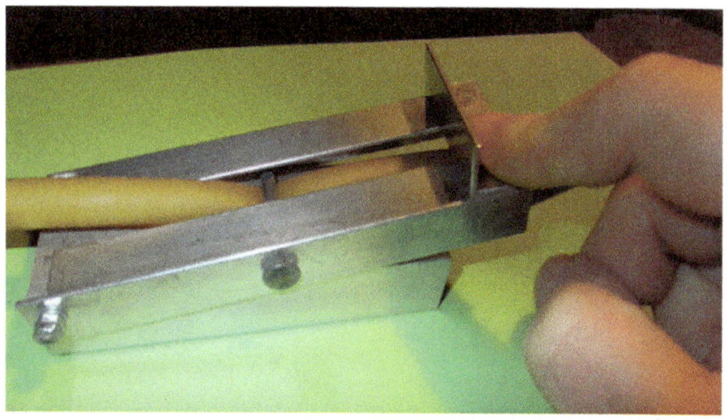

Figure 3b: A push of the lever compresses the tube with the clip: the fluid doesn't flow anymore. The device is made of aluminum profiles, screwed and steel axes. The set is very easy to construct and has proven very effective.

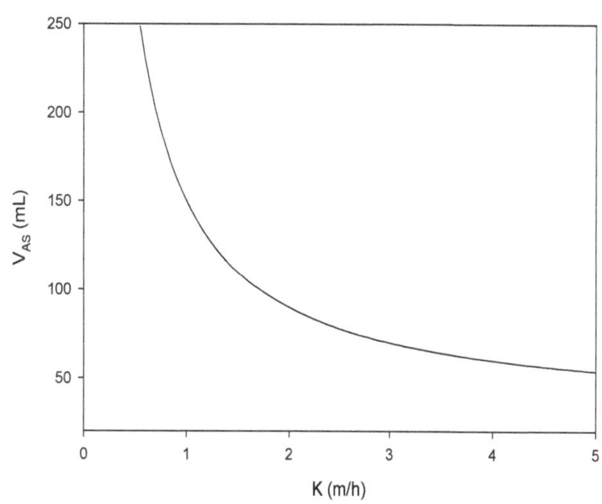

Figure 4: Volume of activated sludge (VAS) in the cylinder as a function of cohesion coefficient according to the "classical" relation (constants: V0=30 mL; Vasc upload velocity=4 m/h).
The figure shows that the sludge volume (and thus its height) in the cylinder is higher when K is small. The low cohesion leads to greater sludge expansion for (Vasc) equal ascending rate.

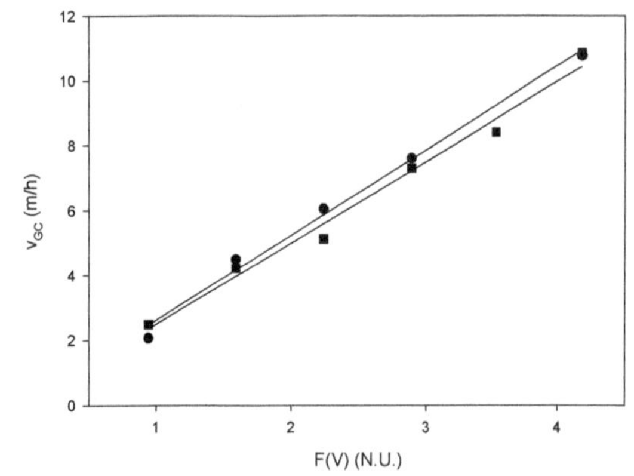

Ascending rate as defined by the Degrémont (1989) relationship.

$$v_{asc} = K\left(\frac{V_{asc}}{V0} - 1\right)$$

This relationship is a straight line allowing to determine the value of the cohesion

Figure 5a: This figure represents the result of the method maintaining the sludge height constant in the graduated cylinder at different ascending rate ($V_{asc}=V_{GD}$). One sees that the straight slopes for AS processed (●) and untreated sludge (■) are almost identical. K(●)=2.50 ± 0.15 m/h ; r=0.99 / K(■)=2.62 ± 0.13 m/h ; r=0.99. coefficient K by linear regression (NU=no units).

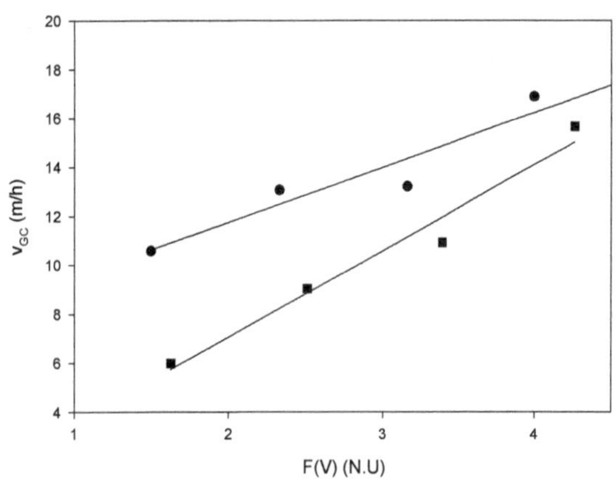

Figure 5b: This figure represents the result of the method bringing the height of the sludge in the graduated cylinder as quickly as possible to different volumes values ($V_{asc}=V_{GD}$). One sees that the straight slopes for treated AS (●) and untreated sludge (■) are now distinct. K(●)=3.53 ± 0.45 m/h ; r=0.97 / K(■)=2.50 ± 0.28 m/h ; r=0.98.

(or may be) related to sludge cohesion. Some chemical [4] or physical [8] processes could be aligned in this concept and do not necessarily use cheaper or safer processes for the environment (as some flocculants). The use of enzymes as clarifier's stabilizers against hydraulic shocks but also as cofactors in sludge treatment appears as a realistic perspective, after an optimization period of the process both in the hydrodynamic field and from the perspective of treatment and recovery of sewage sludge application [9-12].

References

1. Degrémont (1989) Technical Memento water, Lavoisier, Technical and Documentation, Paris 1: 353.
2. Tian Y, Zheng L, Sun DZ (2006) Functions and behaviors of activated sludge extracellular polymeric substances (EPS): a promising environmental interest. J Environ Sci (China) 18: 420-427.
3. Wingender J, Neu TR, Flemmings HC (1999) Microbial extracellular polymeric substances. Characterization, structure and function, Springer-Verlag, Berlin.
4. Boutin J, Combet A, Communal JP (1987) Chlorosulfate basic aluminum, its manufacturing process and its application as a coagulant.
5. Thierie J (2014) A Method for Exploring Suspended Particles Interactions by Means of Optical Density Measurements: Application to Bacterial Floc and Solid Nutrients. J Microb Biochem Technol 6: 279-285.
6. Leonard A (2005) Benchmark "innovative technologies" sludge recovery of STEP in Europe Laboratory of Chemical Engineering. University of Liège.
7. Environment Agency and the Energy Management (1999) ADEME and CEMAGREF Publication, WWTP sludge. Use of organic waste in revegetation, France.
8. Ha JS, Moon JH, Xin Z, Koo KY, Kim YU, et al. (2014) Required Energy Reduction for Treatment of Excess Activated Sludge using Ultrasonic Vibration. Advanced Science and Technology Letters 69: pp.34-38.
9. Fahs A, Quilès F, Jamal D, Humbert F, Francius G (2014) In situ analysis of bacterial extracellular polymeric substances from a Pseudomonas fluorescens biofilm by combined vibrational and single molecule force spectroscopies. J Phys Chem B 118: 6702-6713.
10. McSwain BS, Irvine RL, Hausner M, Wilderer PA (2005) Composition and distribution of extracellular polymeric substances in aerobic flocs and granular sludge. Appl Environ Microbiol 71: 1051-1057.
11. Raynaud M, Vaxelaire J, Olivier J, Dieudé-Fauvel E, Baudez JC (2012) Compression dewatering of municipal activated sludge: effects of salt and pH. Water Res 46: 4448-4456.
12. Jiao Y, Cody GD, Harding AK, Wilmes P, Schrenk M, et al. (2010) Characterization of extracellular polymeric substances from acidophilic microbial biofilms. Appl Environ Microbiol 76: 2916-2922.

Local Reaction Circuits as Motors for the Self-Organization of Multimolecular Systems

Jacques Ricard*

Jacques Monod Institute CNRS, University Paris, Paris, France

Abstract

When a physical system is complex it may possess a behaviour that displays some similarities with true living systems. If we consider, for instance, an enzyme reaction involving the random binding of two substrates A and B to the active site of the enzyme, the correct manner to study this reaction is to derive the rate equation as a combination of elements in interaction, namely the free enzyme E and the two enzyme-substrate complexes, EA and EB. In addition to these steady state concentrations there are local circuits of E, EA and EB. These circuits can be modified by the global system itself in such a way that, for pure physical reasons, the system can favour, for instance, a circuit running to E, to EA or to EB. These circuits exist and their functioning can be modulated by the system considered as a whole.

Keywords: Local reaction circuits; Multimolecular systems

Introduction

Multimolecular systems can spontaneously change their topology according to certain physical laws. Such a situation could be considered a model for the self-organization of prebiotic systems that would not be "living" structures but could possibly possess some properties of living systems. An enzyme catalysed reaction can be considered as a network of enzyme states that mutually interact according to a certain topology, in such a way that the global system possesses some features of living systems. Such situations occur even with relatively simple enzyme systems in such a way that these systems seem to mimic primitive biological processes. As a matter of fact, it is quite possible that life has not appeared at once in multi molecular systems but has been preceded by a number of prebiotic systems that possessed some properties of true living systems.

Local Circuits, Non-equilibrium and the Spontaneous Organization of More Complex Systems

The reasoning that has been followed in this Section can be extended to more complex and realistic situations involving the random binding of two, or several, ligands A and B to an enzyme [1]. As it is assumed that ligand binding takes place randomly, the overall process can be described as shown in Figure 1.

The mathematical treatment of the system upon pseudo-equilibrium conditions leads to the classical equation

$$\frac{v}{[E_o]}=\frac{kK_1K_2[A][B]}{1+K_1[A]+K_3[B]+K_1K_2[A][B]} \qquad (1)$$

Where, as usual, v is the steady state reaction velocity of the process, $[E_o]$ the total enzyme concentration, $[A]$ and $[B]$ the substrate concentrations, K_1, K_2 and K_3 the corresponding equilibrium binding constants, k the catalytic rate constant. Even though this equation fits the steady state rate data, it is not correct and should be considered as an approximation. As a matter of fact, such an enzyme system cannot be in equilibrium and at the same time catalyse a chemical process [2]. Hence this equation that usually fits the experimental data can only be considered as an approximation.

One should expect that the concentration of the various enzyme forms should be dependent upon the various local reaction flows that appear in the system during conversion of the substrate into the products. These reaction flows are basically dependent upon two types of events: substrate binding to and desorption from the enzyme on one hand, catalysis on the other hand [3]. Generation of the free enzyme E through substrate desorption is shown in Figure 2. The overall process involves substrate desorption from EA and EB states. The corresponding expression of the process leading to the free enzyme is then

$$[E]=k_{-1}k_{-2}(k_{-3}+k_4[A])+k_{-3}k_{-4}(k_{-1}+k_2[B]) \qquad (2)$$

If we consider now the formation by desorption of state EA from

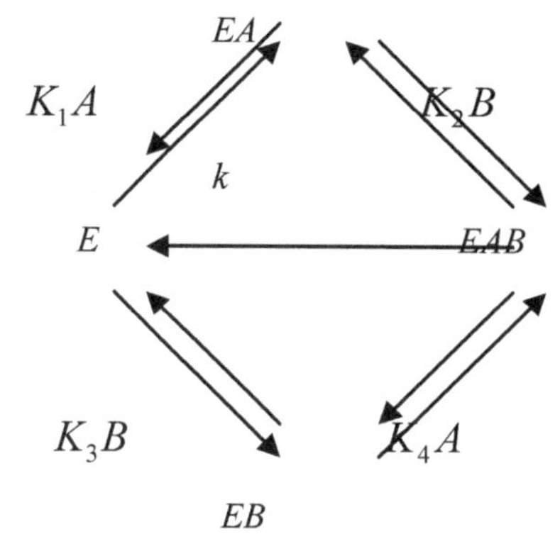

Figure 1: A random binding of two substrates (A and B) to the enzyme *E*. *K1,K2*... are the binding constants of *A, B,*... to the enzyme and *k* the apparent catalytic constant.

***Corresponding author:** Jacques Ricard, Director, Jacques Monod Institute CNRS, University Paris, Paris, France, E-mail: Jkricard@aol.com

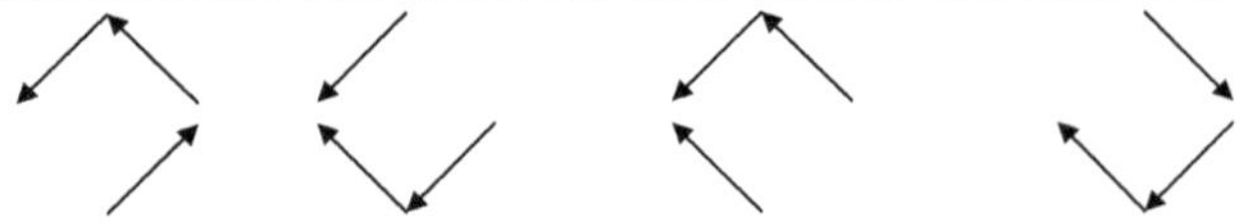

Figure 2: State E can be reached through substrate desorption steps that do not involve catalysis.

This process is described in equation (2).

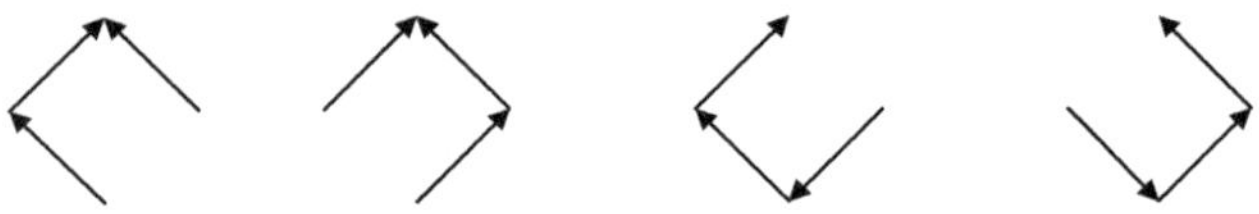

Figure 3: State *EA* can be reached through substrate desorption steps that do not involve catalysis. This process is described in equation (11).

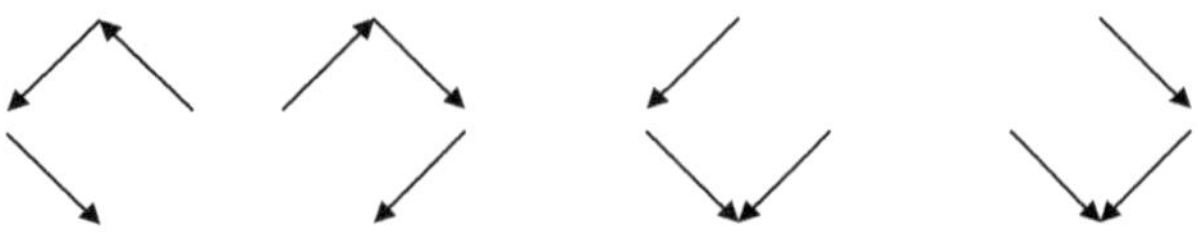

Figure 4: State *EB* can be reached through substrate desorption steps that do not involve catalysis.

This process is described in equation (15).

the other states (Figure 3) the situation is now similar, although with some additional complexities, to the one previously described. For instance, the process described in states 3 and 4 (Figure 3) is similar, with some additional subtleties, to the one previously described. The processes in states 3 and 4 (Figure 3) can be written as

$$[EA_3] = K_1[A](k_{-1}k_{-2}k_{-3}) \tag{3}$$

$$[EA_4] = \frac{k_{-3}K_3[B]k_{-4}K_4[A]k_2[B]}{K_2[B]} = k_{-3}k_{-4}K_1[A]k_2[B]$$

Hence the sum $[EA_3]+[EA_4]$ becomes

$$[EA_3]+[EA_4] = K_1[A]\{k_{-3}k_{-4}(k_{-1}+k_2[B])\} \tag{4}$$

and the sum of the four terms leading to $[EA]$ is now

$$[EA] = K_1[A]\{k_{-1}k_{-2}(k_{-3}+k_4[A]) + k_{-3}k_{-4}(k_{-1}+k_2[B])\} \tag{5}$$

The same reasoning can be followed for the enzyme sates that bind substrate *B* instead of *A*. Thus the process in states 1 and 2 can be expressed as

$$[EB_1] = K_3[B](k_{-1}k_{-2}k_{-3}) \tag{6}$$

And

$$[EB_2] = \frac{k_{-1}k_{-2}K_1[A]K_2[B]k_4[A]}{K_4[A]} = k_{-1}k_{-2}K_3[B]k_4[A] \tag{7}$$

Hence the sum of $[EB_1]+[EB_2]$ can be expressed as

$$[EB_1]+[EB_2] = K_3[B]\{k_{-1}k_{-2}(k_{-3}+k_4[A])\} \tag{8}$$

and the sum of the EB over the four states is now (Figure 4)

$$[EB] = K_3[B]\{k_{-1}k_{-2}(k_{-3}+k_4[A]) + k_{-3}k_{-4}(k_{-1}+k_2[B])\} \tag{9}$$

If now the enzyme has bound both *A* and *B*, the ternary *EAB* state can be generated through the process of Figure 3. Under these conditions one has

$$[EAB] = K_1K_2[A][B]\{k_{-1}k_{-2}(k_{-3}+k_4[A]) + k_{-3}k_{-4}(k_{-1}+k_2[B])\} \tag{10}$$

One can notice that, in this expression, the product K_1K_2 equals that of K_3K_4. This is a consequence of the first principle of thermodynamics. It is important to stress that, in this reasoning, the concentrations of E, EA, EB and EAB are solely obtained through substrate desorption from the enzyme without taking account of the catalytic process (Figure 5).

The common dynamic process that we have found in the genesis of *E, EA, EB, EAB* and which is expressed by relation (17) and (18) can be described by the flows shown in Figures 6-8.

In the processes we have considered so far we have implicitly considered that *E, EA* and *EB* are generated through events that do not involve the catalytic constant *k*. There is no doubt; however, that catalytic constant *k* is, in fact, involved in the processes leading to the reappearance of *E, EA* and *EB* during the catalytic reaction. The steps

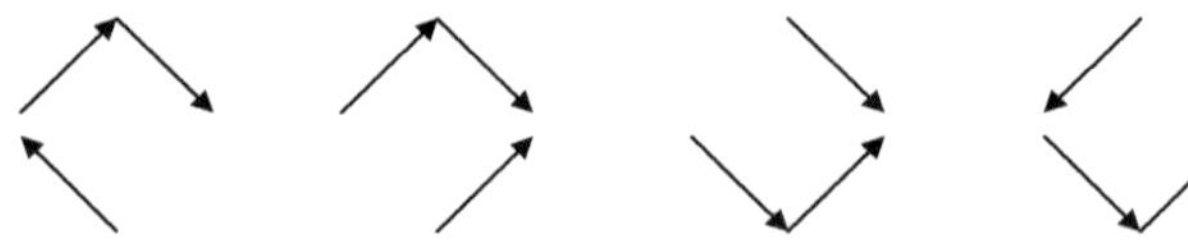

Figure 5: State *EAB* can be reached through substrate desorption steps that do not involve catalysis.

This process is described in equation (16).

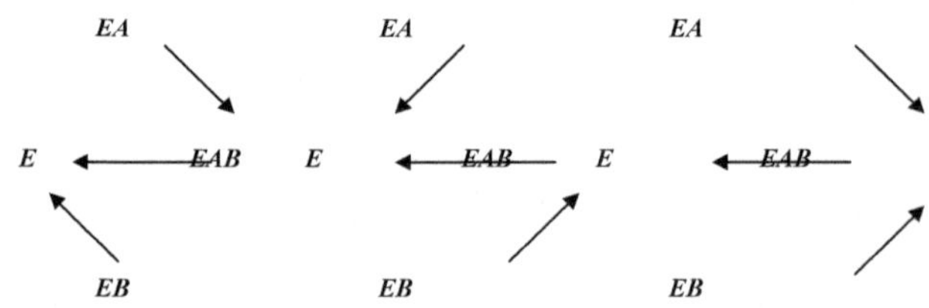

$$v_E = k(k_{-1}+k_2[B])(k_{-3}+k_4[A])$$

Figure 6: "Final" state *E* can be reached from *EAB, EA* and *EB* through a process involving catalysis.

These states can also be involved in a catalytic process leading to the free enzyme state *E*. The corresponding equation is shown in the Figure.

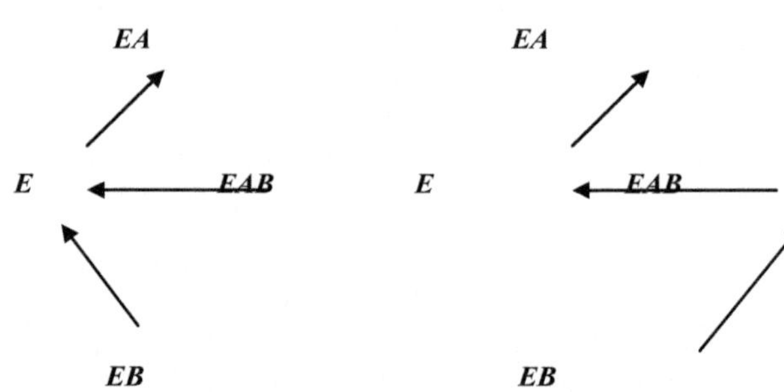

$$u_{EA} = kk_1[A](k_{-3}+k_4[A])$$

Figure 7: "Final" state EA can be reached from E, EB and EAB through a process involving catalysis.

The flux corresponding to this process is described by the equation shown in the Figure.

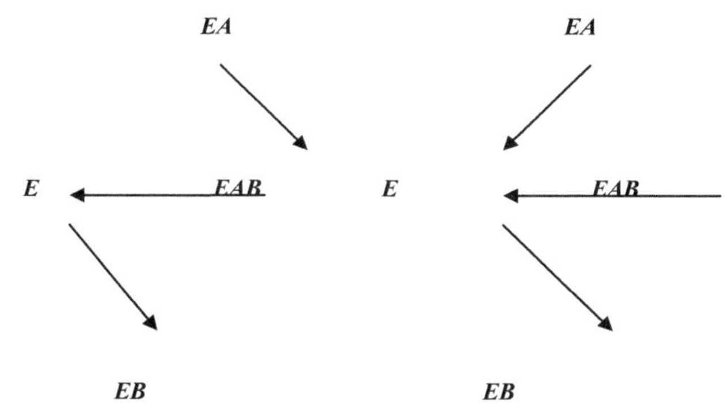

$$u_{EB} = kk_3[B](k_{-1} + k_2[B])$$

Figure 8: "Final" state EB can be reached from EA, E and EAB through a process involving catalysis.

The flux corresponding to this process is described by the equation shown in the Figure.

leading to E and involving the catalytic constant k are shown in Figure 7. The corresponding sum of the flows involving rate constant k is

$$k_{-1}k(k_{-3}+k_4[A]) + kk_2[B](k_{-3}+k_4[A]) = k(k_{-1}+k_2[B])(k_{-3}+k_4[A]) \quad (11)$$

Similarly, the flows leading to EA and to EB are shown in Figures 7,8.

The mathematical expressions of the local flows u_{EA} and u_{EB} are

$$u_{EA} = kk_1[A](k_{-3}+k_4[A]) = kk_{-1}K_1[A](k_{-3}+k_4[A]) \quad (12)$$
$$u_{EB} = kk_3[B](k_{-1}+k_2[A]) = kk_{-3}K_3[B](k_{-1}+k_2[A])$$

and they are described in Figures 7,8. Similarly the mathematical expression of the flow u_E is (equation 11)

$$u_E = k(k_{-1}+k_2[B])(k_{-3}+k_4[A]) \quad (13)$$

The concentrations of E, EA, EB and EAB can then be expressed as

$$[E] = \{k_{-1}k_{-2}(k_{-3}+k_4[A]) + k_{-3}k_{-4}(k_{-1}+k_2[B])\} + k(k_{-1}+k_2[B])(k_{-3}+k_4[A])$$
$$[EA] = K_1[A]\{k_{-1}k_{-2}(k_{-3}+k_4[A]) + k_{-3}k_{-4}(k_1+k_2[B])\} + kk_{-1}K_1[A](k_{-3}+k_4[A]) \quad (14)$$
$$[EB] = K_3[B]\{k_{-1}k_{-2}(k_{-3}+k_4[A]) + k_{-3}k_{-4}(k_{-1}+k_2[B])\} + kk_{-3}K_3[B](k_{-1}+k_2[B])$$
$$[EAB] = K_1K_2[A][B]\{k_{-1}k_{-2}(k_{-3}+k_4[A]) + k_{-3}k_{-4}(k_{-1}+k_2[B])\}$$

If we write these expressions relative to E_{nc}, the free enzyme resulting from substrate desorption i.e.

$$E_{nc} = k_{-1}k_{-2}(k_{-3}+k_4[A]) + k_{-3}k_{-4}(k_{-1}+k_2[B]) \quad (15)$$

One has

$$\frac{[E]}{E_{nc}} = 1 + \frac{k(k_{-1}+k_2[B])(k_{-3}+k_4[A])}{k_{-1}k_{-2}(k_{-3}+k_4[A]) + k_{-3}k_{-4}(k_{-1}+k_2[B])}$$

$$\frac{[EA]}{E_{nc}} = K_1[A] + \frac{kk_{-1}K_1[A](k_{-3}+k_4[A])}{k_{-1}k_{-2}(k_{-3}+k_4[A]) + k_{-3}k_{-4}(k_{-1}+k_2[B])} \quad (16)$$

$$\frac{[EB]}{E_{nc}} = K_3[B] + \frac{kk_{-3}K_3[B](k_{-1}+k_2[B])}{k_{-1}k_{-2}(k_{-3}+k_4[A]) + k_{-3}k_{-4}(k_{-1}+k_2[B])}$$

$$\frac{[EAB]}{E_{nc}} = K_1K_2[A][B]$$

The $u's$ that appear in these expressions are called the "relative local reactions flows". They are defined as

$$u'_E = \frac{k(k_{-1}+k_2[B])(k_{-3}+k_4[A])}{k_{-1}k_{-2}(k_{-3}+k_4[A]) + k_{-3}k_{-4}(k_{-1}+k_2[B])}$$

$$u'_{EA} = \frac{kk_{-1}K_1[A](k_{-3}+k_4[A])}{k_{-1}k_{-2}(k_{-3}+k_4[A]) + k_{-3}k_{-4}(k_{-1}+k_2[B])} \quad (17)$$

$$u'_{EB} = \frac{kk_{-3}K_3[B](k_{-1}+k_2[B])}{k_{-1}k_{-2}(k_{-3}+k_4[A]) + k_{-3}k_{-4}(k_{-1}+k_2[B])}$$

The u's defined in expressions (23) represent the "relative reaction flows" involving the catalytic step and running to the free enzyme (u'_E), the EA complex (u'_{EA}) or the EB complex (u'_{EB}). The reaction flows are expressed relative to the substrate desorption leading to the free enzyme.

Let us consider now an enzyme reaction involving two substrates A and B, the corresponding steady sate rate equation assumes the form

$$\frac{v}{[E]_0} = \frac{k[EAB]/E_{nc}}{\frac{[E]}{E_{nc}} + \frac{[EA]}{E_{nc}} + \frac{[EB]}{E_{nc}} + \frac{[EAB]}{E_{nc}}} \quad (18)$$

Moreover one has

$$\frac{[EAB]}{E_{nc}} = K_1K_2[A][B]$$

$$\frac{[E]}{E_{nc}} = 1 + u'_E$$

$$\frac{[EA]}{E_{nc}} = K_1[A] + u'_{EA} \quad (19)$$

$$\frac{[EB]}{E_{nc}} = K_3[B] + u'_{EB}$$

Hence equation (18) can be rewritten as

$$\frac{v}{[E]_0} = \frac{kK_1K_2[A][B]}{1 + u'_E + K_1[A] + u'_{EA} + K_3[B] + u'_{EB} + K_1K_2[A][B]} \quad (20)$$

The same reasoning can be applied to probabilities and lead to similar results, for instance

$$p(A,B) = \frac{K_1K_2[A][B]}{1 + K_1[A] + K_3[B] + K_1K_2[A][B] + u'_E + u'_{EA} + u'_{EB}} \quad (21)$$

or $p(A)$, namely

$$p(A) = \frac{K_1[A](1+K_2[B]) + u'_{EA}}{1 + K_1[A] + K_3[B] + K_1K_2[A][B] + u'_E + u'_{EA} + u'_{EB}} \quad (22)$$

In rate equation (20), or in probabilities (21), (22) the u' appear in the denominator only, or both in the numerator and the denominator. Moreover in rate equation (20) it appears that, owing to the u' values, the catalytic constant k appears both in the numerator and the denominator of equations (22)

Information in Stable and Unstable Equilibrium Systems

Let us consider, for instance, a protein that binds two ligands, A and B, randomly. In the absence of any reaction between A and B, or in the presence of a slow reaction between these ligands in such a way that the system is in equilibrium, or close to an equilibrium, the probability that the protein has bound a ligand [4], A for instance, is

$$p(A) = \frac{K_1[A] + K_1K_2[A][B]}{1 + K_1[A] + K_3[B] + K_1K_2[A][B]} \tag{23}$$

In this expression K_1, K_2, K_3 are equilibrium binding constants of *A* and *B* to the protein. But one can define another probability, the probability, for instance, that the protein binds *A* after it has already bound *B*. One can define this conditional probability as

$$p(A|B) = \frac{K_4[A]}{1 + K_4[A]} \tag{24}$$

One can then define from these two probabilities, $p(A)$ and $p(A|B)$, an information, consumed or generated, i.e.

$$i(A:B) = \log p(A|B) - \log p(A) \tag{25}$$

If $i(A:B) > 0$, the interaction between *A* and *B* generates information, and conversely it consumes information if $i(A:B) < 0$. These results are indeed valid for stable systems, i.e. for systems that do not evolve with time. But if the concentrations of *A* and *B* vary, expression (23) becomes

$$p(A) = \frac{K_1[A] + K_1K_2[A][B] + u'_{EA}}{1 + K_1[A] + K_3[B] + K_1K_2[A][B] + u'_E + u'_{EA} + u'_{EB}} \tag{26}$$

and the corresponding expression of $p(A|B)$ assumes the form

$$p(A|B) = \frac{[EAB]}{[EB] + [EAB]} = \frac{K_4[A]}{1 + K_4[A] + u'_{EB} / K_3[B]} \tag{27}$$

or

$$p(A|B) = \frac{K_4[A]}{1 + K_4[A] + u^{*'}_{EB}} \tag{28}$$

with $u^{*'}_{EB} = u'_{EB} / K_3[B]$. The expression of $p(A|B)/p(A)$ can be written as

$$\frac{p(A|B)}{p(A)} = \frac{K_4[A]}{1 + K_4[A] + u^{*'}_{EB}} \frac{1 + K_1[A] + K_3[B] + K_1K_2[A][B] + u'_E + u'_{EA} + u'_{EB}}{K_1[A] + K_1K_2[A][B] + u'_{EA}} \tag{29}$$

equivalent to

$$\frac{p(A|B)}{p(A)} = \frac{K_4(A) + K_1K_2[A][B] + K_1K_4[A]^2 + K_1K_2K_4[A]^2[B] + N(u)}{K_1[A] + K_1K_2[A][B] + K_1K_4[A]^2 + K_1K_2K_4[A]^2[B] + D(u)} \tag{30}$$

One can note that no emergence can appear in such a system if $p(A) > p(A|B)$ which means that $K_1[A] > K_4[A]$ and $D(u) > N(u)$. The second condition is of particular interest for it shows whether or not the local reaction flows can generate information in the global system. One can write

$$N = K_4[A](1 + K_1[A] + K_3[B] + K_1K_2[A][B] + u'_E + u'_{EA} + u'_{EB})$$

$$D = (1 + K_4[A] + u^{*'}_{EB})(K_1[A] + K_1K_2[A][B] + u'_A) \tag{31}$$

and the difference *N-D* assumes the form

$$N - D = K_4[A] - K_1[A] + K_4[A]u'_E - K_1[A]u^{*'}_{EB} + K_4[A]u'_{EB} - K_1K_2[A][B]u^{*'}_{EB} - u'_{EA}(1 + u^{*'}_{EB}) \tag{32}$$

One can notice that the functions u'_{EA}, u'_{EB} and $u^{*'}_{EB}$ can be expressed as a function of u'_E. One finds

$$u'_{EA} = \frac{k_1[A]}{k_{-1} + k_2[B]} u'_E$$

$$u'_{EB} = \frac{k_3[B]}{k_{-3} + k_4[A]} u'_E \tag{33}$$

$$u^{*'}_{EB} = \frac{k_{-3}}{k_{-3} + k_4[A]} u'_E$$

It follows that

$$K_4[A]u'_E - K_1[A]u^*_B = (K_4[A] - \frac{K_1[A]k_{-3}}{k_{-3} + k_4[A]}) \tag{34}$$

Similarly one has

$$K_4[A]u'_{EB} - K_1K_2[A][B]u^{*'}_{EB} = K_3K_4[A][B]k_{-3}(\frac{1}{k_{-3} + k_4[A]} - \frac{1}{k_{-3} + k_4(A)})u'_E = 0 \tag{35}$$

Hence the expression of $N - D$ is

$$N - D = K_4[A] - K_1[A] + (K_4[A] - \frac{K_1[A]k_{-3}}{k_{-3} + k_4[A]})u'_E - u'_B\ (1 + \frac{k_{-3}}{k_{-3} + k_4[A]} u'_E) \tag{36}$$

It follows from these expressions, which can be positive or negative, that the difference $N - D$ can also be positive or negative. Hence one can conclude that the system can consume, or generate information.

Discussion

In classical studies of enzyme kinetics it is implicitly, or explicitly, considered that the conversion of *EAB* into *EB, EA* and *E* are generated by simple desorption of the substrates *A* and *B* from the enzyme's active site. What the present studies have shown is that the situation is far more complex. Regeneration of *E, EA* and *EB* is, to a certain extent, effected by simple desorption of the substrates *A* and *B* from the enzyme. But the present studies have also shown the existence of a far more complex system in which the regeneration of *E, EA* and *EB* is dependent upon circuits, or functions, we have called u'_E, u'_{EA}, u'_{EB}. if a local circuit possesses the term u'_{EA} this means that this circuit is leading to the *EA* state. If, conversely, the local circuit contains the u'_{EB} function, this will favour the orientation of the flow toward the *EB* state. Last, but not least, if the local flow, or circuit, contains the u'_E, or the $u^{*'}_{EB}$, term this will tend to orient the flow toward the *E* state.

Another interesting result which is offered by the present mathematical analysis is that if $p(A) > p(A|B)$ then $h(A|B) > h(A)$ and the system generates some information. One can notice that depending on the sign of $\left(K_4[A] - \frac{K_1[A]k_{-3}}{k_{-3} + k_4[A]}\right)u'_E$ and $K_4[A] - K_1[A]$ then states *EB* or *EA* predominate. Hence, in the first case, the "u'_{EA} flow" is active and in the second case it is u'_{EB} which is more active. All these results are consistent with the view that multimolecular networks are highly organized structures that can spontaneously generate this organization. One can, in particular, notice that depending on the signs of $\left(K_4[A] - \frac{K_1[A]k_{-3}}{k_{-3} + k_4[A]}\right)u'_E$ and $K_1[A] - K_4[A]$ then states *EA* or *EB* predominate. If state *EA* predominates this means that the 'u'_{EA} circuit' predominates as well whereas in the second case it is the 'u'_{EB} flow' which is more active. These results are consistent with the view that the "u' circuits" are highly organized structures that can generate a novel organization. On could feel that the system "chooses" between two different circuits, namely the 'u'_{EA} *flow*'' and the '$u^{*'}_{EB}$ *flow*'. In fact, it is not a choice but the random interaction of the system with u'_{EA} or with u'_{EB} that makes the decision.

References

1. Dixon M, Webb EC, Thornec CJR, Tipton KF (1979) Enzymes (3rd edition). Longman Group.
2. Mahler HR, Cordes EH (1966) Biological Chemistry. Harper and Row.
3. Laidler KJ, Bunting P (1973) The Chemical Kinetics of Enzyme Action (1st edition). Clarendon Press Oxford.
4. Klipp E, Libermeister W, Wierling C, Kowald A, Lehrach H, et al. (2010) Systems Biology: a textbook. Wiley-Blackwell.

Reduced Hydrophobicity of the Minor Groove Intercalation Loop is Critical for Efficient Catalysis by Cold Adapted Uracil-DNA N-Glycosylase from Atlantic Cod

Elin Moe[1]*, Netsanet Gizaw Assefa[1], Ingar Leiros[1], Kathrin Torseth[2], Arne O Smalås[1] and Nils Peder Willassen[1]

[1]*The Norwegian Structural Biology Centre, University of Tromsø, N-9037 Tromsø, Norway*

[2]*Department of Laboratory Medicine, Children's and Women's Health, Norwegian University of Science and Technology, N-7489 Trondheim, Norway*

Abstract

The minor groove intercalation loop (leucine-loop) in Uracil-DNA N-glycosylases (UNGs) undergoes significant conformational changes upon substrate interaction. Previous studies have shown that the cold adapted Atlantic cod UNG (cUNG) possesses a tenfold higher catalytic efficiency than human UNG (hUNG). A sequence alignment of the two enzymes revealed amino acid substitutions in and near the leucine-loop of cUNG which may increase the elasticity of the loop and thus explain the observed higher catalytic efficiency of cUNG compared to hUNG. In order to investigate this hypothesis we constructed five single and four multiple mutants of cUNG with cod to human UNG substitutions and characterised them by kinetic experiments at three different temperatures (15, 22 and 37°C) and Differential Scanning Calorimetry (DSC). The results showed that the mutations affected the k_{cat} more than the K_M, and mutants with apparent reduced hydrophobic properties (A266T, V267A, A266T/V267A and A266T/V267A/A274V) showed increased or equivalent k_{cat} to cUNG, while mutants with predicted increased hydrophobicity (A274V, H275Y, L279F, A266T/V267A/A274V/H275Y and A266T/V267A/A274V/H275Y/H250Q) showed reduced k_{cat} compared to cUNG. Thus, mimicking the amino acid composition of the leucine-loop in hUNG reduces the activity of cUNG. The DSC experiments were performed in order to identify a potential increase in the overall stability of the mutants as a result of altered hydrophobicity of the leucine-loop. However, the results showed a slight reduction in the overall stability for mutants with increased k_{cat} and increased stability for mutants with reduced k_{cat}. We suggest these results arise from a need to compensate for a potential increased/reduced elasticity of the leucine-loop which is introduced by the amino acid substitutions. This study confirms the suggested importance of the leucine-loop for efficient catalysis of UNG and shows that it is possible to alter the catalytic capacity of this enzyme by manipulating the amino acid composition of this loop.

Keywords: Cold adaptation of enzymes; DNA repair; Uracil-DNA N-glycosylase; Kinetics; Differential scanning calorimetry

Introduction

Uracil-DNA N-glycosylase (UNG) is a DNA repair enzyme involved in the base excision repair pathway (BER) for removal of uracil in DNA. Uracil in DNA may result from deamination of cytosine [1] or misincorporation during replication [2]. The crystal structure of the catalytic domain is known from a wide range of organisms including human [3], herpes simplex virus 1 (HSV-1) [4], *E. coli* [5], Atlantic cod [6], *D. radiodurans* [7], *Vibrio cholerae* [8], Vaccinia virus [9], *Mycobacterium tuberculosis* [10], *Bacillus substilis* (Banos-Sanz, 2013) and *Staphylococcus aureus* [11]. All ten enzymes have similar structures and consist of a classic single domain α/β-fold with a central four stranded parallel and twisted β-sheet surrounded by eight to eleven α-helices. The N- and C-terminals are on opposite sides of the central β-sheet and the active site is located within a positively charged groove at the C-terminal end of the parallel β-sheet.

The crystal structures have also been determined of UNG in complex with DNA from human [12-14], *Vaccinia virus* [15,16] and *D. radiodurans* [17]. These structures have revealed that the uracil is flipped out of the DNA and into the active site upon detection in a process where the enzyme binds the minor groove, compress the phosphate backbone and kink the DNA [13]. Three Ser-Pro-rich loops are involved in this process (named from the hUNG structure): the 4-Pro loop, the Gly-Ser loop and the minor groove intercalation loop (leucine loop). The four serines in these loops form hydrogen bonds to the phosphates 5' and 3' to uracil, and may orient the enzyme correctly for DNA scanning. The initial backbone compression is suggested to be coupled to a minor groove reading head formed by residues in the leucine loop which make water bridged hydrogen bonds to structurally conserved purine N3 sites and widens up the minor groove [13]. When the uracil is detected, it is flipped out of the DNA by further backbone compression, and the resulting DNA stabilised by penetration of a conserved Leucine (Leu272 in hUNG) into DNA minor groove [13]. It is further shown that the leucine plugging of the cavity in DNA increases the lifetime of the extrahelical base [18].

Cold adapted enzymes are characterised by low thermal stability and high catalytic efficiency compared to their mesophilic counterparts. It has been suggested that a consequence of the reduced stability is an increased flexibility which is necessary in order for the enzymes to perform efficient catalysis at low temperatures [19]. No general structural feature seems to account for the increased efficiency and reduced stability of the cold adapted enzymes [20]. However, for some enzymes correlations between cold adaptation and reduced interactions between structural domains or subunits are found. This is

***Corresponding author:** Elin Moe, NorStruct, Department of Chemistry, University of Tromsø, N-9037 Tromsø, Norway, E-mail: elin.moe@uit.no

observed for salmon trypsin [21], subtilisin from the Antarctic *Bacillus* TA39 [22], lactate dehydrogenase from Antarctic notothenoid fishes [23], citrate synthase from an Antarctic bacteria [24], phosphoglycerate kinase from Antarctic *Pseudomonas* sp. TACII18 [25], anionic trypsin from Chum salmon (*Oncorhynvhus keta*) [26] and alkaline phosphatase from Atlantic cod (*Gadus morhua*) [27]. These findings suggest local flexibility as a strategy for cold adaptation. Local flexibility is also suggested to be a strategy for cold adaptation of malate dehydrogenase from *Aquaspirillium arcticum* [28], chitobiase from *Arthrobacter* sp. TAD20 [29], DNA-ligase from *Pseudoalteromonas haloplanktis* [30] and alkaline proteinase from an Antarctic *Pseudomonas* species [31,32] argues that in order to increase the reaction rate of cold adapted enzymes, the best compromise would be to maximise the flexibility in the regions involved in catalysis and maintain or increase the rigidity in regions not involved in catalysis.

Cold adapted UNG has been isolated from the liver of Atlantic cod (*Gadus morhua*) and characterisation experiments revealed that it possesses cold adapted features, like increased catalytic efficiency and reduced temperature optimum for activity, compared to human UNG (hUNG) [33]. The catalytic domain of cod UNG (cUNG) has been produced (in *E. coli*) and purified, and characterisation experiments shows that the activity of cUNG is more thermolabile and has a 10-fold increase in catalytic efficiency at optimum conditions (pH 7.0 and 50 mM NaCl) compared to hUNG [34]. The crystal structure of the catalytic domain of cUNG has been determined at 1.9 Å resolution and a comparative analysis of the structures of cUNG and hUNG suggested that a less stable C-terminal part and an increased electrostatic surface potential near the active site might be responsible for the cold adapted behaviour of the cod enzyme [6]. The importance of an optimised positive electrostatic surface potential for the cold adapted features of cUNG has been confirmed experimentally by mutational and structural analysis [35], and theoretical calculation studies have suggested that cUNG in part uses local flexibility of the minor groove intercalation loop (leucine-loop) as a strategy for cold adaptation [6,36]. Recently the first thermal stability analysis of cUNG and hUNG, performed by DSC, was published and shows that the thermal melting temperature (T_m) of cUNG is 9°C lower than for hUNG [37]. Thus the overall stability of cUNG is lower than for hUNG and suggests that it is a more elastic molecule which is able to perform rapid conformational changes during catalysis at low temperatures. Lately the crystal structure and biophysical analysis of the complex of cUNG-Ugi (uracil DNA glycosylase inhibitor) further confirmed the importance of the optimised electrostatic surface potential for the high catalytic efficiency of cUNG compared to hUNG [38].

This study was initiated in order to investigate the hypothesis that a more dynamic leucine-loop of cUNG might contribute to the high catalytic efficiency of cUNG compared to hUNG in addition to the previously observed optimised electrostatic surface potential. Thus, nine mutants of cUNG were constructed, with cod to human UNG amino acid substitutions, and analysed by enzyme kinetics and DSC. In summary our results support the hypothesis that the high catalytic efficiency of the cold adapted UNG can in part be explained by a more dynamic leucine loop compared hUNG. However, the DSC experiments showed an unexpected reduced overall stability of the mutants with increased k_{cat} and increased overall stability for the mutants with reduced k_{cat}. We believe that this effect originates from a need to compensate for the increased/reduced rigidity which is introduced in the leucine-loop by the substitutions.

Materials and Methods

Materials

E. coli TOP10 [F - *mcr*A (*mrr-hsd*RMS-*mcr*BC) 80*lac*Z M15 *lac*X74 *deo*R *rec*A1 *ara*D139 (*ara-leu*) 7697 *gal*U *gal*K *rps*L (Str R) *end*A1 *nup*G] was purchased from Invitrogen (Carlsbad, CA). *E. coli* NR8052 [Δ(*pro-lac, thi, ara, trpE9777, ung1*)], purified recombinant human UNG (UNGΔ84) and the gene encoding the catalytic domain of human UNG was kindly provided by Dr. Hans E. Krokan, Institute for Cancer Research and Molecular Biology, Norwegian University of Science and Technology.

Construction of mutants

The following mutants were constructed: A266T, V267A, Double (A266T/V267A), A274V, Triple1 (A266T/V267A/A274V), H275Y, Leuloop (A266T/V267A/A274V/H275Y), L279F and Triple2 (A274V, H275Y, L279F) The template for construction of the mutants was the gene encoding the catalytic domain of recombinant cod UNG with three additional codons (MEF) and two altered Arg codons (from AGA to CGT) in the N-terminal part of the sequence (cUNGΔ81) [34]. The Triple1, Leuloop and H275Y mutants were constructed as previously described [35]. The A266T, V267A, Double, A274V, L279F and Triple2 were constructed using the **QuickChange® Site-Directed Mutagenesis Kit** according to the manual from the manufacturer (Stratagene, CA). The primers used for the mutagenesis were as indicated in Table 1. The downstream primers are complementary and reverse of the primers listed in Table 1. The DNA sequencing was performed using BigDye® Terminator v3.1 Cycle Sequencing Kit (Applied Biosystems).

Expression

Large-scale expression was performed in 2 L baffled Erlenmeyer flasks using a Multitron Incubator shaker (Infors HT, Bottmingen, Switzerland). Fifty ml pre-cultures of *E. coli NR8052* transformed with the *pTrc99A* plasmid containing genes encoding the mutants were used for inoculation of 0.5 L of LB-medium supplemented with 0.1% glucose and 100 μg/ml of ampicillin. The cells were grown at 37°C until OD_{600}

Mutant	5'-primer
A266T	CATGTCTTGCAAACCGTTCATCCATCTC
V267A	CATGTCTTGCAAGCTGCGCATCCATCTCCTTTG
Double (A266T/V267A)	CATGTCTTGCAAACCGCGCATCCATCTCCTTTG
A274V	CCTTTGTCTGTGCATCGTGGGTTCCTTGGTTG
Triple1 (A266T/V267A/A274V)	GCAAACCGCGCATCCATCTCCTTTGTCTGTGCATCGTG
H275Y	GTCTGCTTATCGTGGGTTC
Leuloop (A266T/V267A/A274V/H275Y)	CTTGCAAACCGCGCATCCATCTCCTTTGTCTGTGTATCGTG
L279F	CCTTTGTCTGCTCATCGTGGGTTCTTTGGTTG
Triple2 (A274V/H275Y/L279F)	CCTTTGTCTGTGTATCGTGGGTTCTTTGGTTG

Table 1: Primers used for construction of mutants.

was approximately 2.0. The expression was induced by adding IPTG to 1 mM concentration 30 minutes after lowering the temperature to 30°C or 20°C and the cells were harvested 5 hours or 16 hours after induction, respectively, by centrifugation.

Protein purification

The protein purification was performed using the Äkta Explorer Purification System (Amersham Biosciences, Uppsala, Sweden). Harvested cells from 4 L culture were re-suspended in 150 ml extraction buffer [25 mM Tris/HCl, 10 mM NaCl, 1 mM EDTA, 1% (v/v) glycerol, pH 8.0] and 4 Complete Mini EDTA-free Protease Inhibitor Cocktail tablets from Roche Applied Science (Indianapolis, USA) dissolved in 4 ml distilled water were added. The cells were disrupted on ice by ultrasonication using Sonics VC Ultrasonic Processor (Sonics & Materials, Inc., Newtown, USA) for one hour (9.9 seconds on, 9.9 seconds off) at 40% amplitude. The cell extract was centrifuged and the supernatant was collected. Thereafter, 2% (w/v) protamine sulfate was added to the final concentration of 0.25% (v/v) and incubated on ice for 15 min with stirring. The solution was centrifuged at 20,000 rpm for 1 hour at 4°C, and the supernatant was collected. The pH of the cell extract was adjusted to 7.5 followed by filtration using a 0.2 μm syringe filter (Millipore, Billerica, USA). The filtrate was then applied onto a Q-Sepharose FF column coupled with an SP Sepharose FF column, both equilibrated in 25 mM Tris/HCl, 10 mM NaCl, 1 mM EDTA, 1% (v/v) glycerol, pH 7.5]. The rest of the purification steps were performed as described above. Fractions containing UNG were pooled and concentrated using Amicon Centrifugal Filter Units (molecular weight cut off, 10 kDa) from Millipore.

Activity measurements

Nick translated calf thymus DNA (Sigma-Aldrich, Steinheim, Germany) with deoxy[5-^{3}H]uridine 5'-triphosphate (Amersham Biosciences, Uppsala, Sweden) was used as substrate. Preparation of substrate and measurement of UNG activity was performed as previously described [33]. The standard assay procedure is in short: The enzyme was diluted in cold dilution buffer (5 mM Tris-HCl, 10 mM NaCl, 1% glycerol, pH 8.0). The activity was measured in a total volume of 20 μl in 70 mM Tris-HCl, 10 mM NaCl, 1 mM EDTA, pH 8.0, 100 μg/ml BSA and 230 ng substrate (^{3}H-dUMP DNA). The reaction mixture was incubated for 10 minutes at 37°C, and terminated by addition of 20 μl ice cold single stranded calf thymus DNA (1 mg/ml) and 500 μl 10% TCA. After incubation on ice for 15 minutes, free ^{3}H uracil was separated from precipitated material by centrifugation at 16 000 g for 10 minutes, and analysed using a liquid scintillation counter. One unit enzyme is defined as the amount of enzyme required for release of 1 nmol acid soluble uracil per minute at 37°C.

Kinetics

K_M and k_{cat} were measured in the presence of eight different [^{3}H]-dUMP substrate concentrations in the range of 0.56 -4.5 μM at 15°C, 22°C and 37°C. The amounts of UNG used were different for each mutant at the different temperatures, giving cpm values between 500 - 5000. Assays were performed in 25 mM Tris-HCl, 50 mM NaCl, 1 mM EDTA and 100 μg/ml BSA, pH 7.5 at 37°C. The calculation of the kinetic constants was performed by the enzyme kinetics module in SigmaPlot (SPSS Inc., Chicago, IL).

Differential scanning calorimetry (DSC)

DSC experiments were conducted on a Nano-Differential Scanning Calorimeter III, model CSC6300 from Calorimetry Sciences Corporation (Lindon, USA). Preparations of the native enzymes were dialyzed overnight at 4°C against 1 L of dialysis buffer [25 mm Hepes, 10 mM NaCl, 1 mM EDTA, 1% (v/v) glycerol, pH 7.4) in a Pierce Slide-A-Lyzer dialysis cassettes from Thermo Fisher Scientific Inc (Rockford, USA) with a 10 kDa cut-off followed by filtration using a 0.2 μm syringe filter (Millipore, Billerica, USA).

The dialysis buffers were used as reference buffers in the DSC runs. Reference buffers and samples were degassed for 10 minutes before loading into the DSC cells. The temperature scans were carried out at a constant pressure of 3 atm in a range 10 or 20 to 70°C with a heating or cooling rate of 1°C /min. Refolding experiments were performed for each protein by a backward cooling scan after unfolding and by incubating the unfolded proteins at 4°C for 1 hour before initiating a new forward scan. CpCalc software supplied with the instrument was utilized to generate molar heat capacity values. The experimental transition curves were fitted on a two-state transition model.

Results and Discussion

The minor groove intercalation loop (leucine-loop) undergoes significant conformational changes upon binding of UNG to substrate DNA and brings His268 within hydrogen binding distance of the uracil O_2, to accomplish the formation of the uracil recognition pocket. It also stabilises the distorted DNA when uracil is flipped into the specificity pocket [13,17,18] It is thus important that the loop consists of amino acids which allow rapid conformational changes upon interaction with DNA. A sequence alignment of cod and human UNG (Figure 1) revealed differences in the amino acid composition close to and within the leucine-loop which suggest improved dynamic properties of the cUNG loop.

Two of the residues which anchor the leucine-loop to the β4 strand in hUNG, Thr266 and Ala267, is substituted by Ala and Val in cUNG. In addition are Val274, Tyr275 and Phe279 substituted by Ala, His and Leu in cUNG, respectively. Based on this observation, nine different mutants with both single and multiple substitutions were constructed and characterised with regard to catalytic efficiency and temperature stability (Tables 2 and 3). The positions of the substituted amino acids are shown in Figure 2.

The mutants were produced in large scale and purified following the procedures described earlier and diafiltrated. The yield varied, but was typically between 2-4 mg for each mutant. After purification, the kinetic constants, k_{cat} and K_M were determined for the mutants at 15°C, 22°C and 37°C using the optimal buffer conditions for cUNG (pH 7.5 and 50 mM NaCl). In addition, thermal stability was measured by DSC.

The A266T, V267A and A266T/V267A mutants shows similar and higher k_{cat} compared to cUNG

An analysis of the enzyme kinetics shows that the mutations affected the k_{cat} (turnover rate) more than the Michaelis-Menten constant (K_M), and in the case of the A266T mutant, the k_{cat} is similar to cUNG at all three temperatures and only slightly increased at 37°C. For the A267V mutant the k_{cat} value is similar to cUNG at 37°C, while it is reduced with approximately 50% at 22 and 15°C (Table 3). However, when these two mutations are combined in the A266T/V267A double mutant, the k_{cat} is increased with 40% compared to cUNG at 37°C, while it is slightly decreased with 14 and 2% compared to cUNG at 22 and 15°C, respectively (Table 3).

As mentioned above, A266 and V267 anchor the leucine-loop to the β4 strand in UNG and the substitutions to Thr and Ala, respectively,

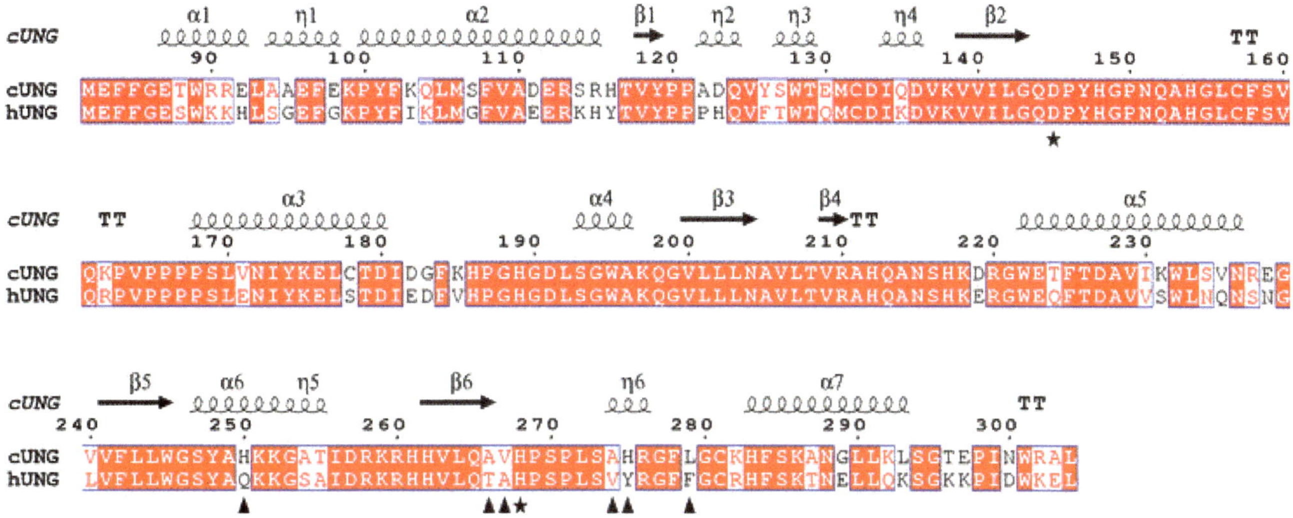

Figure 1: Sequence alignment of cUNG (GI:7413914) and hUNG (GI:14767852). The structural elements of cUNG (PDB ID: 1OKB) are included above the sequence of cUNG. Residues that might improve the flexibility of the Leucine-loop (266, 267, 274, 275 and 279) of cUNG are marked with ▲ and the catalytic residues D145 and H268 are marked with *.

UNG	T (°C)	V_{max} (nmol/min/mg)		k_{cat} (min^{-1})	K_M (µM)		k_{cat}/K_M (min^{-1}µM^{-1})
cUNG [35]	37	49135	± 1626	1241	0.8	± 0.09	1499
	22	24439	± 555	617	0.8	± 0.06	795
	15	11064	± 374	279	0.5	± 0.07	607
A266T	37	53467	± 1079	1352	0.6	± 0.04	2282
	22	24994	± 464	632	1.0	± 0.05	641
	15	10006	± 170	253	0.7	± 0.04	389
V267A	37	47476	± 1363	1197	0.5	± 0.06	2507
	22	15218	± 313	384	0.6	± 0.04	685
	15	5867	± 113	148	0.3	± 0.03	500
Double (A266T/V267A)	37	68898	± 1455	1740	0.7	± 0.05	2500
	22	20891	± 320	528	0.8	± 0.04	686
	15	10805	± 249	273	0.5	± 0.05	549
A274V	37	37379	± 611	944	1.1	± 0.05	830
	22	11746	± 244	297	0.8	± 0.06	381
	15	8991	± 286	227	1.3	± 0.11	174
Triple1 (A266T/V267A/A274V)	37	61843	± 2879	1562	0.7	± 0.12	2199
	22	30544	± 733	771	0.8	± 0.07	929
	15	14550	± 509	367	0.7	± 0.09	496
H275Y [35]	37	23590	± 859	596	0.6	± 0.09	1017
	22	11974	± 553	302	0.6	± 0.11	509
	15	6166	± 384	156	0.7	± 0.16	224
Leuloop (A266T/V267A/A274V/H275Y)	37	26603	± 1293	672	0.7	± 0.12	1018
	22	10415	± 282	263	0.5	± 0.06	548
	15	6007	± 273	152	0.5	± 0.11	292
L279F	37	31674	± 321	800	0.5	± 0.02	1459
	22	11944	± 145	302	0.5	± 0.03	611
	15	7977	± 252	201	0.6	± 0.08	342
Triple2 (A274V/H275Y/L279F)	37	30569	± 676	713	0.8	± 0.06	919
	22	7551	± 160	176	0.8	± 0.05	215
	15	4352	± 79	102	0.7	± 0.04	146
hUNG [35]	37	25347	± 1256	647	2.1	± 0.23	309
	22	12530	± 740	320	2.4	± 0.29	135
	15	5580	± 399	142	2.2	± 0.34	66

Table 2: Kinetic constants of the interactions of cUNG and mutants with substitutions in the Leucine-loop with uracil ($3H^+$ labelled) incorporated calf thymus DNA. The assay was performed at 15, 22 and 37°C in 25 mM Tris/HCl (pH 7.5 at 25°C), 50 mM NaCl, 1 mM EDTA. The V_{max} and K_m were extracted by fitting the velocity data obtained at 8 different substrate concentrations on the *Michaelis-Menten* equation using the enzyme kinetics module in Sigma Plot. The standard error (SD) for the fit is indicated.

UNG	37°C	22°C	15°C
cUNG	0	0	0
A266T	-9	-2	9
V267A	4	38	47
Double (A266T/V267A)	-40	14	2
A274V	24	52	19
Triple1 (A266T/V267A/A274V)	-26	-25	-32
H275Y	52	51	44
Leuloop (A266T/V267A/A274V/H275Y)	46	57	46
L279F	36	51	28
Triple2 (A274V/H275Y/L279F)	43	71	63
hUNG	48	48	49

Table 3: Reduction (%) in k_{cat} at different temperatures for cUNG mutants and hUNG relative to cUNG.

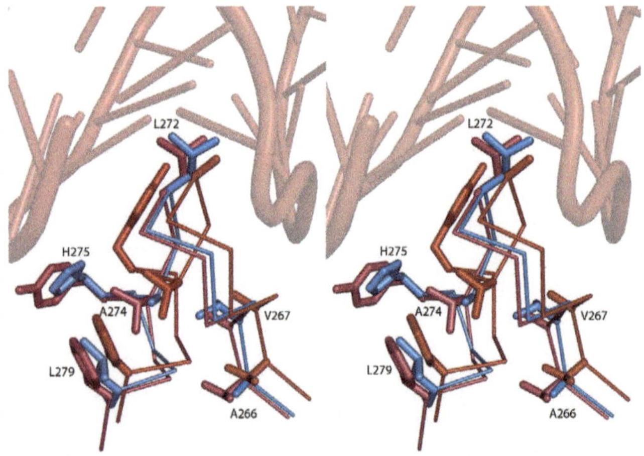

Figure 2: Stereoview of superimposed amino acid residues in the Leucine-loop of cUNG (A266, V267, L272, A274, H275 and L279) in blue (PDB ID: 1OKB), hUNG (T266, A267, L272, V274, Y275 and F279) in pink (PDB ID: 1AKZ) and hUNG-DNA (PDB ID: 2SSP – L272 is substituted by A) in red. The labels indicate the cUNG residues.

probably allow more elastic movements of the leucine-loop which give rise to a more efficient interaction with the damaged DNA. Why the k_{cat} is reduced by approximately 50% for the V267A mutant compared to cUNG at lower temperatures is difficult to explain. However, since electrostatic interactions are more strengthened at low temperatures, substituting the Valine with Alanine may affect one of the two observed hydrogen bonds between His268 and Gln144 or Leu272 and Arg276 in the leucine-loop in such a way that they get stronger, and thus reduces the dynamic properties of the loop and the k_{cat} at 22 and 15°C.

The H275Y substitution severely affects the k_{cat}

The k_{cat} of the V274A, H275Y and L279F mutants are strongly reduced compared to cUNG with most effect on the H275Y mutant which has an approximate reduction of 50% in k_{cat} at all temperatures (Table 3), and indicates that His275 is very important for the high catalytic efficiency of cUNG. This has previously been observed through theoretical calculation studies [36], where one of the experiments showed that the side chain of Arg276 is closer to DNA in a cUNG-DNA complex than in hUNG-DNA, which is caused by a hydrogen bond between His275 and the O´5 atom on the adenine base opposite of the uracil pulling the DNA strand closer to the enzyme. From the structure of cUNG, it does not seem like His275 is able to interact strongly with DNA, since it is pointing out from the minor groove of DNA (Figure 2). However the crystal structure of hUNG in complex with DNA, shows that the side chain of Tyr275 has changed conformation upon binding of DNA, pointing into the minor groove of DNA (Figure 2), and even though the crystal structure of cUNG in complex with DNA has not been determined, we assume that the same will happen with His275. This movement was observed in the model made of cUNG in complex with DNA in the theoretical study [30].

Thus His275 in cUNG probably interacts more strongly with DNA compared to Tyr275 in hUNG, and explains why the catalytic efficiency is severely reduced for the H275Y mutant (Table 2). However, the H275Y mutant shows no significant difference in K_M compared to cUNG, rather the reduction in catalytic efficiency is caused by a reduction in k_{cat}. One explanation for this could be that the His275 residue binds stronger to the transition state than to the ground state, affecting k_{cat} instead of K_M [39]. A strong interaction between the leucine-loop in cUNG and DNA is supported by a mutational study of Arg276 in hUNG, which shows its importance for the activity of hUNG due to favourable interactions with DNA [40].

The multiple mutants confirm the importance of the H275Y substitution

When we combine the A274V, H275Y and L279F mutations in the A274V/H275Y/L279F mutant, the k_{cat} value is even more reduced than for the H275Y mutant at 22 and 15°C, showing an approximate reduction of k_{cat} of 70 and 60% compared with cUNG, respectively, while it is slightly higher than for the H275Y single mutant at 37°C (Tables 2 and 3). The main contributor to the reduced k_{cat} is obviously the H275Y mutation, however it seems like the two other mutations has additional effects. Looking into the structures of cUNG and hUNG again (Figure 2), it seems likely that Val274, Tyr275 and Phe279 generates a hydrophobic cluster in the leucine-loop which may rigidify the loop. The equivalent residues in cUNG are Ala274 and Leu279 in addition to His275 which probably allow more elastic movements of the loop and thus a more efficient interaction with DNA (Figure 2). The V274, Y275 and F279 in the A274V/H275Y/L279F mutant probably generate a similar hydrophobic cluster in the leucine-loop, thus making the loop more rigid and preventing it from rapid conformational changes upon binding to DNA. Results from a molecular dynamic (MD) simulation of cUNG supports this theory showing reduced dynamic properties of the leucine-loop in hUNG compared to cUNG [36]. The same study showed that substituting only His275 with Tyr results in a reduced flexibility of the leucine-loop compared to native cUNG.

The high k_{cat} of the A266T/V267A/A274V mutant is difficult to explain but might be caused by an increased elasticity which is allowed by the A266T and V267A. However, by adding the H275Y substitution in the A266T/V267A/A274V/H275Y mutant, the k_{cat} is reduced. As seen before, the H275Y single mutant has severe effect on k_{cat} of cUNG alone, and is probably causing the k_{cat} to drop also for this multiple mutant, due to less efficient interactions with the DNA.

The substitutions affect the stability of cUNG in an unexpected manner

In order to analyse whether the amino acid substitutions in the leucine-loop had any effect on the overall stability of the protein, T_m was determined of the mutants by DSC. For every experiment in the order of 1-2 mg of protein was used, and all the measurements were performed by increasing the temperature by 1°C every minute until the melting curve was completed. The reversibility of the unfolding was tested by slowly cooling down the protein solution in the sample cell and re-running the DSC experiment, however we were not able to detect any new melting curves, thus the unfolding was irreversible. This was confirmed by visual inspections of the protein solutions after the experiment which showed white protein aggregates, and is in accordance with previous results from thermal stability analysis of cUNG and hUNG [37].

The results from the DSC experiments show that T_m was only slightly decreased or increased for the mutants compared to cUNG (Figure 3). The mutants with T_m lower than cUNG were A274V (43.4°C), H275Y (43.8°C), Triple2 (44.0°C), Leuloop (44.4°C) and Triple1 (44.6°C) while mutants with higher T_m were Double (45.3°C), A266T (46.1), L279F (46.5°C) and V267A (46.5°C) (Table 4). The differences in T_m between native cUNG and the mutants is not large, with only a 1.7 and 1.3°C reduction in T_m for the A274V and H275Y substitutions respectively and a 1.0 and 1.6°C increase for the A266T and V267A substitutions. Such subtle changes in T_m caused by mutations analyzed by DSC is a common phenomenon and have been reported for other cold adapted enzymes like α-amylase from *Pseudoalteromonas haloplanktis* (AHA) [41] and chitinase from the Antarctic *Arthrobacter Sp. TAD20* [42].

Our hypothesis for this experiment was that the substitutions introduced in the leucine-loop would alter the rigidity of the loop and increase or reduce the overall stability accordingly. According

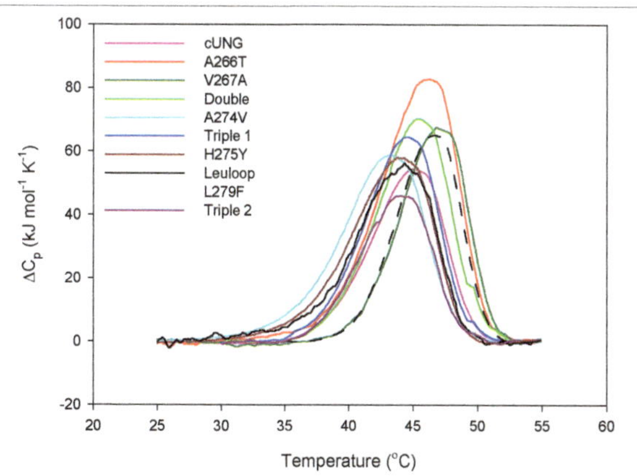

Figure 3: DSC unfolding curves of cUNG and nine cUNG mutants with amino acid substitutions in the Leucine-loop. The transition curves were obtained at 1°C/min scan rate in a buffer [25 mM Hepes, 10 mM NaCl, 1 mM EDTA, 1% (v/v) glycerol, pH 7.4]. The thermograms are baseline-subtracted and normalized for protein concentration.

UNG	ΔH_{cal} (kJ/mol)	T_m (°C)
cUNG	351	45.1
A266T	564	46.1
V267A	406	46.7
Double (A266T/V267A)	489	45.3
A274V	443	43.4
Triple1 (A266T/V267A/A274V)	472	44.6
H275Y	460	43.8
Leuloop (A266T/V267A/A274V/H275Y)	448	44.4
L279F	420	46.5
Triple2 (A274V/H275Y/L279F)	350	44.0

Table 4: Thermodynamic stability parameters of native and mutant cUNG in Hepes buffer by DSC. The heating rate was 1°C per min. The experiments were performed at least twice, and resulted in reproducible T_m values irrespective of the protein concentration used. The molar heat capacity data extracted from experimentally obtained transition curves were deconvoluted on a two-state transition model using CpCalc software. The unfolding was irreversible.

to the hypothesis about an inverse relationship between activity and stability [23], we expected to observe reduced stability of the mutants with increased elasticity, and increased stability of the mutants with reduced elasticity. However, as observed from the results, this is not the case. Instead the mutants with increased hydrophobicity (A274V, H275Y, Triple1, Triple2 and Leuloop) possessed reduced stability, while mutants with reduced hydrophobicity (A266T, V267A, Double and L279F) possessed increased stability. Thus, the mutations affected the overall stability of the protein, however it seems like the changes can only be related to overall instead of local structural rearrangements which occur due to the substitutions and has to be evaluated separately from the changes in activity which are observed of the mutants.

Free energy (ΔG) of hydration explains unexpected stability results

The free energy (ΔG) of hydration is generally unfavourable for hydrophobic amino acids in contact with water at higher temperatures and destabilises proteins [43]. This effect is probably causing the reduced thermostability which is observed of the A274V, H275Y, Triple1, Triple2 and Leuloop mutants (Figure 3).

This theory also counts for the observed increased stability of the A266T, V267A and Double mutant, which probably is caused by the less unfavourable effect the reduced hydrophobicity has on the free energy (ΔG) of hydration.

Some of the activity measurements have been performed at 37°C, and as we can see from the DSC experiments (Figure 3), it seems like the unfolding process has started for both A274V and H275Y at temperatures below this temperature. Based on this result it is tempting to explain the reduced k_{cat} of these mutants with reduced overall stability at 37°C instead of an increased rigidity of the leucine-loop. However, the activity measurements are performed in the presence of DNA, which most likely increase the overall stability of the protein at the same time as it induce conformational changes of the loops and residues in the loops (e.g., Tyr275 in hUNG) which are involved in catalysis [12,13]. This is also well supported by results from a heat inactivation experiment we previously performed of the H275Y mutant, which showed an apparent increased stability of the H275Y mutant compared to cUNG upon substrate binding [35].

The ΔH_{cal} is increased for all the mutants compared to cUNG except Triple2, indicating that an increased amount of energy is needed to break the intramolecular interactions during the unfolding process of the proteins and thus an enhanced stability of the mutants. These results contradict the T_m values showing that several of the mutants possess reduced stability compared to cUNG. However, ΔH_{cal} is also a measure of energy which takes into consideration hydration effects upon unfolding in addition to bond breaking, and due to the aggregation of the protein we would thus be careful to draw any conclusion regarding the ΔH_{cal} of our mutants.

Stability/flexibility and activity relationships

The high catalytic activity of cold adapted enzymes is explained by an increased flexibility (and thus reduced stability) of the proteins which will allow rapid conformational changes at low temperatures [23]. There is growing evidence based on mutagenesis studies that increased catalytic activity of cold adapted enzymes is not always associated with reduced stability, and reduced stability is not always indicative of increased molecular flexibility [44]. Such delicate balance between activity and stability have been reported for subtilisin S41 from the Antarctic *Bacillus TA41* [45], chitinase from the Antarctic *Arthrobacter*

Sp. TAD20 [42] and alkaline phosphatase (AP) from the Antarctic strain TAB5 [46]. This has also been demonstrated in our study which shows that variants with reduced k_{cat} compared to the wild type enzyme were not found to be more stable, and vice versa. A274V, H275Y, Leuloop, Triple1 and Triple2 were characterized by reduction of both enzymatic activity and stability while an improvement of both parameters was observed for A266T, V267A and Double. The only mutant which behaves according to the simple inverse relationship between activity and stability is the L279F mutant who possesses a 1°C increase in T_m and a 40% reduction in activity at 37°C compared to cUNG. Since we do not have the structure of this mutant, the explanation for this behaviour will only be speculative, but through its location at the edge of the leucine-loop (Figure 2) it might generate stabilising interaction with His275 in the loop making it less prone to hydration by surrounding water molecules, and increase the stability of the leucine-loop and thus the overall stability of the enzyme. The features of the Triple1 mutant are also different from the others by possessing both increased stability and activity. A similar result is observed for a mutant of subtilisin from the Antarctic bacillus TA39 [22] and reveals interesting options for industrial application of cold adapted enzymes.

Through theoretical calculation studies it has been suggested that the leucine-loop is more flexible in cUNG than in hUNG, which in combination with an optimized electrostatic surface potential, probably explains the increased catalytic efficiency of cUNG compared to hUNG [36]. We have through this study shown that an increased hydrophobicity of the leucine-loop reduces the activity of cUNG. However, the expected increased rigidity of the loop is not reflected in an increased overall stability, instead showed mutants with expected increased stability reduced stability and vice versa. This corresponds to the findings published by suggesting that a maximised flexibility in regions involved in catalysis in combination with an increased overall rigidity in other parts of the protein is a good compromise for increasing the reaction rate of cold adapted enzymes [32]. Thus, we propose that the increased hydrophobicity which has been introduced in the leucine-loop mutants resulted in an increased rigidity of the loop which has been compensated for by reducing the overall stability of the protein and vice versa.

Conclusion

The minor groove intercalation loop (leucine-loop) is very important for efficient removal of uracil in DNA by bringing the catalytic residue His268 within hydrogen binding distance of uracil, accomplishing the formation of the substrate binding pocket and stabilising distorted DNA when uracil is flipped into the specificity pocket [13,18]. Here, we have studied how substitutions of hydrophobic residues close to and in the leucine-loop affect the activity and overall stability of Atlantic cod UNG (cUNG). Based on our results we suggest that the reduced hydrophobicity of the leucine-loop in cUNG, increases its dynamic properties and thus optimizes the interaction between the leucine-loop and damaged DNA compared to hUNG. Our results confirm the importance of the leucine-loop for catalysis by UNG, and show how the capacity of this enzyme can be altered by manipulating the amino acid content of this loop.

Acknowledgements

We thank the Research Council of Norway, the National Functional Genomics Program (FUGE) and the University of Tromsø for funding this project.

References

1. Lindahl T, Nyberg B (1974) Heat-induced deamination of cytosine residues in deoxyribonucleic acid. Biochemistry 13: 3405-3410.
2. Tye BK, Nyman PO, Lehman IR, Hochhauser S, Weiss B, et al. (1977) Transient accumulation of Okazaki fragments as a result of uracil incorporation into nascent DNA. Proc Natl Acad Sci USA 74: 154-157.
3. Mol CD, Arvai AS, Slupphaug G, Kavli B, Alseth I, et al. (1995) Crystal structure and mutational analysis of human uracil-DNA glycosylase: structural basis for specificity and catalysis. Cell 80: 869-878.
4. Savva R, McAuley-Hecht K, Brown T, Pearl L (1995) The structural basis of specific base-excision repair by uracil-DNA glycosylase. Nature 373: 487-493.
5. Saikrishnan K, Bidya Sagar M, Ravishankar R, Roy S, Purnapatre K, et al. (2002) Domain closure and action of uracil DNA glycosylase (UDG): structures of new crystal forms containing the Escherichia coli enzyme and a comparative study of the known structures involving UDG. Acta Crystallogr D Biol Crystallogr 58: 1269-1276.
6. Leiros I, Moe E, Lanes O, Smalås AO, Willassen NP, et al. (2003) The structure of uracil-DNA glycosylase from Atlantic cod (Gadus morhua) reveals cold-adaptation features. Acta Crystallogr D Biol Crystallogr 59: 1357-1365.
7. Leiros I, Moe E, Smalås AO, McSweeney S (2005) Structure of the uracil-DNA N-glycosylase (UNG) from Deinococcus radiodurans. Acta Crystallogr D Biol Crystallogr 61: 1049-1056.
8. Raeder IL, Moe E, Willassen NP, Smalås AO, Leiros I (2010) Structure of uracil-DNA N-glycosylase (UNG) from Vibrio cholerae: mapping temperature adaptation through structural and mutational analysis. Acta Crystallogr Sect F Struct Biol Cryst Commun 66: 130-136.
9. Schormann N, Grigorian A, Samal A, Krishnan R, DeLucas L, et al. (2007) Crystal structure of vaccinia virus uracil-DNA glycosylase reveals dimeric assembly. BMC Struct Biol 7: 45.
10. Kaushal PS, Talawar RK, Varshney U, Vijayan M (2010) Structure of uracil-DNA glycosylase from Mycobacterium tuberculosis: insights into interactions with ligands. Acta Crystallogr Sect F Struct Biol Cryst Commun 66: 887-892.
11. Wang HC, Hsu KC, Yang JM, Wu ML, Ko TP, et al. (2014) Staphylococcus aureus protein SAUGI acts as a uracil-DNA glycosylase inhibitor. Nucleic Acids Res 42: 1354-1364.
12. Slupphaug G, Mol CD, Kavli B, Arvai AS, Krokan HE, et al. (1996) A nucleotide-flipping mechanism from the structure of human uracil-DNA glycosylase bound to DNA. Nature 384: 87-92.
13. Parikh SS, Mol CD, Slupphaug G, Bharati S, Krokan HE, et al. (1998) Base excision repair initiation revealed by crystal structures and binding kinetics of human uracil-DNA glycosylase with DNA. EMBO J 17: 5214-5226.
14. Parikh SS, Walcher G, Jones GD, Slupphaug G, Krokan HE, et al. (2000) Uracil-DNA glycosylase-DNA substrate and product structures: conformational strain promotes catalytic efficiency by coupled stereoelectronic effects. Proc Natl Acad Sci USA 97: 5083-5088.
15. Schormann N, Banerjee S, Ricciardi R, Chattopadhyay D (2015) Binding of undamaged double stranded DNA to vaccinia virus uracil-DNA Glycosylase. BMC Struct Biol 15: 10.
16. Burmeister WP, Tarbouriech N, Fender P, Contesto-Richefeu C, Peyrefitte CN, et al. (2015) Crystal Structure of the Vaccinia Virus Uracil-DNA Glycosylase in Complex with DNA. J Biol Chem 290: 17923-17934.
17. Pedersen HL, Johnson KA, McVey CE, Leiros I, Moe E, et al. (2015) Structure determination of uracil-DNA N-glycosylase from Deinococcus radiodurans in complex with DNA. Acta Crystallogr D Biol Crystallogr 71: 2137-2149.
18. Jiang YL, Kwon K, Stivers JT (2001) Turning On uracil-DNA glycosylase using a pyrene nucleotide switch. J Biol Chem 276: 42347-42354.
19. Hochachka PW, Somero GN (1984) Biochemical adaptation. Princeton University Press, Princeton, NJ, USA.
20. Smalås AO, Leiros HK, Os V, Willassen NP (2000) Cold adapted enzymes. Biotechnol Annu Rev 6: 1-57.
21. Smalås AO, Heimstad ES, Hordvik A, Willassen NP, Male R, et al. (1994) Cold adaption of enzymes: structural comparison between salmon and bovine trypsins. Proteins 20: 149-166.
22. Narinx E, Baise E, Gerday C (1997) Subtilisin from psychrophilic antarctic bacteria: characterization and site-directed mutagenesis of residues possibly involved in the adaptation to cold. Protein Eng 10: 1271-1279.
23. Fields PA, Somero GN (1998) Hot spots in cold adaptation: localized increases

in conformational flexibility in lactate dehydrogenase A4 orthologs of Antarctic notothenioid fishes. Proc Natl Acad Sci USA 95: 11476-11481.

24. Russell RJ, Gerike U, Danson MJ, Hough DW, Taylor GL, et al. (1998) Structural adaptations of the cold-active citrate synthase from an Antarctic bacterium. Structure 6: 351-361.

25. Bentahir M, Feller G, Aittaleb M, Lamotte-Brasseur J, Himri T, et al. (2000) Structural, kinetic, and calorimetric characterization of the cold-active phosphoglycerate kinase from the antarctic Pseudomonas sp. TACII18. J Biol Chem 275: 11147-11153.

26. Toyota E, Ng KK, Kuninaga S, Sekizaki H, Itoh K, et al. (2002) Crystal structure and nucleotide sequence of an anionic trypsin from chum salmon (Oncorhynchus keta) in comparison with Atlantic salmon (Salmo salar) and bovine trypsin. J Mol Biol 324: 391-397.

27. Asgeirsson B, Nielsen BN, Højrup P (2003) Amino acid sequence of the cold-active alkaline phosphatase from Atlantic cod (Gadus morhua). Comp Biochem Physiol B Biochem Mol Biol 136: 45-60.

28. Kim SY, Hwang KY, Kim SH, Sung HC, Han YS, et al. (1999) Structural basis for cold adaptation. Sequence, biochemical properties, and crystal structure of malate dehydrogenase from a psychrophile Aquaspirillium arcticum. J Biol Chem 274: 11761-11767.

29. Lonhienne T, Zoidakis J, Vorgias CE, Feller G, Gerday C, et al. (2001) Modular structure, local flexibility and cold-activity of a novel chitobiase from a psychrophilic Antarctic bacterium. J Mol Biol 310: 291-297.

30. Georlette D, Damien B, Blaise V, Depiereux E, Uversky VN, et al. (2003) Structural and functional adaptations to extreme temperatures in psychrophilic, mesophilic, and thermophilic DNA ligases. J Biol Chem 278: 37015-37023.

31. Aghajari N, Van Petegem F, Villeret V, Chessa JP, Gerday C, et al. (2003) Crystal structures of a psychrophilic metalloprotease reveal new insights into catalysis by cold-adapted proteases. Proteins 50: 636-647.

32. Lonhienne T, Gerday C, Feller G (2000) Psychrophilic enzymes: revisiting the thermodynamic parameters of activation may explain local flexibility. Biochim Biophys Acta 1543: 1-10.

33. Lanes O, Guddal PH, Gjellesvik DR, Willassen NP (2000) Purification and characterization of a cold-adapted uracil-DNA glycosylase from Atlantic cod (Gadus morhua). Comp Biochem Physiol B Biochem Mol Biol 127: 399-410.

34. Lanes O, Leiros I, Smalås AO, Willassen NP (2002) Identification, cloning, and expression of uracil-DNA glycosylase from Atlantic cod (Gadus morhua): characterization and homology modeling of the cold-active catalytic domain. Extremophiles 6: 73-86.

35. Moe E, Leiros I, Riise EK, Olufsen M, Lanes O, et al. (2004) Optimisation of the surface electrostatics as a strategy for cold adaptation of uracil-DNA N-glycosylase (UNG) from Atlantic cod (Gadus morhua). J Mol Biol 343: 1221-1230.

36. Olufsen M, Smalås AO, Moe E, Brandsdal BO (2005) Increased flexibility as a strategy for cold adaptation: a comparative molecular dynamics study of cold- and warm-active uracil DNA glycosylase. J Biol Chem 280: 18042-18048.

37. Assefa NG, Niiranen L, Willassen NP, Smalås A, Moe E (2012) Thermal unfolding studies of cold adapted uracil-DNA N-glycosylase (UNG) from Atlantic cod (Gadus morhua). A comparative study with human UNG. Comp Biochem Physiol B Biochem Mol Biol 161: 60-68.

38. Assefa NG, Niiranen L, Johnson KA, Leiros HK, Smalås AO, et al. (2014) Structural and biophysical analysis of interactions between cod and human uracil-DNA N-glycosylase (UNG) and UNG inhibitor (Ugi). Acta Crystallogr D Biol Crystallogr 70: 2093-2100.

39. Olufsen M, Smalås AO, Brandsdal BO (2008) Electrostatic interactions play an essential role in DNA repair and cold-adaptation of uracil DNA glycosylase. J Mol Model 14: 201-213.

40. Chen CY, Mosbaugh DW, Bennett SE (2004) Mutational analysis of arginine 276 in the leucine-loop of human uracil-DNA glycosylase. J Biol Chem 279: 48177-48188.

41. D'Amico S, Gerday C, Feller G (2001) Structural determinants of cold adaptation and stability in a large protein. J Biol Chem 276: 25791-25796.

42. Mavromatis K, Feller G, Kokkinidis M, Bouriotis V (2003) Cold adaptation of a psychrophilic chitinase: a mutagenesis study. Protein Eng 16: 497-503.

43. Murphy KP (2001) Stabilization of protein structure. Methods Mol Biol 168: 1-16.

44. Huston AL, Haeggström JZ, Feller G (2008) Cold adaptation of enzymes: structural, kinetic and microcalorimetric characterizations of an aminopeptidase from the Arctic psychrophile Colwellia psychrerythraea and of human leukotriene A(4) hydrolase. Biochim Biophys Acta 1784: 1865-1872.

45. Miyazaki K, Wintrode PL, Grayling RA, Rubingh DN, Arnold FH (2000) Directed evolution study of temperature adaptation in a psychrophilic enzyme. J Mol Biol 297: 1015-1026.

46. Koutsioulis D, Wang E, Tzanodaskalaki M, Nikiforaki D, Deli A, et al. (2008) Directed evolution on the cold adapted properties of TAB5 alkaline phosphatase. Protein Eng Des Sel 21: 319-327.

Energetic and Thermodynamic Analysis of Adsorption Isotherm Type VI of Xenon on Graphite Nanotubes

Manel Ben Yahia[1,2*], Mohamed Ben Yahia[1], Fatma Aouaini[1] and Abdemottaleb Ben Lamine[1]

[1]*Department of Physics, Faculty of Science, King AbdulAziz University, Rabigh, Saudi Arabia*
[2]*Laboratory of Quantum Physics, Faculty of Science, Monastir University, Monastir, Tunisia*

Abstract

New theoretical expressions for the modeling of adsorption isotherms of Xenon on Graphite at 110 K have been established. The establishment of these new expressions is based on statistical physics formalism and some working assumptions. This method allowed estimation of physico-chemical parameters in the theoretical model. The parameters intervening in the adsorption process have been deduced directly from experimental adsorption isotherm by numerical simulation. The proposed models allow a good correlation to Type VI experimental isotherms. We mainly introduce three parameters affecting the adsorption process, namely, the density of Xenon receptor sites NM, the number of molecules per site n and the Xenon adsorption energy. Then we apply the model to calculate thermodynamics functions which govern the adsorption mechanism such as entropy, free enthalpy and internal energy.

Keywords: Adsorption isotherm; Statistical physics; Modeling; Thermodynamic potential

Introduction

The study of adsorption properties of carbon nanotubes has several advantages. From a fundamental point of view, their tubular structure associated with their low dimensionality offers the possibility to study the effects of confinement and curvature on physical properties. As for applications related to the phenomenon of physical adsorption (physisorption), several studies dealing with the possibility of using these materials in storage systems, separation or gas detection [1]. On the other hand, the physisorption is particularly suitable for characterizing surface of the nanotubes. This technique is often implemented for the study of solid powders. The surface is generally characterized in terms of surface area and porosity. When the surface is uniform, that is to say as close as possible to a crystal plane without defects, the adsorption is carried out by condensation of successive monomolecular layers giving rise to a succession of steps of the isotherm adsorption. This type of adsorption has been extensively studied on graphite substrate [2]. The adsorption on the nanotubes leads also to isothermal operation [3,4], which allows to identify the fractions of uniform surface and get information on the nature of the adsorption sites and the method of attachment of the molecules of the latter. The term "surface defects" is a very vague term, used in the case of adsorption to designate heterogeneities may lead to inhomogeneous distribution of the energies of adsorption sites of a given surface. This leads to changes in all the adsorption properties described above on uniform surfaces, but to varying degrees according to the property studied and especially the nature of these heterogeneities.

In our work, we are interested in modeling the adsorption isotherm of Xenon on graphite at 110 K [5], this isotherm is similar to type six in the IUPAC classification [6]. Type VI isotherms are rare and indicate strong interactions adsorbate-adsorbent [7]. In this paper, we present the methodology used to find new and simple theoretical expression that is able to describe and fit well the adsorption isotherm of Xenon on Graphite Nanotubes. The study is composed of two parts: in the first one, we use a statistical physics treatment to establish the expression of an adequate model giving the adsorbed quantity against the pressure. In the second part, we apply the analytical expression to fit experimental data taken from literature. The adsorption isotherm of Xenon on graphite at 110 K achieved by ref. [5] is depicted in Figure 1. Then we apply the model to interpret the adsorption stereographic and energetic and to calculate thermodynamics functions which govern the adsorption mechanism such as entropy, the free enthalpy and the internal energy.

General statistical physics treatment

Adsorption involves an exchange of particles from the free state to the adsorbed one. Its investigation in statistical physics cannot be performed without employing the grand canonical ensemble to take account of the particle number variation. So a chemical potential is introduced in the adsorption process [8]. In order to treat adsorption problems using statistical physics methods, we will proceed with some assumptions as the basis of an initial postulate. As a first approximation, the molecules of Xenon are treated as an ideal gas [8], since the mutual interaction between the adsorbed molecules will be neglected. We consider that a variable number of molecules are adsorbed onto N_M receptor sites located on a unit mass of the adsorbent. The adsorption reaction of gas molecules (X_e) onto receptor sites (S) should include a stoichiometric coefficient n as the following equation:

$$nX_e + S \Leftrightarrow \left(X_e\right)_n S \tag{1}$$

The parameter n is an average number, which can be an integer or not, greater or smaller than 1. If n is greater than 1, it represents the number of molecules anchored on one site, it is a multi-molecular adsorption. If n is less than 1, it would represent "the fraction of molecule per site" [8,9], it is a multi-anchorage adsorption.

***Corresponding author:** Manel Ben Yahia, Department of Physics, Faculty of Science, King AbdulAziz University, Rabigh 21911, Saudi Arabia
E-mail: manelbenyahia@ymail.com

In our treatment, our starting point is the grand canonical partition function describing the microscopic of the adsorption process. The receptor site is supposed to be empty or occupied by one or more molecules. So, we define the state of occupation number N_i that expressed the situation in which this receptor site is placed [9,10]:

$$z_{gc} = \sum_{N_i} e^{-\beta(\varepsilon_i-\mu)N_i} \quad (2)$$

where ε_i is the receptor site adsorption energy, μ is the chemical potential, N_i is the receptor site occupation state, and β is defined as $1/k_B T$, where k_B is the Boltzmann constant and T is the absolute temperature. The total grand canonical partition function related to N_M receptor sites per surface unit, which we assume independent and identical, is equal:

$$Z_{gc} = z_{gc}^{N_M} \quad (3)$$

This total grand canonical partition function led to determine the average site occupation number which can be written [9,11]:

$$N_o = k_B T \frac{\partial Ln\left(z_{gc}\right)^{N_M}}{\partial \mu} \quad (4)$$

When the thermodynamic equilibrium is reached according to the relation (1), the equality related to the chemical potential can be written as $\mu_m = \frac{\mu}{n}$, where is the chemical potential of adsorbed molecule, n is the number or fraction of molecules per site, and μ_m is the chemical potential of the gaseous molecule. The chemical potential can be written as [12,13]:

$$\mu_m = k_B T Ln\left(\frac{N}{z_{Tr}}\right) \quad (5)$$

Where z_{Tr} is the translation partition function which can be written as [8,10]:

$$z_{Tr} = V\left(\frac{2\pi m k_B T}{h^2}\right)^{\frac{3}{2}} = \frac{V}{\Lambda^3} \quad (6)$$

With m being the adsorbed molecule mass, V being the volume of the gas, h Planck's constant and Λ is the Broglie thermal wavelength given by $\frac{h}{\sqrt{2\pi m k_B T}}$.

The translation partition function per unit of volume can be also expressed as function of the saturated vapor pressure and the vaporization energy by the relation [9,10]:

$$\frac{1}{\Lambda^3} = \beta\left(\frac{2\pi m k_B T}{h^2}\right)^{\frac{3}{2}}\left(k_B T\right)e^{\frac{-\Delta E^V}{RT}} e^{\frac{\Delta E^V}{RT}} = \beta P_{VS} e^{\frac{\Delta E^V}{RT}} \quad (7)$$

Where ΔE^V the molar vaporization energy and P_{VS} is the saturated vapor pressure.

Assume V to be 1 liter, the translational partition function of an Xe molecule at T=110 K is 10^{30}. This means that 10^{30} quantum states are thermally accessible to the molecular system.

If there are n molecule(s) anchored per site, the total number of adsorbed molecules is given by:

$$N_a = nN_0 \quad (8)$$

In the following sections, we will establish analytical expression corresponding to the adsorption isotherm of Xenon on graphite on the basis of statistical physics treatment. We present the adequate grand canonical partition function and afterward the obtained model. We will discuss the role of the physicochemical parameters.

Theoretical model

Monolayer model: To develop the expression of the adsorbed quantity according to the pressure, we consider that a variable number of molecules N are adsorbed onto three types of independent sites. The number of the first type is N_{1M} sites, the number of the second type is N_{2M} and the number of the third type is N_{3M}. We suppose that the adsorption onto the first type is carried out with the energy $(-\varepsilon_1)$, with the energy $(-\varepsilon_2)$ onto the second type and with the energy $(-\varepsilon_3)$ onto the third type. In this case, we can write the total grand canonical partition function as follows:

$$Z_{gc} = \left(\left(z_{1gc}\right)^{N_{1M}}\left(z_{2gc}\right)^{N_{2M}}\left(z_{3gc}\right)^{N_{3M}}\right) \quad (9)$$

With z_{1gc}, z_{2gc} and z_{3gc} being the partition functions of one site of the three types. The grand canonical partition function in this case, for each type of receptor site, has the form:

$$z_{1gc} = \sum_{N_i=0,1} e^{-\beta(-\varepsilon_1-\mu_1)N_i} = 1+e^{\beta(\varepsilon_1+\mu_1)} \quad (10a)$$

$$z_{2gc} = \sum_{N_i=0,1} e^{-\beta(-\varepsilon_2-\mu_2)N_i} = 1+e^{\beta(\varepsilon_2+\mu_2)} \quad (10b)$$

$$z_{3gc} = \sum_{N_i=0,1} e^{-\beta(-\varepsilon_3-\mu_3)N_i} = 1+e^{\beta(\varepsilon_3+\mu_3)} \quad (10c)$$

We obtain the expression of the adsorbed against the pressure:

$$N_a = \frac{n_1 N_{1M}\left(\frac{p}{p_1}\right)^{n_1}}{1+\left(\frac{p}{p_1}\right)^{n_1}} + \frac{n_2 N_{2M}\left(\frac{p}{p_2}\right)^{n_2}}{1+\left(\frac{p}{p_2}\right)^{n_2}} + \frac{n_3 N_{3M}\left(\frac{p}{p_3}\right)^{n_3}}{1+\left(\frac{p}{p_3}\right)^{n_3}} \quad \text{(Model 1)}$$

With: $p_i = \frac{k_B T}{\Lambda^3} e^{-\beta\varepsilon_{im}}$

In Model 1, there are nine fitting parameters: p_1, p_2, p_3, n_1, n_2, n_3, N_{1M}, N_{2M} and N_{3M}. The correlation between experimental data and the analytical model is given by the correlation adjustment coefficient R^2=0.995 which is a measure of the correctness of the fit. The fitting parameters of Model 1 are presented in Table 1.

Triple layer model: If we suppose that the site can be empty or occupied by one molecule adsorbed with the energy $(-\varepsilon_1)$, the second molecule adsorbed with the energy $(-\varepsilon_2)$ and the third molecule adsorbed with the energy $(-\varepsilon_3)$.

The grand canonical partition function is written as:

p_1(Pa)	p_2(Pa)	p_3(Pa)	n_1	n_2	n_3	N_{1M}(mmol/g)	N_{2M}(mmol/g)	N_{3M}(mmol/g)
0.002	10.28	570	1.13	1.22	5.66	0.69	1.19	0.66

Table 1: Adjustment parameter values corresponding to the fitting Model 1.

$$z_{gc} = 1 + e^{\beta(\varepsilon_1+\mu)} + e^{\beta(\varepsilon_1+\varepsilon_2+2\mu)} + e^{\beta(\varepsilon_1+\varepsilon_2+\varepsilon_3+3\mu)} \quad (11)$$

We obtain the adsorbed quantity versus the pressure which is written:

$$N_a = \frac{nN_M\left(\left(\frac{p_1}{p}\right)^n + 2\left(\frac{p_1}{p}\right)^n\left(\frac{p_2}{p}\right)^n + 3\left(\frac{p_1}{p}\right)^n\left(\frac{p_2}{p}\right)^n\left(\frac{p_3}{p}\right)^n\right)}{1+\left(\frac{p_1}{p}\right)^n + \left(\frac{p_1}{p}\right)^n\left(\frac{p_2}{p}\right)^n + \left(\frac{p_1}{p}\right)^n\left(\frac{p_2}{p}\right)^n\left(\frac{p_3}{p}\right)^n} \quad \text{(Model 2)}$$

In Model 2, there are five fitting parameters: p_1, p_2, p_3, N_M and n. The correlation between experimental data and the analytical model is given by the correlation adjustment coefficient $R^2=0.983$ which is a measure of the correctness of the fit. The fitting parameters of Model 2 are presented in Table 2.

Multilayer model: We consider now that the type VI isotherm is due to multilayer adsorption process. So, there is complete formation of monomolecular layers before progressing to a subsequent layer which commonly associated with non-porous or macroporous adsorbents. The isotherm presents three partial saturations and a condensation phenomenon occurs at high pressure. To choose adequate model, we suppose that the layers are formed one after the other with different energy. We suppose that L_1 layers of adsorbed molecules are formed with the first energy ($-\varepsilon_1$), L_2 layers with the energy ($-\varepsilon_2$), L_3 layers with the energy ($-\varepsilon_3$) and infinite layers are formed with the energy ε_4. The grand canonical partition function is written as:

$$z_{gc}(T,\mu) = \sum_{N_i=0}^{L_1} e^{-\beta(-\varepsilon_1-\mu)N_i} + \sum_{N_i=L_1+1}^{L_1+L_2} e^{-\beta(-L_1\varepsilon_1-(N_i-L_1)\varepsilon_2-N_i\mu)} + \sum_{N_i=L_1+L_2+1}^{\infty} e^{-\beta(-L_1\varepsilon_1-L_2\varepsilon_2-(N_i-L_1-L_2)\varepsilon_3-N_i\mu)} \quad (12)$$

To determine the adsorbed quantity versus pressure we use the definition of the average site occupation number given by relation (8). We take $p_1 = p_{vs}e^{(\Delta E_1^a)/RT}$, $p_2 = p_{vs}e^{(\Delta E_2^a)/RT}$, $p_3 = p_{vs}e^{(\Delta E_3^a)/RT}$ three energetic parameters, with ΔE_1^a ΔE_2^a ΔE_3^a are the molar adsorption energies and p_{vs} is the saturated vapor pressure of the gas. Finally the adsorbed quantity N_a has the following expression:

$$N_a = \frac{nN_M\left(f_1(p,p_1)+f_2(p,p_1,p_2)+f_3(p,p_1,p_2,p_3)\right)}{\frac{\left(1-\left(\frac{p}{p_1}\right)^{n(L_1+1)}\right)}{1-\left(\frac{p}{p_1}\right)^n} + \frac{\left(\frac{p}{p_1}\right)^{nL_1}\left(\frac{p}{p_2}\right)^n\left(1-\left(\frac{p}{p_2}\right)^{nL_2}\right)}{1-\left(\frac{p}{p_2}\right)^n} + \frac{\left(\frac{p}{p_1}\right)^{nL_1}\left(\frac{p}{p_2}\right)^{nL_2}\left(\frac{p}{p_3}\right)^n}{1-\left(\frac{p}{p_3}\right)^n}} \quad \text{(Model 3)}$$

with f_1, f_2 and f_3 are functions of the pressure and the parameters p_i such as:

$$f_1(p,p_1) = \frac{-\left(\frac{p}{p_1}\right)^{n(L_1+1)}(L_1+1)}{1-\left(\frac{p}{p_1}\right)^n} + \frac{\left(\frac{p}{p_1}\right)^n\left(1-\left(\frac{p}{p_1}\right)^{n(L_1+1)}\right)}{\left(1-\left(\frac{p}{p_1}\right)^n\right)^2}z \quad (13)$$

$$f_2(p,p_1,p_2) = \frac{\left(\frac{p}{p_1}\right)^{nL_1}\left(\frac{p}{p_2}\right)^n(L_1+1)\left(1-\left(\frac{p}{p_2}\right)^{nL_2}\right)}{1-\left(\frac{p}{p_2}\right)^n} - \frac{\left(\frac{p}{p_1}\right)^{nL_1}\left(\frac{p}{p_2}\right)^n\left(\frac{p}{p_2}\right)^{nL_2}L_2}{1-\left(\frac{p}{p_2}\right)^n} + \frac{\left(\frac{p}{p_1}\right)^{nL_1}\left(\frac{p}{p_2}\right)^{2n}\left(1-\left(\frac{p}{p_2}\right)^{nL_2}\right)}{\left(1-\left(\frac{p}{p_1}\right)^n\right)^2} \quad (14)$$

p_1(Pa)	p_2(Pa)	p_3(Pa)	n	N_M(mmol/g)
0.007	10.38	564.78	1.90	0.7

Table 2: Adjustment parameter values corresponding to the fitting Model 2.

p_1(Pa)	p_2(Pa)	p_3(Pa)	n	N_M(mmol/g)	L_1	L_2
0.005	10.22	566.29	1.52	0.92	1.3	6.4

Table 3: Adjustment parameter values corresponding to the fitting Model 3.

$$f_3(p,p_1,p_2,p_3) = \frac{\left(\frac{p}{p_1}\right)^{\hbar_1}\left(\frac{p}{p_2}\right)^{\hbar_2}\left(\frac{p}{p_3}\right)^n(L_1+L_2+1)}{1-\left(\frac{p}{p_3}\right)^n} + \frac{\left(\frac{p}{p_1}\right)^{\hbar_1}\left(\frac{p}{p_2}\right)^{\hbar_2}\left(\frac{p}{p_3}\right)^{2n}}{\left(1-\left(\frac{p}{p_3}\right)^n\right)^2} \quad (15)$$

In the expression of adsorbed quantity given by Model 3, there are seven parameters which are divided in two classes, the first is a steric aspect such as the number of molecules per site n and the density of receptor sites N_M, and the number of layers L_1, L_2, the second aspect is energetic one given by the parameters p_1, p_2 and p_3 which are function of the adsorption energy. The correlation between experimental data and the analytical Model 3 is given by the correlation adjustment coefficient $R^2=0.992$. The fitting parameters of Model 3 are presented in Table 3.

Fitting results and interpretation: We have adjusted the experimental data with the expression of the established models (Model 1, Model 2 and Model 3). We note that the experimental curve shows a high correlation with the proposed model when the coefficient of determination R^2 is close to 1 and which is given by numerical simulation. It is true that in general, the coefficient of determination R^2 is improved when the number of fitting parameters increases which is the case of our adjustment. Indeed, these parameters reflect physical magnitudes. This means that the adsorption phenomenon is better described when a maximum of physical magnitudes are included, thus the space of description of the phenomenon will be a more complete space. By fitting the experimental isotherm we have found that the adjustment coefficient is close to 0.995. The fitting of the experimental isotherm of Xe on graphite at 110 K demonstrates that the proposed Model 1 present a good correlation with the experimental curve.

In comparison with the other models, the Model 1 presents the best values of R^2 which are close to the unit. Then we retain this model as an adequate model to characterize the adsorption of Xenon on graphite.

The parameters of the model give information about the number of molecules per site [10,11], the density of receptor sites and the adsorption energies. We summarize these parameters in two categories [10,11]: on the one hand a steric (n) parameter reflecting the properties of anchorage between adsorbed molecules and receptor sites related to the geometry of molecule, on the other hand an energetic parameter giving the adsorption energy which characterizes the binding between the molecules and receptor site. The parameters of model such as number of molecules per site (n), the density N_M and the energetic coefficients p_1, p_2 and p_3 are presented in Table 1. The adsorption reaction (equation 1) expresses that the parameter n is a stoichiometric coefficient; it represents the number of molecules per site given in the expression of the model. It is also a steric coefficient which gives information on the position of the molecule at the adsorbent surface [9,11].

Energies of adsorption

Using the values of the dimensionless parameters p_1, p_2 and p_3 and the pressure vapor saturation p_{vs}, we calculate the molar adsorption energies ΔE_1^a, ΔE_2^a and ΔE_3^a:

$$\Delta E_i^a = \Delta E^V - RTLn\left(\frac{p_i}{p_{VS}}\right) \quad (16)$$

With: i=1, 2 and 3.

The adsorption energies of Xenon adsorbed on graphite at 110 K are presented in Table 4. The first energy gives information about the interaction between molecules and the surface, whereas the others represent the molecule-molecule interaction. The values of these energies showed that binding between the molecules and the surface takes place via a physical adsorption such as Vander Waals binding. The second and the third energies corresponding to the molecule-molecule interaction are inferior to the first energy.

Adsorption entropy

In this section we present a detailed calculation of the value of the adsorption entropy. This value can be calculated on the basis of statistical physics. We start from the definition of the grand potential using the grand canonical partition function [14]:

$$J = -k_B T \ln Z_{gc} = -\frac{\partial \ln Z_{gc}}{\partial \beta} - TS_a \tag{17}$$

We can obtain the expression of the entropy then:

$$TS_a = -\frac{\partial \ln Z_{gc}}{\partial \beta} + k_B T \ln Z_{gc} \tag{18}$$

$$\frac{S_a}{k_B} = -\beta \frac{\partial \ln Z_{gc}}{\partial \beta} + \ln Z_{gc} \tag{19}$$

Finally the adsorption entropy S_a has the following expression:

$$\frac{S_a}{k_B} = \left[N_{1M} \ln\left(1+\left(\frac{p}{p_1}\right)^{n_1}\right) + N_{2M} \ln\left(1+\left(\frac{p}{p_2}\right)^{n_2}\right) + N_{3M} \ln\left(1+\left(\frac{p}{p_3}\right)^{n_3}\right) - \left(\frac{N_{1M}\left(\frac{p}{p_1}\right)^{n_1} \ln\left(\frac{p}{p_1}\right)^{n_1}}{\left(1+\left(\frac{p}{p_1}\right)^{n_1}\right)} + \frac{N_{2M}\left(\frac{p}{p_2}\right)^{n_2} \ln\left(\frac{p}{p_2}\right)^{n_2}}{\left(1+\left(\frac{p}{p_2}\right)^{n_2}\right)} + \frac{N_{3M}\left(\frac{p}{p_3}\right)^{n_3} \ln\left(\frac{p}{p_3}\right)^{n_3}}{\left(1+\left(\frac{p}{p_3}\right)^{n_3}\right)} \right) \right] \tag{20}$$

We notice, Figure 2 that for the isotherm studied; the entropy has three different behaviors below and above the particular points p_1, p_2 and p_3. Indeed, the entropy increases with the adsorbed quantity before the half-saturation and decreases after this particular point. This is explained by the fact that when the adsorbed quantity is inferior to the one of p_1, the molecule has various possibilities to choose a receptor site to be adsorbed and therefore the disorder increases at the surface with the adsorbed quantity. After this point, the adsorbent molecule has low probability to choose adsorbent site since the surface tends toward the saturation and therefore tends toward the order. The entropy can reach zero when the saturation is reached.

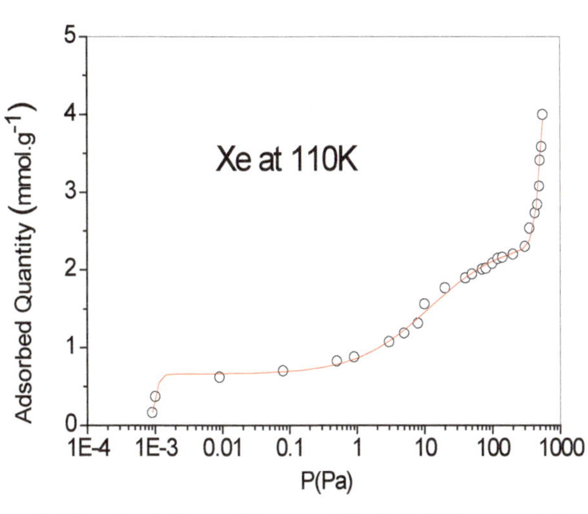

Figure 1: Adsorption isotherm of Xe on graphite at 110 K. The open circles are the experimental data and the line is the simulation data.

ΔE_1^a (kJ/mol)	ΔE_2^a (kJ/mol)	ΔE_3^a (kJ/mol)
12.36	9.70	9.24

Table 4: Adsorption energies of Xenon/graphite at 110 K.

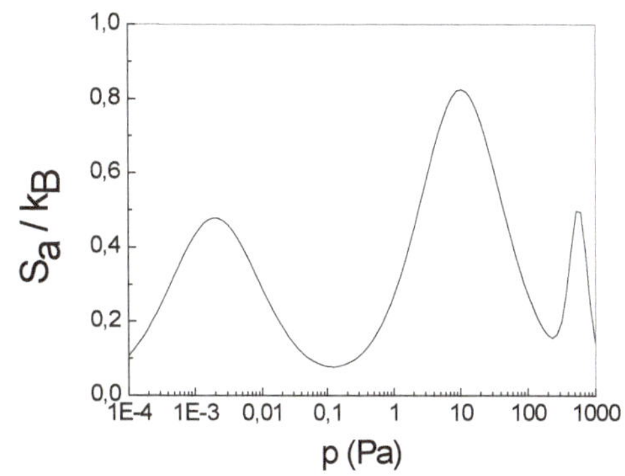

Figure 2: Evolution of the entropy versus the pressure.

Free enthalpy

The free enthalpy described the spontaneity of the system. It is given by the following relation [14]:

$$G = \mu n N_0 = \mu N_a \tag{21}$$

From Eq. (5), μ can be written in the variable p as $\mu = k_B T \ln \beta p \Lambda^3$.

Where μ is the chemical potential of adsorbed molecule and N_a is is the adsorbed quantity.

The free enthalpy G has the following expression:

$$G = k_B T \ln \beta p \Lambda^3 \left(\frac{n_1 N_{1M}}{1+\left(\frac{p_1}{p}\right)^{n_1}} + \frac{n_2 N_{2M}}{1+\left(\frac{p_2}{p}\right)^{n2}} + \frac{n_3 N_{3M}}{1+\left(\frac{p_3}{p}\right)^{n3}} \right) \tag{22}$$

We notice, Figure 3 that for the isotherm studied, the free enthalpy is always negative, what translates that the reaction of adsorption evolves spontaneously.

Internal energy

The expression of internal energy is given by the following relation [14] as:

$$E_{\text{int}} = -\frac{\partial \ln Z_{gc}}{\partial \beta} + \frac{\mu}{\beta}\left(\frac{\partial \ln Z_{gc}}{\partial \mu}\right) \tag{23}$$

The internal energy E_{int} has the following expression:

$$E_{\text{int}} = k_B T \ln \beta p \Lambda^3 \left(\frac{n_1 N_{1M}}{1+\left(\frac{p_1}{p}\right)^{n_1}} + \frac{n_2 N_{2M}}{1+\left(\frac{p_2}{p}\right)^{n2}} + \frac{n_3 N_{3M}}{1+\left(\frac{p_3}{p}\right)^{n3}} \right) - k_B T \left(\frac{N_{1M} \ln\left(\frac{p_1}{p}\right)^{n_1}}{\left(1+\left(\frac{p_1}{p}\right)^{n_1}\right)} + \frac{N_{2M} \ln\left(\frac{p_2}{p}\right)^{n_2}}{\left(1+\left(\frac{p_2}{p}\right)^{n_2}\right)} + \frac{N_{3M} \ln\left(\frac{p_3}{p}\right)^{n_3}}{\left(1+\left(\frac{p_3}{p}\right)^{n_3}\right)} \right) \tag{24}$$

We notice, Figure 4 that for the alloy studied, the internal energy is always negative and the increase of the pressure always pulls a rise in module of the internal energy of system.

Conclusion

Using the grand canonical ensemble in statistical physics and

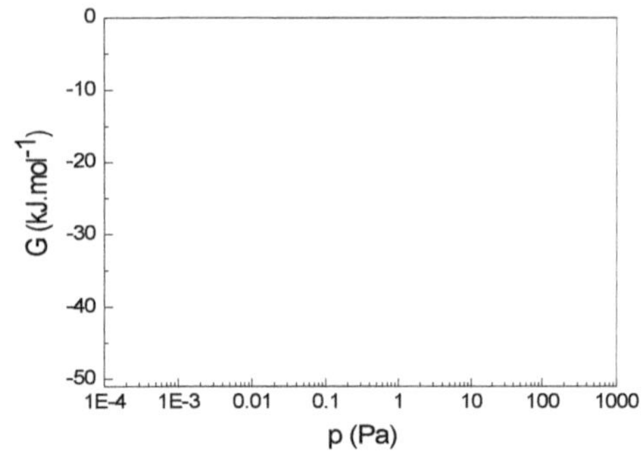

Figure 3: Evolution of the free enthalpy versus the pressure.

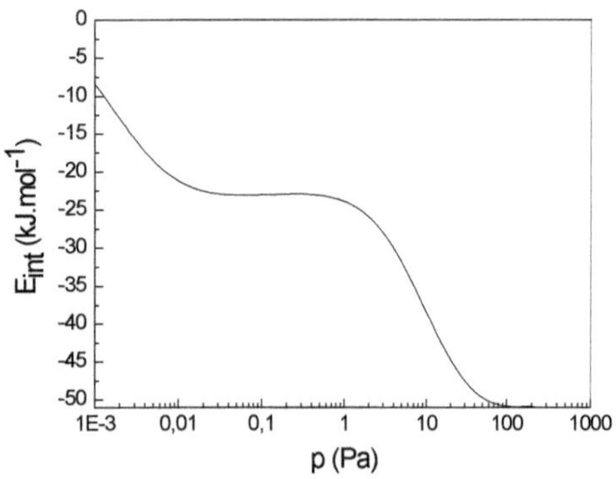

Figure 4: Evolution of the internal energy versus the pressure.

applying some simplifying approaches, we have established new theoretical expressions to fit the adsorption isotherm of Xenon on graphite at 110 K. These models are thermodynamically consistent and the physicochemical parameters intervening in the adsorption process are taken account in the theoretical treatment. It was found that the proposed statistical treatment is sufficiently flexible to give a good representation of experimental data. Thanks to the theoretical modeling; we can well understand the behavior of the adsorption isotherms of Xenon and thus the corresponding adsorption phenomenon at microscopic level. Form the experimental isotherm we can determine the physicochemical parameters related to the adsorption process. All the values of these parameters, in the various equations already determined, may be obtained by fitting procedures and deduced for such considered experimental isotherm.

References

1. McRae E, Varlot K, Dupont-Pavlovsky N, Duval X (2001) Carbon nanotubes: some potential applications related to adsorption on, confinement within and functionalisation of the graphene surface. Le vide 300: 235-250.
2. Thomy A, Duval X (1969) Adsorption de molécules simples sur graphite. J Chim Phys 66: 1966-1973.
3. Muris M, Dufau N, Bienfait M, Dupont-Pavlovsky N, Grillet Y, et al. (2000) Methane and krypton adsorption on single-walled carbon nanotubes. Langmuir 16: 7019-7022.
4. Masenelli-Varlot K, Dupont-Pavlovsky N, McRae E (2002) Comparative adsorption of simple molecules on carbon nanotubes. Dependence of the adsorption properties on the nanotube morphology. Surf Sci 196: 209-215.
5. Moulay-Rachid BABAA (2004) Contribution à l'étude de l'adsorption physique de gaz sur les nanotubes de carbone mono- et multiparois'. Thesis, University Henri Poincaré, Nancy-I.
6. IUPAC (1999) Commission on Colloid and Surface Chemistry Including Catalysis. Pure Appl Chem 57: 603-619.
7. Rouquerol F, Rouquerol J, Sing K (1999) Adsorption by powders and porous solids. Academic press, London, UK.
8. Ben Lamine A, Bouazra Y (1997) Application of statistical thermodynamics to the olfaction mechanism. Chem Senses 22: 67-75.
9. Khalfaoui M, Knani S, Hachicha M, Ben Lamine A (2003) A New theoretical expressions for the five adsorption type isotherms classified by BET based on statistical physics treatment. Journal of Colloid and Interface Science 263: 350-356.
10. Khalfaoui M, Baouab MHV, Gauthier R, Ben Lamine A (2002) Statistical Physics Modelling of Dye Adsorption on Modified Cotton. Adsorption Science & Technology 20:17-31.
11. Khalfaoui M, Baouab MHV, Gauthier R, Ben Lamine A (2002) Dye Adsorption by Modified Cotton. Steric and Energetic Interpretations of Model Parameter Behaviours. Adsorption Science & Technology 20: 33-47.
12. Diu B, Guthmann C, Lederer D, Roulet B (1989) Physique Statistique.Hermann, Paris, France.
13. Knani S, Mathlouthi M, Ben Lamine A (2007) Modeling of the Psychophysical Response Curves Using the Grand Canonical Ensemble in Statistical Physics. Food Biophysics 2: 183-192.
14. Couture L, Zitoun R (1992) Physique statistique. Ellipses 375.

A Theoretical Study of CO Adsorption on Pt-Me (Me- Fe, Co, Ni) Nanoclusters

Korniy S*, Pokhmurskii V and Kopylets V

Karpenko Physico-Mechanical Institute of the National Academie of Sciences of Ukraine, 5 Naukova Str., 79060 Lviv, Ukraine

Abstract

We used DFT calculations to study interaction of carbon monooxide with platinum binary nanoclusters of core-shell structure $Pt_{42}X_{13}$ (where X - transition metals Fe, Co, Ni). Bond strength was calculated for CO adsorption on a platinum monolayer as well as and regularities of the core (Fe, Co, Ni) of binary nanocluster influence on geometry and energy behaviors of the adsorption. On the ground of data obtained we proposed a mechanism of CO interaction with platinum binary nanoclusters. The mechanism is based either on electronic properties changes for surface platinum atoms under the influence of our core or on distribution changes for active adsorption centers in dependence on the core nature.

Keywords: Platinum binary nanoclusters; Adsorption; Carbon monoxide; Density functional theory method; Bond energy

Introduction

In catalytic electrodes, manufactured for low temperature fuel cells, to decrease their prices, binary nanoclusters PtX (X - Cr, Fe, Co, Ni and others) are often used. According to literature [1-3], such binary nanoclusters may possess higher catalytic activity, than pure metals. However, if hydrogen (obtained by hydrocarbon conversion) is used as a fuel, effective work of catalysts is dramatically decreased due to poisoning of nanoclusters by carbon monoxide admixtures. Thus, although high initial catalytic properties of such binary nanoclusters, their stability and corrosive resistance dramatically drops. Even tiny amounts of CO poison platinum anode after blocking active centers and reaction depletion. Many publications of the last years were devoted to the problem of adsorption and oxidation of CO on anodic catalysts for fuel cells [4-6]. These works were directed on search of experiments and theoretical data to propose and understand the mechanisms of CO adsorption and its poisoning of different nanocatalysts. Calculations help to ‘realize’ reaction conditions in a more simple way as compared with experiments [7]. Some problems may be solved through understanding of micro-mechanisms of atom and molecular interaction with nanocatalysts, when theoretical mechanisms for adsorption are understood. Besides proposition of structural and electron reasons of poisoning may be accepted [8-10]. But theoretical approaches in modern modeling often give doubt information, whereas the methods themselves cannot describe nanocluster properties in a correct way due to neglecting of quantum nature of interaction in such systems. Quantum-chemical methods describe surface structure, sizes and composition of nanoclusters in thin details, which sufficiently changes both catalytic activity and corrosive resistance of nanocatalysts. The aim of such investigation for practice is atomic and molecular mechanism of CO interaction with functional platinum catalysts, which help optimized choice of surface morphology of the surface and nature of binary nanocluster catalyst for fuel cells, which are stable to CO. In this occasion an estimation of micro- mechanisms of CO location on surfaces and effects of electronic structure of CO on metal nanoparticles. First principles studies of additional elements influence on CO adsorption on Pt alloys confirmed [11] the role of stresses, induced by such elements, when electronic states on platinum are changed, followed by CO adsorption energy decrease. Further studies [12] confirmed Pt stability to CO poisoning as a result of adsorption energy decrease through CO depletion from platinum active centers. In the present work the process of CO adsorption on platinum binary nanoclusters Pt_nX_m (X - transition metals Fe, Co, Ni) was studied and their resistance to CO poisoning in the environment of low temperature fuel cells was estimated on the basis of quantum-chemical energy parameters comparison.

Computational Details

Platinum binary nanocluster were simulated on the basis of Pt real crystal lattice (f.c.c. with lattice parameter 3.92 A), when atoms were closed parked. From many possible structures one of the most parked is a cubo-octhaedron Pt_{55}, (Figure 1a), constructed from eight elementary f.c.c. lattices of platinum with full symmetry and maximum surface. The choice is motivated by the following reasons: Pt_{55} nanocluster has four coordinate spheres of its surface; the nanocluster is the most stable one due to both quantum-chemical studies [13,14] and mass spectroscopy measurements [15]. Neighbor number in a cubo-octhaedron is the greatest in comparison with other structures and causes its stability.

Our structures were formed as «core-shell» particles, where inner core had atoms of a transition metal (secondary component of the nanocluster), while the shell in one atomic layer was built of platinum as an active catalyst (Figure 1b). Our own [16] and other author [17] studies showed, that the core influence sufficiently on the nanocluster catalytic and corrosive behaviors. Particularly, nanoclusters PtCo with the core-shell structure possesses enhanced catalytic activity and stability to oxide formation during their electro-reduction on cathodes of low temperature fuel cells. Enhanced activity in PtCo obeys to reaction ability of the nanocluster surface to oxygen reduction as well as to the strength of chemo-sorption bonds of the surface with atomic oxygen. Besides, a reactive center position in three-fold sites at the surface should be centered by Co atom in the sub layer. On the other hand similar nanoclusters may possess higher stability in aggressive sour environment, which causes electro-chemical corrosion of a catalyst material.

***Corresponding author:** Korniy S, Karpenko Physico-Mechanical Institute of the National Academie of Sciences of Ukraine, 5 Naukova Str., 79060 Lviv, Ukraine E-mail: kornii@ipm.lviv.ua

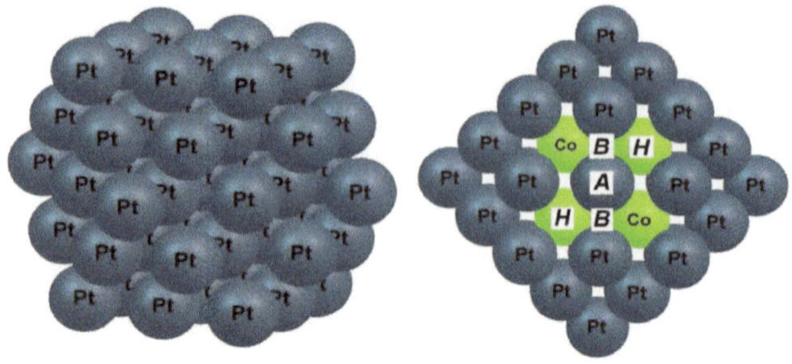

Figure 1: Cubo-octhaedral nanoclusters of platinum Pt_{55}.(a) and binary $Pt_{42}Co_{13}$ (b): A - atop position, B –bridge position, H - hollow position.

In the present work we studied processes of platinum stable nanoclusters and binary nanoclusters Pt_nX_m (X - Fe, Co, Ni; n=42; m=13) interaction with carbon monoxide. The nanoclusters were of the core-shell structure.

Electronic structure of geometry optimized binary nanoclster was calculated in the frame of the NWChem 6.1.1 [18] program with the use DFT method in GGA approximation [19] for exchange-correlation functional PW91 and double-ζ basis set with DZVP polarization [20]. Oxygen and carbon had the 6-31 G** basis sets. The bond energy of CO molecule with a nanocluster surface was found as a change of total energy of our nanocluster system during the molecule approach to the surface. Thus, potential curves were obtained with minima of initial and final stable states and maxima of transition states. Bond energy *H* was a difference between total energy of the system nanocluster-adsorbate and sum of total energies of nanocluster and adsorbate $H=W_{Pt(PtX)CO}-(W_{Pt(PtX)}+W_{CO})$.

Results and Discussion

Firstly we have calculated geometry and electronic properties of a single carbon monoxide molecule. Molecular orbital theory predicts electronic structure of the molecule for ten valence electrons as $[(\sigma_s)^2(\sigma_s*)^2(\pi_x)^2(\pi_y)^2(\sigma_z)^2]$ [21]. CO parameters are as follows: bond order 3, bond length 1.13 Å, ionization potential 14.1 eV, bond energy 256 kcal/mole, vibration frequency 2150 cm-1. Carbon monoxide forms coordination bonds with transition atoms metals by its carbon atom, whereas these metals are able to σ- and π-interaction [22]. We calculated bond distance between C and O in the molecule by different methods (Table 1). Variation in this approximation with different basis sets leads to a little deviation from experiments for bond distance in CO, even for less exact semi-empirical method PM6.

During calculations of CO interaction with pure and binary platinum nanocluster Pt_{55} only one or two molecules of CO were considered (Figure 2) for surface plane (100) with 9 atoms. Calculations were performed in a perpendicular direction of the newly formed bond, as well as different sites - atop (A), bridge (B) and hollow (H). Adsorption energy, distance of CO from the surface, charges on atoms and distance in the molecule were calculated. All six surfaces (100) of our binary nanoclusters were covered with CO molecules in symmetric sites (see Figure 1b): atop - 6 molecules, bridge and hollow- 12 molecules. After that our nanocluster obtained a high symmetry structure, available for optimization. Binding energies were calculated as being related to values per neighbor atoms, similar to the methodology of the paper [23,24]. Bond energy was calculated as follows. For atop sites it was determined by $H=[W_{Pt\text{-}CO}-(W_{Pt}+W_{CO})]/6$, while for bridge and hollows sites $H=[W_{Pt\text{-}CO}-(W_{Pt}+W_{CO})]/12$. Thus, bond energies were obtained as related to a single molecule of carbon monoxide. As we see (Figure 3), the most favorable site for CO on Pt_{55} nanocluster is the hollow one, whereas adsorption heat is the biggest for it. It should be noted, that calculated heat of adsorption is in frames of experimental values of adsorption heats for CO on Pt(111) [25].

It should be noted that contradiction experiments exists regarding Pt - CO bond energies and their consideration under low coverage. Some authors accept atop positions as prevailing [26], other authors accept hollows ones [27]. Such discrepancy is explained by different experimental methods, used in determination of adsorption parameters for different conditions (temperature, pressure, surface coverage). Adsorption energy drops to the values of 1.3-1.6 eV. Calculation results showed prevailing adsorption sites as hollow ones [28]. Calculated bond energies, present here, lie in the frames of experimental values of CO adsorption on Pt (111), and coincide with results of others.

The most stable adsorption of CO was observed in the hollow site, as is evidently from the shortest distance from the surface 1.418 A (Table 2), as well as decrease for CO atomic distance in 10%. Thus, in this site strong bond of CO to the surface occurs without dissociation of the molecule. Really platinum carbonyl compounds have larger bond distances Pt - CO (in decimals) because of more accurate approximation [29,30], used in these studies. Moreover carbonyls with a few CO ligands should enhance bond distances (as compared with low chemisorption), due to steric interactions of these ligands. During adsorption of CO charge on carbon atom growth up in comparison

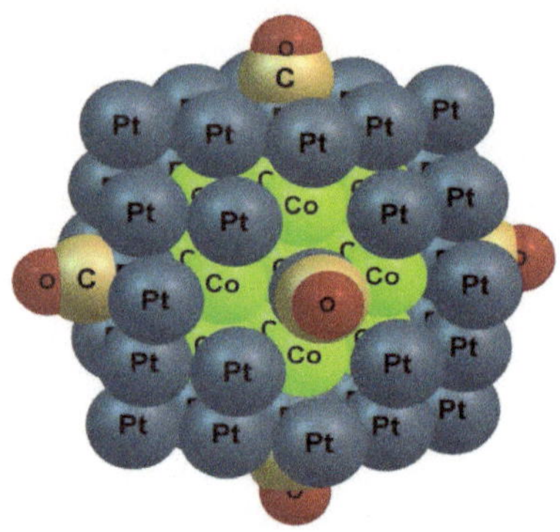

Figure 2: Adsorption of CO molecule on the surface of binary platinum nanocluster $Pt_{42}Co_{13}$.

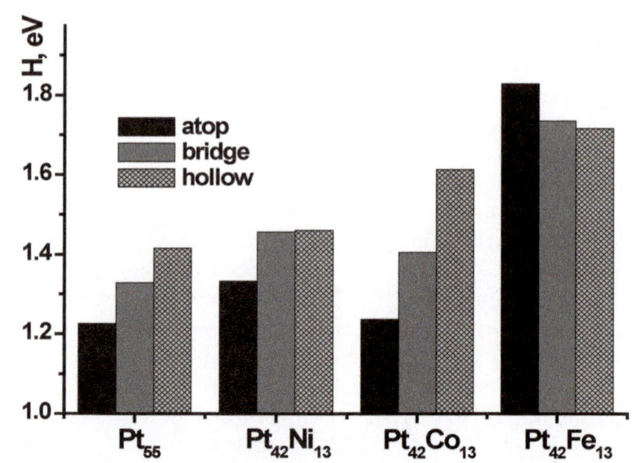

Figure 3: The Calculated bond energy for CO adsorption on nanoclusters.

	Method of calculation					Experiment [23]
	PM6	MP2/6-31G*	MP4/6-311+G(2d, 2p)	B3LYP/ LANL2DZ	B3LYP/6-311+G(2d, 2p)	
d_{CO}, Å	1.1396	1.1511	1.1458	1.1665	1.1256	1.13

Table 1: Calculated distance (d) in molecule CO in different methods.

		atop	bridge	hollow
Pt_{55}	D, A	1.563	1.542	1.418
	d, A	1.132	1.125	1.112
	q, e	0.446	0.481	0.526
$Pt_{42}Ni_{13}$	D, A	1.606	1.533	1.362
	d, A	1.135	1.148	1.169
	q, e	0.542	0.466	0.803
$Pt_{42}Co_{13}$	D, A	1.645	1.505	1.323
	d, A	1.286	1.264	1.328
	q, e	0.224	0.450	0.436
$Pt_{42}Fe_{13}$	D, A	1.334	1.420	1.635
	d, A	1.115	1.126	1.120
	q, e	0.561	0.628	0.634

Table 2: Calculated geometry of CO adsorption on the surface (111) of nanoclusters, D – distance (C–Pt) of the molecule to the surface, q – charge on atom C, d – atomic distance in CO.

with that of free molecule CO. So, there are no contradiction to the statement [22] that in the case of CO adsorption on homogenous metallic surfaces bond strength depends sufficiently on delocalization of electron charge from filled $d\pi$-orbitals of the metal to empty π^*-orbital of CO. This leads to reverse transition of the charge due to donor-acceptor bond with strengthening of the π-bond.

Therefore CO bonding on the surface of platinum nanocluster occurs through donor-acceptor interaction through carbon atom with formation of π- and σ-bonding. To estimate influence of the second component in platinum nanoclusters on its activity in the CO environment, we calculated CO interaction with the surface (111) of binary nanoclusters with core-shell structures. Three kinds of nanocluster $Pt_{42}Co_{13}$, $Pt_{42}Fe_{13}$, $Pt_{42}Ni_{13}$ were considered.

Our results confirm influence of binary nanocluster core on surface interaction of Pt atoms with CO molecule. Thus adsorption heat growth for it is observed in a row $Pt_{55} \leq Pt_{42}Ni_{13} < Pt_{42}Co_{13} < Pt_{42}Fe_{13}$ (Figure 3). Obviously, nanocluster ability for CO bonding determinate electron density of the 5*d*-orbital in connection with surface geometry (dimension effects) and electron interaction of Pt layer on the surface with the core (the ligand effect). Further CO π^*-orbital participation in bond formation depends on energy level gaps of the metal and this orbital. Nearest levels form stronger bonds. Calculated ionization potential for CO π^*-orbital is 5.5 eV, while for metals experimental work functions are as follows: Fe - 7.63 eV, Co - 7.8 eV, Ni - 7.9 Pt - 9.0 eV [31]. The higher the value of work function, the lower bond strength in Pt - CO as a result of increasing energy gap between π^*-level in CO and metal orbital levels.

The strongest bond of CO was observed with the surface of $Pt_{42}Fe_{13}$ at atop positions, oppose to hollow site position of other nanoclusters. The shortest bond length indicates about strong bond. Bond length increases in a row $Pt_{42}Fe_{13} < Pt_{42}Ni_{13} < Pt_{42}Co_{13}$, so semi-dissociative state exists for platinum-nickel $Pt_{42}Ni_{13}$ and full dissociation was found for platinum-cobalt $Pt_{42}Co_{13}$. Platinum resistance to CO poisoning depends on adsorption heats, which show the degree of adsorption centers blocking. The presence of nickel has no sufficient effect on platinum shell, while iron or cobalt in the core decreases platinum resistance to CO poisoning. Such a behavior of $Pt_{42}Ni_{13}$ during its interaction with CO may be described by ligand mechanism of nickel influence on bonding in Pt-CO through induced charges in electronic structure of platinum surface atoms. In this case π-bonds emerge between surface platinum atoms and neighbors in sublayer due to partly filled d-orbitals of metal. During CO adsorption a concurrence exits regarding d-electrons of surface platinum atom between CO and nickel atom. This is confirmed by essential increase in calculated electron density and Muliken charge on carbon in CO-$Pt_{42}Ni_{13}$, than in CO-Pt_{55}.

In prevailing situations heat grows up for CO adsorption on binary nanoclusters, compared with pure platinum nanoclusters. Thus, the core induces poisoning of the surface. For $Pt_{42}Fe_{13}$ a prevailing adsorption of CO is observed in atop sites. This indicates that the core of Fe atom changes electronic behaviors of the catalytic active metal platinum through collective (morphologic) mechanism, when additional component in binary nanoclusters charges distribution of active centers on the surface and allows catalytic reactions to occur. Such a reaction is electrooxidation of hygrogen in bridge and hollow sites.

During interaction of surface platinum with nanocluster core a decrease is in overlapping of sublayer d-orbitals (Fe, Co, Ni) and surface (Pt) layers in a sequence Ni>Co>Fe. As a result most surface d-orbitals become free in platinum atoms, which are able to form bonds both with CO and H atoms. Thus, in the initial step during binary nanocluster interaction with CO and H_2 under low coverage of the surface in all adsorption sites a ligand sphere occurs in metal-environment interface. As a result, a strong chemisorptions of H_2 occurs on neighbor free atop positions due to the conjugation effect.

Conclusions

1. A model of core-shell binary nanocluster $Pt_{42}X_{13}$ (X - Fe, Co, Ni) poisoning by carbon monoxide was proposed in environment of low temperature fuel cells. The model is grounded on adsorption behavior calculations for CO interaction with nanocluster surfaces and an ability of the nanoclusters to strong bond formation with carbon monoxide in dependence on core nature and adsorption sites.

2. Analysis of calculated adsorption heats, adsorption distances, distances in molecules, electronic charges and densities in the system Me-CO lead to establishment of main physico-chemical behaviors of Fe, Co and Ni influence (core of binary nanoclusters) on their interaction with CO. Especially transition metals Fe, Co and Ni sufficiently influence on formed bond behaviors on the platinum surface layer mainly by two mechanisms: ligand (electronic) mechanism, when Fe, Co and Ni change electronic properties of Pt as catalytically active metal, and collective (morphological) mechanism, where additional component of binary nanocluster changes distribution of active sites on Pt surface with new catalytic reaction ways.

3. It was shown, that Pt_{55} has an ability to lower catalytic properties through poisoning by carbon monoxide through formation of strong chemo-sorption bond with the highest adsorption heat, which increases for binary nanoclusters in the sequence $Pt_{55} \leq Pt_{42}Ni_{13} < Pt_{42}Co_{13} < Pt_{42}Fe_{13}$ due to electronic interactions between 5 d-orbital of transition metals and π^*-orbital in CO for increasing ionization potential in the row Fe<Co<Ni.

References

1. Wenzhen L, Qin X, Yushan Y (2010) Nanostructured Pt–Fe/C cathode catalysts for direct methanol fuel cell: The effect of catalyst composition. Int J Hydrogen Energy 35: 2530-2538.
2. Antolini E, Lopes T, Gonzalez ER (2008) An overview of platinum-based catalysts as methanol-resistant oxygen reduction materials for direct methanol fuel cells. J Alloys and Compounds 461: 253-262.
3. Chien-Te H, Jia-Yi L (2009) Fabrication of bimetallic Pt–M (M = Fe, Co, and Ni) nanoparticle/carbon nanotube electrocatalysts for direct methanol fuel cells. J Power Sources 188: 347-352.
4. Zhang S, Yuan XZ, Hin JNC, Wang H, Andreas Friedrich K, et al. (2009) A review of platinum-based catalyst layer degradation in proton exchange membrane fuel cells. J Power Sources 194: 588-600.
5. Wentao W (2009) The effect of internal air bleed on CO poisoning in a proton exchange membrane fuel cell. J Power Sources 191: 400-406.
6. Chen QS, Sun SG, Zhou ZY, Chen YX, Deng SB (2008) CoPt nanoparticles and their catalytic properties in electrooxidation of CO and CH(3)OH studied by in situ FTIRS. Phys Chem Chem Phys 10: 3645-3654.
7. Bi W, Fuller TF (2008) Modeling of PEM fuel cell Pt/C catalyst degradation. J Power Sources 178: 188-196.
8. Goddard III W, Merinov B, Van Duin A, Jacob T, Blanco M, et al. (2006) Multi-paradigm multi-scale simulations for fuel cell catalysts and membranes. Molecular Simulation 32: 251-268.
9. Greeley J, Mavrikakis M (2006) Near-surface alloys for hydrogen fuel cell applications. Catalysis Today 111: 52-58.
10. Geschke D, Bastug T, T Jacob, Fritzsche S, Sepp WD, et al. (2001) Adsorption of CO on cluster models of platinum (111): A four-component relativistic density-functional approach. Phys Rev B 64: 235411.
11. Koper MT, Shubina TE, van Santen RA (2002) Periodic Density Functional Study of CO and OH Adsorption on Pt−Ru Alloy Surfaces: Implications for CO Tolerant Fuel Cell Catalysts. J Phys Chem 106: 686-692.
12. Ali S (2012) Interaction of carbon monoxide with bimetallic Co–Pt clusters: A density functional theory study. Comp Mater Sci 58: 77-86.
13. Francesca B, Riccardo F (2005) Structural properties of nanoclusters: Energetic, thermodynamic, and kinetic effects. Rev Mod Phys 77: 371.
14. Barcaro G, Fortunelli A, Polak M, Rubinovich L (2011) Patchy multishell segregation in Pd-Pt alloy nanoparticles. Nano Lett 11: 1766-1769.
15. Yen TW, Lai SK (2015) Use of density functional theory method to calculate structures of neutral carbon clusters Cn ($3 \leq n \leq 24$) and study their variability of structural forms. J Chem Phys 142: 084313.
16. Pokhmurskii V, Korniy S, Kopylets V (2011) Computer simulation of binary platinum-cobalt nanoclusters interaction with oxygen. J Cluster Sci 22: 449-458.
17. Rossi G, Rapallo A, Mottet C, Fortunelli A, Baletto F, et al. (2004) Magic polyicosahedral core-shell clusters. Phys Rev Lett 93: 105503.
18. Valiev M, Bylaska EJ, Govind N, Kowalski K, Straatsma TP, et al. (2010) NWChem: A comprehensive and scalable open-source solution for large scale molecular simulations. Comput Phys Commun 181: 1477-1489.
19. Keal TW, Tozer DJ (2005) Semiempirical hybrid functional with improved performance in an extensive chemical assessment. J Chem Phys 123: 121103.
20. Perdew JP, Chevary JA, Vosko SH, Jackson KA, Pederson MR, et al. (1992) Atoms, molecules, solids, and surfaces: Applications of the generalized gradient approximation for exchange and correlation. Phys Rev B Condens Matter 46: 6671-6687.
21. George B (1964) Molecular Orbital View of Chemisorbed Carbon Monoxide. J Phys Chem 68: 2772-2777.
22. Doyen G, Ertl G (1974) Theory of carbon monoxide chemisorption on transition metals. Surf Sci 43: 197-229.
23. Roberts MW, McKee CS (1978) Chemistry of the metal-gas interface. Oxford University Press, Oxford.
24. Paz-Borbon LO, Johnston RL, Barcaro G, Fortunelli A (2009) Chemisorption of CO and H on Pd, Pt and Au nanoclusters: a DFT approach. Eur Phys J D 52: 131-134.
25. Markovic NM, Jr Ross PN (2002) Surface science studies of model fuel cell electrocatalysts. Surf Sci Rep 45: 117-229.
26. Yeo YY, Vattuone L, King DA (1997) Calorimetric heats for CO and oxygen adsorption and for the catalytic CO oxidation reaction on Pt{111}. J Chem Phys 106: 392.
27. Ogletree DF, Van Hove MA, Somorjai GA (1986) LEED intensity analysis of the structures of clean Pt(111) and of CO adsorbed on Pt(111) in the c(4 × 2) arrangement. Surf Sci 173: 351.
28. Feibelman PJ, Hammer B, Norskov JK, Wagner F, Scheffler M, et al. (2001) The CO/Pt(111) Puzzle. J Phys Chem 105: 4018-4025.
29. Okabayashi T, Yamamoto T, Okabayashi EY, Tanimoto M (2011) Low-energy vibrations of the group 10 metal monocarbonyl MCO (M = Ni, Pd, and Pt): rotational spectroscopy and force field analysis. J Phys Chem A 115: 1869-1877.
30. Liu Z, Xie H, Qin Z, Cong R, Wu X, et al. (2012) Vibrationally resolved photoelectron imaging of platinum carbonyl anion Pt(CO)(n)- (n = 1-3): experiment and theory. J Chem Phys 137: 204302.
31. Kittel C (1996) Introduction to Solid State Physics. 7th edn, Wiley.

12

Silica Supported Heteropoly Catalysts for Oxidation of Methacrolein to Methacrylic Acid

Lilong Zhou[1,2], Lei Wang[1*], Hui Wang[1], Yunli Cao[1,2], Ruiyi Yan[1] and Suojiang Zhang[1*]

[1]Beijing Key Laboratory of Ionic Liquids Clean Process, Key Laboratory of Green Process and Engineering, State Key Laboratory of Multiphase Complex Systems, Institute of Process Engineering, Chinese Academy of Sciences, Beijing 100190, PR China
[2]Chinese Academy of Sciences, University of Chinese Academy of Sciences, Beijing 100049, PR China

Abstract

To prepare efficient supported heteropoly catalysts for oxidation of methacrolein to methacrylic acid, $CsH_3PMo_{11}VO_{40}$ was supported on SBA-15 and compared with other supports, such as Y zeolite, KIT-6, and porous SiO_2. Nano-particles of $CsH_3PMo_{11}VO_{40}$ were loaded on the inner and outer side surface of the SBA-15 support and were decomposed after calcination and catalytic reaction. SBA-15 supported $CsH_3PMo_{11}VO_{40}$ with the loading amount of 80 wt% showed the best performance and catalytic activity was twice of that of $CsH_3PMo_{11}VO_{40}$. The results of XPS, ^{29}Si and ^{31}P NMR spectra showed that the decomposition of the supported catalysts was caused by the replacement of P in the Keggin structure by Si during calcination and reaction. The low selectivity was detected and due to the formation of MoO_3 and V_2O_5 from the decomposition of the supported $[PMo_{11}VO_{40}]^{4-}$, and further interaction between $CsH_3PMo_{11}VO_{40}$ and SBA-15 was prevented by the formation of a layer of MoO_3 and V_2O_5 between the heteropoly compounds and support. Compared with supported $H_4PMo_{11}VO_{40}$, the addition of Cs improved thermal stability of the supported heteropoly catalysts. This study shows that the key point to prepare efficient and thermal stable silica supported heteropoly catalysts is to avoid the reaction between heteropoly compounds and silica supports.

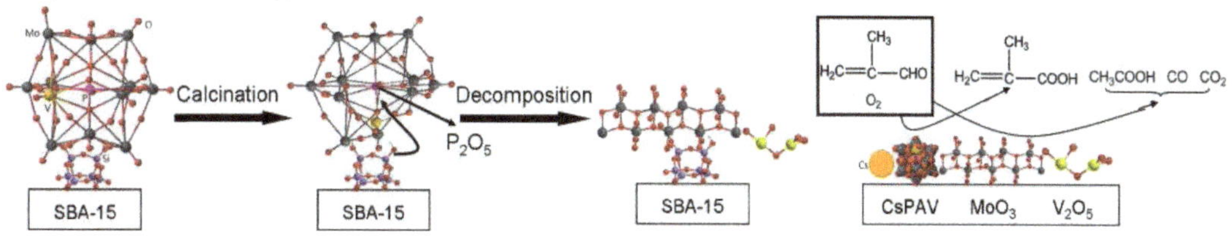

Keywords: Heteropoly catalysts; Support; Selective oxidation; Methacrolein; Methacrylic acid

Introduction

Heteropoly catalysts have attracted a lot of attention in oxidation, acid and shape-selective catalysis [1-5]. The oxidation of methacrolein (MAL) to methacrylic acid (MAA) is a successful example of industrial applications of heteropoly catalysts [6-8]. Due to the high cost and low turnover frequency of this kind of catalysts, researches on supported heteropoly catalysts for higher catalytic efficiency are always attractive.

Silica is a kind of cheap, useful, and thermal stable supports, and some heteropoly acids have been supported on silica [9-19]. Kanno et al. found that SiO_2 supported $H_4PMo_{11}VO_{40}$ (HPAV) had good catalytic activity for oxidation of MAL to MAA, especially when using aminated SiO_2 as the support, on which the TOF was enhanced 5 times. Thermal stability of this kind of supported heteropoly acids was poor as HPAV was decomposed to MoO_3 and V_2O_5 [15,16]. Similarly, the SiO_2 supported $H_3PMo_{12}O_{40}$, reported by Gomez-Sainero, showed low catalytic activity due to the loss of Brönsted acids sites, and $H_3PMo_{12}O_{40}$ was decomposed to MoO_3. To solve this problem, ZrO_2 was added as a co-catalyst, and catalytic activity was obviously improved [12]. $H_4SiW_{12}O_{40}$ and $H_3PW_{12}O_{40}$ were supported on porous SiO_2. However, these two heteropoly acids reacted with silica and formed $[{\equiv}SiOH_2]_3PW_{12}O_{40}$ and $[{\equiv}SiOH_2]_xH_{4-x}SiW_{12}O_{40}$, which leaded to the decomposition of the Keggin structure due to the existence of over acidified protons [13]. Ballarini prepared heteropoly acids in SiO_2 by co-gelation procedure. The catalysts showed poor thermal stability due to the replacement of P by Si as center atoms [14]. Silica-based ordered mesoporous molecule sieves, such as SBA-15, MCM-41 and KIT-6 were also utilized as support for heteropoly acids. These supported catalysts showed better catalytic activity than the bulk ones. However, the supported heteropoly acids have the same drawbacks as those catalysts supported on SiO_2, e.g., low thermal stability [9,10]. So it is necessary to find a way to modify silica and improve the thermal stability ofsupported heteropoly catalysts. The addition of counter ions with large volumes, such as K^+, Cs^+, NH_4, is an effective way to improve thermal stability of heteropoly compounds [20]. The addition of counter ions, such as Cs^+, also can improve the catalytic activity for the oxidation of MAL to MAA and increase the selectivity for MAA [21,22]. Therefore, suitable counter ions are needed to prepare thermal stable supported heteropoly catalysts. In this study, four different silica-based supports (i.e., porous SiO_2, SBA-15, KIT-6 and

***Corresponding authors:** Lei Wang, Beijing Key Laboratory of Ionic Liquids Clean Process, Key Laboratory of Green Process and Engineering, State Key Laboratory of Multiphase Complex Systems, Institute of Process Engineering, Chinese Academy of Sciences, Beijing 100190, PR China
E-mail: lwang@ipe.ac.cn

HY zeolite) with different pore sizes were used as the supports to load $H_4PMo_{11}VO_{40}$, and Cs^+ was added as the counter ion. The preparation condition, counter ion content and loading amount were optimized to get thermally stable supported heteropoly catalysts. The decomposition mechanism of the supported heteropoly catalysts during calcination and oxidation of MAL to MAA were studied in detail.

Materials and Methods

Preparation of catalysts

$H_4PMo_{11}VO_{40}$ (HPAV) was prepared from MoO_3 (Sinopharm Chemical Reagent Co., Ltd.), V_2O_5 (Tianjin Guangfu Fine Chemical Research Institute) and 85% phosphoric acid (Sinopharm Chemical Reagent Co., Ltd.) according to published procedures [3]. Briefly, 5 g MoO_3 and 0.287 g V_2O_5 were added to 100 mL deionized water, which was subsequently heated to reflux. Then, 0.364 g of 85% phosphoric acid was added to the mixture, which was refluxed with vigorous stirring for 5 h. A deep-orange solution then formed. The insoluble solid was filtered, and the solution was evaporated at 80°C, obtaining a bright orange solid. Porous SiO_2 was purchased from Sinopharm Chemical Reagent Co. Ltd., SBA-15, KIT-6 and HY zeolite were prepared following the published procedures [23-25], and the preparation processes were described in Supporting Information. The structures of the obtained supports were confirmed by FT-IR, Raman, BET, XRD and SEM. The supported HPAV was prepared by the following procedure. 1 g HPAV was dissolved in 10 mL deionized water, then 1 g supports (porous SiO_2, SBA-15, KIT-6 or Y zeolite) were added to the HPAV solution, and the mixture was vibrated at 40°C for 12 h. After the vibration, the mixture was dried in an oven at 80°C for 12 h. The yellow powder was the prepared supported HPAV. Supported $CsH_3PMo_{11}VO_{40}$ was prepared by the following way. 0.11 g $CsNO_3$ was dissolved into 10 mL deionized water, and 1 g SBA-15 was added to the solution, then the mixture was vibrated at 40°C for 12 h. After then, the mixture was dried in oven at 80°C for 12 h and SBA-15 with $CsNO_3$ was prepared. 1 g HPAV was dissolved in 10 mL deionized water, and it was dripped into SBA-15 with $CsNO_3$, then the mixture was vibrated at 40°C for 12 h. After the vibration, the mixture was dried in oven at 80°C for 12 h. The bright yellow powder was the prepared catalysts containing Cs. $CsH_3PMo_{11}VO_{40}$ supported on other supports were prepared following the similar processes and were named as CsPAV/SiO_2, CsPAV/SBA, CsPAV/KIT and CsPAV/Y. Catalysts with different supporting amounts were named like this, 0.5CsPAV/SBA, which means 0.5 times of SBA-15 weight of catalysts was supported on SBA-15. All catalysts were calcined at 360°C in a flow of air at a rate of 80 mL·min^{-1}.

Characterization of the catalysts

Fourier transform infrared (FT-IR) spectroscopy was performed on an FT-IR spectrometer (Nicolet 380, Thermal Electron Corporation) with anhydrous KBr as standard. Raman spectroscopy was conducted by a laser Raman spectrometer (HR800 UV, Horiba Scientific Co. Ltd) with a 100 mV laser, equipped with a wavelength of 514 nm and CCD detector. Ultraviolet-visible (UV-vis) diffuse reflectance spectroscopy was recorded on a UV-vis spectrophotometer (UV-2550, Shimadzu Corporation). Powder X-ray diffraction (XRD) was performed on an X-ray diffractometer (Bruker D8 Advance X-ray powder diffractometer). Samples (about 10 mg) were analyzed by conducting thermogravimetry/differential thermal (TG/DTA, analysis using a simulaneous DTA-TG apparatus DTG-60H, Shimadzu, Japan), which allowed both TGA and DSC curves to be recorded simultaneously. The samples were heated from room temperature to 650°C at heating rate of 5°C/min. The oxidation and acid properties of catalysts were tested by hydrogen temperature programmed reduction and ammonia temperature programmed desorption on AutoChem II Chemisorption Analyzer (Micromeritics). The microcosmic pattern of the catalysts was observed by scanning electron microscope (SEM, SU8020, Hitachi Electronic Electric Appliance Company, Japan) and field emission electron microscope (TEM, JEM-2100F, JEOL, Japan). The X-ray photoelectron spectroscopy (XPS) spectra were recorded by a Kratos Axis Ultra DLD spectrometer with monochromatic Al Kα radiation. The surface area and pore structure of the catalysts were tested by Surface area and Pore Analyzer (Mocromeritics ASAP 2460, US). ^{29}Si NMR and ^{31}P NMR were performed on Bruker Avance III (400M).

Determination of catalytic activity

Catalytic oxidation of MAL was performed in a fix bed reactor under 310°C at atmosphere pressure. A total of 0.6 mL catalyst was loaded into the fix bed reactor and the contact time was about 2 s. Before the detection, temperature was increased to 310°C, then a reactant mixture of MAL (4.4 vol.%), O_2 (11.1 vol.%), H_2O (17.8 vol.%) and N_2 (balance) was fed into the reactor to start the reaction. The reaction products were collected and analyzed using GC (SP-7890, Lunan Ruihong Chemical Engineering Instrument Co. Ltd) equipped with a flame ionization detector and a TDX-01 packed column. The catalytic activity of the catalysts was evaluated by TOF, which was calculated using Formula 1:

$$\mathrm{TOF} = \frac{F(MAL) \times Conversion \times Selectivity}{n(catalysts)} \qquad \text{Formula 1}$$

where F(MAL) was the molar flow rate of methacrolein (MAL) (mol·min^{-1}), Conversion was the conversion of MAL, Selectivity was the selectivity of methacrylic acid (MAA) and n (Catalysts) was the mole amount of heteropoly catalysts (the supports was not included).

Results and Discussion

Characterization of the catalysts

To illustrate the changes of structure and chemical properties of the catalysts during calcination and reaction, FT-IR, Raman, XRD, BET, TG-DTA, XPS, ^{29}Si NMR and ^{31}P NMR were performed.

FT-IR: Figure 1 showed the FT-IR spectra of uncalcined, calcined and used CsPAV and supported catalysts with different supporting amounts. The four characteristic vibration bands (1062-1065 cm^{-1}, vs(P-O); 963-966 cm^{-1}, vs(M=O); 865-872 cm^{-1}, vs(Mo-Ob-Mo); 777-797 cm^{-1}, vs(Mo-O_c-Mo) of heteropoly compounds were shown in these spectra [26-28]. The three different vibration bands [1080, 812 and 470 cm^{-1}, vs(Si-O)] of supports were also present in the spectra [29]. Compare with the bulk CsPAV, the vibration bands of Keggin structure in the supported catalysts shifted to high wavenumber due to the interaction between CsPAV and supports. The weak vibration appeared around 1000 cm^{-1} may be attributed to Si-O in Keggin structure which was caused by the replacement of P by Si [14]. To prove this replacement, XPS, ^{29}Si and ^{31}P NMR were used to characterize the catalysts, which would be discussed in section 3.1.8. The spectra showed that CsPAV was well supported on the supports. A new band appeared at 1034 cm^{-1} in the spectra of the calcined CsPAV corresponding to V-O caused by the migration of V from Keggin structure to the secondary structure of catalysts [8]. And no other obvious changes can be found in these FT-IR spectra of the supported catalysts. It demonstrates that Keggin structure was well maintained after calcination. No obvious changes were found in the spectra of the used catalysts. The catalysts with different loadings showed almost the same spectra before calcination, after calcination and reaction. However, there were

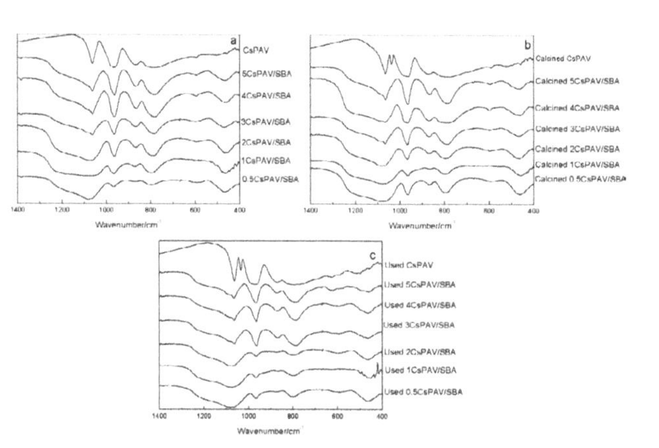

Figure 1: FT-IR spectra of uncalcined, calcined and used supported catalysts with different supporting amount
(a) FT-IR spectra of uncalcined catalysts; (b) FT-IR spectra of catalysts which were calcined at 350°C for 12 h; (c) FT-IR spectra of catalysts which have been used in reaction.

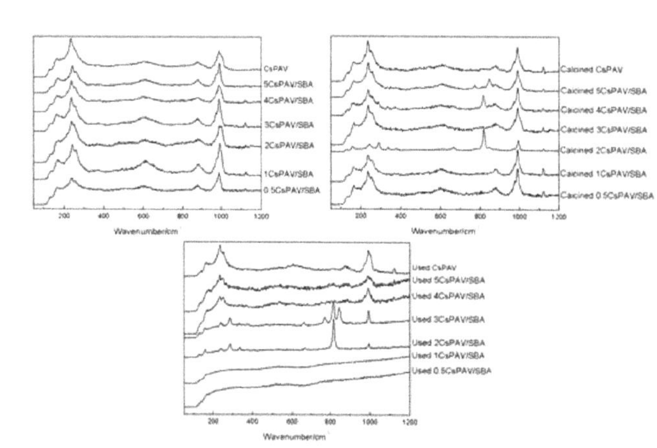

Figure 2: Raman spectra of uncalcined, calcined and used supported catalysts with different supporting amounts.

also some differences between the spectra of catalysts with different loadings. With the increase of loading amount, the vibration bands of Keggin structure shifted to low wavenumber (1081 to 1062 cm^{-1}, vs(P-O); 965 to 963 cm^{-1}, vs(M=O); 868 to 865 cm^{-1}, vs(Mo-O_b-Mo)); 797 to 781 cm^{-1}, vs(Mo-O_c-Mo) because of the decrease of interaction intensity between HPAV and support. To enhance thermal stability of the supported catalysts, Cs contents and loading amount were tuned. In the spectra of HPAV/SBA (Figure S1, Supporting Information), some vibration bands (994 cm^{-1}, $vs_{Mo=O}$; 589 cm^{-1}, $vs_{Mo-O-Mo}$) belonging to MoO_3 were shown. It suggests that HPAV was decomposed during the preparation process. HPAV might react with silica species, which led to the decomposition of the catalysts [15]. After calcination and reaction, another vibration band of MoO_3 was shown at 869 cm^{-1}, and the other two bands became stronger, which suggests the further decomposition of the catalysts. For catalysts with Cs, only a weak vibration at 589 cm^{-1} was observed. After calcination and reaction, no other changes could be detected by the FT-IR, indicating that the addition of Cs could improve thermal stability of supported catalysts.

Raman: In Figure 2, the typical main vibration bands (625 cm^{-1}, Mo-O_c-Mo; 876-909 cm^{-1}, Mo-O_b-Mo; 984 cm^{-1}, Mo-O-P; 1002 cm^{-1}, Mo=O_d; 230-250 cm^{-1}, the deformation vibration of the terminal M=O groups and the entire framework of Keggin structure.) of the catalysts with Keggin structure were shown [8,30]. The vibration band appearing at 1120 cm^{-1} could be assigned to the Si-O band in the supports. After calcination, the vibration bands at 250 and 984 cm^{-1} in the spectra of 0.5CsPAV/SBA and 1CsPAV/SBA transformed from a shoulder to a single band because of loss of crystalline water and partial decomposition of Keggin structure during calcination [30]. For other catalysts, vibration bands at 230-250 and 825 cm^{-1}, which belong to MoO_3, were shown, indicating that the supported catalysts were decomposed significantly during calcination. Decomposition of the catalysts was not observed as the vibration bands of MoO_3 didn't appear in the spectra of 0.5CsPAV/SBA and 1CsPAV/SBA, while the vibration bands of MoO_3 in the spectra of 2CsPAV/SBA and 3CsPAV/SBA indicated decomposition of these catalysts. Integral Keggin structures were detected in the spectra of 4CsPAV/SBA and 5CsPAV/SBA. It indicates that more supported CsPAV was decomposed in the catalysts with more loading amount than the ones with less loadings. The vibration band of Si-O became too strong after calcination because of the interactions between catalysts and supports during calcination. After reaction, the spectra of CsPAV remained the same. While the spectra of the support became very strong, and almost no vibration bands of Keggin structure were detected. The vibration bands emerged at 825 and 1002 cm^{-1} can be assigned to Mo-O in MoO_3 in the spectra of used 2CsPAV/SBA and 3CsPAV/SBA. It seems that the supported catalysts were almost completely decomposed after reaction. The vibration bands of MoO_3 was supposed to be detected in the spectra of used 0.5CsPAV/SBA and 1CsPAV/SBA, yet only strong vibration bands of supports were detected. The vibration bands of the decomposed catalysts might be clouded by the vibration bands of supports. Vibration bands, which belong to Keggin structure can be found in the spectra of 4CsPAV/SBA and 5CsPAV/SBA, which indicates that with the increase of the loading amount, the surface of the catalysts would not interact with the support directly and decomposition can be avoided.

XRD: XRD patterns of the catalysts (2θ=10.6°, 18.4°, 21.2°, 23.8°, 26°, 30.2°, 32.2°, 35.6°, 38.9°, 43.2°, 47.4°, 51.2°, 55°, 61.8°) showed the typical cubic crystalline of cesium molybdovanadophosphoric hydrate ($Csh_3PMo_{11}VO_{40}\cdot nH_2O$) (Figure 3). The peak at 8° could be assigned to HPAV. No peaks related to MoO_3 or $H_4PMo_{11}VO_{40}$ were found in the patterns of the catalysts before calcination, suggesting that HPAV did not completely react with Cs^+ to form CsPAV. A new peak emerged at 27.3° in the patterns of calcined supported catalysts which could be assigned to MoO_3. It indicates that the supported catalysts were partially decomposed after calcination. The peak at 8° belonging to HPAV disappeared because of the decomposition of HPAV. The diameters of crystal were calculated by Scherrer formula, $D_{hkl}=k\lambda/\beta\cos\theta_{hkl}$ (where k is shape factor, cubic is 0.9; λ is the wavelength of X-ray (nm); β is FWHM (radian); θ_{hkl} is diffraction angle (radian).). The diameters of the catalysts increased with the increase of loading amount of CsPAV (20, 21.5, 22, 39, 43.9 and 44.7 nm, respectively), which indicates that CsPAV crystal grew larger and lager with the increase of loading amount of CsPAV. With the loading amount increasing to 2, MoO_3 patterns emerged at 10°, and peaks of MoO_3 became stronger and stronger with the increase of loading amount of CsPAV. It indicates that Keggin structure was decomposed during the preparation process, and more MoO_3 formed with the increase of loading amount. After calcination and reaction, the patterns of MoO_3 became more obvious, which showed that catalysts were further decomposed during the calcination and reactions. No obvious changes in the patterns of catalysts were observed after reaction. The sizes of the supported catalysts changed

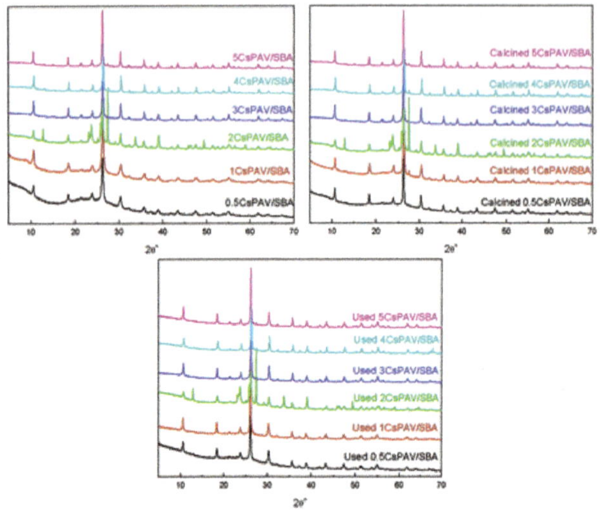

Figure 3: XRD patterns of uncalcined, calcined and used supported catalysts with different supporting amounts.

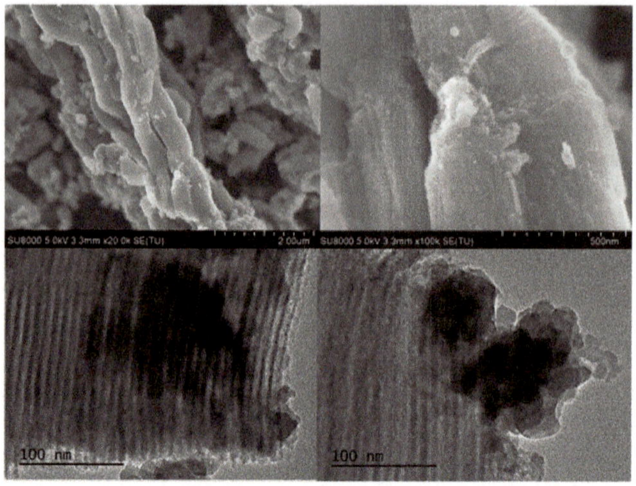

Figure 4: SEM (Top) and TEM (Bottom) images of CsPAV/SBA.

little after calcination and catalytic reaction, suggesting that most of Keggin structures kept integral after reaction. The XRD patterns of HPAV/SBA were obviously different from Cs$_x$PAV/SBA (Figure S2). Some peaks appeared at 10.5°, 19.1°, 25° and 28.9°, which showed that HPAV were well dispersed on SBA-15. The wide peaks showed that the diameter of HPAV was very small. HPAV might disperse on SBA-15 by a single molecule, because the surface area of SBA-15 used in this experiment was about 1000 m^2/g, while the diameter of a single HPAV molecule is about 1 nm [15,16], and HPAV/SBA was 1 g HPAV supported on 1 g SBA-15, while 4 g HPAV could form a single molecule layer on 1 g SBA-15. The patterns of the catalysts with Cs are similar. The main peaks could be assigned to CsPAV,and the peaks of HPAV (8°) were also shown in the patterns of CsPAV/SBA and Cs2PAV/SBA. And it disappeared in the patterns of Cs$_3$PAV/SBA, which demonstrated that all PMo$_{11}$VO^{4-} reacted with Cs$^+$ during preparation of Cs$_3$PAV/SBA. The sizes of Cs$_{1\sim3}$PAV crystals on the support are almost the same (21-22 nm). The supported HPAV was decomposed after calcination, so the peaks of MoO$_3$ were obviously observed. HPAV in Cs$_x$PAV/SBA was also decomposed during calcination. The peaks at 23.5° and 24.9° show that some integral Keggin structures remained after calcination. The peak of MoO$_3$ (27.3°) was also found in the patterns of CsPAV/SBA and Cs$_2$PAV/SBA. No peaks of MoO$_3$ were found in Cs3PAV/SBA. So the appearance of MoO$_3$ was due to the decomposition of HPAV in the supported catalysts, indicating that increasing of Cs content can improve thermal stability of the catalysts.

SEM and TEM: SEM and TEM could reveal the state of the supported catalysts. Figure 4 shows that the bulk CsPAV has no particular shape and size. The images of supported catalysts showed that all the supports kept integral after preparation and calcination. CsPAV was clustered on the surface of the supports in the sizes of 30-150 nm which were in accordance with the calculated results of XRD. Although the supports were different, CsPAV showed the same shape. The CsPAV might form from the following process. Firstly, CsNO$_3$ was uniformly dispersed on the supports. After the HPAV solution dripped onto the supports with CsNO$_3$, some Cs$^+$ was added into the solution, which could reacted with PMo$_{11}$VO$_{40}$, forming CsPAV on the outer surface of the supports. Some PMo$_{11}$VO$_{40}$ might come into the pore of the supports and react with Cs$^+$ to form CsPAV on the pore wall. With the increase of loading amount, CsPAV particles on the surface of SBA-15 increased obviously. When the loading amount was increased to 75%, almost all the holes in SBA-15 were filled by CsPAV. With the loading amount further increasing, the surface of SBA-15 was all covered by CsPAV. The CsPAV particles also clustered by themselves. No independent HPAV particles could be found in HPAV/SBA (Figure S3), indicating that HPAV was well dispersed on the SBA-15. TEM images shows that CsPAV particles were supported on the surface of the carriers. Some CsPAV might enter the pore of SBA-15. TEM images also show that CsPAV cumulated in the entrance of the channels.

N$_2$-sorption: Figure 5 showed that the surface area and pore diameter of SBA-15 were about 1017.8 m^2·g^{-1} and 6.6 nm, respectively. The nitrogen absorption isothermal lines of SBA-15 show that it was a kind of mesporous material with some micropores which were indicated in the part of nitrogen absorption isothermal lines below 0.1. After CsPAV being loaded on SB A-15, the surface area, pore volume and the ratio of micropores decreased with the increase of loading amount of CsPAV. The surface area decreased from 1017.8 to 70.0 m^2·g^{-1}. As mentioned above, some CsPAV entered the pores of SBA-15. The average diameter of CsPAV was higher than 20 nm. So the decrease of surface area was due to the clog of CsPAV in the entrances and channels of pores. When the supporting amount increased to 3, the micropores almost disappeared and some new pores with average diameter of 3.8 nm were formed. As can be seen in the Nitrogen absorption isothermal lines and pore width distribution of CsPAV, its pores were accumulation ones with wide diameter distribution. So these pores with average diameter of 3.8 nm were formed when the CsPAV molecules entered the pores of supports. With the further increase of loading amount, the ratio of pores with average diameter of 3.8 nm increased obviously, and the hysteresis loops of 4HPAV/SBA and 5HPAV/SBA were divided into two parts. It suggests that with the increase of loading amountof CsPAV, more catalysts entered the pores of supports. Some pores with wide diameter distribution were also formed in 4HPAV/SBA and 5HPAV/SBA which indicates the accumulation of CsPAV crystals on the outer surface of the supports.

TG-DTA: TG-DTA was performed to figure out the changes during calcination and thermal stability of the supported catalysts. The TG curves of SBA-15 showed a two-stage loss (Figure 6). The first one between 100 and 150°C can be assigned to loss of absorbed water, which indicated that about 7 wt% waters existed in SBA-15. The second stage could be attributed to the loss of crystalline water. The TG curves of HPAV/SBA exhibited the first one at 150-200°C,

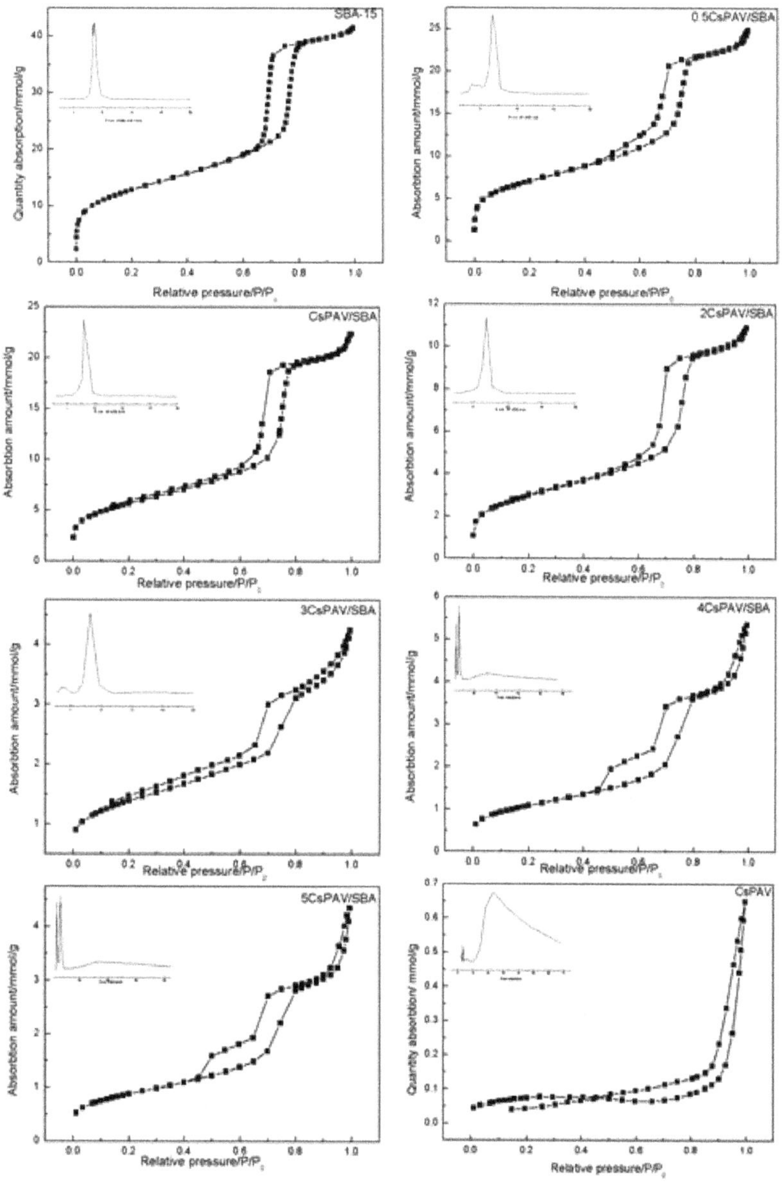

Figure 5: Nitrogen absorption isothermal lines and pore width distribution of supported catalysts with different supporting amount.

which was higher than that of SBA-15. The reason was that besides the absorbed water in SBA-15 and HPAV, there also existed crystalline water in HPAV which needed higher temperature to be desorbed. The second weight loss could be distributed to the decomposition of HPAV, which was completely decomposed at about 400°C. Compared with bulk HPAV, the temperature for the complete decomposition of the supported HPAV was 30°C lower. The result shows that dispersion of HPAV as a single molecule layer made its thermal stability decrease, which might be due to that there was not enough water or other counter ions in the secondary structure to support Keggin structure. Compared with HPAV/SBA and SBA-15, the weight loss of CsPAV/SBA was not that obvious. The first weight loss could be ascribed to the absorbed water (about 5%). Then the crystalline water lost slowly. It's hard to figure out the crystalline water content on TG curves. CsPAV/SBA was decomposed completely at about 500°C which was about 100°C higher than HPAV/SBA, suggesting that the addition of Cs could significantly enhance the final decomposition temperature of the supported catalysts. With the increase of loading amount of CsPAV, the absorbed water content (about 3.5%) further decreased because of the decrease of surface area. 3CsPAV/SBA was completely decomposed at about 450°C. The results show that the addition of Cs in these supporting catalysts could obviously enhance thermal stability of the supported heteropoly catalysts.

XPS: XPS was preformed to study the chemical changes of the catalysts during calcination and reaction. The chemical states of O, Mo, P, V and Si in the catalysts were detected and analyzed. O atoms in the catalysts play an important role in the selective oxidation of MAL to MAA [16]. The oxide species in the supported catalysts were analyzed

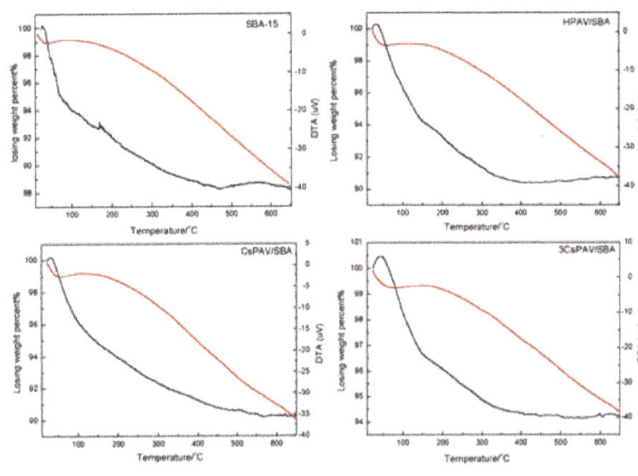

Figure 6: TG-DTA curves of supported catalysts and SBA-15.

by XPS, and two main kinds of oxide species were found (Figure 7). One was at 533 ± 0.25 eV which can be assigned to O species in SiO_2, and the other was at 530 ± 0.2 eV assigned to O species in the supported CsPAV. The binding energy of these two O species in HPAV/SBA increased after calcination and decreased after catalytic reaction (Table S1). The increase of binding energy of O species was caused by the oxidation and decomposition during calcination, and the decrease of binding energy of O was due to the reduction by MAL during reaction. The decreased amount of O species in HPAV detected by XPS may be due to the decomposition of the supported HPAV. These two O species in CsPAV/SBA showed the same tendency as HPAV/SBA due to the existence of HPAV in CsPAV/SBA. There were also two different O species in CsPAV (531.10~531.17 and 532.02~532.20 eV). These two O species showed different energy binding tendencies from the supported catalysts during calcination and reaction. The reduced one was the dominant species. The binding energy decreased after calcination and increased slightly after reaction, and the amount of reduced O species decreased after calcination and reaction. These changes were caused by the partial decomposition of catalysts during calcination and reaction, such as the migration of V atoms from the primary structure to secondary structure.

XPS spectra showed that two different Mo species existed in the catalysts. The binding energy of first one was 233 ± 0.3 eV and that for the other was 232 ± 0.2 eV.

These two can be assigned to Mo species in Keggin structure in different chemical states. The chemical states of Mo in the catalysts were Mo^{5+} and Mo^{6+}. Compared with the binding energy of HPAV/SBA (233.30 eV), that of Mo in CsPAV/SBA (233.12 eV), CsPAV (233.17 eV) and 3CsPAV/SBA (233.20 eV) was lower (Table S2). It suggests that some supported HPAV was decomposed into MoO_3 during preparation process, while the supported CsPAV was kept integral. The addition of Cs can prevent the decomposition of the catalysts during preparation process, and no Mo in low chemical state was found in CsPAV, calcined CsPAV and 3CsPAV/SBA. It suggests that the reduced chemical state of heteropoly compound was caused by the interaction between catalysts and supports. The amount of reduced Mo in HPAV/SBA and CsPAV/SBA decreased after calcination due to the oxidation of Mo^{5+} by O_2 during calcination. The amount of reduced Mo in all catalysts increased after reaction because of the oxidation of MAL on Mo-O species. No Mo species which belongs to MoO_3 was detected by XPS and it may due to the reduction of Mo by MAL during reaction. Compared with CsPAV/SBA, less reduced Mo was found in 3CsPAV/SBA. It indicates

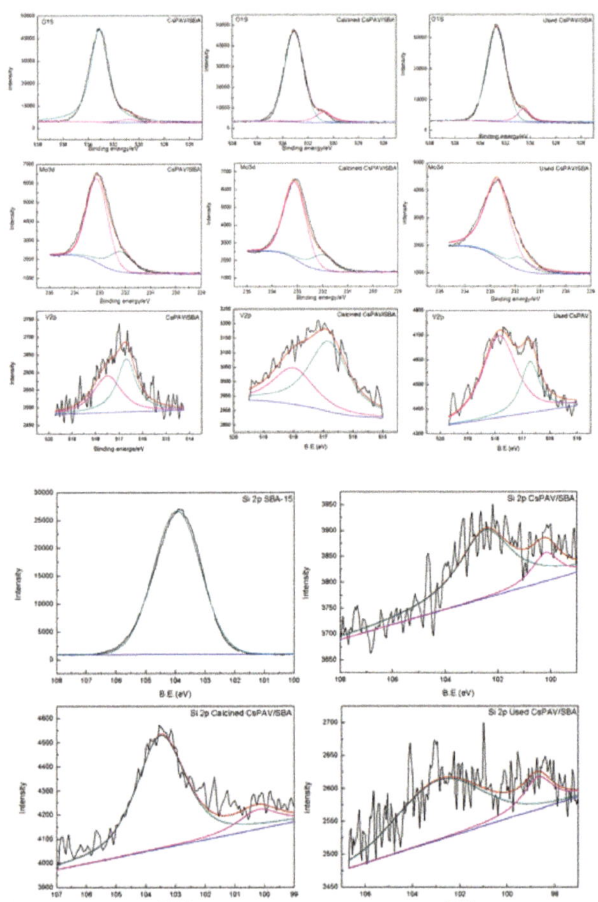

Figure 7: XPS spectra of O 1s, Mo 3d, V 3d, Si 2p in uncalcined, calcined, and used CsPAV/SBA.

that increasing loading amount can reduce the interaction between the catalysts and the supports.

Vanadyl species are also active species for this reaction [8]. XPS results showed that both V^{5+} (517.52~517.92 eV) and V^{4+} (516.65~516.98 eV) existed in the catalysts. The binding energy of V in the supported CsPAV and HPAV was also lower than that in bulk CsPAV, and more V^{5+} existed in bulk CsPAV than the supported CsPAV and HPAV. It also suggests that loading heteropoly compounds on SBA-15 can reduce their chemical states (Table S3). The amount of V^{4+} was decreased after calcination due to the oxidation by O_2 in air. The amount of V^{5+} in the used HPAV/SBA increased after reaction due to the decomposition of the catalysts during reaction. The decrease of V^{5+} in CsPAV/SBA after reaction suggested that the addition of Cs had improved thermal stability of the supported heteropoly catalysts and prevented the formation of V_2O_5.

Only one peak can be found at 103.8 eV in the Si spectra of SBA-15 which can be assigned to the silica species in silica dioxide (Figure 7; Table S4), suggesting that only one kind of silica species was on the surface of SBA-15. After CsPAV being supported on SBA-15, the binding energy of silica shifted to 102.5 eV, and a new peak appeared at 100.2 eV. The changes of the binding energy showed that the supported CsPAV strongly interacted with silica species on the surface. The new emerged peak can be assigned to silica species surrounded by Mo atoms formed from the replacement of centre P atom by Si in Keggin structure. After calcination, the ratio of this kind of silica species decreased which suggested that some of the new formed Keggin structures with Si as centre atoms decomposed. The binding energy of the other silica species also increased to 103.5 eV which suggested that CsPAV interacted with the support and then decomposed during calcination. The binding energy of both of the two silica species decreased after catalytic reaction. The silica species emerged at 98.7 eV can be assigned to $MoSi_X$, indicating that the decomposed catalysts continued to react with the support during the catalytic reaction, and the rest of Keggin structures with Si as centre atoms were completely decomposed during the catalytic reaction. The XPS data suggests that the thermal stability of the catalysts decreased because of the interactions between the catalysts and the support which led the decomposition of the catalysts during calcination and reaction.

^{31}P NMR and ^{29}Si NMR: ^{29}Si and ^{31}P NMR spectra were collected to figure out the mechanism of decomposition of the supported CsPAV on SBA-15. Only one peak was shown in the ^{31}P NMR spectra of CsPAV and the supported catalysts (Figure 8), suggesting that all the Keggin structures on the catalysts had the same state. However, the XRD patterns showed that some HPAV existed in the CsPAV/SBA. So there should be at least 2 peaks in the ^{31}P NMR spectra of CsPAV/SBA. The only peak in ^{31}P NMR spectra suggested that HPAV was well dispersed in the CsPAV. The chemical shift of ^{31}P NMR spectra of bulk CsPAV (-5 ppm) was different from that of HPAV/SBA (-4.5 ppm) due to the replacement of protons by Cs^+ in the catalysts [29-31]. After it was supported on SBA-15, the peak shifted upfield (CsPAV/SBA, -4.9 ppm; 3CsPAV/SBA, -4.8 ppm) was found, indicating that CsPAV could interact with SBA-15 during preparation. After calcination, the peak of P in CsPAV moved downfield (-5.2 ppm) because of the lost of crystalline water, but it shifted upfield for the other three supported catalysts, implying that the interactions between the catalysts and SBA-15 became stronger after calcination. The chemical shift of P in CsPAV was -5 ppm after the catalytic reaction due to the acquisition of crystalline water from the feed fluid. The peaks of P in the supported catalysts shifted downfiled after calcination and reaction, which indicated that Keggin structure interacting with SBA-15 may decompose after the catalytic reaction. ^{29}Si NMR was used to prove the results from XPS data and find out how the interactions occurred between the supported HPAV, CsPAV and SBA-15 (Figure 9). Two major peaks were found in the ^{29}Si NMR spectra of the supported catalysts which can be assigned to (SiO)3=Si-OH (-102.5 ppm) and (SiO)3=Si-O-Si= (-110 ppm) [32,33]. These two peaks belonged to Si atoms in SBA-15. The new peak at -91 ppm can be found in the spectra of HPAV/SBA and calcined catalysts, which can be assigned to Si atoms in the Keggin structure. It indicated that SBA-15 interacted with HPAV during preparation and calcination, and the reaction between SBA-15 and CsPAV only happened during calcination. The results were in agreement with other characteristic methods previously mentioned. The spectra suggested that some Si atoms entered Keggin structure of CsPAV and replaced P atoms. After reaction, disappearance of this peak suggested that the center atom of Keggin structures was replaced by Si.

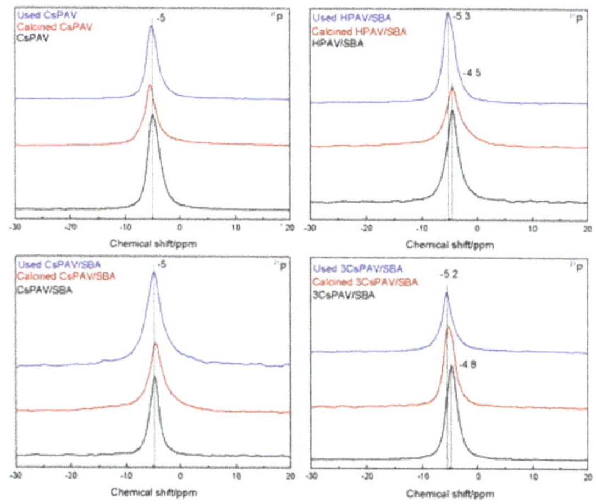

Figure 8: 31P NMR spectra of uncalcined, calcined and used CsPAV and supported catalysts.

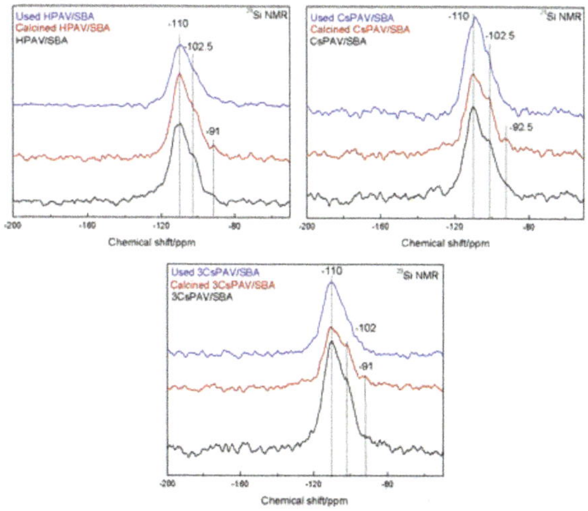

Figure 9: 28Si NMR spectra of uncalcined, calcined and used CsPAV and supported catalysts.

Reaction results

The effects of supports with different pore sizes: The above prepared catalysts were used to catalyze the oxidation of MAL to MAA. The catalytic activity of bulk CsPAV increased from 0.0016 to 0.019 min^{-1} with the reaction temperature increasing from 290 to 330°C and then decreased slightly (0.018 min^{-1}) at 350°C (Figure 10). The activity of CsPAV/SBA increased from 0.013 to 0.026 min^{-1} with temperature increasing from 290 to 320°C and then decreased to 0.005 min^{-1} when temperature reached to 350°C. Catalytic activity of CsPAV/KIT increased from 0.008 to 0.017 min^{-1} (290-310°C) and then decreased to 0.004 min^{-1} (350°C). The catalytic activity of CsPAV/SiO_2 decreased from 0.012 to 0.004 min^{-1} with reaction temperature increasing from 290 to 350°C, and the activity of CsPAV/Y decreased from 0.005 to 0.0015 min^{-1}. The supported CsPAV showed higher catalytic activity than bulk CsPAV at 290°C. The selectivity of MAA on bulk CsPAV increased from 6.6% to 58.2% when the temperature increased from 290 to 330°C and declined to 47.1% at 350°C. For the supported catalysts, selectivity was much lower than bulk CsPAV (except at 290°C) and decreased with increase of reaction temperature (25% to 2.1% on CsPAV/SBA, 18.2% to 4% on CsPAV/KIT, 16.8% to 3.5% on CsPAV/SiO_2 and 6.2% to 2.5% on CsPAV/Y). The highest MAA selectivity by using the supported catalysts was obtained on CsPAV/SBA, 25%. So SBA-15 was chosen for further study, and the catalytic activity was tested at 320°C. Bulk CsPAV was bright yellow before calcination and became chartreuse after calcination and turned to be dark green after catalytic reaction. The color changes of the catalysts may be due to the changes in chemical states of the elements. The green color might be caused by the reduction of Mo and V in the catalyst. For the supported catalysts, they had the same color as bulk CsPAV be-

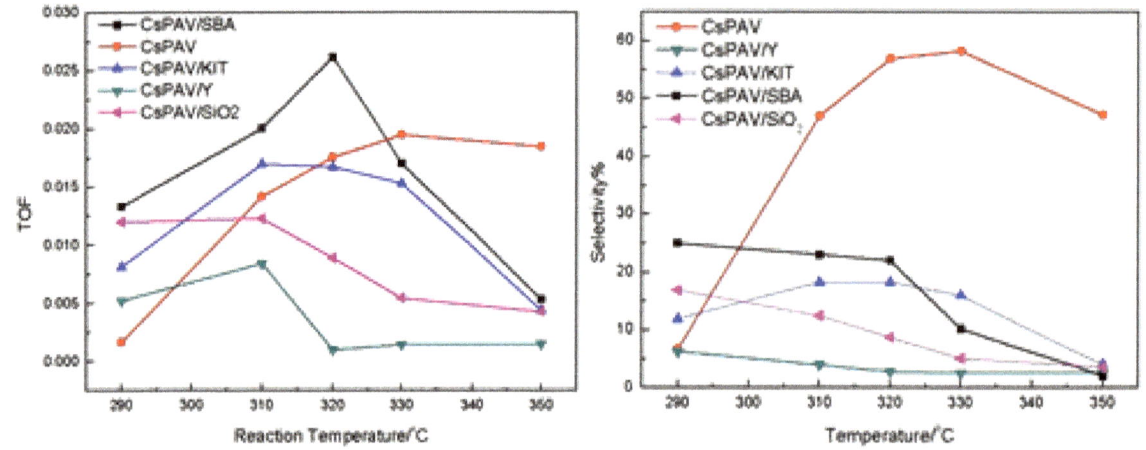

Figure 10: The catalytic activity of CsPAV supported on different carriers in the oxidation of MAL to MAA (the unite of TOF is min^{-1}, and the loading amount was 50%).

fore and after calcination. They became brown after catalytic reaction, which may be caused by the decomposition of the supported CsPAV. So the decrease of catalytic activity and selectivity may be due to the decomposition of the catalysts. To enhance thermal stability of the catalysts, Cs content and loading amount were changed and the influence on catalytic activity investigated.

The supported catalysts with different Cs amount: The addition of Cs in the heteropoly catalysts could improve thermal stability and MAA selectivity [7,20]. Different amounts of Cs were added to the catalysts. The Brønsted acidity is necessary for the oxidation of MAL to MAA, so at least a proton should be left per heteropoly anion. The results showed that the addition of Cs decreased the catalytic activity (TOF), especially when the Cs content was higher than 1 per heteropoly anion (Tables 1 and 2). However, the selectivity of MAA was increased with the addition of Cs and it decreased significantly when the Cs content was higher than 1 per heteropoly anion. The results indicated that the preferential Cs content was 1 per heteropoly anion, and strong Brønsted acidity was needed for this reaction.

Catalysts with different loading amount: The activity increased from 0.026 to 0.042 min^{-1} with the loading amount of CsPAV increasing from 1 to 3 per gram of SBA-15, and then decreased to 0.023 min^{-1} on 5CsPAV/SBA, while the MAA selectivity showed a different trend, which increased from 25% to 44% with the increase of loading amount of CsPAV (Figure 11). Oxidation of MAL to MAA occurred on the active sites on the surface of catalysts. With the increase of loading amount of CsPAV, the sizes of CsPAV crystals became larger, which might show the similar properties with bulk CsPAV. So catalysts with high loading amount of CsPAV showed the similar selectivity as bulk CsPAV and the increase of loading amount could improve the selectivity. Compared with bulk CsPAV, the catalytic activity was improved more than twice on 3CsPAV/SBA. The suitable CsPAV loading amount of this catalyst was 80 wt%.

The catalytic life of the supported catalysts: Figure 12 showed that the catalytic activity kept steady in 12 hours, which increased slightly from 0.026 to 0.029 min^{-1} within the first 6 hours. Then it hold steady in next 6 hours. The selectivity increased from 27.6% to 37.1% with time

Catalysts	BET surface area/ $m^2 \cdot g^{-1}$	Pore volume/ $cm^3 \cdot g^{-1}$	Average Pore width/ nm
SBA-15	1017.8	1.307	6.6
0.5CsPAV/SBA	569.4	0.822	6.4
1CsPAV/SBA	450.5	0.734	7.01
2CsPAV/SBA	236.7	0.349	6.9
3CsPAV/SBA	108.5	0.121	6.6
4CsPAV/SBA	86.2	0.187	7.33
5CsPAV/SBA	70.0	0.149	7.4
CsPAV	5.8	0.021	16.40

Table 1: BET surface area, pore volume and pore width of the supported catalysts with different supporting amounts.

Catalysts	TOF/min^{-1}	Selectivity%
HPAV/SBA	0.028	20.4
CsPAV/SBA	0.026	25
Cs_2PAV/SBA	0.007	7.1
Cs_3PAV/SBA	0.002	4.5

Table 2: The catalytic activity and selectivity of supporting catalysts with different Cs contents at 320°C.

going on, and then waved around 35%. No obvious deactivation was observed, which indicated that the addition of Cs could enhance thermal stability ofcatalysts.

Discussion

According to the characterization results, changes of supported catalysts occurred during preparation, calcination and reaction. Firstly, Cs^+ was well dispersed on the supports. After the Cs^+ -containing supports were immersed into the HPAV solution, $PMo_{11}VO_{40}$ reacted with Cs and precipitated on the outer and inner surface of the supports. Due to the quick reaction of $PMo_{11}VO_{40}^{4-}$ and Cs^+, the reaction between $PMo_{11}VO_{40}^{4-}$ process. If HPAV was supported directly on the supports, $PMo_{11}VO_{40}^{4-}$ would react with SiO_X species on the surface of the supports and led to decomposition of HPAV during drying.

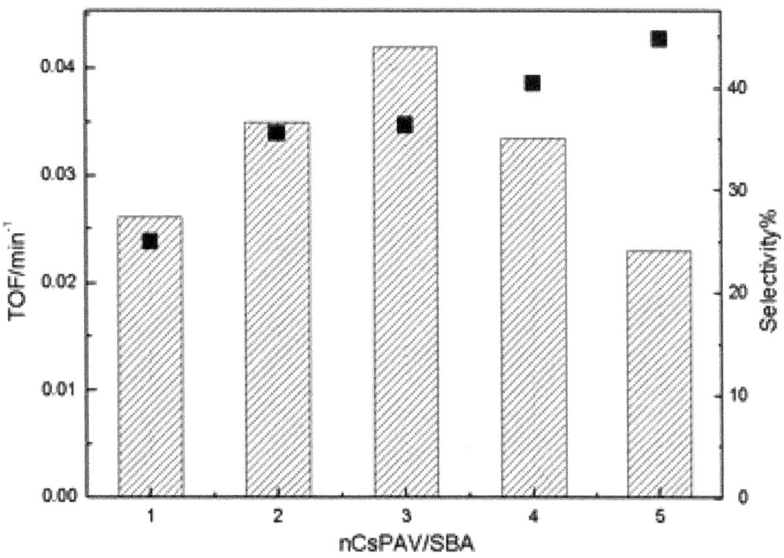

Figure 11: The catalytic activity and MAA selectivity on the supported catalysts with different CsPAV loading amounts at 320°C (TOF, bar; Selectivity, cubic point).

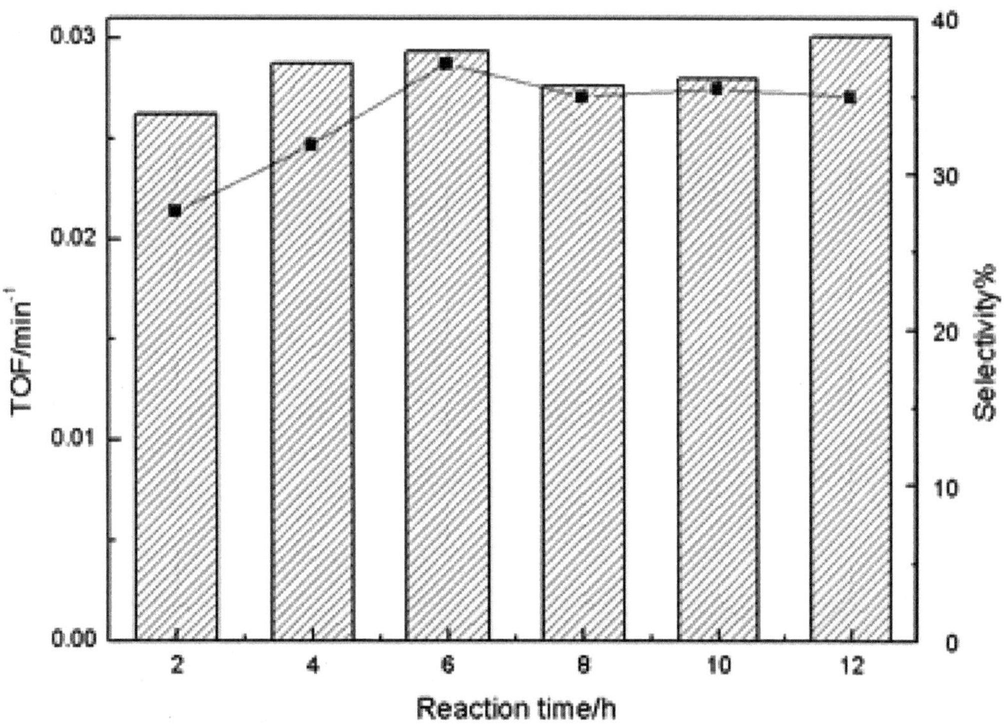

Figure 12: The catalytic activity and selectivity of 3CsPAV/SBA at 320°C with time on stream (TOF, bar; Selectivity, cubic point).

Without the counter cations, HPAV in single molecule layer can easily interact with silica species and was decomposed by heating. Unlike HPAV dispersing on the supports in a single molecule layer, CsPAV clustered in small crystalline in the size of about 20 nm on the surface of the supports, and some catalysts molecules entered channels of the supports. With the support of other heteropoly anions, counter cation (mainly Cs^+) and crystalline water, the catalysts showed better thermal stability than supported HPAV. However, there was some HPAV existed in the supported CsPAV, and it reacted with SiO_X species on the surface during calcination and reaction. Si replaced P as the center atom in the heteropoly anions in Keggin structure, which caused the decrease of thermal stability and catalytic activity of the supported catalysts. The supported HPAV was decomposed into MoO_3 and V_2O_5 which had good oxidization ability during calcination and reaction. So there were at least 3 kinds of oxidation catalysts on the supports, CsPAV, MoO_3 and V_2O_5. According to the results of XRD, the crystalline sizes of MoO_3 and V_2O_5 were rather small (smaller than 100 nm), so they would show very good catalytic activity. CsPAV is a good catalyst for oxidation of MAL to MAA, while MAL was oxidized to acetic acid and CO_X on MoO_3 and V_2O_5. That's why MAA selectivity on the supported catalysts was lower than bulk CsPAV.

With the increase of loading amount of CsPAV, the surface of SBA-15 was covered by CsPAV, and MoO_3 as well as V_2O_5 were not exposed directly to the feed. So MAL was oxidized on CsPAV, and the MAA selectivity was improved. The decomposed $PMo_{11}VO_{40}^{4-}$ formed a layer of oxides between CsPAV and SBA-15 to avoid the further reaction between them during reaction and decomposition of supported CsPAV, so the catalytic activity and selectivity were steady with the reaction going on. In general, supporting heteropoly catalysts on SiO_2 can reduce the size of the catalysts and improve the catalytic activity. The key point to prepare efficient supported heteropoly is to avoid the reaction between the catalysts and SiO_2 supports and one way for that is to use a layer to separate catalysts and supports.

Conclusions

$CsH_3PMo_{11}VO_{40}$ was supported on different kinds of SiO_2 supports. The one supported on SBA-15 showed the best catalytic performance in the oxidation of MAL to MAA. $CsH_3PMo_{11}VO_{40}$ formed clusters on the outer and inner surfaces of SBA-15. Compared with $H_4PMo_{11}VO_{40}$ /SBA-15, the addition of Cs^+ could improve thermal stability of the catalysts. The most suitable supporting amount of CsPAV was 80 wt%. The catalytic activity was improved twice than bulk $CsH_3PMo_{11}VO_{40}$, but selectivity was low. The low selectivity was caused by MoO_3 and V_2O_5 formed from the decomposition of the supported $H_4PMo_{11}VO_{40}$ and $CsH_3PMo_{11}VO_{40}$. The decomposition of $CsH_3PMo_{11}VO_{40}$ and $CsH_3PMo_{11}VO_{40}$ was due to the replacement of P by Si in Keggin structure. The decomposed catalysts formed an oxides layer between integral $CsH_3PMo_{11}VO_{40}$ and supports which prevented any further reaction between $CsH_3PMo_{11}VO_{40}$ and supports or decomposition of $CsH_3PMo_{11}VO_{40}$. The key point to prepare efficient and thermal stable supported heteropoly is to avoid the reaction between the catalysts and silica supports.

Acknowledgements

The authors gratefully acknowledge the financial support of National Key Projects for Fundamental Research and Development of China (2016FB0600903) "Strategic Priority Research Program" of the Chinese Academy of Sciences (XDA07070600), National Key Projects for Fundamental Research and Development of China (2016YFB0601303), and National Science Funds for Distinguished Young Scholar (21425625).

References

1. Makoto M, Noritaka M, Koichi K, Atsushi K, Yasuo K, et al. (1982) Catalysis by heteropoly compounds. (III). The structure and properties of 12-heteropolyacids of molybdenum and tungsten (H3PMo12-xWxO$_{40}$) and their salts pertinent to heterogeneous catalysis. Bull Chem Soc Jpn 55: 400-406.
2. Mizuno N, Misono M (1998) Heterogeneous Catalysis. Chem Rev 98: 199-218.
3. Makoto M (2001) Unique acid catalysis of heteropoly compounds (heteropolyoxometalates) in the solid state. Chem Commun 1141-1152.
4. Martin L, Jean-Marc MM (2000) Effect of iron counter-ions on the redox properties of the Keggin-type molybdophosphoric heteropolyacid Part I. An experimental study on isobutane oxidation catalysts. Appl Catal A-Gen 200: 89-101.
5. Toshio O (2002) New catalytic functions of heteropoly compounds as solid acids. Catal Today 73: 167-176.
6. Yasuo K, Kanji S, Makoto M, Yukio Y (1982) Catalysts by heteropoly Compounds IV. Oxidation of methacrolein to methacrylic acid over 12-molybdophosphoric acid. J Catal 77: 169-179.
7. Marosi L, Otero Areán C (2003) Catalytic performance of Csx(NH4) yHzPMo12O40 and related heteropolyacids in the methacrolein to methacrylic acid conversion: in situ structural study of the formation and stability of the catalytically active species. J Catal 213: 235-240.
8. Lilong Z, Lei W, Suojiang Z, Ruiyi Y, Yanyan D (2015) Effect of vanadyl species in Keggin-type heteropoly catalysts in selective oxidation of methacrolein to methacrylic acid. J Catal 329: 431-440.
9. Kala Raj NK, Deshpande SS, Ingle RH, Raja T, Manikandan P (2004) Heterogenized molybdovanadophosphoric acid on amine-functionalized SBA-15 for selective oxidation of alkenes. Catal Lett 98: 217-224.
10. Popa A, Sasca V, Halasz J (2008) Catalytic properties of molecular sieves MCM-41 type doped with heteropolyacids for ethanol oxidation. Appl Surf Sci 255: 1830-1835.
11. Spojakina AA, Kostova NG, Sow B, Stamenova MW, Jiratova K (2001) Thiophene conversion and ethanol oxidation on SiO2-supported 12-PMoV-mixed heteropoly compounds. Catal Today 65: 315-321.
12. Gomez Sainero LM, Damyanova S, Fierro JLG (2001) Methanol oxidation over ZrO2–SiO2 supported phosphomolybdic acid. Appl Catal A-Gen 208: 63-75.
13. Legagneux N, Basset JM, Thomas A, Lefebvre F, Goguet A, et al. (2009) Characterization of silica-supported dodecatungstic heteropolyacids as a function of their dehydroxylation temperature. Dalton Trans: 2235-2240.
14. Ballarini N, Candiracci F, Cavani F, Degrand H, Dubois JL, et al. (2007) The dispersion of Keggin-type P/Mo polyoxometalates inside silica gel, and the preparation of catalysts for the oxidation of isobutaneto methacrolein and methacrylic acid. Appl Catal A-Gen 325: 263-269.
15. Mitsuru K, Toshiya Y, Wataru N, Ken O, Yuichi K (2010) Catalytic oxidation of methacrolein to methacrylic acid over silica-supported 11-molybdo-1-vanadophosphoric acid with different heteropolyacid loadings. J Catal 273: 1-8.
16. Mitsuru K, Yu-ki M, Toshiya Y, Toshio H, Wataru N, et al. (2011) 11-Molybdo-1-vanadophosphoricacid H4PMo11VO40 supported on ammonia-modified silica as highly active and selective catalyst for oxidation of methacrolein. Catal Commun 13: 59-62.
17. Makoto H, Kazushi A (2001) Reaction of butane to isobutane on silica-supported 12-tungstosilicic acid. Green Chem, 3: 170-172.
18. Jean-Michel T, Christelle M, Katharina B (1996) A new method to prepare silica supported heteropolyanion catalysts Formation on the silica surface of calcium and magnesium salts of phosphomolybdic acid, H3PMo12O40. Appl Catal A-Gen 138: L1-L6.
19. Yihang G, Kexin L, Xiaodan Y, James H C (2008) Mesoporous H3PW12O40-silica composite: Efficient and reusablesolid acid catalyst for the synthesis of diphenolicacid from levulinic acid. Appl Catal B-Envir 81: 182-191.
20. Marchal-Roch C, Laronze N, Villanneau R, Guillou N, Tézé A, et al. (2000) Effects of NH4+, Cs+, and H+ Counterions of the Molybdophosphate Anion in the Oxidative Dehydrogenation of Isobutyric Acid. J Catal 190: 173-181.
21. Marosi L, Otero Areán C (2003) Catalytic performance of Csx(NH4) yHzPMo12O40 and related heteropolyacids in the methacrolein to methacrylic acid conversion: in situ structural study of the formation and stability of the catalytically active species. J Catal 213: 235-240.
22. Deuber LM, Gaube W, Martin FG, Hibst H (1996) Effects of Cs and V on

Heteropolyacid Catalysts in Methacrolein Oxidation. 11th international Congress on Catalysis - 40th Anniversary Studies in Surface Science and Catalysis, 101: 981-990.

23. Dongyuan Z, Qisheng H, Jianglin F, Bradley FC, Galen DS (1998) Nonionic triblock and star diblock copolymer and oligomeric surfactant syntheses of highly ordered, hydrothermally stable, mesoporous silica structures. J Am Chem Soc 120: 6024-6036.

24. Beck JS, Vartuli JC, Roth WJ, Leonowicz ME, Kresge CT, et al. (1992) A new family of mesoporous molecular sieves prepared with liquid crystal templates. J Am Chem Soc 144: 10834-10843.

25. Michael GB, Laurel M, David EWV (1982) Flemington NJ. US Patent, US4309313.

26. Shin RM, Takao M, Isao O, Kenji H (1997) Preparation of encaged heteropoly acid catalyst by synthesizing 12-molybdophosphoric acid in the supercages of Y-type zeolite. Appl Catal A-Gen 165: 219-226.

27. Carmen S, Wolfgang FH (2000) Modification of faujasites to generate novel hosts for "ship-in-a-bottle" complexes. Catal Today 60: 193-207.

28. Heng Z, Ruiyi Y, Li Y, Yanyan D, Lei W, et al. (2013) Investigation of Cu- and Fe-Doped CsH3PMo11VO40 Heteropoly Compounds for the Selective Oxidation of Methacrolein to Methacrylic Acid. Ind Eng Chem Res 52: 4484-4490.

29. Karthikeyana G, Pandurangan A (2009) Heteropolyacid (H3PW12O40) supported MCM-41: An efficient solid acid catalyst for the green synthesis of xanthenedione derivatives. J Mole Catal A-Chem 311: 36-45.

30. Claude RD, Ahmed A, Mohaned MB, Suzanne L, Michel F (1996) Catalysis by 12-Molybdophosphates: 1. Catalytic Reactivity of 12-Molybdophosphoric Acid Related to its Thermal Behavior Investigated through IR, Raman, Polarographic, and X-ray Diffraction Studies: A Comparison with 12-Molybdosilicic Acid. J Catal 164: 16-27.

31. Kei I, Akiko O, Hiroshi K, Makoto M (1998) Catalysis by heteropoly compounds Part 39. The structure and redox behaviour of vanadium species in molybdovanadophosphoric acid catalysts during partial oxidation of isobutene. J Chem Soc Faraday Trans 94: 1765-1770.

32. Kausik M, Anirban G, Rajiv K (2002) Heteropolyacids aided rapid and convenient syntheses of highly ordered MCM-41 and MCM-48: exploring the accelerated process by 29Si MAS NMR and powder X-ray diffraction studies. Chem Commun, pp: 2404-2405.

33. Li W, Chen S, Yu J, Fang D, Ren B, et al. (2016) In-situ synthesis of interconnected SWCNT/OMC framework on silicon nanoparticles for high performance lithium-ion batteries. Green Energy & Environment 1: 91-99.

Evaluation of Adding Carbon Tetrachloride as Propulsion to the Thermal Cracking Reactor due to the Amount of Formed Coke in Different Coil Outlet Temperatures (COT)

Afshin Davarpanah*

Department of Petroleum Engineering, Science and Research Branch, Islamic Azad University, Tehran, Iran

Abstract

Nowadays, polyvinyl chloride (PVC) was the most usual plastic. Thereby, to make polyvinyl chloride its monomer must be produced firstly that was called vinyl chloride (VCM). It was severely endothermic reaction that was done in an ethylene dichloride ethylene dichloride thermal cracking reactor within temperature range of 680-758°K and pressure range of 2500000 Pascal, thus this cracking reaction changed into hydrochloric acid and VCM. In production unit, monomeric chloride had the main and principal role as the core of the process of thermal cracking that occurs in the furnace. Increasing of wall temperature cause to boil gas mixture and causing pyrolysis reactions. Regarding to the simulation results showed that number of pyrolysis produced composition have maximum concentration in the length of reactor that illustrated these compositions participated in secondary reactions. Furthermore, by increasing the amount of coil outlet temperatures, the amount of formed coke will be increased. If Carbon tetrachloride considered as the chlorine radicals, it has an important role as the motivator in the cracking procedures, radicals causing an enhancement in VCM production.

Keywords: Carbon tetrachloride; Conversion factor; Coke; Coil outlet temperatures; Thermal cracking

Introduction

Design and Simulation of thermal cracking reactor because of the chemical reactions in the processes and heat transfer in the furnace are very complex. Therefore, this process from the chemical reactions engineering perspective and from the perspective of transfer phenomena is highly regarded. Computational Fluid Dynamics (CFD) tools help us to investigate the development of detailed models of these reactors and by understanding the basic principles, the optimized design of reactor and by saving the cost and time performance of this equipment in a virtual laboratory can be done. In this research, investigate the proposed model of EDC cracking process in an industrial furnace and by operational data from a petrochemical unit to qualify the model and remove the reductions. To begin, by previous researches that worked on this specific occasion were collected. Then they are studied and investigated. Moreover, the reactor mathematical models were provided and solved by numerical software. Thereby, the operation of the reactor at different operating conditions investigated and sensitivity analysis of effective parameters will be done. As e result, this work could design new reactors or increasing the optimization capacity of existing reactors.

Molecular cracking

Molecular cracking was a process that used in the petrochemical industry and to reduce their hydrocarbons molecular weight by breaking their bonds. This process is the main principal of converting crude oil to useful fuels such as gasoline, diesel, jet fuel and kerosene. Thermal cracking, catalyst cracking, hydrocracking and cracking by water vapour were the most useful cracking processes in the industry. This process was done in the high temperature and pressure and without presence of catalyzer or in the low temperature and pressure with presence of catalyzer.

Geometry of thermal cracking reactor

In EDC thermal cracking process, coils containing process fluid, inside a box called furnace, are positioned horizontally. Coils placement inside the furnace shown in Figure 1 [1]. That positioned on the on both sides of the wall furnace, by burning gases of these furnaces the heat produced, that the heat transferred to the coils by convection or radiation model (Table 1). The number of intended coils to simulate 11 rows of pipes (Inconel 600) with overall length of 95 meters and length of each row tubing is 13.5 meters. Inner diameter of coils is 0.134 meter the distance between centers of the two tubes is 0.32 meter (Figures 2 and 3).

Mass balance equation in the software proposed in form of equation 1

$$\frac{\partial(\rho y_i)}{\partial t}+\frac{\partial(\rho u_i y_i)}{\partial x_i}=-\frac{\partial J_i}{\partial x_i}+R_i+S_i \qquad (1)$$

ρ=Density

t=time

u_i=Potential energy

J_i=diffusion term

S_i=heat at point i

R_i=input heat to the reactor

Phrase J_i reflects is the term infiltration, which is not used in this simulation [2,3]. The energy equation in the Fluent software as follows:

***Corresponding author:** Afshin Davarpanah, Department of Petroleum Engineering, Science and Research Branch, Islamic Azad University, Tehran, Iran

E-mail: afshindpe@gmail.com

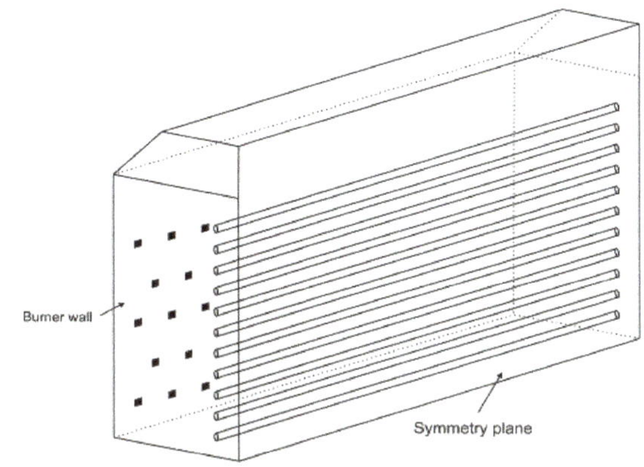

Figure 1: Placement of coil in the furnace.

Figure 2: Schematic of one row of the coil in the software.

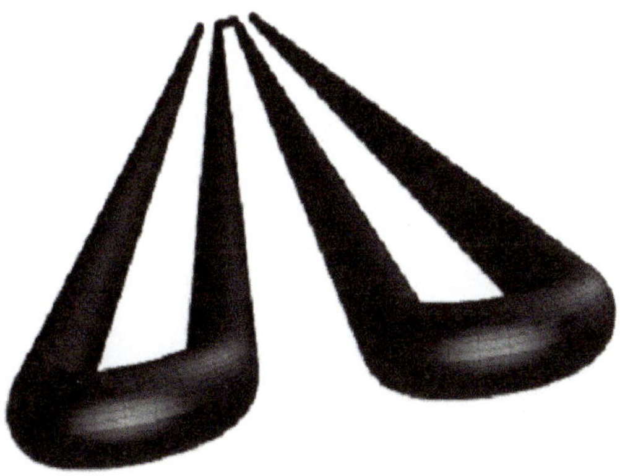

Figure 3: Prospective a row of coil.

$$\frac{\partial}{\partial t}(\rho E)+\frac{\partial}{\partial x_i}(u_i(\rho E+p))=\frac{\partial}{\partial x_i}\left(K_{eff}\frac{\partial T}{\partial x_i}-\sum_j h_j J_j+u_i(\tau_{ij})_{eff}\right)+S_h \quad (2)$$

ρ=Density

p=pressure

t=time

T=Temperature

u_i=input Potential energy

u_j=output Potential energy

T_{ij}=torque

J_j=Output diffusion term

s_i=input thermal of chemical reactions

S_h=thermal of chemical reactions

K_{eff}=the effective conductivity

E=total energy

K_t+K=K_{eff} is the effective conductivity, K_t turbulent thermal conductivity, according to the model used is the perturbation. The first three terms on the right side of the equation as an expression of directed energy transfer, distribution of components and dispersion of viscosity. S_h is thermal of chemical reactions [4-6]. E is expressed as follows:

$$E=h-\frac{p}{\rho}+\frac{u_i^2}{2} \quad (3)$$

u_i=input Potential energy

p=pressure

ρ=Density

E=total energy

u_i=input Potential energy

h=input heat

Mass continuity equation in FLUENT software is in form of eqn 4:

$$\frac{\partial \rho}{\partial t}+\frac{\partial \rho u_i}{\partial x_i}=0 \quad (4)$$

Above equation was the general form of the mass balance equation for compressible and non-compressible flows.

$$\frac{\partial(\rho u_i)}{\partial t}+\frac{\partial(\tilde{n}u_i u_j)}{\partial x_i}=-\frac{\partial \rho}{\partial x_i}+\frac{\partial \tau_{ij}}{\partial x_j}+\rho g_i+F_i \quad (5)$$

In the above equation, P static pressure, τ_{ij} tensor stress, ρg_i and F_i physical forces of gravity and physical external factors are in i direction. Tensor stress is as follows:

$$\tau_{ij}=\left[\mu\left(\frac{\partial u_i}{\partial x_j}+\frac{\partial u_j}{\partial x_i}\right)\right]-\frac{2}{3}\mu\frac{\partial u_i}{\partial u_i}\delta_{ij} \quad (6)$$

μ the molecular viscosity and the second semester is the influence of volumetric expansion [3].

Standard k-ε model

This was a comprehensive model that account as the simplest turbulence model. Two educational turbulent models enables solving the two transfer equations that separately determined turbulent length and time scale. Standard k-ε model is useful for powerful engineering calculations. Powerful of this equation, economical and acceptable accuracy for a wide range of turbulent flow as a result of its popularity in industrial flow and heat transfer simulation. This model is a semi-empirical model. Standard k-ε model, was a model based on transfer equations for kinetic turbulent energy and dissipation rate (ε). In achieving the standard k-ε model assumes that the fully turbulent flow and molecular viscosity effects have been ignored. Because of the strengths and weaknesses of the equation to improve its performance, k-ε realizable and k-ε RNG model was introduced. K and ε is obtained by solving simultaneously the following two equations:

Reactions	A[(cm³/mole) n⁻¹s⁻¹]	n	E (kJ/mole)
Radical reaction chain initialization			
EDC → R_1+R_2	5.9 × 1015	1	342
CCl_4 → $R_{1+}R_8$	2.2 × 1012	1	230
Radical reaction chain propagation			
EDC+R_1 → HCl+R_3	1.3 × 1013	2	7
EDC+R_5 → VCM+R_3	1.2 × 1013	2	34
EDC+R_2 → EC+R_3	1.0 × 1012	2	42
EDC+R_4 → 1,1+R_3	5.0 × 1011	2	45
EDC+R_6 → 1,1,2+R_3	2.0 × 1011	2	48
EDC+R_7 → 1,1,1,2+R_3	1.0 × 1011	2	56
EDC+R_8 → CHC_{l3}+R_3	1.0 × 1012	2	63
VCM+R_1 → R_4	9.1 × 1010	2	0
VCM+R_1 → HCl+R_5	1.2 × 1014	2	56
VCM+R_5 → CP+R_1	5.0 × 1011	2	31
VCM+R_4 → $C_4H_6H_2$+R_1	2.0 × 1010	2	30
VCM+R_2 → EC+R_5	3.0 × 1011	2	61
R_3 ↔ VCM+R_1	2.1 × 1014	1	84
R_5 ↔ C2H2+R_1	5.0 × 1014	1	90
R_6 ↔ Di+R_1	2.0 × 1013	1	70
R_7 ↔ Tri+R_1	2.5 × 1013	1	70
2C_2H_2+R_5 → C_6H_6+R_1	1.0 × 1014	2	20
EC+R_1 → HC_l+R_2	1.7 × 1013	1	4
1,1+R_1 → HCl+R_4	1.2 × 1013	1	6
1,1,2+R_1 → HCl+R_6	1.7 × 1013	1	15
1,1,1,2+R_1 → HCl+R_7	1.7 × 1013	2	17
$CHCl_{3+}R_1$ → HCl+R_8	1.6 × 1013	2	14
CCl_4+R_5 → Di+R_8	5.0 × 1011	2	33
CCl_4+R4 → 1,1,2+R8	1.0 × 1012	2	33
CCl_4+R_5 → 1,1,1,2+R_8	5.0 × 1011	2	33
Radical reaction chain termination			
$R_{2+}R_1$ → VCM+HCl	1.0 × 1013	2	13
R_3+R_1 → Di+HCl	1.0 × 1013	2	12
R_6+R_8 → Di+CCl_4	1.0 × 1013	2	13
C_2H_2+2R1 → 2C+2HCl	1.6 × 1014	2	70

Table 1: Energy of EDC thermal cracking reaction [2].

$$\frac{\partial}{\partial t}(\rho K)+\frac{\partial}{\partial x_i}(\rho K v_i)=-\frac{\partial}{\partial x_j}\left[\left(\mu+\frac{\mu_t}{\sigma_k}\right)\frac{\partial k}{\partial x_j}\right]+G_k+G_b-\rho\varepsilon-Y_M+S_K \quad (7)$$

$$\frac{\partial}{\partial t}(\rho\varepsilon)+\frac{\partial}{\partial x_i}(\rho\varepsilon v_n)=-\frac{\partial}{\partial x_j}\left[\left(\mu+\frac{\mu_t}{\sigma_\varepsilon}\right)\frac{\partial \varepsilon}{\partial x_j}\right]+C_{1\varepsilon}G_k\frac{\varepsilon}{k}-C_{2\varepsilon}\frac{\varepsilon^2}{k}+S_\varepsilon$$

G_k=turbulent kinetic energy generated due to the gradient of speed.

G_b=turbulent kinetic energy generated due to the buoyancy forces.

Y_M=Share Swinging delay in the compressibility turbulence for a total loss of (j)

$C_{1\varepsilon}$=constant parameter of model=1.44

$C_{2\varepsilon}$=1.92

σ_k=turbulent Prantel numbers for k=1

σ_ε=turbulent Prantel number for ε=1.3

S_K=term defined source for k

S_ε=term defining source for ε

μ_t=turbulence viscosity that can be obtained from the following equation:

$$\mu_t=\rho C_\mu\frac{K^2}{\varepsilon} \quad (9)$$

RNG k-ε model

This model is based on statistical methods, this model is the same as the standard k-ε model but with the following modifications: 1- An additional term (ε) has been added that significantly increases accuracy for flow to layer velocity. 2- The effect of the rotational movement (vortex motion) included in turbulences and increases accuracy for rotary flow. 3- An analytical formula for turbulent Prantel numbers are provided, as standard k-ε model considers a fixed amount. 4- While the standard k-ε model is the model for high Reynolds numbers, the theory of RNG for analysis of a differential equation for the turbulent viscosity can be provided that it considered the effects of low Reynolds. Effective use of this feature depends on proper behavior of the region near the wall of furnace. K-ε realizable model.

This model have limitations on Reynolds stresses that match with the physics of turbulent flows. This is possibly not in the standard k-ε model and not in the k-ε RNG model. In places where a large curvature in flow lines, vortex and rotation of the screw. K-ε RNG and k-ε realizable models have significant progress towards standard k-ε model. Preliminary studies show that the k-ε realizable model between k-ε models have best performance for confirmation numbers of complex flows and segregated flows. For all of this cases the performance of this model is better than the standard k-ε model. As it was observed that in the standard k-ε model, turbulent viscosity achieved the following equation:

$$\mu_t=\rho C_\mu\frac{K^2}{\varepsilon} \quad (10)$$

That C_μ is constant, but in the k-ε realizable C_μ achieved by the following equation:

$$C_\mu=\frac{1}{A_o+A_s\dfrac{KU^*}{\varepsilon}} \quad (11)$$

$$U^*=\sqrt{S_jS_j+\tilde{\Omega}_j\tilde{\Omega}_j} \quad (12)$$

$$\tilde{\Omega}_j=\Omega_j-2\varepsilon_{jk}\omega_k \quad (13)$$

$$\Omega_j=\bar{\Omega}_j-\varepsilon_{jk}\omega_k \quad (14)$$

$\bar{\Omega}_j$ Average rate of tensor with rotational speed is ω_k constants are defined below.

$$A_o=4/04,\ A_s=\sqrt{6}\cos\phi$$

$$\phi=\frac{1}{3}cos^{-1}\left(\sqrt{6}\frac{S_jS_{jk}S_{ki}}{\left(S_jS_j\right)^{3/2}}\right) \quad (15)$$

$$S_j=\frac{1}{2}\left(\frac{\partial u_j}{\partial x_i}+\frac{\partial u_i}{\partial x_j}\right) \quad (16)$$

As it's be seen C_μ placed a function of the average strain, rotational speeds, the angle speed of rotation systems, and turbulent fields are (k and ε) (Table 2).

Wall of furnace

Profile of heat was applied in the walls. A total of eleven wall was located and each row of coil has a heat profile of its own. To enter the thermal flux resulting from furnaces for applying to the coils level use from proposed profiles were provided by Kegard in 2007. The length of each row of coils can be categorized Profile provided by the amount of heat to be entered in the software (Figure 4).

Methodology of the Work (Simulation)

Boundary conditions of cracking reactor

Inlet flow: Input for fluid flow due to the mass flow rate and the percentage of each of the compounds at the entrance of the inlet is used for the mass flow condition that account as boundary condition (Table 3).

Output flow: The output flow used for pressure boundary condition. The boundary conditions are summarized in Table 4. Investigate the conversionfactorandtheamountofformedcokeduetothedifferentCOTs. In this part of the research in different COTs EDC conversion factor and the amount of formed coke has been investigated. The amount of COT according to recent issues in review of EDC cracking processes to the values of 760, 770, 780 and 790 are being selected. How to achieve this profile to be the first calculated $\frac{COT_{desired}}{COT_{base}}$ and then by using the proportion a variable factor is described. As a result, the new temperature profile in a particular COT due to the new data was achieved. In this method, input temperature to the coil doesn't change and base COT temperature multiply in the $\frac{COT_{desired}}{COT_{base}}$ and the other base profile points due to eq.17 and eq.18 multiply in a constant parameter to obtain new Profile points.

$$Weight\ factor(x) = \frac{Temperature(x) - Temperature(l=0)}{Temperature(l=150) - Temperature(l=0)} \quad (17)$$

$$T(x) = Weight\ factor(x) \times \frac{T(l=150)_{desired}}{774.1} \times T_{base} \quad (18)$$

The results of these calculations can be seen in Figure 5, COT different temperature profiles have been obtained. Figure 6 indicates the EDC conversion factor in different COTs, as the figure shows at higher conversion factor it will be higher in the COTs. But according to the Figure 7 if the amount of COT is higher, the amount of formed coke also will be higher. So by regarding the simulation results it can be said that for optimum economic performance of furnace, the temperature profile of the fluid is very important.

Investigating of adding carbon tetrachloride as propulsion

By adding this compound to the feed in addition to increase conversion factor, it has other effects on the process. Due to increasing number of chlorine radicals in the process, number of unwanted compounds and Acetylene production and as a result coke production has been increased. Creating unwanted compounds in the next stages, in addition to VCM treating problem, also an important loss costly. As shown in Figure 8 conversion factor increases with increasing carbon tetrachloride, on the other hand in the Figure 9 formed coke, of the increases in carbon tetrachloride. So regarding to the simulation results it can be said that for optimum economic performance of furnace, the temperature profile of the fluid is very important. Therefore, optimize the amount of added carbon tetrachloride to the feed for economic performance of thermal cracking unit is very important. In this amount of propulsion conversion factor increases as an allowable limit and reactor performance due to forming coke and unwanted compounds are economical. It can be seen in the Figure 5.

Conclusion

K-ε realizable model is the most appropriate model for calculation because of these reasons: [7,8]

Temp (K)	Transfer Coefficient (W/m.K)
373	15.9
573	19
753	21.2
973	25.7
Density (Kg/m³)	8470
Heat capacity (J/Kg.k)	444

Table 2: Thermo-physical properties of coils [4].

Feed mass flow rate	Kg/s 6.086
Mass percentage of Input EDC	0.9707
Mass percentage of Input VCM	0.0184
Mass percentage of HCL input	0.0109

Table 3: Output fluid conditions.

Location	Type of boundary condition
Walls	Heat flux
Input gas	Input mass rate
Output flow rate	Output pressure

Table 4: Applied boundary conditions in the software.

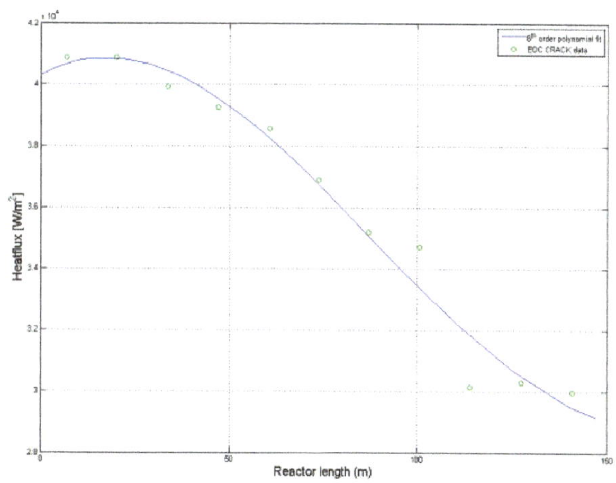

Figure 4: Thermal profiles calculated by Kegard.

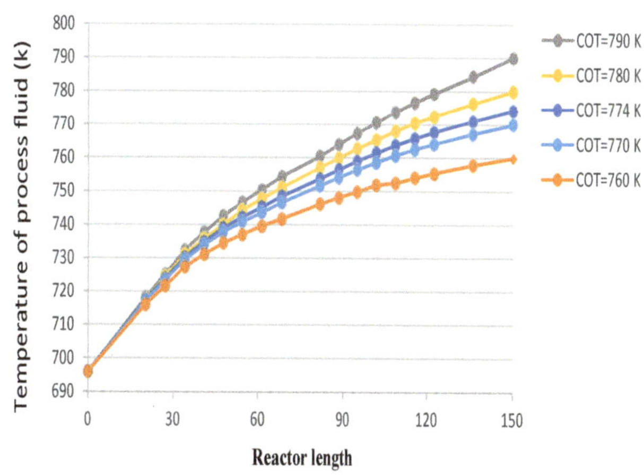

Figure 5: Process fluid temperature profile along the reactor.

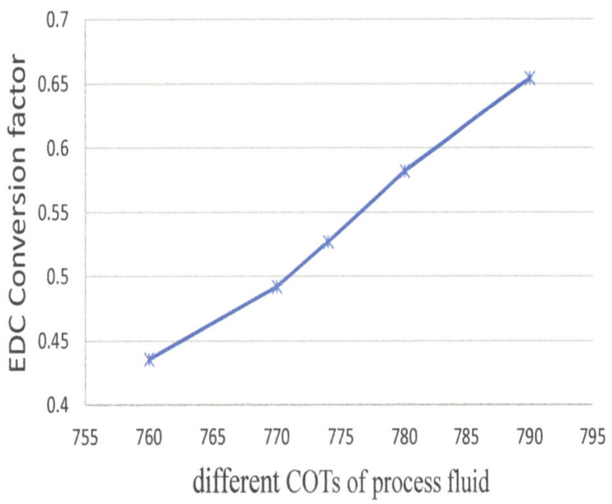

Figure 6: Process fluid temperature profile along the reactor.

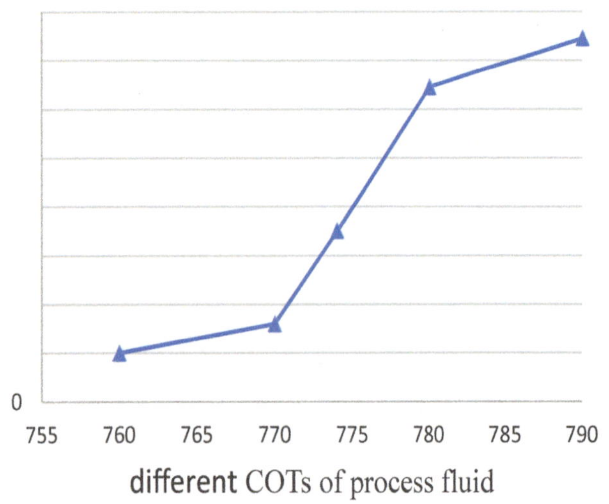

Figure 7: Process fluid temperature profile along the reactor.

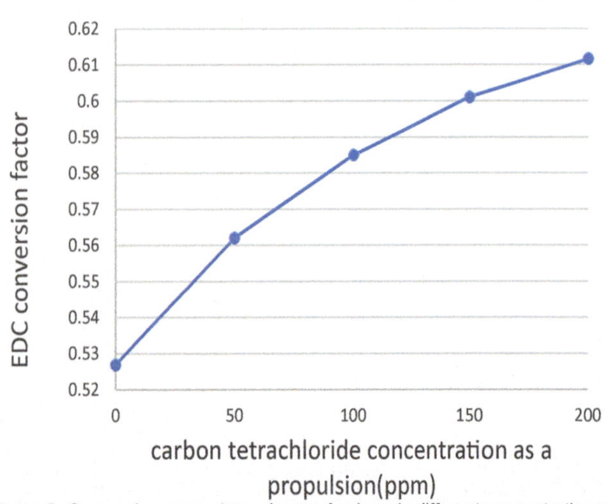

Figure 8: Conversion percentage change Ay.dy.sy in different concentrations of carbon tetrachloride.

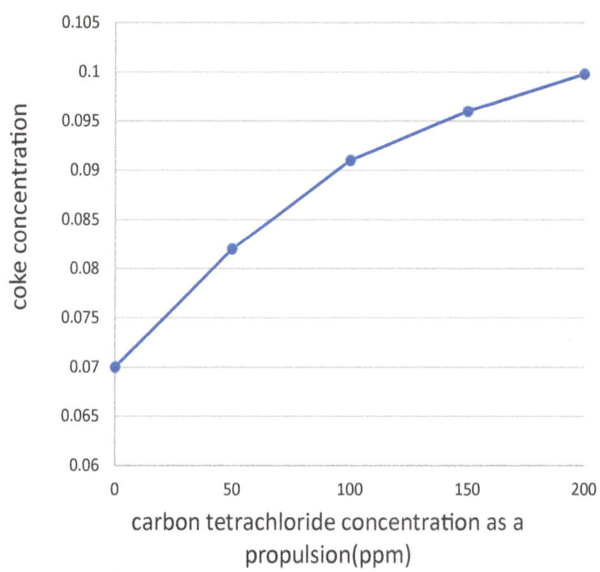

Figure 9: Change the concentration of coke in different concentrations of carbon tetrachloride.

1- This version contains a new formula for turbulence viscosity.

2- A modified transferring equation for scattering rate, ε, can be considered. Therefore, this model is an exact equation for fluctuations transferring has been made.

In this research, effect of added carbon tetrachloride as a propulsion has been investigated. According to result of this research and other issues, if the amount of this compound does not reach more than 100 ppm, conversion factor will be increased. If this amount will be more than 100 ppm it's not economically suggested.

References

1. Kaggerud TH (2007) Modeling an EDC Cracker using Computational Fluid Dynamics (CFD). p: 70.
2. Li C, Hu G, Zhong W, He W, Du W, et al. (2013) Coke Deposition Influence Based on a Run Length Simulation of a 1,2-Dichloroethane Cracker. Industrial & Engineering Chemistry Research 52: 17501-17516.
3. Fluent AN (2011) ANSYS FLUENT User's Guide. Ansys Inc., p: 2427.
4. Berreni M, Wang M (2011) Modelling and dynamic optimisation for optimal operation of industrial tubular reactor for propane cracking. Computer Aided Chemical Engineering 29: 955-959.
5. Fogler HS (1999) Element of chemical reaction engineering.
6. Bird RB, Stewart WE, Lightfoot EN (2007) Transport Phenomena. John Wiley & Sons. New York, USA.
7. Davis JR (2000) Nickel, Cobalt and their Alloys. ASM International.
8. Berreni M, Wang M (2011) Modelling and dynamic optimization of thermal cracking of propane for ethylene manufacturing. Computers & Chemical Engineering 35: 2876-2885.

14

Catalyst and Catalysis for Fischer-Tropsch Synthesis: A Comparative Analysis of Iron and Cobalt Catalysts on SBA-15

Mala A Sainna[1]* and Abba MK[2]

[1]*Manchester Institute of Biotechnology and School of Chemical Engineering and Analytical Science, The University of Manchester, 131 Princess Street, Manchester M1 7DN, United Kingdom*

[2]*Department of Petroleum Engineering, Faculty of Engineering and Engineering Technology, Abubakar Tafawa Balewa University Bauchi, PMB 0248, Yelwa Road, Bauchi State, Nigeria*

Abstract

The influence of Cobalt and iron loading, cobalt and iron precursor, and the effect of promoters(Mg and Mn) on the physico-chemical properties as well as the catalytic properties and structural appearance of mesopourous Co/SBA-15, Fe/SBA-15 and bimetallic Fe, Co/SBA-15 catalysts for Fischer-Tropsch (FTS) reaction is investigated. The various prepared catalysts were characterized by X-ray diffraction (XRD), scanning electron micropscope (SEM), Energy dispersion X-ray diffraction (EDX) and Temperature programmed reduction (TPR). For the Co/SBA-15 and Fe/SBA-15 prepared and promoted with Magnesium and manganese (Mg, Mn), the dispersion of the metal loaded (Fe, Co) increases significantly when compared to the un-promoted Co/SBA-15 and Fe/SBA-15, and the dispersion of metal loaded (Co, Fe) was found to be higher with Mg than Mn promoter. The bimetallic catalyst was also found to have a higher dispersion of the metals than a single metal loaded catalyst (Co/SBA-15, Fe/SBA-15). Promoting the catalyst with Mn increases dispersion but it was said to decrease reducibility leading to a catalyst that is less reactive than the un-promoted one [1-3]. The mesopourous catalyst prepared did not show any crystal growth in their meso-channels which propose the dispersion of the active metals on the mesopore and the formation of tiny crystals (nanocrystal) undetectable by the X-ray diffractogram. The absence of an intense peak of the crystals (Co_3O_4, Fe_3O_4) at an expected angle might be due to the existence of ions of cobalt and iron which might be existing as a nano-cluster or due to the disintegration of the mesopores structures.

Keywords: Fischer-Trosch; Catalyst; Heterogeneous catalyst; Preparation; Characterization

Introduction

The search for alternatives to non-renewable energy, such as crude oil and to provide a sulphur free fuel is becoming an interesting field of study to scientist and researchers all over the world. Hydrogenation of carbon monoxide (CO) with several catalysts such as, nickel, cobalt, and iron to produce mostly alkanes, alkenes and oxygen containing compounds as by-products is known as Fischer- Tropsch Synthesis (FTS). The mechanism for the reaction is predicted to follow a stepwise chain growth of hydrocarbons, according to Schulz-Flory Distribution [4-13]. Currently, the main commercial interest in Fischer-Tropsch (F-T) process is the production of high quality sulphur-free synthetic diesel fuels from natural gas. The formation of hydrogen and carbon monoxide ($CO+H_2$) commonly referred to as synthesis gas, involves several processes such as dry reforming of methane, partial oxidation of methane, steam reforming and water gas shift reaction. This process serves as routes to producing fuel precursors and chemicals used in refineries and petrochemical plants making the F-T process viable [9-13]. The renewed interest in F-T synthesis, is probably not because of the abundant natural gas in the globe, but because of the global demand for fuel supplies and environmental policies placed on the quality of fuel emitted to the environment. While the concept of hydrogen fuel and biomass remains an important option for fossils fuel, synthetic fuel produced by the F-T process is currently promoted as the fuel for the future. However, the maximum potential of the F-T process is yet to be achieved. This is because, the cost of running an F-T plant is expensive compared to the direct production of gasoline and diesel via crude oil. Consequently, research and huge investments in Europe, Asia, Africa and the U.S to improve the amount of C_2-C_4 olefins is proliferated. The research is aimed at optimizing the selectivity of cracking high-boiling waxy hydrocarbons that can be converted through cracking processes to any desired hydrocarbon of any length [14-18]. The F-T reaction is highly exothermic and very slow. Catalyst is therefore required to increase the rate of reaction and reduce the energy consumed during the reaction. Gleaves et al. [19] used and modified several heterogeneous catalysts by promoting the catalyst with mostly metals from Group VIII. For example, coke deposition on Ru, Fe, Co, Rh, Pd, Pt, and Ir affects the mechanism [17-19]. Roa et al. [20] illustrated that the deposition of coke on metal surfaces are as a result of disproportionation of CO and cracking of hydrocarbons during the reaction. This can de-activate the catalyst and reduce product yield. Meanwhile, Basini et al. [21], tried to understand the mechanism of F-T reaction on different catalyst optimizing steam/carbon and oxygen/carbon as feed stock. They suggested that, majority of group VIII metals are catalytically active for the reaction, but their resistance to coke deposition on catalyst performance decreases across the periodic table. Whereas, noble metals, such as Pt, and Rh, have been reported to be more resistant to coke deposition and used as promoters [22], however, they are very expensive and not used on large scale, as compared to nickel, cobalt and iron which are widely used in large scale (Scheme 1).

In this report, we thoroughly investigated and rationalized the influence of Cobalt and iron loading, cobalt and iron precursor, and the effect of Mg and Mn promoters on the physico-chemical properties as well as the catalytic properties and structural appearance of mesoporous

***Corresponding author:** Mala A Sainna, Manchester Institute of Biotechnology and School of Chemical Engineering and Analytical Science, The University of Manchester, 131 Princess Street, Manchester M1 7DN, United Kingdom, E-mail: alhaji.mala@manchester.ac.uk

$CH_4+1/2O_2$	→	$CO+2H_2$	ΔH=38 kJ /mole	(1)
CH_4+CO_2	→	$2CO+2H_2$	ΔH=-247kJ/mol	(2)
$CO+H_2O$	→	CO_2+H_2	ΔH=41 kJ /mol	(3)
CH_4+H_2O	→	$2CO+3H_2$	ΔH=-206 kJ /mol	(4)

Scheme 1: The synthesis gas production is performed by different processes, this processes includes partial oxidation of methane (1), Dry reforming of methane (2), Water gas shift reaction (3) and Steam methane reforming (4) [8].

Co/SBA-15, Fe/SBA-15 and bimetallic Fe, Co/SBA-15 catalysts for Fischer-Tropsch synthesis (FTS). The various catalyst prepared within this report is characterized using XRD, SEM, EDX, and TPR which is thoroughly discussed in the result section of the report.

Experimental Methods

Sol-Gel method

The mesoporous silica support (SiO_4) was prepared using procedure of Sol-Gel method as reported in previous work [22]. Reagents and solvents were purchased from commercial sources and purified by standard procedures before use.

SBA-15

The preparation of SBA-15 support is synthesised as follows: 8 grams of tri-block copolymer P123 ($EO_{20}PO_{70}EO_{20}$ also known as the poly (ethylene glycol)-poly (propylene glycol)-poly (ethylene-glycol)) is measured using an electronic weighing balance and dissolved in 240 grams of de- ionized (DI) water in a polyethylene bottle. The solution in the polyethylene bottle was placed in a glass bowl containing water and kept on a hot plate at 40°C for 6 hours under stirring at 545 rpm, until the whole solute P123 is completely dissolved, leading to a clear solution. 40 ml of 37% HCl is then slowly added using 50 ml measuring cylinder by dropping on the solution to reduce the pH of the solution. 10 munities later, 18.5 ml Tetraethylorthosilicate (TEOS) is slowly dropped in (approximately 30 drop per minute) the mixture was kept for 24 hrs at 40°C under stirring condition for Hydrolysis and condensation reactions to complete. A white precipitated solution was formed which was then transferred into an oven at 95°C for 24 hours to age without stirring. The white solute precipitate was then filtered, washed sufficiently with approximately 600 ml of DI water to remove the template (P123) and afterwards the white solute powder was dried in an oven at 60°C for 24 hours using a glass watch vessel. The white powdered solute was removed from the oven and transferred into a crucible for calcinations under static condition for 6 hours at 550°C by setting the crucible to increase temperature to 5°C per minute and after heating for 6 hours it was cooled at 10°C per minute and finally the pure mesoporous silica was called the SBA-15 [21-25].

Metal loading by wet impregnation

Preparation of cobalt active catalyst by wet impregnation method: In wet impregnation, the cobalt loading on the channels of the mesoporous silica support is defined as the weight percentage of elemental Co in the catalysts. The equations below signifies the steps to follow in obtaining the percentage of the cobalt precursor used for the cobalt loading and the amount of the cobalt loaded in the channels of the SBA-15 silica support used; i.e.,

$$weight\ of\ cobalt = \frac{Molar\ weight\ of\ Cobalt \times weight\ of\ Cobalt\ nitrate}{Relative\ molecular\ weight\ of\ Cobalt\ nitrate} \quad (1)$$

To identify the weight of cobalt in a given weight of support, the weight of cobalt obtained using the equation above divided by the summation of the weight of a given support and the weight of cobalt obtained using eqn. (1) above, multiplied by 100 to obtain the percentage of the cobalt loaded in the channels of the given support. It is simplified as shown below;

$$weight\ of\ Cobalt\ in\ given\ weight\ of\ support\ \% = \frac{Weight\ of\ cobalt \times 100}{Weight\ of\ support + weight\ of\ cobalt} \quad (2)$$

The concentration of the precursor used in the DI water in which it is dissolved is denoted as the equation below;

$$M_1V_1 = M_2V_2 \quad (3)$$

Where;

M_1=Molarity of water

V_1=Volume of water

M_2=Molarity of salt precursor

V_2=Volume of salt precursor

The equation below is used to measure the concentration of the salt precursor used and the DI water or absolute alcohol;

$$Concentration\left(\frac{Mol}{dm3}\right) = \frac{Mass}{Molarmass} \quad (4)$$

The preparation of Co/SBA-15 by wet impregnation is conducted by the addition of 10 ml of absolute alcohol (ethanol) with 2 grams of $Co(NO_3)_2.6H_2O$ in a round bottom flask with a short condenser at the top of the flask to condense the evaporated alcohol in the solution and the flask is kept in a bowl containing water, the mixture is stirred for at least 60 munities under 50°C on a hot plate to dissolve the salt precursor thoroughly (however the amount of the salt precursor was added according to the percentage of the metal required within the channels of the SBA-15 silica support), then 0.5 grams of the calcined mesoporous SBA-15 silica is added gently on the salt solution and allowed to stir for another 120 minutes at 50°C. The solution is left to cool till it is at room temperature then 10 ml of DI water is added to reduce the viscosity of the solution. The solution is then filtered and without washing the filtrate it is then dried in a vacuum oven for 60 minutes at a temperature of 120°C and a pressure of 800 psi. After drying, the sample is calcined at a temperature of 400°C for 4 hours by setting the crucible to a heating rate of 10°C per minutes and later cooled at 10°C per munities, to obtain the mechanical required structure of the catalyst. And finally the sample obtained is known as the Co/SBA- 15 active catalyst.

Preparation of cobalt active catalyst with promoter using wet impregnation method: The preparation of active catalyst with promoter using wet impregnation method is conducted by the addition of 10 ml of absolute alcohol with 2 grams of the salt precursor together with 1 gram of Magnesium nitrate. Add both precursors in a round bottom flask containing 10 ml of absolute alcohol, stir for 1 hour at 50°C on a hot plate then add 0.5 grams of the calcined SBA-15 silica and allow to stir for another 2 hours, allow the solution to cool to room temperature then filter. Dry the filtrate in a vacuum heat for 1 hour at 12°C and calcine for 4 hours at 400°C with a heating rate of 10°C per minute followed by a cooling rate of 10°C per minute. The active catalyst obtained is expected to have a better dispersion as well as activity and selectivity due to the promoter within the channels of the mesoporous SBA-15 catalyst support. The amount of the metal deposited on the channels is measured by using X- ray diffractions (XRD) [26-51].

Preparation of bimetallic active catalyst by co-impregnation method: The bimetallic active catalyst (Co-Fe/SiO_2) is prepared by the addition of 1.3 grams of cobalt nitrate hexahydrate, which is equivalent to 35% theoretically, and 1.2 grams of iron (III) nitrate which is also equals to 36% and the two precursors are stirred to dissolved in absolute alcohol under 50°C for 1 hour followed by the addition of 0.5 grams of the calcined SBA-15 silica slowly in the homogeneous solution. The solution is allowed to stir at 50°C for 2 more hours which is then filtered and dried under 120°C for 1 hour in a vacuum oven. The dried sample is calcined for 4 hours at 400°C by the same heating and cooling rate as explained in the above procedure, and an orange powder is obtained.

Results and Discussion

Energy dispersive X-ray spectroscopy results

The energy dispersive x-ray spectroscopy (EDX) is an investigative method used in the chemical characterization or elemental analysis of the sample prepared. We utilized it to display the various samples of the catalyst prepared which were supported on silica SBA-15 were shown by elemental composition to note the amount of metal loaded (Fe, Co, Mg and Mn) after preparation of the samples using wet impregnation method. All the samples were prepared using DI water as a solvent instead of ethanol to dissolve the metal precursor and the concentration ranging from 0.46M to 0.49M. The samples characterized by means of EDX were recognized as x-SBA/15-n; where x refers to the percentage of the metal loaded in a sample, and the n is the various numbers of the samples prepared (Figures 1-7 and Tables 1-7).

Scanning Electron Microscope (SEM) images evaluation

The scanning electron microscope is basically used to determine the samples composition, surface topology and some other properties like the electrical conductivity when electrons from the SEM interact with the atoms that unite to form the sample [52]. The Figures 8 and 9 below indicate the cobalt supported catalyst and iron supported

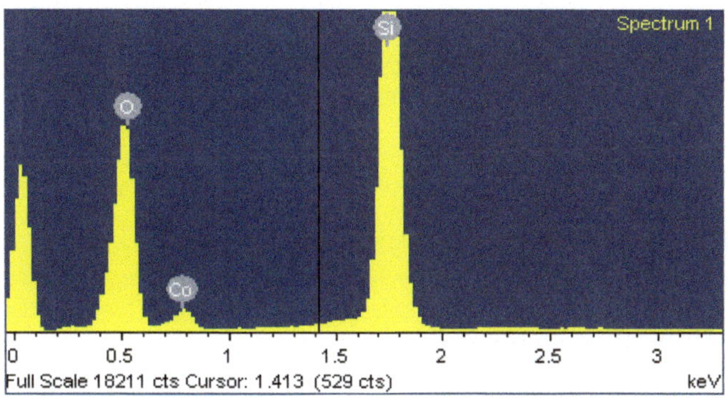

Figure 1: Spectrum of elemental composition of Calcined mesoporous 15.60% Co/SBA-15 silica.

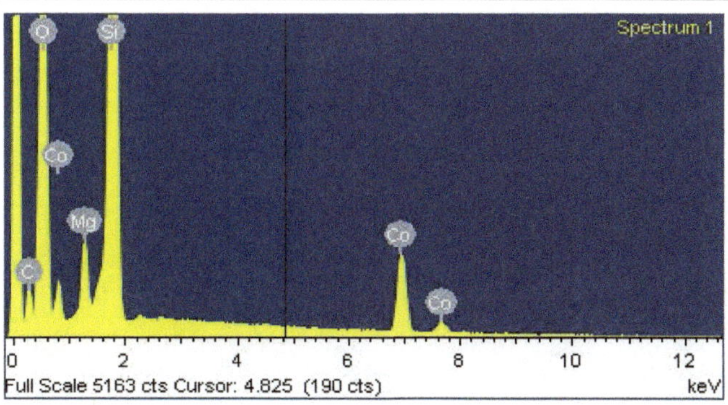

Figure 2: Spectrum of elemental composition of 6.73% Co/SBA-15 promoted with 1.38% Mg.

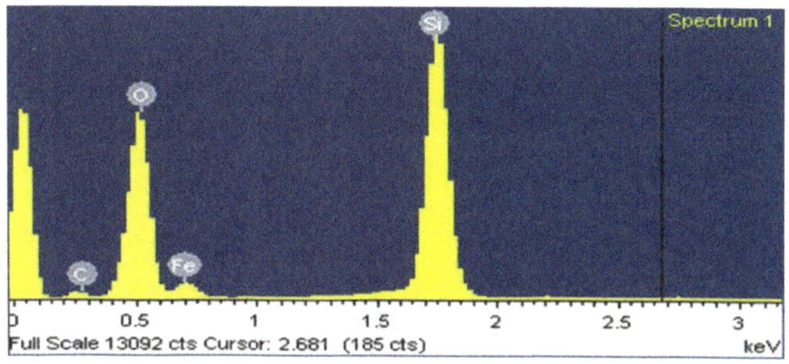

Figure 3: Spectrum of elemental composition of 29.19% Fe loading on silica SBA-15.

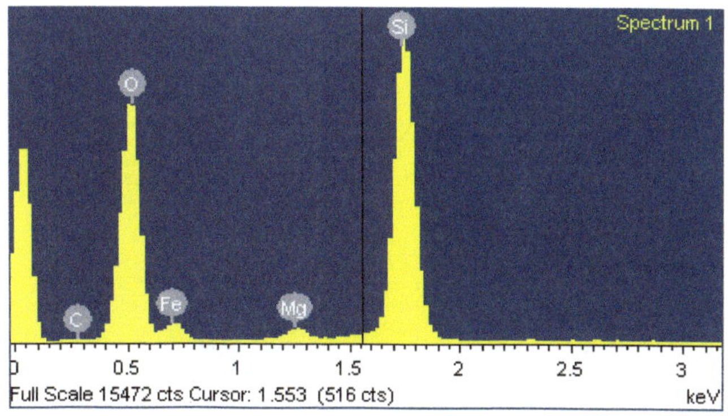

Figure 4: Spectrum of elemental composition of 28.82% Fe/SBA-15 promoted with 1.02% Mg.

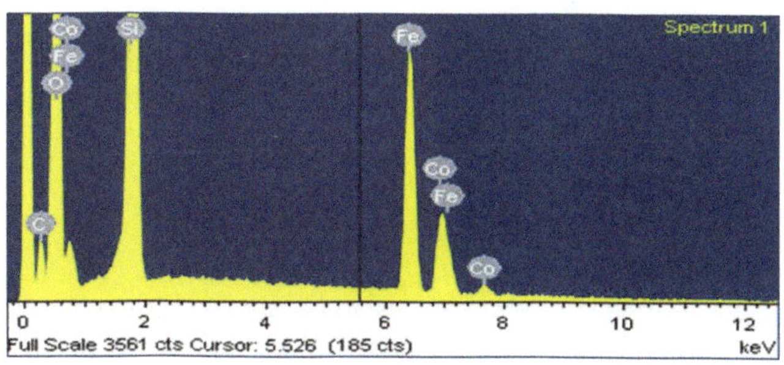

Figure 5: Spectrum of elemental composition of bimetallic loading comprising of 14.37% Fe and 5.41% Co supported on SBA-15 silica.

Element	Weight%	Atomic%
O K	47.90	65.68
Si K	36.50	28.51
Co K	15.60	5.81
Total	100.00	100.00

Table 1: Results showing weight% and the atomic% of 15.60% cobalt loading on SBA-15.

Element	Weight%	Atomic%
C K	11.30	17.87
O K	47.70	56.64
Mg K	1.38	1.08
Si K	32.89	22.24
Co K	6.73	2.17
Total	100.00	100.00

Table 2: Results of 6.73% Co/SBA-15 loading and promoted with 1.38% Mg.

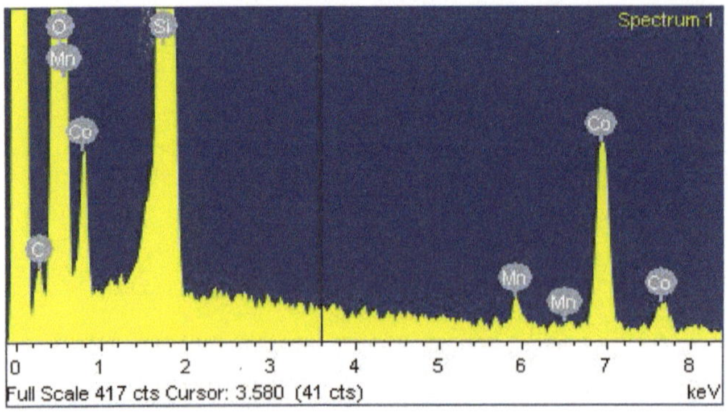

Figure 6: Spectrum illustrating the elemental composition of Co/SBA-15 promoted with 0.61% Mn.

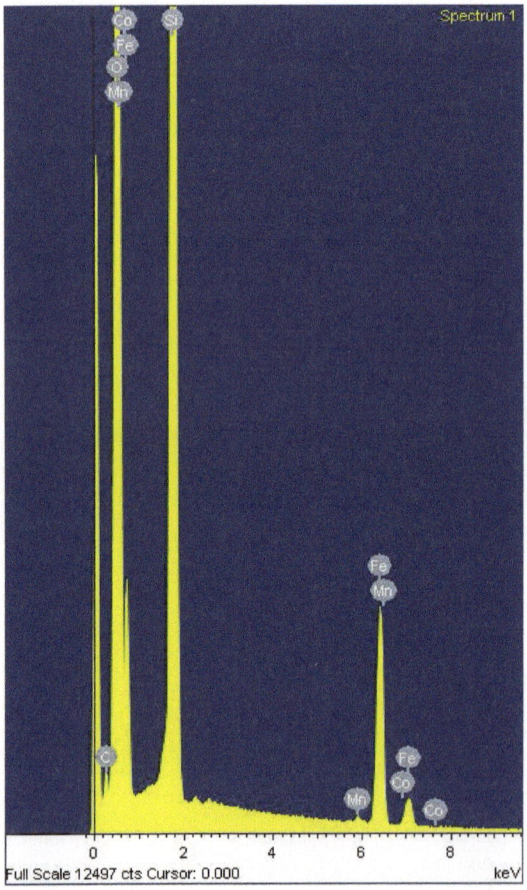

Figure 7: Elemental composition of 33.48% Fe/SBA-15 promoted with 0.50% Mn.

catalyst respectively after calcination; As can be observed from both images of the prepared catalyst at a magnification of 4.5 μm above, it can be seen that the catalysts were agglomerated and this is simply due to the fact that both catalyst lost the components of the precursors like the nitrates, water, the Tri-block copolymer used during preparation [poly (ethylene glycol)-poly (propylene glycol)-poly (ethylene-glycol)]. The absence of these residues of these components is as a result of the calcination of the prepared catalysts at 400°C as explained earlier. Even though the image of Fe/SBA-15 shows some of the mesoporous structures where broken, it is obvious that the pores are not blocked by some surfactants to reduce the active metal dispersion on the support. Therefore, calcination of the catalyst before and after impregnation will increase surface area of the catalyst.

X-Ray Diffraction (XRD) spectroscopy result evaluation

X-ray diffraction spectroscopy pattern of an un-promoted as well as the magnesium and manganese promoted iron and cobalt catalysts prepared are calcined at 400°C together with those of the bulk Co_3O_4 spinel's phase are presented below. All samples were prepared using cobalt nitrate and iron (III) nitrates precursors. It is apparent from the samples diffraction peaks, all cobalt catalysts showed reflection characteristics of the Co_3O_4 spinel for the cobalt catalyst by observing an intense peak which is usually expected at 36.90°; the peak becomes narrower as the percentage of the cobalt increases indicating an increase of the mean Co_3O_4 particles. Although the intensity of the peaks

Element	Weight%	Atomic%
C K	6.18	11.88
O K	36.97	53.33
Si K	27.65	22.72
Fe K	29.19	12.06
Totals	100.00	100.00

Table 3: Results of 29.19% Fe/SBA-15 loading.

Element	Weight%	Atomic%
C K	3.43	6.65
O K	39.68	57.85
Mg K	1.07	1.02
Si K	27.01	22.43
Fe K	28.82	12.04
Totals	100.00	100.00

Table 4: Results of 28.82% Fe/SBA-15 loading promoted with 1.07% Mg.

Figures 8: SEM image of mesoporous 15.60% Co/SBA-15.

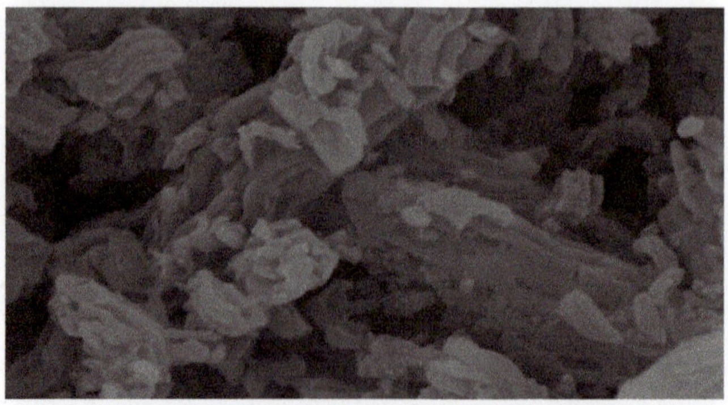

Figures 9: SEM image of mesoporous 29.19% Fe/SBA-15.

Element	Weight%	Atomic%
C K	12.46	21.32
O K	39.65	50.93
Si K	28.11	20.57
Fe K	14.37	5.29
Co K	5.41	1.89
Totals	100.00	100.00

Table 5: Results of bimetallic loading of 14.37% Fe and 5.41% Co supported on SBA-15 silica.

Element	Weight%	Atomic%
C K	8.85	14.34
O K	48.76	59.3
Si K	34.05	23.59
Mn K	0.61	0.22
Co K	7.73	2.55
Totals	100.00	100.00

Table 6: Results of 7.73% Co SBA-15 silica promoted with 0.61% Mn.

Element	Weight%	Atomic%
C K	3.62	7.26
O K	38.06	57.25
Si K	24.33	20.85
Fe K	33.48	14.42
Mn K	0.50	0.22
Totals	100.00	100.00

Table 7: Results of 33.48% Fe SBA-15 silica promoted with 0.50% Mn.

are much lower than expected for some of the cobalt loaded catalyst, which could be due to the dispersion of the metal oxides (Co_3O_4, CoO and Fe_3O_4) on the surface of the SBA-15 silica rather than within the SBA-15 silica as desired. On the other hand, by promoting the Co/SBA-15 silica with Magnesium, it significantly increases the dispersion of the cobalt and the iron. Although, the increase in dispersion and the presence of Mn favors a strong interaction of the cobalt oxides with the support and possibly MnO phase can lead to a drastic catalyst deactivity of up to 50% as compared with the unpromoted one [53]. In the case of the iron catalyst, the wide angle shows a very sharp and unexpectedly long peak which is also an indication that some of the irons were dispersed on the surface of the SBA-15 silica support (Figure 10).

The wide angle diffractogram above indicate the cobalt peak within the different mesoporous silica SBA-15 prepared and impregnated with different percentage of cobalt nitrate, the peaks of Co_3O_4 crystals were obvious in each of the catalyst diffractogram which is at an angle of 36.5° as expected [52]. The peak for the catalyst with 15.60% Cobalt loading is higher among the rest with a lower cobalt loading; it is arranged according to (Series 4) 15.60% Co, (Series 3) 6.73% Co, (Series 2) 7.73% Co, (Series 1) 5.41 Co in a bimetallic catalyst. The catalyst with lower loading has less intense peak as compared to the catalyst with higher loading. According to Martinez et al. [52], the characteristics peak of Co_3O_4 by the diffractogram is not very obvious at 36.5° at a lower Co loading. The mesoporous structures seems to be dependent on the quantity of the metal loading on the support has been observed after calculating the crystallite sizes of each of the samples prepared using shearers equation. The table below shows the average crystallite diameters of each of the sample prepared (Table 8).

The table above indicates the metal loading expected theoretically, what was obtained after preparation and the crystallite size diameter of each of the samples prepared after calculating using the shearer's equation. The percentage of the reduced catalyst and the percentage of metal dispersion on the mesoporous SBA-15 silica support are also calculated using the interatomic distance equations [38-40]. As can be observed from the Table, the samples with higher metal loading has the larger particle size, this confirms the dependent of the metal loading on the particle size (Figure 11).

The above diffractogram compares the different iron catalyst supported with SBA-15 silica with the iron loaded at different proportion. The Fe_2O_3 peak is expected at an angle of 24.3°, 33.4°, and 35.6° as reported [53]. The diffractogram is arranged according to (Series 4) 29.19% Fe, (Series 3) 28.82% Fe, (Series 2) 33.48% Fe, and (Series 1) 14.37% Fe for the bimetallic loading. It is important to understand the fact that larger metal oxide crystallite diameter as compared to the pore diameter of the SBA-15 silica support will lead to the dispersion of the crystal on the surface of the SBA-15 silica and not within the meso-channels of the support. This can be identified by observing the XRD of the catalyst by having rough surface with unclear peaks as can be observed from the XRD patterns of "Series 4" above. For a very well dispersion of the metal on the support, three well resolved diffraction peaks of d_{100}, d_{110}, and d_{200} associated with p6mm hexagonal symmetry can be obtained showing that the mesostructure was highly ordered. The crystallite size of the SBA-15 silica support is usually around 6-15 nm [31], however the diffractogram clearly showed the dispersion of the metals on the surface of the support which means that the average diameter of the crystal is larger than that of the support which leads to the dispersion of the metal on the surface. A very narrow and highly intense peak was also observed on the 29.19% Fe/SBA-15 silica which is also due to the dispersion of the metal on the surface of the support [54] (Figure 12).

Temperature programmed reduction evaluation

Figure 13 below shows the Temperature programmed reduction (TPR) peaks of cobalt oxides in an atmosphere of hydrogen which is recorded using the CATLAB-PCS (Micro-reactor with integrated mass spectrometer) [34] at 500°C (at an ascending heating rate of 15°C per minute). The first series (Series 2) on the graph for Hydrogen shows two peaks where the reduction of the cobalt oxide crystals occurred from Co_2O_3 to CoO and the second peak is a reduction of CoO to Co-zero which is in the reduced state. This peaks were proven by the

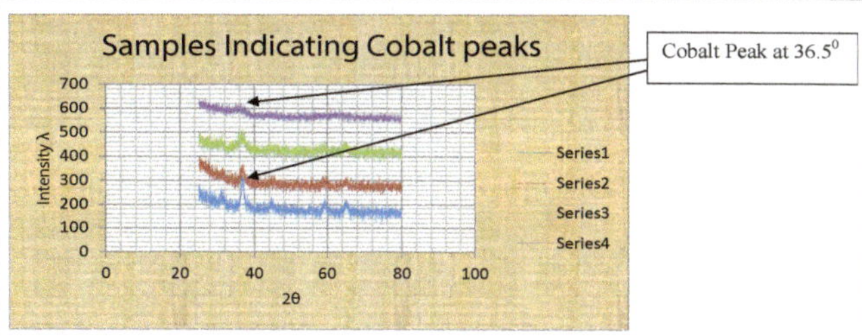

Figure 10: Wide angle diffractogram indicating cobalt peaks of various samples at an angle of 36.5° as expected.

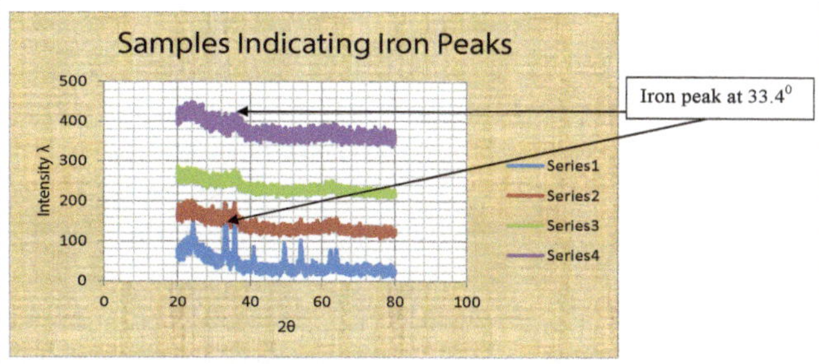

Figure 11: Wide angle diffractogram indicating Iron peaks of various samples at an angle of 33.4° as expected.

Catalyst	(wt%) of metal loading expected Theoretically	(wt%) of metal loaded practically	Average Crystal diameter (nm)	(Wt%) Reduced metal catalyst	(Wt%) Metal dispersion
Co/SBA-15 silica	35	15.60	13.2	9.9	9.6
Co/SBA-15 Silica promoted with Mg	34	6.73	8.7	6.5	14.8
Co/SBA-15silica promoted with Mn	30	7.73	10.5	7.9	12.2
Fe/SBA-15 silica	35	29.19	15.3	11.5	8.3
Fe/SBA-15silica Promoted with Mg	35	28.82	8.3	6.3	15.2
Fe/SBA-15 silica promoted with Mn	35	33.48	13.2	9.9	9.6
Fe-Co/SBA-15 silica	Fe 20, Co 10	Fe 14.37, Co 5.41	8.1	6.1	15.7

Table 8: Table indicating the percentage of metal loading expected theoretically and the percentage obtained practically together with the average particle sizes for each of the catalyst, the percentage of reduced metal catalyst and the percentage metal dispersion.

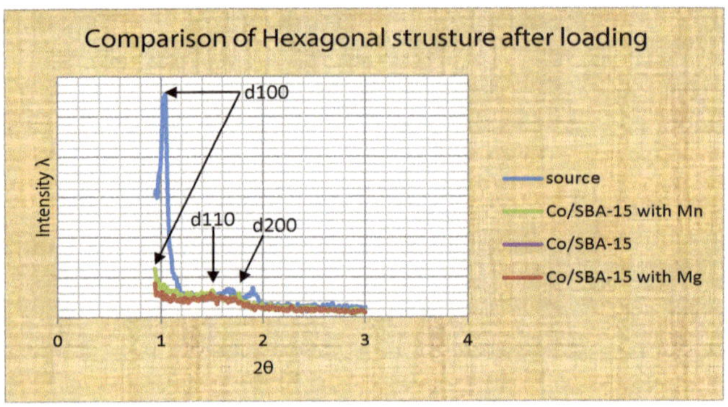

Figure 12: Comparison of Hexagonal structures after loading of Co/SBA-15 at different quantity.

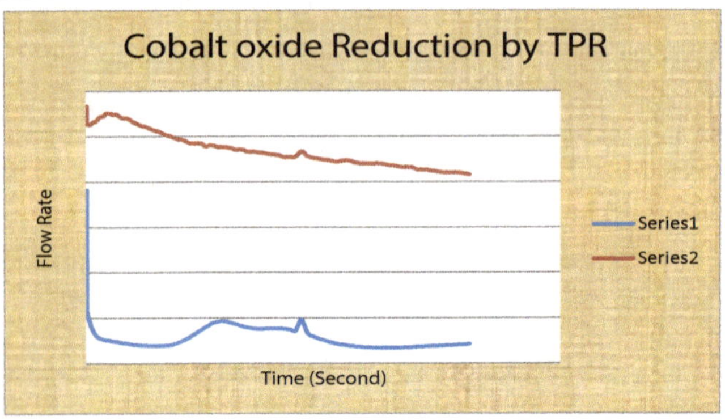

Figure 13: Graph of Cobalt oxide reduction by Temperature Programmed Reduction technique using CATLAB-PCS Microreactor. The Graph is showing the peaks at a reduction of Co_2O_3 to CoO (Series 1), and a reduction of CoO to Co-zero.

peaks (Series 1) observed by formation of water at an align position to the peaks formed by hydrogen, the first reduction occurred at a temperature less than 350°C and the second reduction occurred at a temperature less than 450°C as can be observed from the graph above. The temperature programmed reduction techniques sometimes gives quantitative and qualitative picture of the catalyst surface reducibility as well as its sensitivity to promoters chemical changes or metal/ support interactions (Figure 13).

Conclusions

The seven different samples of Fischer Tropsch catalyst where prepared and the physico-chemical properties of their mesoporous SBA-15 supported format was investigated as a function of the metal loading, percentage dispersion of the metal on the support, effect of magnesium (Mg) and manganese (Mn) promoters, the average crystal diameter in each of the sample. Calcined samples of Co/SBA-15 prepared from Co (II) nitrate hexahydrate with cobalt loading of 15.60 wt% practically, another 6.73 wt% Co/SBA-15 promoted with Mg, 7.73 wt% Co/SBA-15 promoted with Mn where all characterized using XRD, EDX/SEM and TPR and it was concluded as follows;

- The average crystal diameter is observed to be larger in iron catalyst and this could be due to the fact that Fe metal has higher percentage loading on the support than the cobalt; this indicates the dependence of metal loading on the average crystal diameter. i.e., the higher the metal loading on the support, the larger the average crystal diameter.
- According to the observation on the wide angle pattern, it can be considered that the strong interaction of the support and the metal catalyst loaded can lead to high dispersion of the metal catalyst loaded within the meso-channels of the support, and this will lead to a weak peak on the wide angle diffractogram of the samples.
- The dispersion of the active metal sizes of Cobalt and iron samples increases with the addition of promoter, however it increases most with the magnesium promoter (Mg) which has the highest dispersion of 14.8 wt% on 6.73% Co/SBA-15, and 15.2 wt% on 28.82% Fe/SBA-15 as compared to manganese promoter with 12.2 wt% dispersion on 7.73% Co/SBA-15 and 9.6 wt% dispersion on 33.43% Fe/SBA-15. However, an increase in Mn promoter favors a strong interaction of the metal oxide with the support and perhaps with the MnO component generating a catalyst that maybe about 50% less active than the un- promoted catalyst sample [52].
- The dispersion was also observed to be high with the bimetallic 14.37% Fe-5.41% Co/SBA- 15 catalyst with the active metal dispersion on the support of 15.7% was observed.
- It should be well-known that the absence of intense reflection at expected peaks by the wide angle XRD in an active metal catalyst suggest the formation of either small metal crystal particles that is undetectable by the X-ray diffractogram, or it means the disintegration of the mesoporous structure. Likewise for the small angle, the absence of the peaks at d_{100}, d_{110} and d_{200} might be due to the collapse of the hexagonal structure of the support as observed from the XRD pattern of the small angles XRD result [52-56].

Acknowledgements

MAS and AMK thanks the Petroleum Technology Development Fund for a studentship.

References

1. Zhao Y, Klie RF (2011) J Underg Research 25: 25-28.
2. Wilson TP, Paul HK, Ellgen PC (1981) J Cata 69: 193-201.
3. Knop-Gericke A, Havecker M, Schedel-Niedrig T, Schlogl R (2000) Top Catal 10: 187.
4. Kobayashi M, Kobayashi H (1972) Application of transient response method to the study of heterogeneous catalysis: I. Nature of catalytically active oxygen on manganese dioxide for the oxidation of carbon monoxide at low temperatures. J Cataly 27: 100-107.
5. Kobayashi M, Kobayashi H (1972) Application of transient response method to the study of heterogeneous catalysis: II. Mechanism of catalytic oxidation of carbon monoxide on manganese dioxide. J Catal 27: 108-113.
6. Kobayashi M, Kobayashi H (1972) Application of transient response method to the study of heterogeneous catalysis: III. Simulation of carbon monoxide oxidation under an unsteady state. J Catal 27: 114-119.
7. Dekker FHM, Nazloomian J, Bliek GA, Kapteijn F, Moulijn JA, et al. (1997) J Appl Catal A 151: 247-266.
8. Olea M, Kunitake M, Shido T, Iwasawa Y (2001) TAP study on CO oxidation on a highly active Au/Ti $(OH)_4$* catalyst. Phys Chem Chem Phys 3: 627-631.
9. Buyevskaya OV, Wolf, D, Baerns M (1994) Rhodium-catalyzed partial oxidation of methane to CO and H_2. Transient studies on its mechanism. Catal Lett 29: 249-260.
10. Schuurman Y, Mirodatos C (1997) Uses of transient kinetics for methane activation studies. Appl Catal A 151: 305-331.
11. Aparicio LM (1997) Transient isotopic studies and microkinetic modeling of methane reforming over nickel catalysts. J Catal 165: 262-274.
12. Heitnes Hofstad K, Hoebink JHBJ, Holmen A, Marin GB (1998) Partial oxidation of methane to synthesis gas over rhodium catalysts. Catal Today 40: 157-170.
13. Berger-Karin C, Sebek M, Pohl MM, Bentrup U, Kondra-tenko VA, et al. (2012) Chem Cat Chem 4: 1368-1375.
14. Keiski RL, Salmi T, Niemisto P, Ainassaari J, Pohjola V (1996) Stationary and transient kinetics of the high temperature water-gas shift reaction. J Appl Catal A 137: 349-370.
15. Kalamaras CM, Panagiotopoulou P, Kondarides DI, Efstathiou AM (2009) Kinetic and mechanistic studies of the water–gas shift reaction on Pt/TiO_2 catalyst. J Catal 264: 117-129.
16. Kala-maras CM, Petallidou KC, Efstathiou AM (2013) The effect of La 3+-doping of CeO 2 support on the water-gas shift reaction mechanism and kinetics over Pt/Ce1- x La x O_2-δ. Appl Catal B 136: 225-238.
17. Wainwright MS, Hoffman TW (1977) The oxidation of ortho-xylene on vanadium pentoxide catalysts. I. Transient kinetic measurements. J Chem Eng 55: 552-556.
18. Hinz A, Nilsson B, Andersson A (2000) Simulation of transients in heterogeneous catalysis: a comparison of the step-and pulse-transient techniques for the study of hydrocarbon oxidation on metal oxide catalysts. Chem Eng Sci 55: 4385-4397.
19. Gleaves JT, Ebner JR, Kuechler TC (1988) Temporal analysis of products (TAP)-a unique catalyst evaluation system with submillisecond time resolution. Catal Rev 30: 49-116.
20. Gleaves JT, Yablonsky G, Zheng X, Fushimi R, Mills PL (2010) Temporal analysis of products (TAP)-recent advances in technology for kinetic analysis of multi-component catalysts. Journal of Molecular Catalysis A: Chemical 315: 108-134.
21. Angeli SD, Mon teleone G, Giaconia A, Lemonidou AA (2014) State-of-the-art catalysts for CH_4 steam reforming at low temperature. Int J Hydrogen Energy 39: 1979 -1997.
22. Duarte RB, Nachtegaal M, Bueno JMC, van BoKhoven JA (2012) Understanding the effect of Sm_2O_3 and CeO_2 promoters on the structure and activity of Rh/ Al2O3 catalysts in methane steam reforming. J Catal 296: 86-98.
23. Wei JM, Iglesia E (2004) Mechanism and Site Requirements for Activation and Chemical Conversion of Methane on Supported Pt Clusters and Turnover Rate Comparisons among Noble Metals. J Phys Chem B 108: 4094-4103.
24. Wei JM, Iglesia E (2004) Structural requirements and reaction pathways in methane activation and chemical conversion catalyzed by rhodium. J Catal 225: 116-127.
25. Wei JM, Iglesia E (2004) Isotopic and kinetic assessment of the mechanism of reactions of CH 4 with CO_2 or H_2O to form synthesis gas and carbon on nickel catalysts. J Catal 224: 370-383.
26. Craciun, R, Daniell, W, Knozinger, H (2002) The effect of CeO 2 structure on the activity of supported Pd catalysts used for methane steam reforming. Appl Catal A 230: 153-168.
27. Sainna MA, Singh D, Kumar D, de Visser SP (2015) A Trimetal Carbene with Reactivity Reminiscent of Fischer–Tropsch Catalysis. Organometallics 34: 1651-1660.
28. Halabi MH, De Croon MHJM, van der Schaaf J, Cobden PD, Schouten JC, et al. (2010) Intrinsic kinetics of low temperature catalytic methane-steam reforming and water-gas shift over Rh/Ceα Zr1− $αO_2$ catalyst Appl Catal A General 389: 80-91.
29. Halabi MH, De Croon MHJM, van der Schaaf J, Cobden PD, Schouten JC, et al. (2010) Low temperature catalytic methane steam reforming over ceria–zirconia supported rhodium. Applied Catalysis A: General 389: 68-79.
30. Kurungot S, Yamaguchi T (2004) Stability improvement of Rh/γ-Al2O3 catalyst layer by ceria doping for steam reforming in an integrated catalytic membrane reactor system. Catal Lett 92: 181-187.
31. Quan ZA, Qin YN, Liu C (1991) Promoting effect of cerium oxide in supported nickel catalyst for hydrocarbon steam-reforming. Appl Catal 70: 1-8.
32. Van Santen RA (2009) Complementary structure sensitive and insensitive catalytic relationships. Acc Chem Res 42: 57-66.
33. Duarte RB, Safonova OV, Krumeich F, Makosch M, Van Bokhoven JA (2013) Oxidation State of Ce in CeO2-Promoted Rh/Al2O3 Catalysts during Methane Steam Reforming: H2O Activation and Alumina Stabilization. ACS Catal 3: 1956-1964.
34. Chin YH, Buda C, Neurock M, Iglesia E (2013) Consequences of metal-oxide interconversion for C-H bond activation during CH4 reactions on Pd catalysts. J Am Chem Soc 135: 15425-15442.
35. Buyevskaya OV, Walter K, Wolf D, Baerns M (1996) Primary reaction steps and active surface sites in the rhodium-catalyzed partial oxidation of methane to CO and H_2. Catal Lett 38: 81-88.
36. Wang DZ, Dewaele O, DeGroote AM, Froment GF (1996) Reaction mechanism and role of the support in the partial oxidation of methane on Rh/Al_2O_3. J Catal 159: 418-426.
37. Shaikhutdinov S, Heemeier M, Baumer M, Lear T, Lennon D, et al. (2001) Structure-Reactivity Relationships on Supported Metal Model Catalysts: Adsorption and Reaction of Ethene and Hydrogen on Pd/Al_2O_3/NiAl (110). J Cata l 200: 330-339.
38. Chou P, Vannice MA (1987) Calorimetric heat of adsorption measurements on palladium. I. Influence of crystallite size and support on hydrogen adsorption. J Catal 104: 1-16.
39. Bernal S, Calvino JJ, Cifredo GA, Laachir A, Perrichon V, et al. (1994) Influence of the Reduction/Evacuation Conditions on the Rate of Hydrogen Spillover on Rh/CeO_2 Catalysts. Langmuir 10: 717-722.
40. Bernal S, Calvino JJ, Cifredo GA, Rodriguez-Izquierdo JM, Perrichon V, et al. (1992) Reversibility of hydrogen chemisorption on a ceria-supported rhodium catalyst. J Catal 137: 1-11.
41. Gritschneder S, Reichling M (2007) Structural elements of CeO_2(111) surfaces. Nanotechnology 18: 1-6.
42. Yang ZX, Wang Q, Wei S, Ma D, Sun Q (2010) The Effect of Environment on the Reaction of Water on the Ceria(111) Surface: A DFT+U Study. J Phys Chem C 114: 14891-14899.
43. Molinari, M, Parker, S. C, Sayle, D. C, Islam, M. S. Water Adsorption and Its Effect on the Stability of Low Index Stoichiometric and Reduced Surfaces of Ceria. J. Phys. Chem. C 2012, 116, 7073 - 7082.
44. Norrish K, Taylor RM (1962) Quant Diffrac Ana, pp: 98-109.
45. Sainna MA, Kumar S, Kumar D, Fornarini S, Crestoni ME, et al. (2015) A comprehensive test set of epoxidation rate constants for iron(IV)–oxo porphyrin cation radical complexes. Chem Sci 6: 1516-1529.
46. Peri JB (1965) Infrared and Gravimetric Study of the Surface Hydration of γ-Alumina. J Phys Chem 69: 211-219.

47. Hass KC, Schneider WF, Curioni A, Andreoni W (2000) First-Principles Molecular Dynamics Simulations of H_2O on α-Al_2O_3 (0001)J Phys Chem B 104: 5527-5540.

48. Fathi M, Monnet F, Schuurman Y, Holmen A, Mirodatos C (2000) Reactive oxygen species on platinum gauzes during partial oxidation of methane into synthesis gas. J Catal 190: 439-445.

49. Wang R, Xu HY, Liu XB, Ge QJ, Li WZ (2006) Role of redox couples of Rh 0/ Rh δ+ and Ce 4+/Ce 3+ in CH_4/CO_2 reforming over Rh–CeO_2/Al_2O_3 catalyst. Appl Catal A 305: 204-210.

50. Meng M, Zha Y, Luo J, Hu T, Xie Y, et al. (2006) A study on the catalytic synergy effect between noble metals and cobalt phases in Ce-Al-O supported catalysts. J Appl Catal A 301: 145-151.

51. Vazquez A, Lopez T, Gomez R, Bokhimi X (2001) Synthesis, characterization and catalytic properties of Pt/CeO_2-Al_2O_3 and Pt/La_2O_3-Al_2O_3 sol–gel derived catalysts. J Mol Catal A 167: 91-99.

52. Okayama S, Haraichi S, Matsuhata H (2005) Reference sample for the evaluation of SEM image resolution at a high magnification--nanometer-scale Au particles on an HOPG substrate. J Electron Microsc (Tokyo) 54: 345-350.

53. Martínez ACL (2003) J Catal, pp: 486-499.

54. Kumar N, Singh P, Kumar S (2006) Physical, X-ray diffraction and scanning electron microscopic studies of uroliths. Indian J Biochem Biophys 43: 226-232.

55. Pattanayak BC (2010) Synthesis and Characterization of Alumina/Iron Oxide Mixed Nanocomposite. Rourkela: National Institute of Technology, Rourkela.

56. Yinyong S, Gen SWP (2006) Appl Catal A, pp: 1-7.

15

Micellar Catalysis on 1,10-Phenanthroline Promoted Chromic Acid Oxidation of Dimethyl Sulfoxide to Dimethyl Sulfone in Aqueous Media at Room Temperature

Rumki N, Aniruddha G, Susanta M and Bidyut Saha*

Homogeneous Catalysis Laboratory, Department of Chemistry, The University of Burdwan, Golapbag, Burdwan, Pin 713104, WB, India

Abstract

The monomeric species of Cr(VI) was found to be kinetically active in the absence of 1,10-phenanthroline (phen) under pseudo-first order conditions, whereas in the 1,10-phenanthroline promoted path, the Cr(VI)-phenanthroline complex undergoes a nucleophilic attack by dimethyl sulfoxide (DMSO) to form a ternary complex which subsequently undergoes a redox decomposition leading to dimethyl sulfone and Cr(III)-phenanthroline complex. A detailed study was done to investigate the effect of cationic surfactant N-cetylpyridinium chloride (CPC), anionic surfactant sodium dodecyl sulphate (SDS) and neutral surfactant (TX-100) on the phen promoted chromic acid oxidation of DMSO in aqueous media at 30°C. The aggregation and morphology of the micelles formed by the surfactants were studied by optical microscope and Transmission Electron Microscope (TEM). 1,10-phenanthroline in association with SDS micelle exhibited almost 70-fold rate enhancements compared to the uncatalyzed reaction path.

Keywords: Dimethyl sulfoxide; Cr(VI); 1,10-Phenanthroline; Sodium dodecyl sulphate; TX-100

Introduction

Surfactants are a surface active agent that shows characteristic behaviors at surfaces and interfaces [1-3]. The driving force for their adsorption at an interface is to lower the interfacial tension (i.e. interfacial free energy per unit area). The fundamental property of surfactant of reducing the interfacial tension is responsible for its use as emulsifier, detergent, foaming agent, wetting agent and so on [4-15]. Surfactant molecules consist of at least two parts, one of them is soluble to a specific fluid (lyophilic part) and the other is insoluble (lyophobic part). The terms are hydrophilic part and hydrophobic part respectively when water is used as the solvent. The concentration at which micelle formation starts is known as critical micelle concentration (CMC) which is an important feature of a surface active agent. Molecular structures of the surfactant, temperature, polarity of the medium are some of the factors on which CMC depends. In polar solvents, above CMC, the surfactant molecules spread their hydrophilic part in the polar medium and their hydrophobic ends being repelled, tend to associate generating similar environment. This type of micelle is generally called normal micelle. The fact is opposite in nonpolar solvent where the hydrophobic tail group is remaining out in the nonpolar medium and hydrophilic head group remains far from the non-polar medium. This type of micelle formation is known as reverse micellization. According to the charges on the hydrophilic head group surface active agents are divided in four categories (i) cationic (ii) anionic (iii) nonionic (iv) zwitterionic. The oxidation of various organic compounds by Cr(VI) was found to be more pronounced in the presence of micellar catalyst and promoters. Micellar catalysis plays an important role in many organic reactions [16,17]. Oxidation is a fundamental reaction in organic synthesis and will possibly play an important role in the development of value-added chemicals from biomass. We used dimethyl sulphoxide (DMSO) as an organic substrate which is oxidized to dimethyl sulfone by the powerful oxidant Cr(VI) consequent to the reduction of non-toxic Cr(III). Chromium has both beneficial and detrimental properties. The two oxidation states of chromium are naturally occurring; one is trivalent chromium and another is hexavalent chromium. The behavior of chromium species depends strongly on their oxidation states: Cr(III) is essential for human nutrients, whereas most hexavalent compounds are toxic, several of which even cause lung cancer. From the standpoint of carcinogenic effect of hexavalent chromium Cr(VI), the kinetic aspects of interaction of Cr(VI) with the different biologically relevant reducing agents have become very significant in recent days to both biochemists and inorganic chemists. Hexavalent chromium used as a strong oxidant in various organic synthesis [18]. But the rate of oxidation of dimethyl sulfoxide with Cr(VI) is very slow. Surfactant micelles and promoters are used to speed up the oxidation of DMSO by chromic acid. DMSO contains two methyl group along with S=O group. Previously, we investigated the mechanistic aspects of Cr(VI) oxidation of dimethyl sulfoxide (DMSO) in aqueous acidic media in the presence and absence of picolinic acid promoter [19,20]. In fact, Cr(VI) can oxidize DMSO slowly without any catalyst only under drastic conditions. The present paper describes the mechanistic aspects of oxidation of DMSO by Cr(VI) in the presence of phenanthroline in different types of micellar media. The major role of micellar media is to organize different chemical reactions as multifunctional nano-reactors. To achieve the target, several kinetic measurements were performed during the oxidation of DMSO. Micellar medium as a nano-reactor in combination with different promoting agents is a very efficient for rapid oxidation of DMSO. The N-hetero-aromatic nitrogen bases are generally used as chelating agents and found to act as an effective oxidation promoters. Different compounds such as picolinic acid, bipyridine, 1,10-phenanthroline, EDTA are used as promoters for Cr(VI) oxidation of organic substrates [1,2,16]. Our present work will be found out the preferable micellar medium that

***Corresponding author:** Bidyut Saha, Homogeneous Catalysis Laboratory, Department of Chemistry, The University of Burdwan, Golapbag, Burdwan, Pin 713104, WB, India, E-mail: b_saha31@rediffmail.com

catalyses the oxidation of dimethyl sulfoxide (DMSO) by Cr(VI) with 1,10-phenanthroline (phen) as promoter. In this study we have selected three different type of surfactants: cationic (cetyl pyridinium choloride, CPC), anionic (sodium dodecylsulfate, SDS), nonionic (octylphenol-polyoxyethylene ether, TX-100). The present paper describes the detailed micellar effect on the title reaction to provide a better insight to the reaction mechanism.

Experimental

Materials and reagents

All chemicals were commercially available and used directly as received from the manufacturer. DMSO (SRL, AR), $K_2Cr_2O_7$ (BDH, AR), H_2SO_4 (BDH, AR), 1,10-phenanthroline (AR, Merck), sodium dodecyl sulfate (AR, SRL, India), N-cetylpyridinium chloride (AR, SRL, India), Triton X-100 (AR, SRL) and all other necessary chemicals used in highest purity available commercially. Solutions were prepared by using doubly-distilled water.

Procedure and kinetic measurements

Solutions of the oxidant and reaction mixtures containing known quantities of the substrate (i.e. DMSO), and promoter (phen) are prepared under the pseudo-first-kinetic conditions $(DMSO)>>(Cr(VI))_T$ and $(phen)>>(Cr(VI))_T$. Acid and other necessary chemicals were separately thermostated (± 0.1°C). The reaction was started by mixing requisite amount of the oxidant with the reaction mixture. Progress of the reaction was monitored by following the decay of Cr(VI) at 450 nm wave length at different time intervals with the UV-VIS (UV-VIS-NIR-3600 (SHIMADZU)) spectrophotometer. Quartz cuvettes of path length 1 cm were used. The pseudo-first-order rate constants (k_{obs}, s^{-1}) were calculated from the slope of plots of $\ln(A_{450})$ versus time (t) which were linear. No interference was observed due to other species at 450 nm. The scanned spectra, spectrum after completion of the reaction and other spectra were recorded with a UV-VIS spectrophotometer. Under the experimental conditions, the possibility of decomposition of the surfactants by Cr(VI) was investigated and the rate of decomposition in this path was kinetically negligible.

Product analysis and stoichiometry

DMSO is quantitatively oxidized to sulfone under the kinetic condition $(DMS)>>(Cr(VI))_T$ and Cr(VI) is finally reduced to a Cr(III)-phrn complex. Completion of the reaction was indicated by the disappearance of Cr(VI) color. A reaction mixture (under the kinetic conditions) containing the substrate and chromic acid was allowed to stand overnight, neutralised with sodium carbonate, and extracted with ether. The solvent was removed after drying (anhydrous Na_2SO_4) and a solid was obtained. Its m.p. (109°C) corresponds to the mp of dimethyl sulfone (Me_2SO_2) (literature reported M.P.=107-109°C) [19]. The (Cr(VI)):(DMSO) ratio was also estimated. The unreacted DMSO was estimated by Ce(IV). The stoichiometry conforms to

$$2HCrO_4^- + 3Me_2S{=}O + 8H^+ \xrightarrow{\text{Pheanthroline}} 2Cr(III) + 3Me_2SO_2 + 5H_2O$$

Results and Discussions

Dependence on $(Cr(VI))_T$

Under the experimental conditions $(DMSO)>>(phen)_T>>(Cr(VI))_T$, both in the presence and the absence of phen, the rate of disappearance of Cr(VI) shows a first-order dependence on $(Cr(VI))_T$. This first-order dependence on Cr(VI) is also maintained in the presence of surfactants [20].

Dependence on $(phen)_T$

The pseudo-first order rate constants (k_{obs}, s^{-1}) were determined from the slope of plots of $\ln(A_{450})$ against time (t) at wavelength 450 nm. The slope (Figure 1) obtained from the plots $\ln(A_{450})$ versus time (t) divided by the time(t=60 sec) using the equation: k_{obs}=slope/time (t) was equal to the pseudo-first order rate constants(k_{obs}, s^{-1}). Time t=60 sec or 1 min in the denominator of the equation has been divided as the total reaction was monitored for several minutes (Figure 1). When k_{obs} versus $(phen)_T$ are plotted, it gives positive intercepts that measures the rate of the relatively slower uncatalyzed path. The pseudo-first order rate constant obtained in the absence of phen under the same conditions is nearly equal to the results measured from the intercept of the plot of k_{obs} versus $(phen)_T$ (Figure 2). In the presence and absence of surfactants the observation is maintained. During the reaction, under the condition $(phen)>>(Cr(VI))_T$, (phen) is reduced due to formation of inert Cr(III)-(phen) complex (Figure 2).

Spectroscopic analysis of the reaction

UV-VIS Spectra: The *in vitro* reduction of Cr(VI) to Cr(III) has been established by the help of UV VIS spectroscopy. The final color of the solution changes to pale blue (λ_{max}=580 nm, and λ_{max}=415 nm Figure 3) in unpromoted path while it turns into a pale violet (λ_{max}=553 nm) solution in phenanthroline mediated promotion. Two peaks are observed in unpromoted pathway due to the transition [21-23] from λ_{max}=580 nm for $^4A_{2g}(F) \rightarrow {}^4T_{2g}(F)$ and λ_{max}=415 nm for $^4A_{2g}(F) \rightarrow {}^4T_{1g}(F)$ of Cr(III) respectively. Interestingly, the spectra of the end solution in unpromoted path seems to be identical with that obtained from the aqueous solution of pure chromic sulphate but a different spectrum obtained in promoted pathway. It seems to contain Cr(III) promoter complex at the end solution in promoted pathway which has been established early [24,25]. A hypsochromic shift is associated for the transition $^4A_{2g}(F) \rightarrow {}^4T_{2g}(F)$ in promoted pathway compared to the final solutions of the unpromoted reaction pathway (Figure 3). The blue shift for the transition $^4A_{2g}(F) \rightarrow {}^4T_{2g}(F)$ indicates that the charge transfer band in the Cr(III) phen complex is due to the presence of strong electron donating atom (three nitrogen donor) in phen. A charge transfer band is associated with the band arising for the transition from $^4A_{2g}(F) \rightarrow {}^4T_{1g}(F)$ which is justified from the fact that a suitable charge transfer transition from the metal to ligand (MLCT) is possible and thus we observed the charge transfer band at lower energy visible range. The vacant π^* (antibonding) M.O. on ligand (phen) switches the electron-transfer from metal to ligand. Thus from the above results it reveals that, a charge transfer band at lower energy should be arising for the formation of a Cr(III) promoter complex at the end solution (Figure 3). A further study was done to investigate whether any reaction intermediate was formed at any time interval. The reaction solution with presence and absence of promoter and surfactant was scanned at a regular time interval in between 250-800 nm to follow the reaction progress. The observed spectrum show that there is a gradual decrease in concentration of Cr(VI) and subsequently appearance of Cr(III) with increase time in unpromoted reaction condition with no distinct isobestic point (Figure 4). From this spectrum (Figure 4), it can only be observed with course of time the rate of absorbance decreases indicating the concentration of Cr(VI) decreases with time. An isobestic point was observed at λ_{max}=527 nm for phen promoted reaction pathway (Figure 5). Isobestic point was shifted when a mixture of three micellar catalysts (CPC, SDS, TX-100) and promoter (phen) are introduced in the reaction solution (Figures 6-8). In this experimental condition the reaction solution was scanned and the observed spectrum gives a single isobestic point. Thus it follows

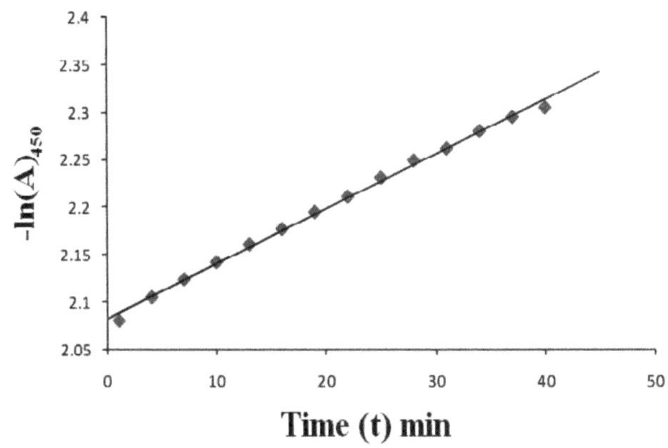

Figure 1: Representative first-order plot to evaluate the pseudo-first-order rate constant (k_{obs}) for phenanthroline-promoted Cr(VI) oxidation of DMSO in aqueous solution at 30°C. $(DMSO)_T$=75 × 10^{-4} mol dm^{-1} ,$(Cr(VI))_T$=5 × 10^{-4} mol dm^{-1}, $(H_2SO_4)_T$=0.5 mol dm^{-1}, $(phen)_T$=50 × 10^{-4} mol dm^{-1}.

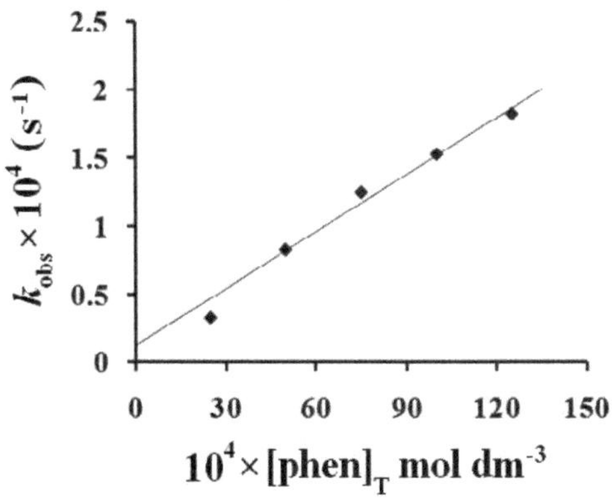

Figure 2: Dependence of k_{obs} on $(phen)_T$ for Cr(VI) oxidation of DMSO in aqueous H_2SO_4 at 30°C. $(DMSO)_T$=75 × 10^{-4} mol dm^{-3}.

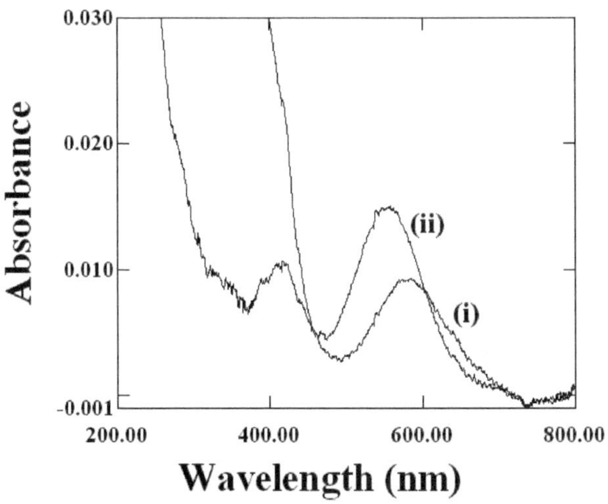

Figure 3: Absorption spectra of **(i)** Unpromoted reaction mixture (after completion of reaction): $(DMSO)_T$=75 × 10^{-4} mol dm^{-3}, $(Cr(VI))_T$=5 × 10^{-4} mol dm^{-3}, $(H_2SO_4)_T$=0.5 mol dm^{-3} (The spectrum of the chromic sulfate is identical with this under the experimental condition). **(ii)** Promoted reaction mixture (after completion of reaction): $(DMSO)_T$=75 × 10^{-4} mol dm^{-3}, $(Cr(VI))_T$=5 × 10^{-4} mol dm^{-3}, $(H_2SO_4)_T$=0.5 mol dm^{-3}, (phen)=25 × 10^{-4} mol dm^{-3}.

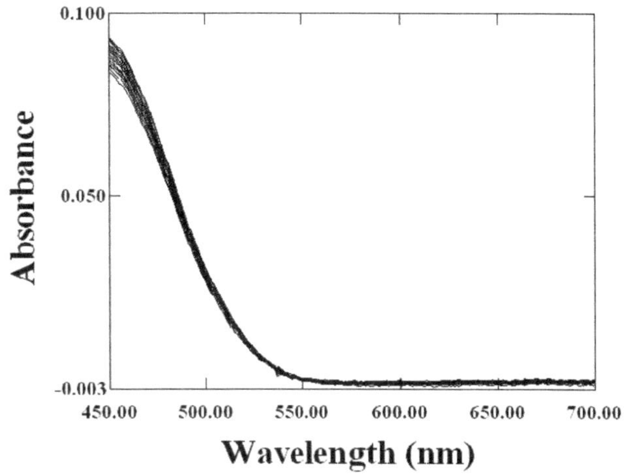

Figure 4: Scanned absorption spectra of the reaction mixture at regular time intervals (15 min). $(DMSO)_T$=75 × 10^{-4} mol dm^{-3}, $(Cr(VI))_T$=5 × 10^{-4} mol dm^{-3}, $(H_2SO_4)_T$=0.5 mol dm^{-3}.

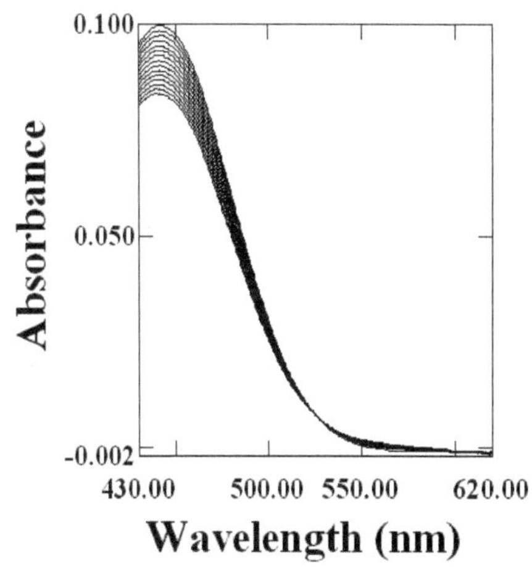

Figure 5: Scanned absorption spectra of the reaction mixture at regular time intervals (3 min). $(DMSO)_T$=75 × 10^{-4} mol dm^{-3}, $(Cr(VI))_T$=5 × 10^{-4} mol dm^{-3}, $(H_2SO_4)_T$=0.5 mol dm^{-3}. $(phen)_T$=25 × 10^{-4} mol dm^{-3} Temp=30°C.

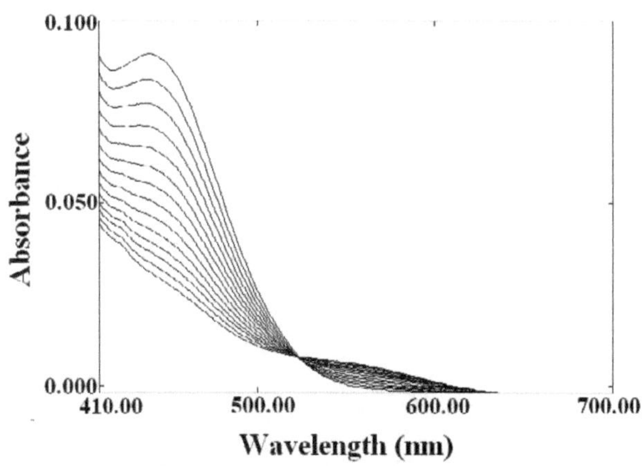

Figure 6: Scanned absorption spectra of the reaction mixture at regular time intervals (3 min). $(DMSO)_T$=75 × 10^{-4} mol dm^{-3}, $(Cr(VI))_T$=5 × 10^{-4} mol dm^{-3}, $(H_2SO_4)_T$=0.5 mol dm^{-3}, $(phen)_T$=25 × 10^{-4} mol dm^{-3} $(SDS)_T$=2 × 10^{-2} mol dm^{-3}. $\lambda_{isobestic}$=525 nm. Temp=30°C.

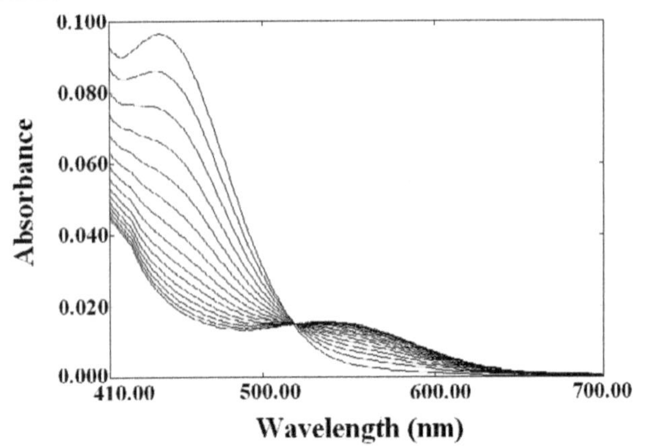

Figure 7: Scanned absorption spectra of the reaction mixture at regular time intervals (3 min). $(DMSO)_T$=75 × 10^{-4} mol dm^{-3}, $(Cr(VI))_T$=5 × 10^{-4} mol dm^{-3}, $(H_2SO_4)_T$=0.5 mol dm^{-3}, $(TX\text{-}100)_T$=2 × 10^{-2} mol dm^{-3}, $(Phen)_T$=25 × 10^{-4} mol dm^{-3} $\lambda_{isobestic}$=517 nm. Temp=30°C.

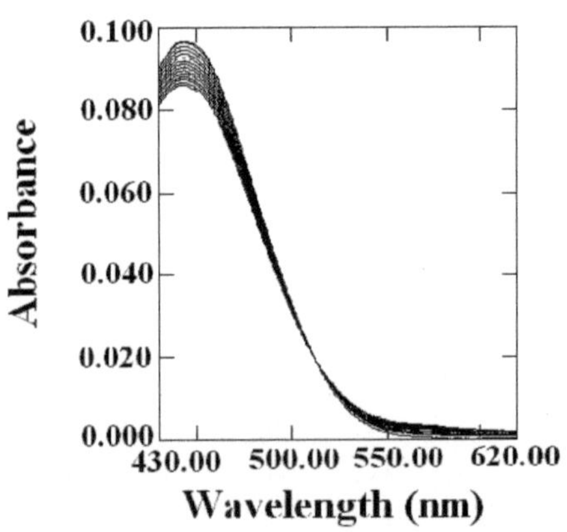

Figure 8: Scanned absorption spectra of the reaction mixture at regular time intervals (3 min). $(DMSO)_T$=75 × 10^{-4} mol dm^{-3}, $(Cr(VI))_T$=5 × 10^{-4} mol dm^{-3}, $(H_2SO_4)_T$=0.5 mol dm^{-3}, $(CPC)_T$=2×10^{-2} mol dm^{-3}, $(Phen)_T$=25 × 10^{-4} mol dm^{-3}, $\lambda_{isobestic}$=512 nm. Temp=30°C.

that there might be short lifetime intermediates Cr(IV) and Cr(V) [26] which can formed with the reduction from Cr(VI) to Cr(III) at very low concentrations and Cr(VI)-phen is the active oxidant in promoted oxidation pathway [23,25,27-31].

Mechanism of the reaction and rate law

From the spectroscopic analysis it clearly indicates that sulfone is the main oxidized product in the reaction. The mechanism of phen promoted chromic acid oxidation of DMSO has been shown in the Scheme 1 given below. The active oxidant Cr(VI)-phen complexes react (supported by Figure 9) with the substrate DMSO to form a ternary complex which undergoes redox decomposition in a rate-limiting step [25,32-34] giving rise to the product sulfone.

Effect of non-functional micellar catalyst on the reaction rate

Partitioning of reactant in promoted path: From the observed data, it is cleared that the rate of oxidation in phen promoted path is highest when it is catalyzed by SDS. The enhancement of rate arises from the preferential partitioning of the positively-charged Cr(VI)-phen (active oxidant, AO^+) complex and neutral substrate (Scheme 1) by electrostatic attraction in the micellar surface [35]. Acceleration of rate is also arises when combination of SDS and the hetero-aromatic nitrogen base (phen) is used (Figure 10a). The Cr(VI)-promoter complex for phen represented as AO^+ (Scheme 1), a cationic complex, has been argued as the active oxidant in the promoted path. The most pronounced acceleration for the DMSO oxidation process is observed in the SDS catalyzed phen promoted path (Tables 1 and 2). The maximum rate enhancement arises from preferential partitioning of the positively-charged Cr(VI)-phen and neutral substrate (Scheme 1, Figure 10b) in the micellar surface. Because of the electrostatic attraction it can preferably distributed in the micellar pseudo-phase (Figure 10a) of the anionic surfactant SDS. Thus in presence of SDS micelle, the oxidation process can propagate in both the micellar pseudo-phase (where both the active oxidant and substrate are preferably concentrated) and aqueous phase to give the observed rate acceleration. A similar phenomenon of smaller rate acceleration has been noticed for combination of TX-100 micelles (Figure 10b) and phenanthroline used. Table 1 show that when phen is used as promoter in the micellar catalyzed path the rate of reaction is greater in SDS compared to TX-100 medium. This phenomenon is due to the larger size of the phen ring. The cationic micelle CPC retards the rate of reaction in presence of promoter also. The positive head groups of CPC micelle repel the positively charged active oxidant in the surface (Figure 10c). So the effective concentration of the reactant in the micellar region decreases and resulting retardation of the rate.

The substrate can make effective collision with the AO^+ and enhancement of rate is produced in SDS. Such enhanced local concentration of reactants increases the observed rate of reaction to a remarkable extent. So the highest acceleration of rate in case of SDS catalyzed phen promoted path is explained by the easy entering of Cr(VI)-phen complex into the relatively hydrophobic core of SDS micelle compared to that of TX-100 micelle. SDS allows maximum number of Cr(VI)-phen compared to TX-100 micelles due to electrostatic and steric factor only. In this route the oxidation rate was almost 70 times enhanced compared to the ordinary uncatalyzed and unpromoted reaction path. So $t_{1/2}$ (half-life) for the reaction is minimum for SDS-phen combined reaction path (Tables 1 and 2). Considering all the above fact it can be concluded that SDS is the most suitable micellar catalyst for this conversion. The overall analysis of our experimental results compared to the literature works are presented in Table 3. From Table 1 it is shown that the rate is accelerated with increasing the concentration of hetero-aromatic nitrogen base promoter

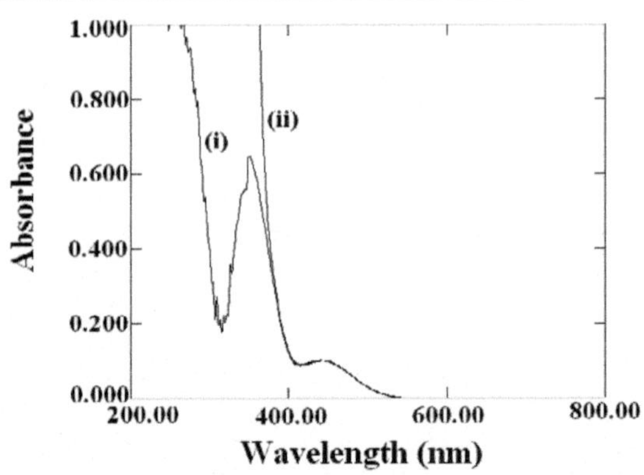

Figure 9: Absorption spectrum of reaction mixture with and without promoter (in absence of substrate): **(i)** $(Cr(VI))_T$=5 × 10^{-4} mol dm^{-3}, $(H_2SO_4)_T$=0.5 mol dm^{-3}. **(ii)** $(Cr(VI))_T$=5×10^{-4} mol dm^{-3}, $(H_2SO_4)_T$=0.5 mol dm^{-3}, $(phen)_T$=25 × 10^{-4} mol dm^{-3} Temp=30°C.

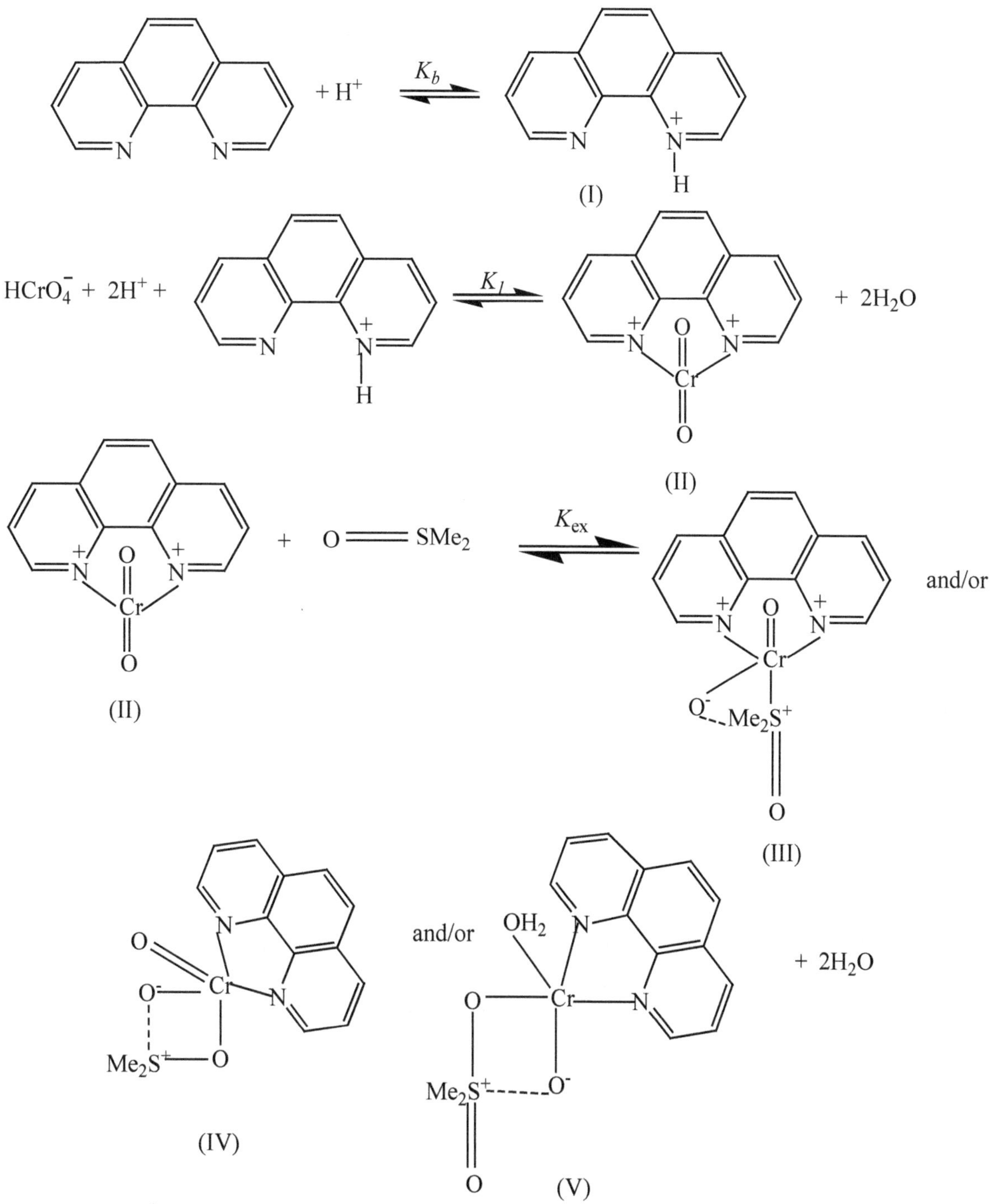

$$(III), (IV), (V) \xrightarrow{K} Me_2SO_2 + Cr^{IV}\text{-phen complex}$$

$$Cr^{IV}\text{-phen complex} + Me_2SO \xrightarrow{fast} Cr^{III}\text{-phen complex} + Me_2SO^{\cdot}$$

$$Cr^{VI}\text{-phen complex} + Me_2SO^{\cdot} \xrightarrow{fast} Cr^{V}\text{-phen complex} + Me_2SO_2$$

$$Cr^{V}\text{-phen complex} + Me_2SO \xrightarrow{fast} Cr^{III}\text{-phen complex} + Me_2SO_2$$

Scheme 1: The mechanism of phen promoted chromic acid oxidation of DMSO.

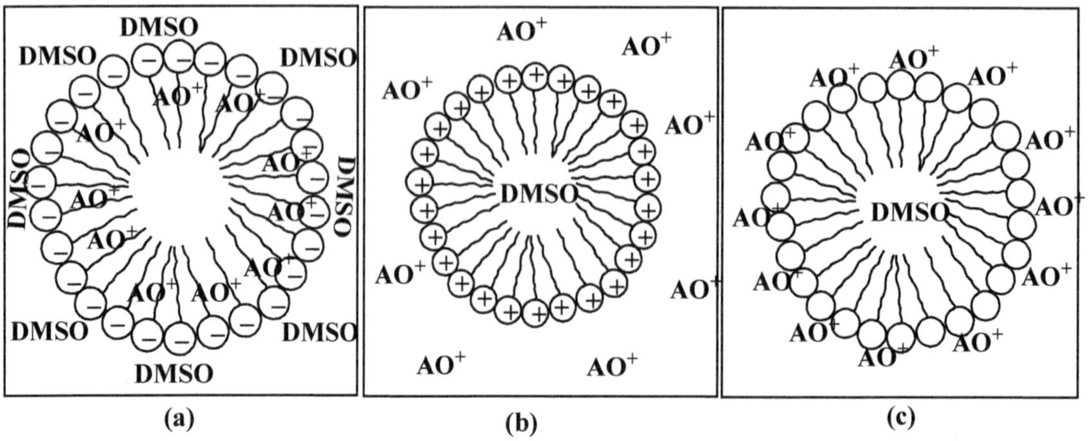

Figure 10: Schematic representation of partitioning of substrate and active oxidant (AO^+=Cr(VI)-promoter complex) in **(a)** Anionic surfactant, **(b)** Neutral surfactant, and **(c)** Cationic surfactant.

Promoter (mol dm^{-3})		Micellar catalyst (mol dm^{-3})		$10^4 \times k_{obs}$ (sec^{-1})	$t_{1/2}$ (hrs)
None		None		0.125 ± 0.01	15.4
1,10 - phenanthroline (phen)	0.0025	None		0.33 ± 0.03	5.83
	0.005			0.833 ± 0.02	2.31
	0.0075			1.33 ± 0.01	1.44
	0.01			1.5 ± 0.04	1.28
	0.0125			1.8 ± 0.02	1.07
1,10 - phenanthroline (phen)	0.0025	SDS	0.02	2 ± 0.01	0.962
	0.005			4 ± 0.03	0.481
	0.0075			5 ± 0.02	0.385
	0.01			7 ± 0.01	0.275
	0.0125			9 ± 0.02	0.214
1,10 - phenanthroline (phen)	0.0025	TX-100	0.02	3.33 ± 0.02	0.57
	0.005			2.83 ± 0.04	0.68
	0.0075			2.5 ± 0.03	0.77
	0.01			2.3 ± 0.04	0.83
	0.0125			1.8 ± 0.01	1.06
1,10 - phenanthroline (phen)	0.0025	CPC	0.02	0.5 ± 0.01	3.85
	0.005			0.67 ± 0.02	2.87
	0.0075			1.0 ± 0.02	1.92
	0.01			0.86 ± 0.03	2.32

$(Cr(VI))_T = 5 \times 10^{-4}$ mol dm^{-3}, $(H_2SO_4)_T = 0.5$ mol dm^{-3}, $(DMSO)_T = 75 \times 10^{-4}$ mol dm^{-3}, Temp=30°C

Table 1: k_{obs} and half life of the reaction in presence and absence of promoter and non-functional micellar catalyst.

Promoter	Micellar catalyst	$k_{eff(promoter)(water)}$	$k_{eff(promoter)(surfactant)}$
1,10-phenanthroline	None	11	-
1,10-phenanthroline	SDS	-	39
1,10-phenanthroline	TX-100	-	19

$(Cr(VI))_T = 5 \times 10^{-4}$ mol dm^{-3}, $(H_2SO_4)_T = 0.5$ mol dm^{-3}, $(DMSO)_T = 75 \times 10^{-4}$ mol dm^{-3}, $(phenantholine)_T = 100 \times 10^{-4}$ mol dm^{-3}, $(SDS)_T = 2 \times 10^{-4}$ mol dm^{-3}, Temp=30°C

Table 2: Kinetic parameters and some representative k_{eff} values for the Cr(VI) oxidation of DMSO in presence of promoter and micellar catalyst (surfactant) in aqueous acidic media.

Entry	Catalyst	Conditions	Reference	10^4 k_{obs} (s^{-1}) (rate constant)	Half life ($t_{1/2}$) (h)
1	$(NaBrO_3)$-$(NaHSO_3)$	$(NaBrO_3)=1 \times 10^{-3}$ mol dm^{-3} $(NaHSO_3)=2 \times 10^{-3}$ mol dm^{-3} Temp=30°C	[30]	1.96	0.982
2	DCSB in 1:1 water acid medium	$(HClO_4)=0.1$ mol dm^{-3} I=0.3 mol dm^{-3} (DCSB)=1 × 10^{-3} mol dm^{-3} Temp=30°C	[31]	1.9	1.01
3	Pd(II)	$(IO_4^-)=1 \times 10^{-4}$ mol dm^{-3}$(OH^-)=0.5$ mol dm^{-3} Pd(II)=1 × 10^{-4} mol dm^{-3} Temp=25°C	[32]	3.8	0.506
4	SDS (catalyst)+phen (promoter)	(phen)=0.0125 mol dm^{-3} (SDS)=0.02 mol dm^{-3}	Curent work	9	0.214

Table 3: A comparative study of different catalysts used in DMSO to dimethyl sulfone conversion.

phenanthroline. As a result of which the half life ($t_{1/2}$) value of the reaction decreases uniformly. SDS enhances the rate of DMSO oxidation (Figure 11) to a great extent compared to the micellar catalyst TX-100. On the other hand the rate is retarded in comparison to the uncatalyzed and unpromoted reaction when the micellar surface is cationic (CPC) (data not shown in Table 1). It was obtained that the combination of phenanthroline promoter and the CPC surfactant decreased the rate constant for the oxidation of DMSO by Cr(VI). In presence of anionic surfactant (SDS) the reaction undergoes completion just 38.52 minute under the kinetic experimental condition. Thus from the data in Table 1 it is shown that reaction is completed in least time and least half time (12.84 min) in presence of SDS at its highest concentration (125×10^{-4} mol dm^{-3}) used. In case of neutral surfactant TX-100, with increasing phenanthroline concentration, the rate of entering large active oxidant (Cr(VI)-phen complex, AO^+) in the micellar core decreases due to the steric effect resulting rate increment to a smaller extent (Figure 12). Here steric factor predominates over electrostatic interaction. When k_{obs} vs $(\text{phenanthroline})_T$ in presence of SDS is plotted the linear plot with positive intercept is obtained (Figure 11). The maximum rate enhancement observed in SDS catalyzed phenanthroline pathway with a rate constant of 9×10^{-4} s^{-1}. The pseudo first order rate constants (k_{obs}, sec^{-1}) are determined from the slope of the plots of $-\ln(A_{450})$ *versus* time (A=absorbance of Cr(VI)) at wavelength 450 nm. The half-life values ($t_{1/2}$) are calculated from the data by the relation $t_{1/2}=\ln 2/k_{obs}$, where k_{obs} is pseudo first order rate constant.

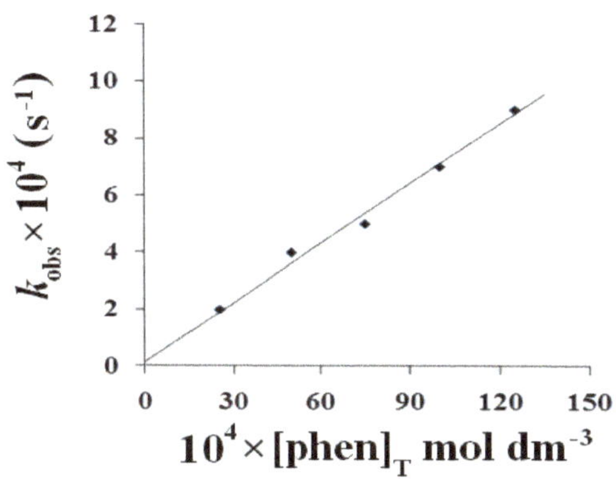

Figure 11: Dependence of k_{obs} on $(\text{phen})_T$ for Cr(VI) oxidation of DMSO in presence of SDS in aqueous H_2SO_4 at 30°C. $(\text{DMSO})_T$=0.375 mol dm^{-3}, $(\text{SDS})_T$=2 × 10^{-4} mol dm^{-3}.

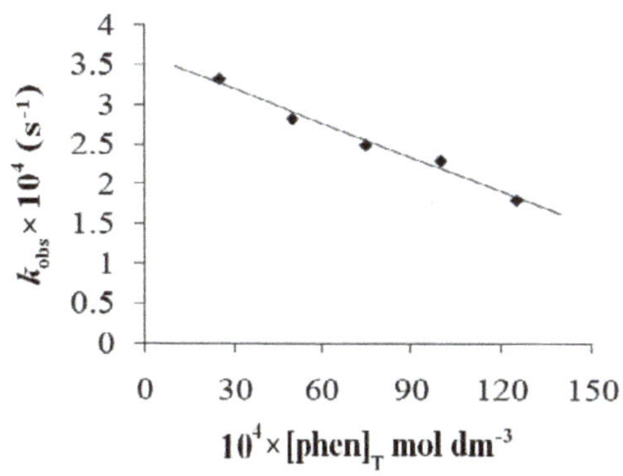

Figure 12: Dependence of k_{obs} on $(\text{phen})_T$ for Cr(VI) oxidation of DMSO in presence of TX-100 in aqueous H_2SO_4 at 30°C. $(\text{DMSO})_T$=0.375 mol dm^{-3}, $(\text{TX-100})_T$=2 × 10^{-4} mol dm^{-3}.

Optical images of micelles

The optical micrographs are taken for mixtures three different surfactants CPC, SDS and TX-100 in 100× magnification in an optical microscope (LEICA DM 1000). The images (Figure 13) represent the evidence of micelle formation in aqueous medium. The formation of micelles in aqueous medium was clearly identified by the optical images found. They are observed in 10 µm scale in an optical microscope. Some of the micelles are attached to the inner walls of the empty vesicles as evident from the optical images.

High resolution transmission electron microscope (HRTEM) images of CPC micelle

The HR-TEM investigation was done at 20 kV acceleration voltages using lacey carbon coated Cu grid of 300 mess size in HR-TEM microscope (JEOL JEM 2100). Samples were prepared by placing sample mixture drops directly on the copper grids using a micropipette. The surfactants present in the aqueous medium were allowed to settle. TEM pictures (Figure 14) are presented which illustrate the multiple morphologies of the aggregates made from the different surfactants CPC, SDS and TX-100.

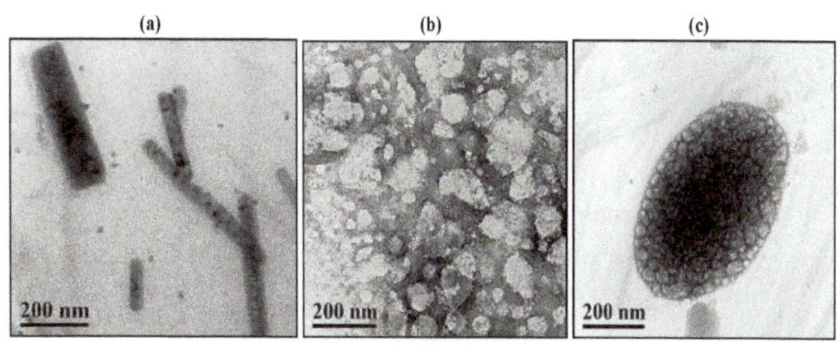

Figure 14: (a) to (c) Different views of TEM images showing micelles formed by CPC, SDS and TX-100 in aqueous medium, (a) (CPC)=1.5 × 10^{-3} mol dm^{-3}, $(\text{SDS})_T$=10 × 10^{-3} mol dm^{-3}, (TX-100)=0.30 × 10^{-3} mol dm^{-3}, Temp=30°C.

Conclusion

The active oxidant Cr(VI)-phen complex is responsible for oxidation of DMSO to sulfone in phen promoted aqueous micelle media. Though the cationic surfactant (CPC) retards the rate of oxidation, the rate is increased in presence of negative surfactant (SDS). Thus it is of growing interest to the chemist as not only an organic transformation is done in aqueous media but also the rate of transformation increases to a greater extent in aqueous micellar media (when the micelle is SDS). So the use of carcinogenic organic solvent may be reduced. Besides DMSO reduces Cr(VI) to Cr(III). Consequently, possibility of water pollution by carcinogenic Cr(VI) from industrial effluent can be reduced. In this study phen is used as promoter and CPC, SDS, TX-100 are used as micelle media. The oxidation product sulfone has been confirmed by

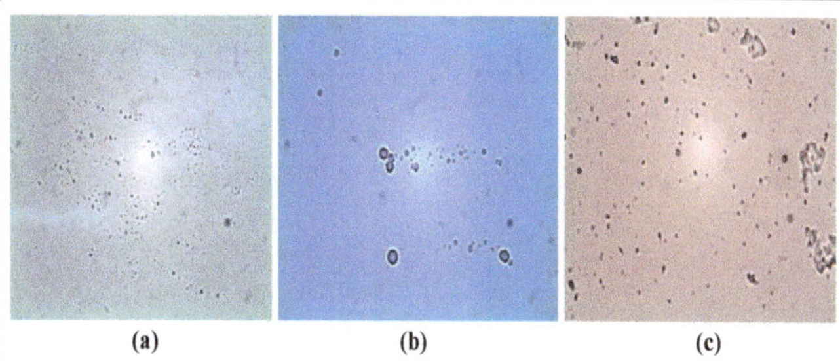

Figure 13: Different views of optical micrograph images of surfactants in water: **(a)** CPC in water **(b)** SDS in water and (c) TX-100 in water.

its boiling point. Combination of SDS and phen accelerates the rate of DMSO oxidation by Cr(VI) up to 70-fold compared to the same reaction in the absence of surfactant.

Acknowledgment

The authors are thankful to CSIR, New Delhi, India and UGC, New Delhi for providing financial help in the form of a project and fellowship.

References

1. Mandal J, Chowdhury KM, Paul KK, Saha B (2010) Kinetics and mechanism of 2, 2'-bipyridine-catalyzed chromium(VI) oxidation of propan-2-ol in the presence and absence of surfactants. J Coord Chem 63: 99-105.
2. Chowdhury KM, Mandal J, Saha B (2009) Micellar catalysis of chromium(VI) oxidation of ethane-1,2-diol in the presence and absence of 2,2'-bipyridine in aqueous acid media. J Coord Chem 62: 1871-1878.
3. Basu A, Ghosh SK, Saha R, Nandi R, Ghosh T, et al. (2011) Effect of some non functional surfactants and electrolytes on the hexavalent chromium reduction by glycerol: A mechanistic study. Tenside Surf Det 48: 453-458.
4. Wadekar SD, Patil SV, Kale SB, Lali AM, Bhowmick DN, et al. (2011) Study of Glycerol Residue as a Carbon Source for Production of Rhamnolipids by Pseudomonas aeruginosa. Tenside Surf Det 48: 16-22.
5. Nowicki J, Sokolowski A, Reska D (2011) Surface Activity and Adsorption Properties of New Perfluorinated Carbohydrate Surfactants. Tenside Surf Det 48 : 11-15.
6. Fanun M, Glatter O (2011) SAXS Study on Azithromycin Loaded Nonionic Microemulsions. Tenside Surf Det 48 : 34-39.
7. Ghosh S, Bhowmick DN, Pratap AP (2011) Application of Neem and Karanjia Oils as Natural Pesticide Microemulsion Systems. Tenside Surf Det 47: 369-375.
8. Wadekar SD, Patil SV, Kale SB, Lali AM, Bhowmick DN, et al. (2010) Structural Elucidation and Surfactant Properties of Rhamnolipids Synthesized by Pseudomonas aeruginosa. Tenside Surf Det 48 : 286-292.
9. Datla VM, Beck K, Shim E, Pourdeyhimi B (2010) Reverse Phase HPLC Analysis of Commercial Surfactants used as Melt Additives. Tenside Surf Det 47: 142-150.
10. Kishore K, Upadhyaya SK (2010) Studies on Acoustic and Thermodynamic Behaviour of Terbium Laurate and Myristate in Mixed Organic Solvents. Tenside Surf Det 47: 184-189.
11. Bajpai D, Tyagi VK (2010) Nonionic Surfactants: An Overview. Tenside Surf Det 47: 190-196.
12. Deb N, Bagchi S, Mukherjee AK (2010) Charge transfer complex formation between TX-100/CCl4 reverse micelle and a series of π-electron acceptors: determination of cmc and aggregation number. Mol Phys 108: 1505-1511.
13. Nevidimov AV, Razumov VF (2009) Molecular dynamics simulations of AOT reverse micelles' self-assembly. Mol Phys 107: 2169-2180.
14. Brodskaya EN, Mudzhikova GV (2006) Molecular dynamics simulation of AOT reverse micelles. Mol Phys 104: 3635-3643.
15. Dubey N (2011) Thermodynamic and FT-IR Study of Micellization of Sodium Dodecylbenzene Sulfonate in Some Simple Alcohols. Chem Eng Commun 198: 1394-1404.
16. Saha R, Ghosh A, Saha B (2011) Micellar catalysis on 1, 10-phenanthroline promoted hexavalent chromium oxidation of ethanol. J Coord Chem 64: 3729-3739.
17. Sar P, Ghosh A, Malik S, Saha B (2015) Sodium dodecylsulphate-catalyzed hetero-aromatic nitrogen base-promoted chromium(VI) oxidation of 2-propenol to 2-propenal in aqueous media. Res Chem Intermed.
18. Sundaram S, Raghavan PS (2011) Chromium-VI Reagents: Synthetic Applications. Springer.
19. Saha B, Islam M, Das AK (2005) Micellar effect on the catalytic co-oxidation of dimethyl sulfoxide and oxalic acid by chromium(VI) in aqueous acid media: A kinetic study. Prog React Kinet Mech 30: 145-156.
20. Das AK, Mondal SK, Kar D, Das M (2001) Micellar Effect on the Reaction of Picolinic Acid Catalyzed Chromium(VI) Oxidation of Dimethyl Sulfoxide in Aqueous Acidic Media: A Kinetic Study. Int J Chem Kinet 33: 173-181.
21. Ghosh SK, Saha R, Ghosh A, Mukherjee K, Saha B (2012) Micellar Catalysis on 1,10-Phenanthroline Promoted Chromic Acid Oxidation of Glycerol in Aqueous Media. Tenside Surf Det 49: 370-375.
22. Meenakshisundaram SP, Gopalakrishnan M, Nagarajan S, Sarathi N (2007) Oxalic acid catalysed chromium (VI) oxidation of some 2-amino-4, 6-diarylpyrimidines. Catal Commun 8: 713-718.
23. Figgis BN (1966) Introduction to ligand fields, Wiley Eastern Limited, New Delhi, India. Pp. 222.
24. Ud-Din K, Hartani K, Khan Z (2002) One-step three-electron oxidation of tartaric and glyoxylic acids by chromium(VI) in the absence and presence of manganese(II). Transition Met Chem 27: 617-624.
25. Jorgensen CK (1964) Absorption Spectra and Chemical Bonding in Complexes, Pergamon Press Ltd: Oxford/London. Pp. 290.
26. Mandal J, Chowdhury KM, Paul KK, Saha B (2008) Kinetics and mechanism of 2,2'-bipyridyl promoted chromic acid oxidation of ethanol and propan-1-ol in aqueous micellar media. Open Catal J 1: 1-5.
27. Khan Z, Masan S, Raju, Ud-Din K (2003) A mechanistic study of the ethylenediaminetetraacetic acid-, 2,2′-bipyridyl-, and manganese(II)-assisted one-step two- and three-electron oxidation of lactic acid by chromium(VI) Transition Met Chem 28: 881-887.
28. Meenakshisundaram S, Markkandan R (2004) 1,10-Phenanthroline catalysis of the chromium(VI) oxidation of alcohols: a kinetic and mechanistic study Trans Met Chem 29: 136-143.
29. Viroopakshappa J, Jagannadham V (2002) Kinetics and mechanism of oxidation of dimethyl sulphoxide by sodium bromate-sodium bisulphite reagent in aqueous medium. Proc Indian Acad Sci (Chem Sci) 114: 149-154.
30. Gowda BT, Jayalakshmi KL, Jyothi K (2003) Kinetics and Mechanism of Oxidation of DimethylSulphoxide by Mono- and Di-Substituted N,N-Dichlorobenzenesulphonamides in Aqueous Acetic Acid. Z Naturforsch 58b: 787-794.
31. Koli BI, Nandibewoor ST (2009) Kinetics and mechanism of palladium(II) catalysed oxidation of dimethyl sulfoxide by alkaline periodate. Indian J Chem 48A: 958-963.
32. Ghosh A, Saha R, Mukherjee K, Ghosh SK, Saha B, et al. (2013) Suitable combination of promoter and micellar catalyst for kilo fold rate acceleration on benzaldehyde to benzoic acid conversion in aqueous media at room temperature: a kinetic approach. Spectrochim Acta A Mol Biomol Spectrosc 109: 55-67.
33. Mukherjee K, Saha R, Ghosh A, Ghosh SK, Saha B, et al. (2013) Combination of best promoter and catalyst for hypervalent chromium oxidation of L-sorbose to lactone of C_5 aldonic acid in aqueous media at room temperature. J Mol Liq 179: 1-6.
34. Mukherjee K, Saha R, Ghosh A, Ghosh SK, Saha B, et al. (2013) Efficient combination of promoter and catalyst for chromic acid oxidation of propan-2-ol to acetone in aqueous acid media at room temperature. Spectrochim Acta A Mol Biomol Spectrosc 101: 294-305.
35. Ghosh A, Saha R, Mukhejee K, Ghosh SK, Bhattacharyya SS, et al. (2013) Selection of suitable combination of nonfunctional micellar catalyst and hetero-aromatic nitrogen base as promoter for chromic acid oxidation of ethanol to acetaldehyde in aqueous medium at room temperature. Int J Chem Kinet 45: 175-186.

16

Cell Disruption and Pressurized Liquid Extraction of Carotenoids from Microalgae

Josef Taucher*, Sascha Baer, Philipp Schwerna, Dominik Hofmann, Magdalena Hümmer, Rainer Buchholz and Anna Becker

Institute of Bioprocess Engineering, Friedrich-Alexander-Universität Erlangen-Nürnberg, Erlangen, Germany

Abstract

Microalgae are very promising organisms for the production of high-value compounds such as carotenoids. Nevertheless, their commercial use is so far hampered by the lack of efficient processes and currently feasible for very few strains. One of the most relevant factors is the high effort for downstream processing, e.g., cell disruption and extraction. Thus, the presented studies were dedicated to investigate and optimize these two steps for carotenoid production with microalgae. Six different cell disruption techniques (high pressure homogenizer, ball mill, Ultra Turrax, repeated freeze and thaw, freeze-drying, ultra-sonication) were compared in lab scale for three species: *Haemotococcus pluvialis*, *Chromochloris zofingiensis* and *Chlorella sorokiniana*. The carotenoid recovery was determined via HPLC-UV/Vis after pressurized solvent extraction. Furthermore, factors influencing the applied extraction methods such as solvent, temperature, duration and number of cycles were optimized in order to reach highest recovery rates. While rough mechanical methods such as ball mill and high pressure homogenizer showed the highest effectivity for cell disruption of all three investigated strains, the influence of non-mechanical methods - i.e., repeated freeze and thaw cycles - on the efficiency of the extraction of astaxanthin, lutein and β-carotene increased reversely proportional to cell size and cell wall rigidity. For *H. pluvialis* repeated freezing and thawing resulted in a factor 240 times lower extraction yield compared to high pressure homogenizer, while both methods were comparable for *C. sorokiniana*. From six tested solvents, dichloromethane resulted in the highest carotene recovery yield, three times higher in comparison to n-hexane. Variation of the extraction temperature from room temperature to 120°C showed an optimum at 60°C. Nearly complete extraction was reached after one cycle of 10 minutes. The data presented here demonstrate the necessity of lab scale optimization of cell disruption and extraction process for future upscaling.

Keywords: *H. pluvialis*; *C. sorokiniana*; *C. zofingiensis*; Carotenoids; Cell disruption; Pressurized liquid extraction; HPLC

Abbreviations: C: Cell Disruption Yield; CDW: Cellular Dry Weight; DCM: Dichloromethane; DMSO: Dimethyl Sulfoxide; HPLC: High Performance Liquid Chromatography; MeOH: Methanol; MTBE: Methyl *tert*-Buthyl Ether; PFD: Photon Flux Density; PLE: Pressurized Liquid Extraction; PSM: Photobioreactor Screening Module; SAG: Culture Collection of Algae at Goettingen University in Germany

Introduction

Microalgae, including cyanobacteria, play an increasing role in science and industry, due to their wide range of commercial and potential novel products [1-3]. They can be cultivated using sunlight (energy source) and waste streams, such as recycled media from other processes (e.g., aquaculture, biogas digestate - nutrient source) and industrial flue gas (anorganic carbon source), that positively contribute to cost reduction and ecological balance [4-9]. One of the most important, and on the market best established, high value products from microalgae are carotenoids. Carotenoids are a group of structurally highly diverse terpenoid pigments (more than 750 have been isolated), which can be divided into carotenes and oxygenated derivatives of carotenes, so called xanthophylls [10-12]. They act as light-harvesting pigments, absorb light in a range of λ=400-550 nm and transfer the light to chlorophyll [12,13]. Furthermore, they stabilize the structure and the functionality of the photosynthetic apparatus [11,12], e.g., by quenching chlorophyll triplet states, dissipating excess energy, or scavenging reactive oxygen species [11,14]. In general, applications of carotenoids can be divided into three main groups: i) as natural dyes in food and feed industry, ii) as feed additives in aquaculture and poultry farming, iii) as well as in the pharmaceutical sector and in cosmetics, due to their antioxidative properties [10,11,15]. The most commonly used carotenoids are β-carotene, astaxanthin and lutein that can be produced by various microalgae such as *Dunaliella salina, Haematococcus pluvialis, Chromochloris zofingiensis, Chlorella sorokiniana*, etc. [1,11,15]. The overall market for carotenoids was estimated to US$ 1.2 billion in 2010 and was predicted to increase up to US$ 1.4 billion in 2018 [15]. Market prices vary from 300 - 3,000 US$·kg^{-1} and 2,500-10,000 US$·kg^{-1} in case of β-carotene and astaxanthin, respectively [16,17].

Although there are many described algae species, until now only few have found application in biotechnological processes (e.g., *H. pluvialis, D. salina, Arthrospira platensis, Chlorella species*) [18]. This is due to still relatively high production costs making them incapable of competing with the chemical industry while only delivering one class of value products [19]. To approach this issue, integrated processes allowing almost complete utilization of the produced biomass, waste streams and nutrients are needed. Derived from the oil-based refinery, in which the raw material is more or less utilized completely, such integrated biotechnical applications are called biorefinery processes. Particularly, an optimal design of the downstream processing is essential for feasibility of such systems as it may contribute up to 90% of the total costs [20].

***Corresponding author:** Josef Taucher, Institute of Bioprocess Engineering, Friedrich-Alexander-Universität Erlangen-Nürnberg, Erlangen, Germany
E-mail: josef.taucher@fau.de

In order to establish an efficient isolation procedure for potential products for the future industrial application, at first an optimization of the extraction in labscale has to be performed for various strains of microalgae. This is not only necessary during the screening phase to assure comparability of the results, but also in later stages of the process design as the collected data and experience can be used for example for upscaling. In case of intracellular products not only extraction parameters such as: selected solvent, temperature, duration, cycle number are crucial but also an efficient cell disruption. At the same time, when considering further upscaling of the process, other aspects such as costs and feasibility for industrial application have to be considered. Thus, the main aim of the work presented here was the optimization of the cell disruption and extraction procedure for carotenoids for future application as an important isolation step of one of the relevant product classes in a microalgae based biorefinery.

Materials and Methods

Strains and media

In this work, three microalgae strains, provided by the Culture Collection of Algae at Goettingen University in Germany (SAG), were used: *Haematocouccus pluvialis* SAG 34-1b, *Chromochloris zofingiensis* SAG 211-14 and *Chlorella sorokiniana* SAG 211-8k. Basal medium (with 1 $g{\cdot}L^{-1}$ peptone and without soil extract) suggested by SAG was used for photoautotrophical cultivation of *H. pluvialis.* In case of *C. zofingiensis* and *C. sorokiniana,* a modified medium of Arnon *et al.* [21] described by Cordero *et al.* [10], was used for photoautotrophical cultivation.

Cultivation

Photobioreactor screening modules (PSM) were used for cultivations under monoseptic conditions, as previously described [22-24]. The aeration was kept constant at 0.45 $L{\cdot}min^{-1}$, enriched with 3% CO_2, and the temperature was 25°C. The light intensity (photon flux density PFD) was measured with a submersible spherical light sensor (US-SQS/L, ULM-500, Walz, Germany) in the center of the PSM filled with water. *H. pluvialis* and *C. sorokiniana* were cultivated with a light intensity of 700 $\mu mol{\cdot}m^{-2}s^{-1}$. In case of *C. zofingiensis* the PFD was increased after six days from 470 up to 700 $\mu mol{\cdot}m^{-2}s^{-1}$ to boost the carotenoid production. All three microalgae were kept in batch mode for two weeks.

Determination of algae carotenoid yield

After two weeks of cultivation, the harvested and freeze-dried biomass was disrupted/homogenized using various methods described below and extracted via pressurized liquid extraction (PLE) at the given conditions. Extracts were evaporated to dryness under the nitrogen and redissolved in 2.5 mL of a mixture of acetone and MeOH (9:1, v:v) before saponification. The latter was started by addition of 0.5 mL of 0.05 M NaOH in MeOH and incubated for 3 h (room temperature, darkness) to eliminate carotenoid esters [25-27]. Next, 3 mL of petroleum ether were added, the mixture was washed with 3 mL 10% NaCl, centrifuged (3,500 rpm, 1 min, 5810R, Eppendorf; Hamburg, Germany) and the aqueous phase was discarded. The washing step was repeated twice. The resulting organic phase was evaporated to dryness under N_2 and the resulting substance was redissolved in 1 mL of MTBE. Extracts were analyzed via HPLC-UV/Vis (Prominence HPLC System, Shimadzu, Japan) according to conditions reported previously by Sander with the following modifications [28]. The chromatographic separation was performed on a reversed phase column (YMC carotenoid C_{30}, 150 × 4.6 mm, 3 μm) with a mobile phase consisting of two solvent mixtures of methanol, MTBE and water (A: 81:15:4, B: 8:89:3, v:v:v) under following gradient conditions. The concentration of 2% B was kept constant for 11 minutes, then increased to 40% B in 7 minutes and held for 6.5 minutes, followed by an increase to 100% B (in 2.5 min, held for 3 min) and successive reequilibration to 2% B (in 3 min, held for 7 min). The flow rate was 1 $mL{\cdot}min^{-1}$, injection volume was 5 μL and column temperature was kept at 22°C. The detection of the analytes was done at 445, 452 and 475 nm for lutein, beta-carotene and astaxanthin, respectively. All these steps were performed in triplicates.

For calibration, a stock solution of astaxanthin (Adipogen International Inc., San Diego, USA) was prepared by dissolving 10 mg in 33.3 mL of DMSO, resulting in a concentration of 300 $\mu g{\cdot}mL^{-1}$. Stock solutions of lutein (CaroteNature GmbH, Ostermundigen, Switzerland) and β-carotene (Sigma Aldrich, St. Lou's, USA) had the same concentration by solving 1 and 10 mg in 3.3 and 33.3 mL of MTBE, respectively. All analytes were calibrated using mixed standards prepared by appropriate dilution of stock solutions in a range of 0.05 to 100 $\mu g{\cdot}mL^{-1}$.

Cell disruption methods

In the scope of this work various cell disruption methods were evaluated based on the optical determination of the disruption yield and total carotenoid yield after pressurized liquid extraction (Dionex ASE 350, 1 cycle, 30 min, 100 bar, 60°C, solvent:acetone:MeOH, 9:1 v:v for *H. pluvialis* and DCM for *C. zofingiensis* and *C. sorokiniana*). The disruption yield C was calculated as ratio of intact and disrupted cells by counting in a Neubauer-improved hemocytometer (equation 1).

$$C[\%] = \frac{cells_{disrupted}}{cells_{total}} \times 100\% = \frac{cells_{disrupted}}{cells_{disrupted} + cells_{intact}} \times 100\%$$

The following disruption methods, performed according to the parameters summed up in Table 1, were compared: labscale high pressure homogenizer (French Pressure Cell Press, SLM instruments, Inc., Urbana, USA), labscale ball mill (TissueLyser LT, Qiagen, Venlo, Netherlands), Ultra Turrax (T 18 Basic, IKA-Werke, Staufen, Germany), repeated freeze and thaw cycles, freeze-drying (Alpha 1-4, Christ, Osterode, Germany) and ultra-sonication (UW 2200, Bandelin electronic, Berlin, Germany). All samples were freeze-dried again before extraction in order to remove water residues and assure comparability.

Cell disruption	Parameters
High pressure homogenizer	Biomass resuspended in 10 mL of H_2O_u, p=1000 bar, cycles n=3
Ball mill	Biomass, frozen at -80°C for 1 h, 1 cycle, 50 Hz, 2 min
Freeze and thaw	Biomass resuspended in 10 mL of H_2O_u, freezing -20°C, thawing at RT for 30 min, cycles n=3
Freeze-drying	Biomass resuspended in 10 mL of H_2O_u frozen at -80°C overnight, P=0.4 mbar, overnight
Ultra-sonication	Biomass resuspended in 10 mL of H_2O_u, maximum intensity, 10 min, one cycle
Ultra Turrax	Biomass resuspended in 10 mL of H_2O_u, 24,000 rpm, 10 min, on ice

Table 1: Parameters for various cell disruption methods tested. Each method was performed with dried biomass of *H. pluvialis* (30 mg), *C. zofingiensis*, *C. sorokiniana* (40 mg).

The methods were tested on freeze-dried biomass of three algae with different morphological characteristics: (I) *H. pluvialis* (30 mg, encysted state, 30-50 µm, rigid cell walls, [29,30], (II) *C. zofingiensis* (30 mg, 2-15 µm, [31]) and (III) C. *sorokiniana* (30 mg, 2-4. µm, [32]). The optical evaluation of the disruption yield was exemplarily applied on *H. pluvialis*.

Optimization of the pressurized liquid extraction method

Selection of the solvent for PLE of carotenoids from algae biomass (30 mg dried biomass, 1 cycle, 30 min, 100 bar, 60°C) was performed based on the experiments with biomass of *H. pluvialis* after Ultra Turrax-disruption (conditions see Table 1) using 6 different solvents: acetone, ethanol, ethyl acetate, n-hexane, dichloromethane and a mixture of acetone:MeOH (9:1, v:v).

Next, the influence of temperature on the efficiency of the carotenoid extraction at first between 20 and 60°C (1 cycle, 30 min, 100 bar, solvent: acetone:MeOH, 9:1, v:v) and later between 60 and 120°C (1 cycle, 30 min, 100 bar, solvent: DCM) was tested with the biomass of *C. zofingiensis* (30 mg, freeze-dried, no disruption) and *H. pluvialis* (40 mg, after Ultra Turrax-disruption).

After appointing the optimum extraction temperature, various extraction durations - 5, 10, 20 and 30 min - were investigated, on the same biomass of *H. pluvialis* that was used before, in order to test the possibility of shortening the extraction time for increased sample throughput. Additionally, the effect of using multiple extraction cycles (4 × 5 min) was evaluated.

Results and Discussion

Cell disruption

A downstream process for recovery of intracellular products in general consists of the following unit operations: harvesting, dehydrating, product isolation (cell disruption and extraction) and product purification [18]. To gain high product yields, both efficient cell disruption and extraction is necessary. For instance in the work of Ceron *et al.* the recovery of lutein from freeze-dried biomass of *S. almeriensis* with various cell disruption methods was up to 60% higher compared to a method without cell disruption [33]. Cell disruption can be performed by a multitude of established methods, generally divided into non-mechanical methods, such as alkaline lysis, organic solvents, freezing, osmotic shock, etc. or mechanical methods, e.g., homogenizers, bead mills, ultra-sonication [18,34].

In the work presented here, four mechanical (ball mill, Ultra Turrax, high pressure homogenizer and ultra-sonication) as well as two non-mechanical methods (repeated freeze and thaw cycles, lyophilization) were used for the three microalgae *H. pluvialis, C. zofingiensis, C. sorokiniana*. To evaluate the efficiency of the mentioned methods an optical determination of the disruption yield C, based on counting disrupted and intact cells after cell disruption, was used (see equation 1). The disruption yield C of freeze-dried *H. pluvialis* biomass is shown in (Figure 1). The highest disruption yield - around 80% - of the thick encysted cell walls was reached by using mechanical methods such as ball mill, Ultra Turrax and high pressure homogenizer, while other mechanical (ultra-sonication) and non-mechanical methods, such as repeated freeze and thaw cycles resulted in disruption yields of ≤ 10%. In the next step the disrupted biomass was used for determination of the total carotenoid extraction yield via HPLC-UV/vis analysis. The obtained total carotenoid yield values are normalized to the highest reached value for each alga (in case of *H. pluvialis* - high pressure homogenizer) (Figure 2). In general, the normalized total carotenoid values are in good agreement with the calculated disruption yield. The highest carotenoid extraction yield was measured after disruption with the high pressure homogenizer (4.21 µg total carotenoid·mg^{-1} dry weight (d.w.), taken as 1 for normalization purpose), ball mill (3.56 µg total

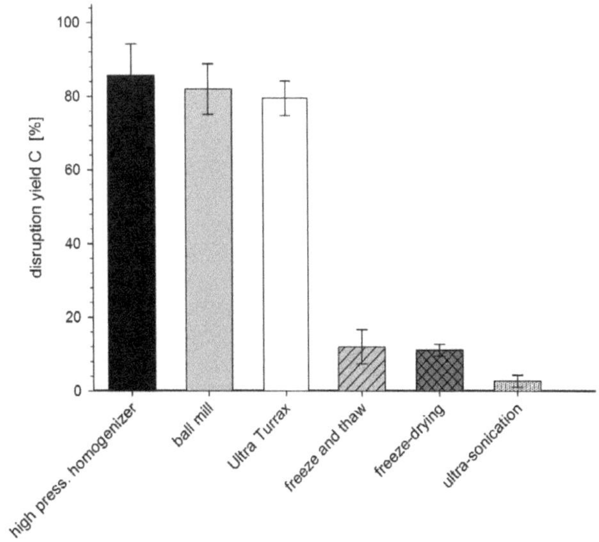

Figure 1: Disruption yield by optical determination of *H. pluvialis* cells. The biomass of *H. pluvialis* (30 mg) was disrupted as described above and intact and disrupted cells were counted in a Neubauer-improved haemocytometer (n=3). The disruption yield was calculated according to formula (1).

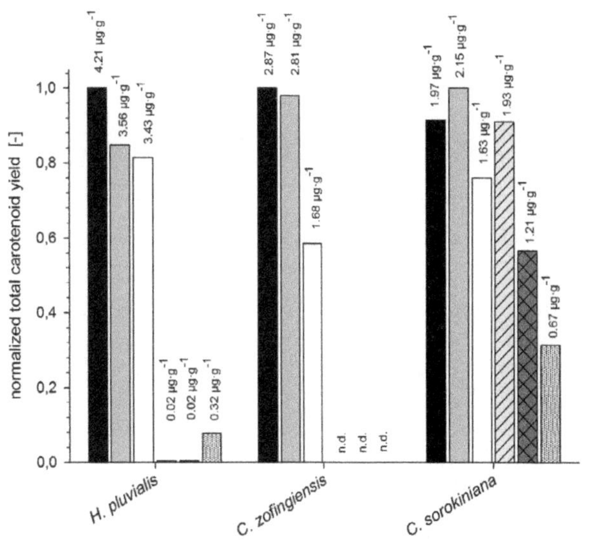

Figure 2: Normalized cumulated carotenoid yield of *H. pluvialis, C. zofingiensis* and *C. sorokiniana* measured by HPLC-UV/Vis resulting from different cell disruption methods (black bar: high pressure homogenizer; light gray bar: ball mill; white bar: Ultra Turrax; dashed bar: repeated freeze and thaw; double dashed bar: freeze-drying; dotted bar: ultra-sonication; n.d.: not determined). Freeze-dried algae biomass (30, 40 and 40 mg for *H. pluvialis, C. zofingiensis* and *C. sorokiniana)* was disrupted as described, extracted via PLE (60°C, 30 min, 100 bar) with acetone:MeOH (9:1, v:v) (*H. pluvialis*) or DCM (*C. zofingiensis, C. sorokiniana*), saponified and analysed by HPLC-UV/Vis on a C30 column. Data was normalized on the highest value reached for each alga.

carotenoid·mg^{-1} d.w.) and Ultra Turrax (3.42 µg total carotenoid·mg^{-1} d.w.). Repeated freeze-and-thaw-cycles resulted in 240-fold lower total carotenoid concentration (0.02 µg total carotenoid·mg^{-1} d.w.) in comparison to high pressure homogenizer. Similar observations for *H. pluvialis* are described in the work of Mendes-Pinto *et al.* In this work mechanical disruption via a cell homogenizer resulted in nearly factor four higher total carotenoid amounts compared to non-mechanical methods like alkaline lysis, spray drying or treatment with base or enzymes. Also autoclaving of the biomass showed nearly the same effectivity as mechanical disruption [35]. However, even though the method used here for the determination of the disruption yield gives a very fast and a relatively good estimation of the disruption effectivity, it can only be performed with the statistically selected parts of the samples and only for some algal species. In case of *H. pluvialis*, the method can be easily applied and gives reliable data due to big cell diameters during encysted stadium [36]. For the other two investigated microalgae *C. zofingiensis* and *C. sorokiniana* with small cell diameters, it is inapplicable. Furthermore, the optical impression of the cell integrity should not lead to the conclusion that carotenoids cannot be extracted, as minor damage might not be visible but might still allow extraction of carotenoids. All in all, this method used here for *H. pluvialis* is a quick and easy one and results in a rough estimation, however due to the above listed limitations, further evaluation based on the total carotenoid yield measured by HPLC is necessary.

In the next step, the best three disruption methods selected in the experiments with *H. pluvialis* were tested on the biomass of *C. zofingiensis* (2-15 µm), showing the best results again for the high pressure homogenizer (2.87 µg total carotenoid·mg^{-1} d.w.) and ball mill (2.81 µg total carotenoid·mg^{-1} d.w.). In contrast to the other two species, for *C. sorokiniana* - the smallest tested cells (2-4.5 µm) - the effectivity of repeated freeze and thaw cycles of the biomass showed comparable results (1.95 µg total carotenoid·mg^{-1} d.w.) to the results reached with the use of high pressure homogenizer (1.97 µg total carotenoid·mg^{-1} d.w.) and ball mill (2.15 µg total carotenoid·mg^{-1} d.w.). Furthermore, also other methods, such as lyophilization and ultra-sonication, showed better results for this species (0.57 and 0.31 normalized carotenoid yield) in comparison to *H. pluvialis* (0.005 and 0.08 normalized carotenoid yield) (Figure 2). This is in good correspondence to the data of Prakabaran *et al.* that showed comparably good results achieved by osmotic shock and bead mill for lipid extraction for *Chlorella sp.* [37]. As a consequence, it can be assumed that the effectivity of non-mechanical disruption methods increases inversely proportionally to the cell size and the rigidity of cell walls of the examined species.

In order to establish an efficient isolation procedure for industrial application, the suitability of the various investigated methods for scale-up has also to be taken into account. Because of high power consumption per unit of biomass volume ultra-sonication and lyophilization are considered to be suitable only for small-scale applications [18,38]. Capital cost for repeated freeze and thaw cycles are relative low compared to the other used methods. However, the fact that this is a very time-consuming as well as the difficulty to maintain the product quality under such conditions has to be addressed. In contrast to this, mechanical methods i.e., high pressure homogenization and bead mills are commercially available and applied [18]. Furthermore, these methods showed the best disruption yields resulting in the highest carotenoid recoveries for all three tested algae. Finally, the right choice of the best cell disruption method strongly depends on the used microalgae, the product of interest as well as the scale of production (quantity and quality) [18,34].

Pressurized liquid extraction of carotenoids

One of the most important factors affecting the extraction yield of carotenoids via PLE is the selection of the suitable extraction solvent (Figure 3). The application of n-hexane for *H. pluvialis* resulted in the isolation of the lowest measured carotenoid amount (1.59 µg·mg^{-1} d.w.), followed by ethyl acetate (1.98 µg total carotenoid·mg^{-1} d.w.), ethanol (2.15 µg total carotenoid·mg^{-1} d.w.), acetone (2.43 µg total carotenoid·mg^{-1} d.w.) and acetone:methanol (9:1, v:v) (3.43 µg total carotenoid·mg^{-1} d.w.). The use of DCM showed the best results with 4.72 µg total carotenoid·mg^{-1} d.w. - a 3-folds higher yield than reached with hexane. However, this general finding should not be generalized for individual carotenoids. In fact, β-carotene gave comparable results for most applied solvents whereas the yield for lutein extraction could be improved by factor 2.9 and for astaxanthin by factor 3.2 by using DCM compared to hexane. This can be explained by the increased solubility of more polar compounds such as astaxanthin or lutein in dichloromethane (i.e., astaxanthin 30 g·L^{-1} in DCM, 0.2 g·L^{-1} in acetone [27]). Increased solubility of the very unpolar β-carotene in hexane (30 mg L^{-1}, described by Craft and Soares) compared to the other solvents did not show significant influence [39]. Due to the low yield in the examined species (0.2 mg L^{-1} in the extract) all tested solvents were able to extract it completely. Jamie *et al.* also reported better extraction yields for PLE extraction from *H. pluvialis* using ethanol in comparison to n-hexane, which is in consent with the results presented above [40]. Mixing dichloromethane with other solvents showed no additional advantages (data not shown here), while Denery showed the best results for a mixture of DCM and methanol (1:3, v:v) for extraction of astaxanthin from *H. pluvialis* compared to acetone, ethanol and mixtures of acetone and methanol (7:3, v:v) [41]. In contrast to this, various mixtures of acetone and methanol are commonly used in literature as extraction solvents [10,25,42]. Furthermore, other extraction media such as oil [43,44] or CO_2 for supercritical fluid extraction [45,46] have also been described in literature but not tested in this work. Further tests are necessary to evaluate these methods.

The first experiments with temperature variations for PLE showed that the yield of the carotenoid extraction can be improved by increasing the temperature from room temperature to 60°C. Therefore, further tests with even higher temperatures were conducted here in order to check if additional improvement is possible, as the solubility itself is a thermodynamic equilibrium, influenced by the temperature. It turned out, that the carotenoid recovery yield for the selected solvent decreased again with temperatures above 80°C, probably due to the thermal destruction of the analytes (Figure 4). At 120°C, the total yield of carotenoids was 35% lower than after extraction at 60°C (3.08 compared to 4.72 µg total carotenoid·mg^{-1} d.w.). The optimum could be found around 60°C. That corresponds well with findings of Mustafa *et al.* who tested PLE with ethanol for α- and β-carotene from carrots showing 42% reduction of β-carotene recovery by using 180°C in comparison to 60°C [47]. Denery *et al.* did not find a reduction of astaxanthin recovery by PLE from *H. pluvialis* using acetone between 20°C and 100°C but a reduction of lutein and total pigment amounts at higher temperatures [41]. In contrast to this, Jaime *et al.* found increasing extraction yields for temperatures up to 200°C for carotenoid extraction with ethanol from *H. pluvialis* [40]. From the somewhat contradicting results reported by the above mentioned groups, it can be assumed that optimum temperature depends on the solvent that is used for the extraction and has to be carefully selected for specific applications. In the case of the work presented here, future tests with smaller temperature intervals might reveal a more exact optimum.

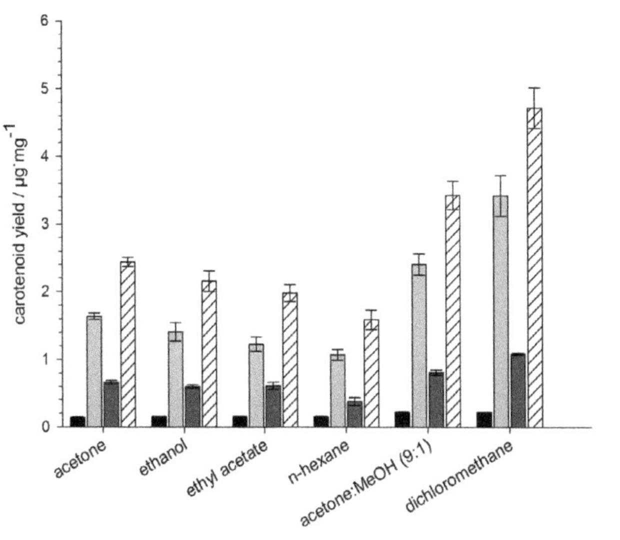

Figure 3: Yield of individual and total carotenoids (black bar: β-carotene; light gray bar: astaxanthin; dark gray bar: lutein; dashed bar: total carotenoids) of *H. pluvialis* measured by HPLC UV/Vis resulting from pressurized liquid extraction with different solvents: 30 mg of freeze-dried *H. pluvialis* biomass were disrupted using Ultra Turrax, extracted via PLE (60°C, 30 min, 100 bar) by six selected solvents, saponified and analysed by HPLC -UV/Vis on a C30 column.

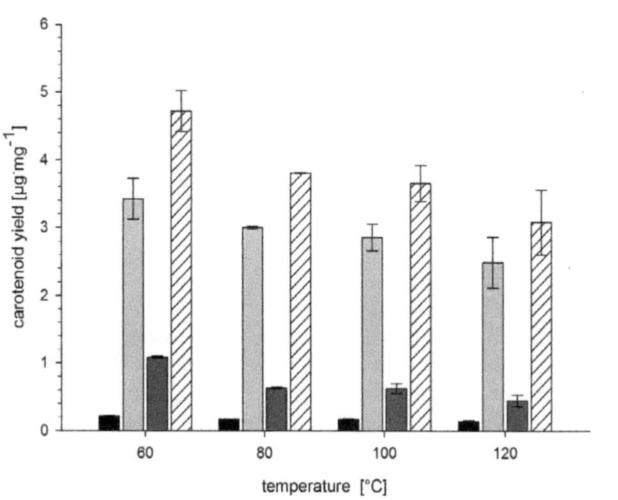

Figure 4: Yield of individual and total carotenoids (black bar: β-carotene; light gray bar: astaxanthin; dark gray bar: lutein; dashed bar: total carotenoids) of *H. pluvialis* measured by HPLC UV/Vis resulting from pressurized liquid extraction with different temperatures: 30 mg of freeze-dried *H. pluvialis* biomass were disrupted using Ultra Turrax, extracted via PLE (30 min, 100 bar, DCM) by 4 different temperatures, saponified and analysed by HPLC -UV/Vis on a C30 column.

The optimum duration of the extraction process depends on the time needed to build up the equilibrium between the analyte concentration in the sample matrix and the solvent. Five minutes showed to be insufficient for carotenoid extraction in the tested setup, resulting in only 79% of the maximal total carotenoid yield (3.74. instead of 4.72 µg total carotenoid·mg^{-1} d.w.) and 58% of the lutein recovery (0.63 instead of 1.08 µg lutein·mg^{-1} d.w.), whereas a nearly complete carotenoid level could be reached applying 10 minutes and more (Figure 5). Implementing additional extraction cycles did not improve the results, as 99% of the analytes were extracted within the first cycle (Figure 6). The carotenoid yield in the extracts of the third and fourth cycle were below the detection limit (0.0125 µg·mg^{-1} for astaxanthin and lutein, 0.025 µg·mg^{-1} for ß-carotene). This suggests that the equilibrium between solvent and biomass is established after 10 minutes and according to the solubility data was also far away from saturation [27,39]. Performing multiple steps is often preferred in literature, e.g., Mustafa *et al.* showed that the optimum PLE of β-carotene with ethanol at 60°C to be five 2 min cycles after 5 min preheating [47]. On the other hand, this highly depends on the solvent and the used matrix.

With the parameters selected (high pressure homogenizer, PLE at 60°C for 10 min, in 1 cycle, solvent: DCM), a total carotenoid yield of 4.78 µg total carotenoid·mg^{-1} d.w. and a total astaxanthin yield of 3.69 µg astaxanthin·mg^{-1} d.w. could be determined for *H. pluvialis*, which is 0.4% of the total cell weight. The values that can be found in literature range from 0.5 to 4% astaxanthin depending on the strain, culture conditions and cultivation mode [31,34,41,42,48]. Values of up to 98 µg·mg^{-1} d.w. have been reported for heterotrophic cultivations [49]. The lower carotenoid yield presented in this study can be easily explained by the fact that no optimization regarding the carotenoid production was performed at this stage of the work. As numerous possible modifications of the process to induce astaxanthin production is described in literature (nitrogen starvation, light stress, salinity, iron addition, strain selection, genetic modifications, heterotrophic growth on acetate), further optimizations of the cultivation is possible and planed for future work [50-56]. High pressure homogenizer and ball mill showed the best disruption for *C. zofingiensis*, the parameters for the PLE were based on the findings regarding the extraction of carotenoids from *H. pluvialis*. Applying an optimized isolation process for carotenoids to the biomass of *C. zofingensis* 1.48 µg lutein·mg^{-1} d.w. and 1.29 µg astaxanthin·mg^{-1} d.w. could be measured. Values reported for this species by Del Campo and Liu range between 4 and 1.3 to 6.5 µg·mg^{-1} d.w. for lutein und astaxanthin, respectively [25,31]. The lower yield for lutein measured in this work can be explained by the harvesting of the biomass at the end of the cultivation while the highest yield was reported for an early stage of cultivation by Del Campo *et al.* [25]. Thus further optimization of the whole production process including cultivation is necessary. Possible approaches were published in literature [50,57,58]. For *C. sorokiniana* the highest yield of lutein (2.08 µg lutein·mg^{-1} d.w.) was measured after the ball mill disruption. In comparison, Cordero *et al.* reported 4.2 µg lutein·mg^{-1} d.w. in wildtype algae and up to 7 µg·mg^{-1} d.w. in genetically modified strains, also suggesting that further optimization of the cultivation process is necessary [10].

Conclusion

Due to the fact that downstream processing is one of the decisive elements, it has to be very carefully optimized in order to reach the maximum recovery yield of the products at lowest possible costs. For this reason, various cell disruption methods and parameters for pressurized liquid extraction of the carotenoids from various algae species were examined in labscale in this work. The influence of the cell size and cell wall properties on the success of cell disruption methods was shown: while mechanical methods, i.e., high pressure homogenizer or ball mill were most efficient for all three tested microalgae, the effectivity of soft disruption increased with decreasing cell size and rigidity. Selecting DCM as the extraction solvent at 60°C improved the extraction yield of total carotenoids. The extraction time could be set to 10 minutes, with no additional extraction cycles, improving

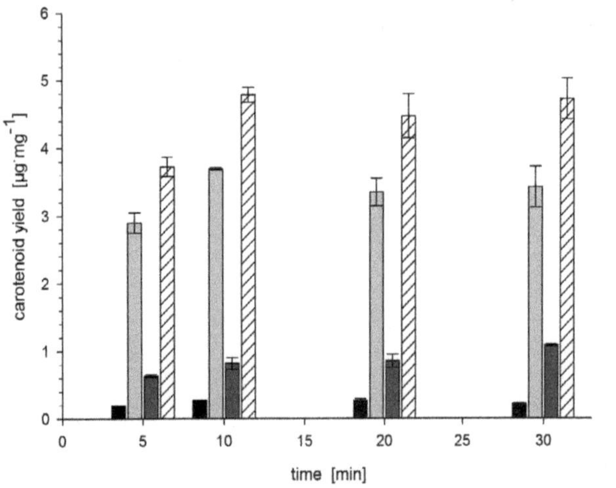

Figure 5: Yield of individual and total carotenoids (black bar: β-carotene; light gray bar: astaxanthin; dark gray bar: lutein; dashed bar: total carotenoids) of *H. pluvialis* measured by HPLC UV/Vis resulting from pressurized liquid extraction with different extraction durations: 30 mg of freeze-dried *H. pluvialis* biomass were disrupted using Ultra Turrax, extracted via PLE (60°C, 100 bar, DCM) for 4 different durations, saponified and analysed by HPLC -UV/Vis on a C30 column.

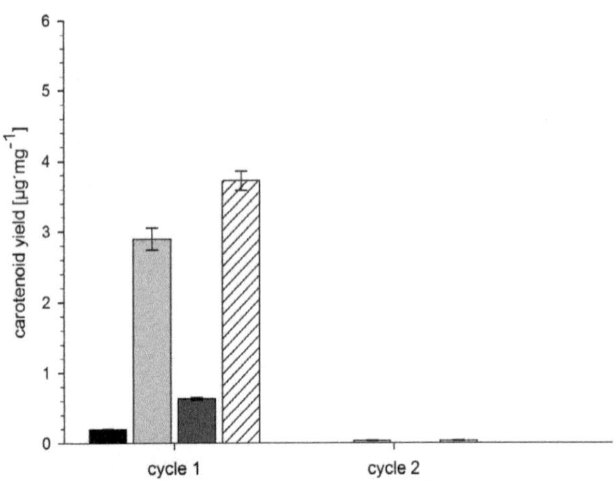

Figure 6: Yield of individual and total carotenoids (black bar: β-carotene; light gray bar: astaxanthin; dark gray bar: lutein; dashed bar: total carotenoids) of *H. pluvialis* measured by HPLC UV/Vis resulting from pressurized liquid extraction with multiple extraction cycles: 30 mg of freeze-dried *H. pluvialis* biomass were disrupted using Ultra Turrax, extracted via PLE (60°C, 10 min, 100 bar, DCM) for 4 different durations, saponified and analysed by HPLC -UV/Vis on a C30 column.

the sample throughput. The best extraction yield for carotenoids from above described species could be achieved by PLE using DCM at 60°C in a 10 min cycle (n=1) with lyophilized and high pressure disrupted biomass. The carotenoid yield was lower than the highest values reported in literature since when this work was carried out the cultivation conditions were not optimized for the highest product yield. Obviously, culture optimization will be as crucial as optimizing the downstream process. Nevertheless, the proposed sample preparation procedure allows reliable carotenoid determination in algae biomass that is necessary for further optimization of the whole process for economic use of microalgae for this application.

Conflict of Interest

The authors have declared no conflicts of interest.

Acknowledgements

The authors are grateful for financial support of the Federal Ministry of Education and Research (BMBF) and the Bavarian State Ministry of the Environment and Consumer Protection, due to which this work could be performed.

References

1. Borowitzka MA (2013) High-value products from microalgae - their development and commercialisation. J Appl Psychol 25: 743-756.
2. Yaakob Z, Ali E, Zainal A, Mohamad M, Takriff MS, et al. (2014) An overview: biomolecules from microalgae for animal feed and aquaculture. J Biol Res (Thessalon) 21: 6.
3. Pulz O, Gross W (2004) Valuable products from biotechnology of microalgae. Appl Microbiol Biotechnol 65: 635-648.
4. Borowitzka MA, Moheimani N (2013) Sustainable biofuels from algae. Mitigation and Adaptation Strategies for Global Change 18: 13-25.
5. Thukral N, Sharma M (2015) Microalgae as Future Fuel: Real Opportunities and Challenges. J Thermodyn Cataly 6.
6. Rodolfi L, Chini Zittelli G, Bassi N, Padovani G, Biondi N, et al. (2009) Microalgae for oil: strain selection, induction of lipid synthesis and outdoor mass cultivation in a low-cost photobioreactor. Biotechnol Bioeng 102: 100-112.
7. Brennan L, Owende P (2010) Biofuels from microalgae - A review of technologies for production, processing, and extractions of biofuels and co-products. Renewable and Sustainable Energy Reviews 14: 557-577.
8. Raja R, Hemaiswarya S, Kumar NA, Sridhar S, Rengasamy R, et al. (2008) A perspective on the biotechnological potential of microalgae. Crit Rev Microbiol 34: 77-88.
9. Bhatnagar A, Chinnasamy S, Singh M, Das KC (2011) Renewable biomass production by mixotrophic algae in the presence of various carbon sources and wastewaters. Applied Energy 88: 3425-3431.
10. Cordero BF, Obraztsova I, Couso I, Leon R, Vargas MA, et al. (2011) Enhancement of lutein production in Chlorella sorokiniana (Chorophyta) by improvement of culture conditions and random mutagenesis. Mar Drugs 9: 1607-1624.
11. Guedes AC, Amaro HM, Malcata FX (2011) Microalgae as sources of carotenoids. Mar Drugs 9: 625-644.
12. Varela JC, Pereira H, Vila M, León R (2015) Production of carotenoids by microalgae: achievements and challenges. Photosynth Res 125: 423-436.
13. Carvalho AP, Silva SO, Baptista JM, Malcata FX (2011) Light requirements in microalgal photobioreactors: an overview of biophotonic aspects. Appl Microbiol Biotechnol 89: 1275-1288.
14. Demmig-Adams B, Adams WW 3rd (2002) Antioxidants in photosynthesis and human nutrition. Science 298: 2149-2153.
15. Sun Z, Liu J, Bi YH, Zhou ZG (2014) Microalgae as the Production Platform for Carotenoids. OMICS Group International.
16. Spolaore P, Joannis-Cassan C, Duran E, Isambert A (2006) Commercial applications of microalgae. J Biosci Bioeng 101: 87-96.
17. Rosenberg JN, Oyler GA, Wilkinson L, Betenbaugh MJ (2008) A green light for engineered algae: redirecting metabolism to fuel a biotechnology revolution. Curr Opin Biotechnol 19: 430-436.
18. Richmond A, Hu Q (2013) Handbook of Microalgal Culture. 2nd edn. John Wiley & Sons, Ltd., Oxford, UK.
19. Olaizola M (2003) Commercial development of microalgal biotechnology: from the test tube to the marketplace. Biomol Eng 20: 459-466.
20. Straathof A (2011) The Proportion of Downstream Costs in Fermentative Production Processes. Moo-Young, Murray. Comprehensive Biotechnology 2: 811-814.

21. Arnon D, McSwain BD, Tsujimoto HY, Wada K (1974) Photochemical activity and components of membrane preparations from blue-green algae. I. Coexistence of two photosystems in relation to chlorophyll a and removal of phycocyanin. Biochim Biophys Acta 357: 231-245.

22. Heining M, Sutor A, Stute SC, Lindenberger CP, Buchholz R, et al. (2015) Internal illumination of photobioreactors via wireless light emitters: a proof of concept. J Appl Psychol 27: 59-66.

23. Klein BC, Walter C, Lange HA, Buchholz R (2012) Microalgae as natural sources for antioxidative compounds. J Appl Psychol 24: 1133-1139.

24. Geier SC, Huyer S, Praebst K, Husmann M, Walter C, et al. (2012) Outdoor cultivation of Chlamydomonas reinhardtii for photobiological hydrogen production. J Appl Psychol 24: 319-327.

25. Del Campo JA, Rodríguez H, Moreno J, Vargas MA, Rivas J, et al. (2004) Accumulation of astaxanthin and lutein in Chlorella zofingiensis (Chlorophyta). Appl Microbiol Biotechnol 64: 848-854.

26. Yuan JP, Chen F (1998) Chromatographic Separation and Purification of trans-Astaxanthinfrom the Extracts of Haematococcus pluvialis. J Agric Food 46: 3371-3375.

27. Johnson EA, An GH (1991) Astaxanthin from Microbial Sources. Crit Rev Biotechnol 11: 297-326.

28. Sander LC, Sharpless KE, Craft NE, Wise SA (1994) Development of engineered stationary phases for the separation of carotenoid isomers. Anal Chem 66: 1667-1674.

29. Kobayashi M, Katsuragi T, Tani Y (2001) Enlarged and astaxanthin-accumulating cyst cells of the green alga Haematococcus pluvialis. J Biosci Bioeng 92: 565-568.

30. Boussiba S (2000) Carotenogenesis in the green alga Haematococcus pluvialis: Cellular physiology and stress response. Physiol Plant 108: 111-117.

31. Liu J, Sun Z, Gerken H, Liu Z, Jiang Y, et al. (2014) Chlorella zofingiensis as an alternative microalgal producer of astaxanthin: biology and industrial potential. Mar Drugs 12: 3487-3515.

32. Lizzul AM, Hellier P, Purton S, Baganz F, Ladommatos N, et al. (2014) Combined remediation and lipid production using Chlorella sorokiniana grown on wastewater and exhaust gases. Bioresour Technol 151: 12-18.

33. Cerón MC, Campos I, Sánchez JF, Acién FG, Molina E, et al. (2008) Recovery of lutein from microalgae biomass: development of a process for Scenedesmus almeriensis biomass. J Agric Food Chem 56: 11761-11766.

34. Del Campo JA, García-González M, Guerrero MG (2007) Outdoor cultivation of microalgae for carotenoid production: current state and perspectives. Appl Microbiol Biotechnol 74: 1163-1174.

35. Mendes-Pinto MM, Raposo M, Bowen J, Young AJ, Morais R, et al. (2001) Evaluation of different cell disruption processes on encysted cells of Haematococcus pluvialis: effects on astaxanthin recovery and implications for bio-availability. J Appl Psychol 13: 19-24.

36. Grewe C (2009) Untersuchungen zurAstaxanthin-Biosynthese in den Grünalgen Scenedesmus sp. und Haematococcus pluvialis.

37. Prabakaran P, Ravindran AD (2011) A comparative study on effective cell disruption methods for lipid extraction from microalgae. Lett Appl Microbiol 53: 150-154.

38. Ranjith Kumar R, Hanumantha Rao P, Arumugam M (2015) Lipid Extraction Methods from Microalgae: A Comprehensive Review. Frontiers in Energy Research 2: 457.

39. Craft NE, Soares JH (1992) Relative Solubility, Stability and Absorptivity of Lutein and @-Carotenein Organic Solvents. J Agric Food Chem 40: 431-434.

40. Jaime L, Rodríguez-Meizoso I, Cifuentes A, Santoyo S, Suarez S, et al. (2010) Pressurized liquids as an alternative process to antioxidant carotenoids' extraction from *Haematococcus pluvialis* microalgae. LWT - Food Science and Technology 43: 105-112.

41. Denery JR, Dragull K, Tang C, Li QX (2004) Pressurized fluid extraction of carotenoids from Haematococcus pluvialis and Dunaliella salina and kavalactones from Piper methysticum. Analytica Chimica Acta 501: 175-181.

42. Sarada R, Vidhyavathi R, Usha D, Ravishankar GA (2006) An efficient method for extraction of astaxanthin from green alga Haematococcus pluvialis. J Agric Food Chem 54: 7585-7588.

43. Nonomura AM (1987) Process for producing a naturally-derived carotene/oil composition by direct extraction from algae.

44. Fernández-Sevilla JM, Acién Fernández FG, Molina Grima E (2010) Biotechnological production of lutein and its applications. Appl Microbiol Biotechnol 86: 27-40.

45. Herrero M, Cifuentes A, Ibañez E (2006) Sub- and supercritical fluid extraction of functional ingredients from different natural sources: Plants, food-by-products, algae and microalgae: A review. Food Chem 98: 136-148.

46. Careri M, Furlattini L, Mangia A, Musci M, Anklam E, et al. (2001) Supercritical fluid extraction for liquid chromatographic determination of carotenoids in Spirulina Pacifica algae: a chemometric approach. J Chromatogr A 912: 61-71.

47. Mustafa A, Trevino LM, Turner C (2012) Pressurized hot ethanol extraction of carotenoids from carrot by-products. Molecules 17: 1809-1818.

48. Cuellar-Bermudez SP, Aguilar-Hernandez I, Cardenas-Chavez DL, Ornelas-Soto N, Romero-Ogawa MA, et al. (2015) Extraction and purification of high-value metabolites from microalgae: essential lipids, astaxanthin and phycobiliproteins. Microb Biotechnol 8: 190-209.

49. Domínguez-Bocanegra AR, Guerrero Legarreta I, Martinez Jeronimo F, Tomasini Campocosio A (2004) Influence of environmental and nutritional factors in the production of astaxanthin from Haematococcus pluvialis. Bioresour Technol 92: 209-214.

50. Orosa M, Valero JF, Herrero C, Abalde J (2001) Comparison of the accumulation of astaxanthin in Haematococcus pluvialis and other green microalgae under N-starvation and high light conditions. Biotechnol Lett 23: 1079-1085.

51. Aflalo C, Meshulam Y, Zarka A, Boussiba S (2007) On the relative efficiency of two- vs. one-stage production of astaxanthin by the green alga Haematococcus pluvialis. Biotechnol Bioeng 98: 300-305.

52. Borowitzka MA, Huisman JM, Osborn A (1991) Culture of the astaxanthin-producing green alga Haematococcus pluvialis 1. Effects of nutrients on growth and cell type. J Appl Psycholy 295-304.

53. Boussiba S, Avigad V (1991) Astaxanthin Accumulation in the Green Alga Haematococcus pluvialis. Plant Cell Physiol 32: 1077-1082.

54. Boussiba S, Bing W, Yuan JP, Zarka A, Chen F, et al. (1999) Changes in pigments profile in the green alga Haeamtococcus pluvialis exposed to environmental stresses. Biotechnology Letters 21: 601-604.

55. Fábregas J, Otero A, Maseda A, Domínguez A (2001) Two-stage cultures for the production of astaxanthin from Haematococcus pluvialis. J Biotechnol 89: 65-71.

56. Cai M, Li Z, Qi A (2009) Effects of iron electrovalence and species on growth and astaxanthin production of Haematococcus pluvialis. Chinese Journal of Oceanology and Limnology 27: 370-375.

57. Sun N, Wang Y, Li YT, Huang JC, Chen F, et al. (2008) Sugar-based growth, astaxanthin accumulation and carotenogenic transcription of heterotrophic Chlorella zofingiensis (Chlorophyta). Process Biochemistry 43: 1288-1292.

58. Wang Y, Peng J (2008) Growth-associated biosynthesis of astaxanthin in heterotrophic Chlorella zofingiensis (Chlorophyta). World J Microbiol Biotechnol 24: 1915-1922.

Solvent Effects on the Extraction of Copper(II) with Lauric Acid

Guerdouh A* and Barkat D

Laboratory of chemical molecular and environment, Department of Industrial Chemistry, Faculty of Science and Technology, Biskra University, 07000 Biskra, Algeria

Abstract

The solvent extraction of copper(II) from nitrate medium with lauric acid (HL) at 25°C is studied as a function of various parameters: pH, concentration of lauric acid and the nature of solvent. The solvent effects on the extraction of copper(II) using polar and nonpolar solvents are treated. Extracted species differs from solvent to solvent: $CuL_2(HL)_2$ for cyclohexane, dichloromethane, toluene or chloroform and CuL_2 for 1-octanol and methyl isobutyl ketone. The extraction constants, percentage extraction (%E) and free energy (ΔG°) are also calculated for different solvents.

Keywords: Solvent extraction; Copper(II); Lauric acid; Solvent effects; Free energy

Introduction

Solvent extraction system is one of the effective techniques used for separation of metal cations from aqueous solutions [1-8]. It is a suitable method for preventing and protecting environment from pollution. The extraction of copper(II) using carboxylic acids has been a subject of much work research [9-13]. Lauric acid was employed as an extractant for the separation of numerous cations, the composition of extracted species and their extraction constants were reported [14-16]. Ghanadzadeh et al. [17] established that lauric acid extracts capably copper(II) from aqueous solutions and it was shown that in the organic phase dimeric complexes of the $(CuR_2.HR)_2$ formula are formed. The solvents have a great importance on the extraction efficiency and distribution ratio due to their polarity [18]. As a consequence, studies of the effect of solvents on the extraction of metal ions have been reported by many researchers [19-23]. Yamada et al. [24] studied the extraction of gallium (III) with decanoic acid in different solvents. They established that the extracted decanoates were more extensively polymerized in the less polar solvent than the more polar solvent. However, Ghebghoub et al. [25] investigated the effect of diluents on the extraction of copper(II) with di(2-ethylhexyl)phosphoric acid. They found that in the extracted species were CuL_2 and $(CuL_2.2HL)$ in polar and nonpolar diluents respectively. The present paper describes the results obtained from the solvent extraction of copper(II) from nitrate medium by lauric acid in several organic solvents. The scope of the work is to determine the stoichiometries coefficients of the extracted species and their equilibrium constants. In addition, the solvent effects on this extraction system are examined and the interaction with several solvents is interpreted.

Materials and Methods

Reagents and solutions

Lauric acid (Biochem), Copper nitrate (Biochem) and sodium nitrate (Biochem) were used without further purification. Chloroform, toluene, dichloromethane, cyclohexane, 1-octanol and methyl isobutyl ketone (MIBK), were employed as the organic solvents after washing several times with distilled water. The ionic strength of the aqueous medium was assumed to be unity ($[NaNO_3]$=1M). Aqueous phase: $[Cu^{2+}]=1.57 \times 10^{-3}$ M; $[NaNO_3]$=1M. Organic phase: [HL]=0.01, 0.02 and 0.04M.

Extraction and analytical procedures

Experiments were carried out by shaking equal volumes (25 ml) of both phases in thermostatted vessels. The time required to reach the equilibrium state was 30 min. The pH of the aqueous phase was adjusted by adding the necessary amount of 0.1M NaOH. Then, after the two phases were separated completely by gravity, concentrations of the metal remaining in the aqueous phase were determined photometrically at 820 nm using a Philips UV-VIS SP6-36. The metal ion concentrations in the organic phase were calculated from the difference between the metal ion concentrations in the aqueous phase before and after extraction. All the experiments were carried out at constant temperature T=25°C.

Results and Discussion

General treatment of extraction equilibrium of Copper (II) with lauric acid

It is well known that carboxylic acids are present as dimeric species in nonpolar solvents such as toluene, hexane or benzene [14,26] and as monomeric species in polar solvents such as: 1-octanol, and 4-methyl-2-pentanone. [26,27]. The extraction process may be described by the following equilibrium in nonpolar solvents:

$$\left(ML_n\left(HL\right)_m\right)_{j,org} + njH^+ \; jM^{n+}_{aq} + j\frac{n+m}{2}\left(HL\right)_{2,org} \xrightleftharpoons{K_{ex1}} \tag{1}$$

Where, equilibrium constant K_{ex1} is defined as Equation (2) and it can be rewritten as (Equation 3) by using the distribution ratio, D of copper (II).

$$K_{ex1} = \frac{\left[(ML_n(HL)_m)_j\right]_{org}\left[H^+\right]^{nj}}{\left[M^{n+}\right]_{aq}^{j}\left[(HL)_2\right]_{org}^{j(n+m)/2}} \tag{2}$$

$$\log D = (j-1)\log[M^{n+}]_{aq} + j\frac{n+m}{2}\log\left[(HL)_2\right]_{org} + (nj)\,pH - \log\alpha_{Cu} + \log j + \log K_{ex1}$$

Where the distribution coefficient, D, is defined as the ratio between the concentration of metal in organic and aqueous phase.

***Corresponding author:** Guerdouh A, Laboratory of chemical molecular and environment, Department of Industrial Chemistry, Faculty of Science and Technology, Biskra University, 07000 Biskra, Algeria
E-mail: pamelguerdouh@yahoo.fr

However, in polar solvents equations (1), (2) and (3) have:

$$jM^{n+}_{aq} + j(n+m)(HL)_{org} \xrightleftharpoons{K_{ex2}} (ML_n(HL)_m)_{j,org} + njH^+ \quad (4)$$

$$K_{ex2} = \frac{\left[(ML_n(HL)_m)_j\right]_{org}\left[H^+\right]^{nj}}{\left[M^{n+}\right]^j_{aq}\left[HL\right]^{j(n+m)}_{org}} \quad (5)$$

$$\log D = (j-1)\log[M^{n+}]_{aq} + j(n+m)\log[HL]_{org} + (nj)\,pH - \log\alpha_{Cu} + \log j + \log K_{ex2} \quad (6)$$

Where, the species in the aqueous and organic phases are designated by the subscripts aq and org respectively. Other parameters are defined as:

$M=Cu^{2+}$, n=cationic charge, m=number of monomeric acids contained in the complex, j=degree of polymerization, $(HL)_2$; (HL)=extractant in dimeric and monomeric form respectively, α_{Cu}=the side reaction coefficient allowing for metal complexation in the aqueous phase.

The equilibrium slope method has been used to analyze the experimental data for each extraction system. The stoichiometries of the extracted species were determined on the basis of some equations derived from Equations. (3) and (6). With a predetermined lauric acid concentration in the organic phase, the relationship $\log D + \log[M^{n+}]_{aq} = f(\log[M^{n+}]_{aq} + npH)$ should yield a straight line of slope j. On the other hand, if only $(ML_n(HL)_m)_j$ is the extractable species in the extraction system and when j and α are equal to unity, the plot of $\log D$ *versus* pH at constant concentration of the extractant should yield a straight line with a slope of n. In addition, the plot of log D *versus* $\log[(HL)]_2$ and log[HL] at constant pH should give a straight line of slope $j(n+m)/2$ and $j(n+m)$ respectively, from which the value of m may be calculated. Therefore, the extraction constants can be also determined by intercept with the axis. The percentage extractions (%E) of Copper (II) can be calculated by:

$$\%E = \frac{D}{D+1} \times 100 \quad (7)$$

Analysis of the extraction equilibrium in nonpolar solvents

In order to investigate stoichiometries of the extracted species in organic phase and to find out which chemical equation applies to the extraction process, a series of Copper(II) extractions were carried out with lauric acid.

First, the degree of polymerization of the extracted copper laurates is given by the slope of the plot (log D + $\log[Cu^{+2}]_{aq}$) versus ($\log[Cu^{+2}]_{aq}$ + 2pH) at a constant $[(HR)_2]$. The results obtained for chloroform, toluene, dichloromethane and cyclohexane are illustrated in Figure 1. It is evident from Figure 1, linear relations are observed with the slope of one (j=1) which indicates that in organic phase monomeric species with lauric acid occur. Second, according to equation (3) the relationships log D versus pH are determined at a constant $[(HR)_2]$ in chloroform, toluene, dichloromethane and cyclohexane. The results are shown in Figures 2a-d linear relations are observed with the slope of two (n=2), confirming that the monomeric extracted species in organic phase is $CuL_2(HL)_m$ ($j = \alpha_{Cu} = 1$).

Finally, a study on the effect of lauric acid concentration on Copper(II) extraction enables us to determine the number of lauric acid molecules involved in the extracted species. This effect was investigated by making a series of copper(II) extractions at various $[(HR)_2]$ dissolved in chloroform, toluene, dichloromethane and cyclohexane. Concentration of dimer was calculated as $[(HL)_2]=[HL]/2$.

According to equation (3) Plots of log D versus $\log[(HL)_2]_{org}$ at constant pH were linear with the slope about of 2, as shown in Figure 3, that is, $(2+m)/2=2$ or m=2. Consequently, the composition of the extracted species is $CuL_2(HL)_2$. A same complex was obtained by Baba et al. [10] in the solvent extraction of copper(II) with 2-butylthiododecanoic acid and 2-bromododecanoic acid in hexane. Therefore, equations (1) and (3) could be rewritten as:

$$Cu^{2+}_{aq} + 2(HL)_{2,org} \xrightleftharpoons{K_{ex1}} CuL_2(HL)_{2,org} + 2H^+ \quad (8)$$

$$\log D = 2\log\left[(HL)_2\right]_{org} + 2\,pH + \log K_{ex1} \quad (9)$$

Percentage extraction (%E) and the distribution coefficient (D) were calculated to determine the extracting capability of lauric acid diluted in cyclohexane, dichloromethane , toluene and chloroform (Tables 1- 4). The distribution coefficient (D) and extraction percentage (%E) increased with increasing pH. The visible spectra of the loaded organic phase were performed in dichloromethane, toluene chloroform, and cyclohexane (Figure 4). Copper laurates shows an absorbance in the 660-678 nm region which indicated to octahedral geometry coordination of the extracted species [12]. In this study, it was found that two molecules of dimeric lauric acid coordinating with the central copper ion, water molecules would complete the coordination sphere so that the Copper ion could obtain a stable structure. On the basis of this interpretation, the plausible structure of the monomeric copper laurates $CuL_2(HL)_2$ in nonpolar solvents may be written as shown in Figure 5. This structure is similar with that obtained by Adjel [28] on the extraction of copper (II) by capric acid in chloroform.

Analysis of the extraction equilibrium in polar solvents

The degree of polymerization of the extracted copper laurates is given by the slope of the plot ($\log D + \log[Cu^{+2}]_{aq}$) versus ($\log[Cu^{+2}]_{aq}$ + 2pH) at a constant [HR]. The results obtained for 1-octanol and MIBK are illustrated in Figure 6. It is evident from Figure 6, linear relations are observed with the slope of one (j = 1) which indicates that in organic phase monomeric species with lauric acid occur.

According to equation (6), the plots of $\log D$ versus pH were determined at a constant [HR] in 1-octanol and MIBK, The results are shown in Figure 7a and Figure 7b, linear relations are observed with the slope of 2 (n=2), confirming that the monomeric extracted species in organic phase is $CuL_2(HL)_m$ ($j=\alpha_{Cu}=1$). In the other hand, the

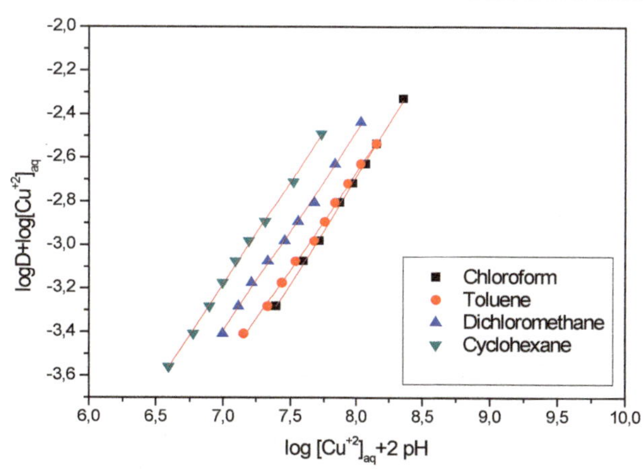

Figure 1: Determination of the degree of polymerization of copper laurates in non polar solvents.

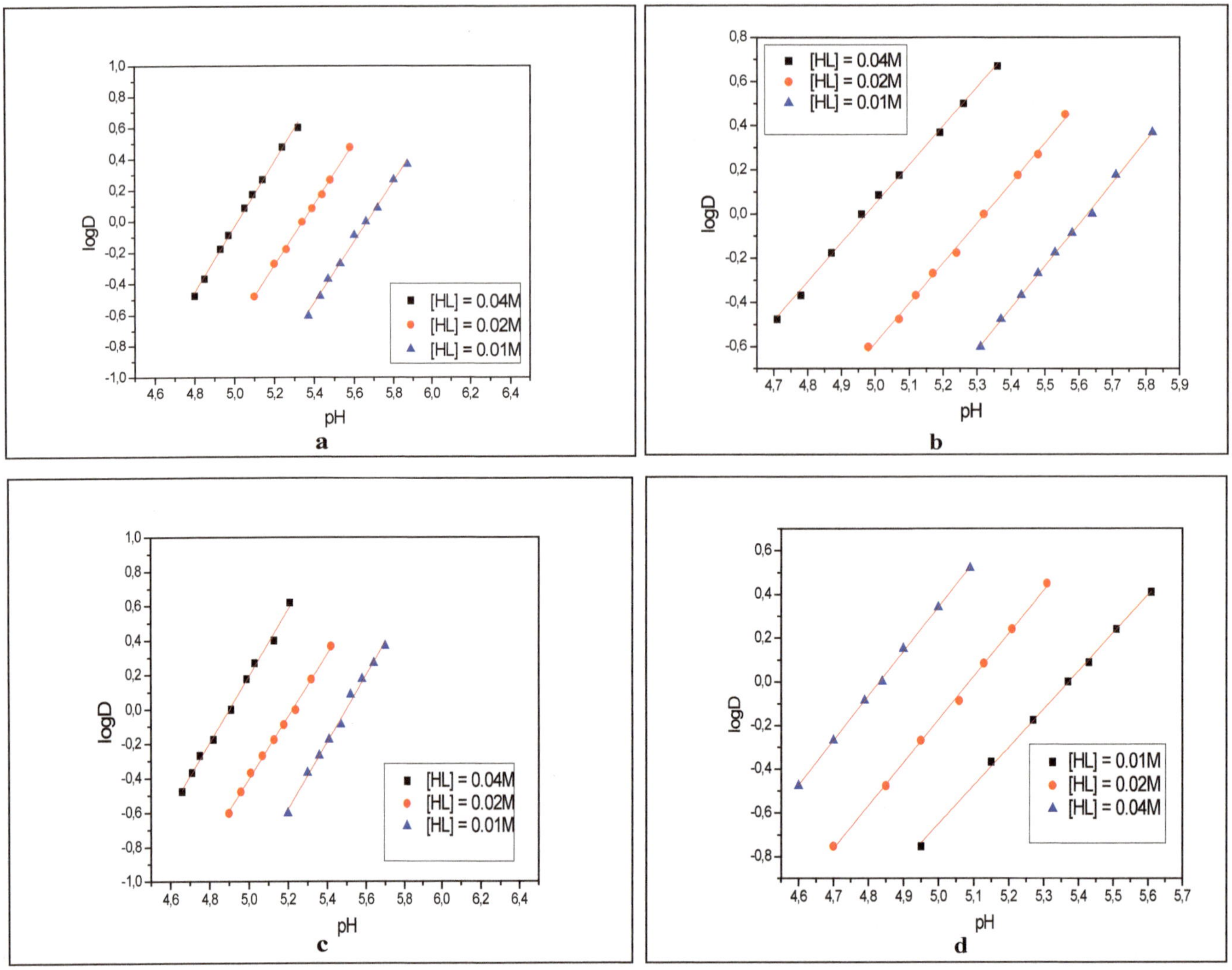

Figure 2: Distribution coefficient log *D* against pH at the extraction of Copper(II) with lauric acid dissolved in nonpolar solvents : a = cyclohexane, b = Dichloromethane, c = chloroform, d = toluene.

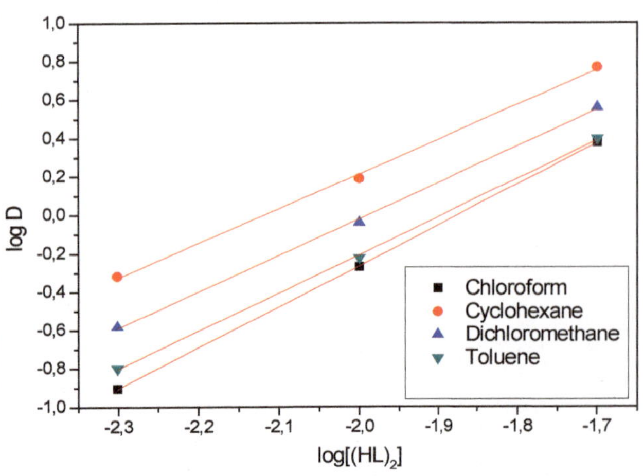

Figure 3: Determination of the number of lauric acid involved in the extracted species at pH=5.2.

effect of lauric acid concentration on Copper (II) extraction enables us to determine the number of lauric acid molecules involved in the extracted species. This effect was investigated by making a series of Copper (II) extractions at various [HR] dissolved in 1-octanol and MIBK.

According to equation (6), Plots of log D versus $\log[\mathrm{HL}]_{org}$ at constant pH were linear with a slope of 2, as shown in Figure 8, that is, $2 + m = 2$ or $m = 0$. Consequently, the composition of the extracted species is CuL_2. This result is in accordance with that reported by Yamada et al. [27]. Therefore, equations (4) and (6) could be rewritten as:

$$\mathrm{Cu}_{aq}^{+} + 2(\mathrm{HL})_{org} \underset{}{\overset{ex2}{\rightleftharpoons}} \mathrm{CuL}_{2,org} + 2\mathrm{H}^{+} \quad (10)$$

$$\log D = 2\log[\mathrm{HL}\ ]_{org} + 2\ \mathrm{pH} + \log\mathrm{K}_{ex2} \quad (11)$$

Percentage extraction (%E) and the distribution coefficient (D) were calculated to determine the extracting capability of lauric acid diluted in 1-octanol and MIBK (Tables 5 and 6). The distribution coefficient (D) and extraction percentage (%E) increased with increasing pH.

The visible spectra of copper laurates in MIBK and 1-octanol showed an absorbance at 668 and 700 nm respectively (Figure 9). This

pH	$[Cu]_{aq}$ x 10^{-3} (mol/l)	$[Cu]_{org}$ x 10^{-3} (mol/l)	*D*	%E
4.53	1.25	0.31	0.24	19.75
4.60	1.18	0.38	0.32	24.77
4.70	1.02	0.54	0.53	34.80
4.79	0.86	0.70	0.81	44.83
4.84	0.78	0.78	0.99	49.84
4.93	0.62	0.94	1.49	59.87
5.04	0.47	1.09	2.32	69.90
5.15	0.31	1.25	3.98	79.93
5.27	0.15	1.41	8.96	89.96

Table 1: The distribution coefficient (D) and extraction percentage (%E) for extraction of copper (II) with lauric acid (0.04M) diluted in cyclohexane. $[Cu]_{aq.initial}$= $1.57x10^{-3}$M.

pH	$[Cu]_{aq}$ x 10^{-3} (mol/l)	$[Cu]_{org}$ x 10^{-3} (mol/l)	D	%E
4.66	1.18	0.38	0.32	24.77
4.71	1.10	0.46	0.42	29.78
4.75	1.02	0.54	0.53	34.80
4.82	0.94	0.62	0.66	39.81
4.86	0.86	0.70	0.81	44.83
4.91	0.78	0.78	0.99	49.84
4.94	0.70	0.86	1.21	54.86
4.99	0.62	0.94	1.49	59.87
5.03	0.55	1.01	1.84	64.89
5.13	0.39	1.17	2.98	74.92
5.24	0.23	1.33	5.64	84.95

Table 2: The distribution coefficient (D) and extraction percentage (%E) for extraction of copper (II) with lauric acid (0.04M) diluted in dichloromethane. $[Cu]_{aq.initial}$= $1.57x10^{-3}$M.

pH	$[Cu]_{aq}$ x 10^{-3} (mol/l)	$[Cu]_{org}$ x 10^{-3} (mol/l)	D	%E
4.71	1.18	0.38	0.32	24.77
4.78	1.10	0.46	0.42	29.78
4.87	0.94	0.62	0.66	39.81
4.92	0.86	0.70	0.81	44.83
4.96	0.78	0.78	0.99	49.84
5.01	0.70	0.86	1.21	54.86
5.07	0.62	0.94	1.49	59.87
5.14	0.55	1.01	1.84	64.89
5.19	0.47	1.09	2.32	69.90
5.29	0.31	1.25	3.98	79.93
5.44	0.15	1.41	8.96	89.96

Table 3: The distribution coefficient (D) and extraction percentage (%E) for extraction of copper (II) with lauric acid (0.04M) diluted in toluene. $[Cu]_{aq.initial}$= $1.57x10^{-3}$M.

result indicated to octahedral geometry coordination of the extracted species [27]. In this study, it was found that two molecules of monomeric lauric acid coordinating with the central Copper ion, water and/or solvent molecules coordinate to copper atoms in place of lauric acid molecules. On the basis of this interpretation, the plausible structure of the monomeric copper laurates CuL_2 in polar solvents may be written as shown in Figure 10. A similar structure was obtained by Yamada et al. [27] on the extraction of Copper (II) with decanoic acid into 1-octanol.

Analysis of the extraction equilibrium in nonpolar and polar solvents

In order to compare the extraction properties of several solvents, a series of Copper (II) extractions were performed using lauric acid 0.02 M dissolved in toluene, dichloromethane, chloroform, cyclohexane, MIBK and 1-octanol. The results obtained are presented in Figure 11 as plots of log D against pH. The relationships have a linear nature of

pH	$[Cu]_{aq}$ x 10^{-3} (mol/l)	$[Cu]_{org}$ x 10^{-3} (mol/l)	D	%E
4.8	1.18	0.38	0.32	24.77
4.85	1.10	0.46	0.42	29.78
4.93	0.94	0.62	0.66	39.81
4.97	0.86	0.70	0.81	44.83
5.05	0.70	0.86	1.21	54.86
5.09	0.62	0.94	1.49	59.87
5.14	0.55	1.01	1.84	64.89
5.24	0.39	1.17	2.98	74.92
5.32	0.31	1.25	3.98	79.93

Table 4: The distribution coefficient (D) and extraction percentage (%E) for extraction of copper (II) with lauric acid (0.04M) diluted in chloroform. $[Cu]_{aq.initial}$= $1.57x10^{-3}$M.

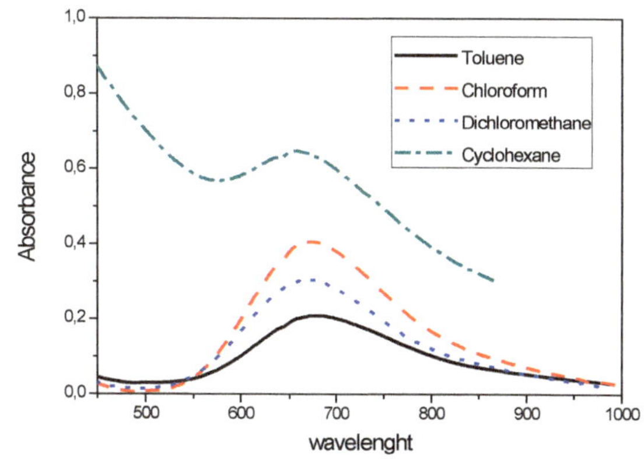

Figure 4: Visible spectra of copper laurates in nonpolar solvents.

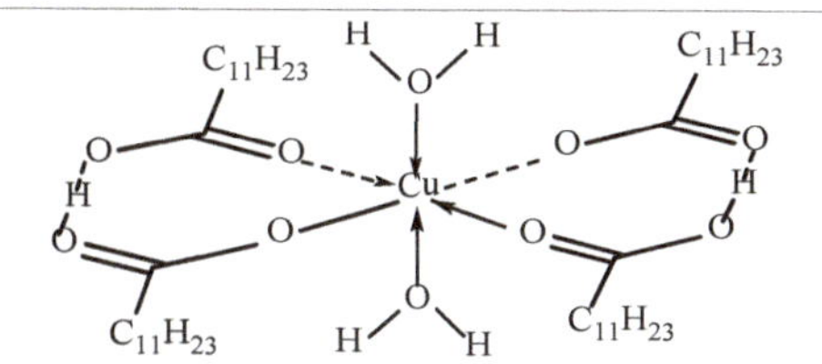

Figure 5: The plausible structure of the complex of copper ion with lauric acid dissolved in nonpolar solvents.

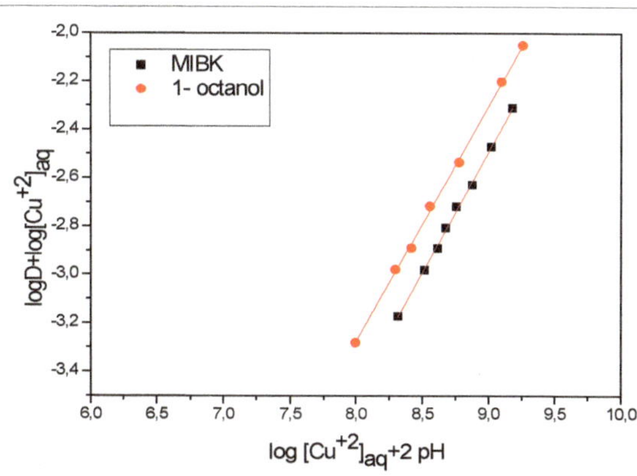

Figure 6: Determination of the degree of polymerization of copper laurates in polar solvents.

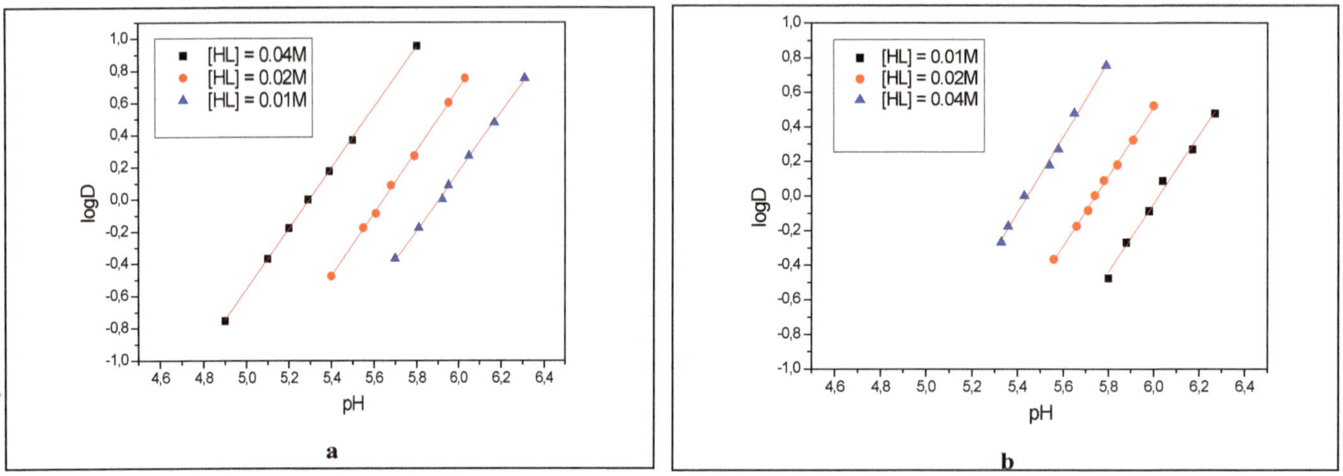

Figure 7: Distribution coefficient log *D* against pH at the extraction of Copper(II) with lauric acid dissolved in polar solvents : a = 1-octanol, b = MIBK. ane, c = chloroform, d = toluene.

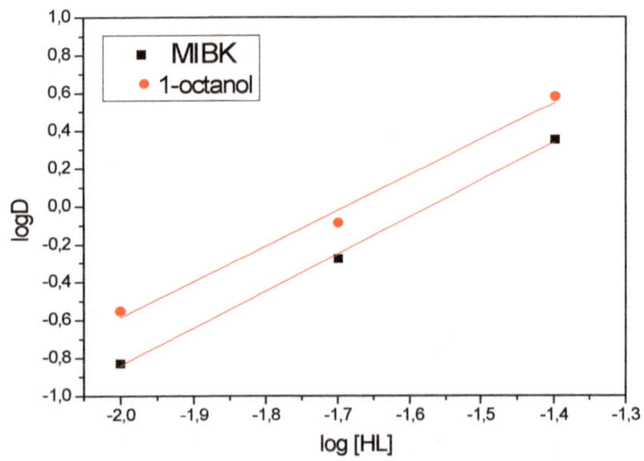

Figure 8: Determination of the number of lauric acid involved in the extracted species at pH=5.6.

pH	$[Cu]_{aq}$ x 10^{-3} (mol/l)	$[Cu]_{org}$ x 10^{-3} (mol/l)	D	%E
4.9	1.33	0.23	0.17	14.73
5.1	1.10	0.46	0.42	29.78
5.2	0.94	0.62	0.66	39.81
5.29	0.78	0.78	0.99	49.84
5.39	0.62	0.94	1.49	59.87
5.5	0.47	1.09	2.32	69.90
5.6	0.31	1.25	3.98	79.93
5.7	0.23	1.33	5.64	84.95

Table 5: The distribution coefficient (D) and extraction percentage (%E) for extraction of copper (II) with lauric acid (0.04M) diluted in 1-octanol. $[Cu]_{aq.initial}$= $1.57x10^{-3}$M.

slopes about to 2. From these results, it was observed that nonpolar solvents showed better performance in Copper (II) extraction than polar solvents. This result is in accordance with that obtained in literature [24]. Also, it is clear that the cyclohexane was as the preferred solvent for copper (II) extraction, may be explained by the total absence of interactions between lauric acid and the solvent. Oukebdane et al. [29] tested the extraction of copper (II) by a mixture of lauric acid and TOPO in different solvents (chloroform, nitromethane and cyclohexane). They observed that cyclohexane as the favored solvent.

On the other hand, the extraction of Copper (II) in dichloromethane is better than in chloroform and toluene, these results can depend on dielectric constants of the solvents. Dichloromethane having high dielectric constants is favored for the extraction of all the metal ions. In addition, the better salvation of the complexes by dichloromethane may be a valuable reason for better extraction [30,31]. The Copper (II) extraction in chloroform is weakest from the stronger interactions between extractant and chloroform in the organic phase, these interactions decreased the activity of lauric acid. However, in polar solvents, 1-octanol and MIBK, it was observed that the distribution coefficient decreased as the polarity of solvents increased. It can be explained that the polar solvents are capable to forming hydrogen bonds with electron donor atoms of lauric acid lead to a reduction in the active concentration of the lauric acid, consequently, poorer extraction of Copper results.

In MIBK, which is a more polar solvent, the extraction proceed rabidly and a lower amount of Copper (II) was extracted than that in 1-octanol, would probably explained by the stronger chemical interaction between MIBK and lauric acid in addition to the solubility of MIBK in aqueous phase. This result is in accordance with that observed by Ghebghoub et al. [25]. While a poor extraction was achieved when using MIBK as solvent on the extraction of Copper (II) by di(2-ethylhexyl)phosphoric acid. Table 7 gives the stoichiometries of the extracted copper species with lauric acid in different solvents and their extraction constants. The log K_{ex} values are higher in nonpolar solvents than in polar solvents. This result can be explained by the extensive solvation of lauric acid in polar solvents [32].

Thermodynamic studies of copper (II) extraction with lauric acid in different solvent

In order to investigate the solvent effects on the thermodynamic parameter of Copper (II) extraction, the free energy (ΔG°) was calculated for different solvents: chloroform, dichloromethane, cyclohexane, toluene, 1-octanol and methyl isobutyl ketone (MIBK) at constant tempetrature T = 298.2 k. The ΔG° is expressed by the following equation [17,33].

$$\Delta G^\circ = -2.303\, RT\, logK_{ex} \quad (12)$$

Where R is the universal gas constant (8.314 J/mol K), T is the thermodynamic temperature and K_{ex} is the extraction constant. The

pH	$[Cu]_{aq}$ x 10^{-3} (mol/l)	$[Cu]_{org}$ x 10^{-3} (mol/l)	D	%E
5.33	1.023	0.54	0.53	34.80
5.36	0.94	0.62	0.66	39.81
5.43	0.78	0.78	0.99	49.84
5.54	0.62	0.94	1.49	59.87
5.58	0.55	1.01	1.84	64.89
5.65	0.39	1.17	2.98	74.92
5.79	0.23	1.33	5.64	84.95

Table 6: The distribution coefficient (D) and extraction percentage (%E) for extraction of copper (II) with lauric acid (0.04M) diluted in MIBK. $[Cu]_{aq.initial}$= 1.57x10^{-3} M.

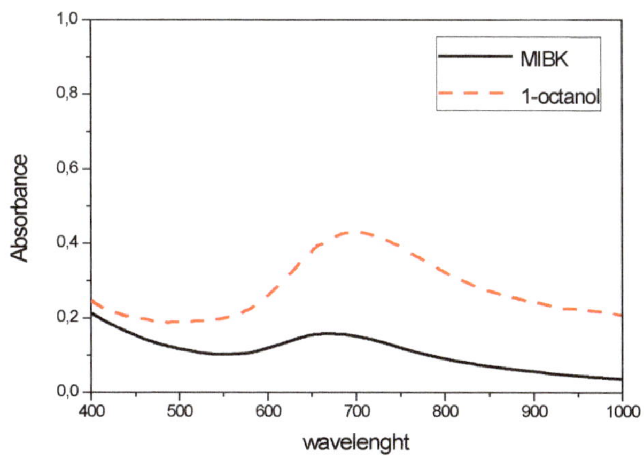

Figure 9: Visible spectra of copper laurates in polar solvents.

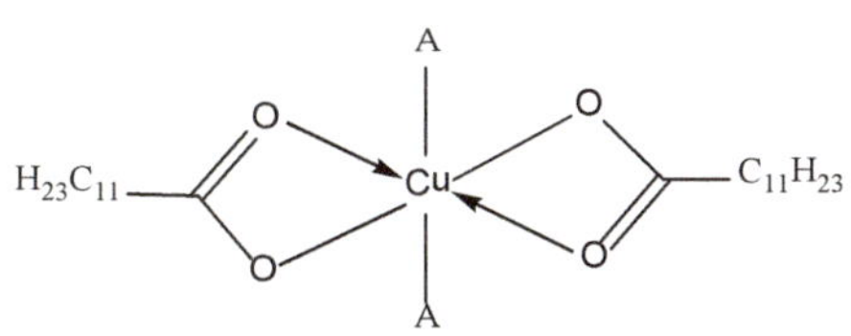

Figure 10: VThe plausible structure of the complex of copper ion with lauric acid dissolved in polar solvents. A: H_2O or 1-octanol or MIBK.

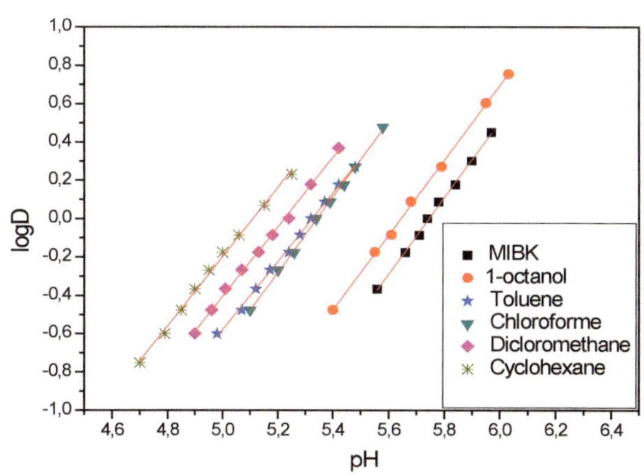

Figure 10: Distribution coefficient log *D* against pH at the extraction of copper(II) with lauric acid dissolved in different solvents. [HL] = 0.02M; [Cu^{2+}] = 1.57 x 10^{-3}M.

Solvant	$logK_{ex}$	ΔG° (kJ/mol)	Extracted species
Cyclohexane	-10.71	61.15	$CuL_2(HL)_2$
Dichloromethane	-11.08	63.26	$CuL_2(HL)_2$
Toluene	-11.31	64.57	$CuL_2(HL)_2$
Chloroform	-11.39	65.03	$CuL_2(HL)_2$
1-octanol	-11.89	67.88	CuL_2
MIBK	-12.14	69.31	CuL_2

Table 7: The extraction constant and free energy (ΔG°) for extraction of copper (II) with lauric acid diluted in several solvents at T= 298.2 K.

results obtained are shown in Table 7. It is evident from Table 7, the free energy (ΔG°) increases with the polarity of solvents. In addition, the (ΔG°) value is positive. The positive value of (ΔG°) indicated that the Copper (II) extraction with lauric acid didn't occurred spontaneously in all solvents examined.

Conclusion

The results obtained in this work have established the feasibility of using lauric acid to remove the heavy metal ions like Copper (II) from aqueous solutions. Lauric acid extracts Copper (II) as a monomeric Copper laurates from nitrate medium according to the following stoichiometric relations.

$$Cu^{2+}_{aq} + 2(HL)_{2,org} \underset{}{\overset{K_{ex1}}{\rightleftharpoons}} CuL_2(HL)_{2,org} + 2H^+ \text{ for nonpolar solvent}$$

$$Cu^{2+}_{aq} + 2(HL)_{org} \underset{}{\overset{K_{ex2}}{\rightleftharpoons}} CuL_{2,org} + 2H^+ \text{ for polar solvent}$$

The choice of solvent, polar or nonpolar, has a greater effect on the extraction efficiency. The nonpolar solvents showed better performance than the polar solvents on the Copper extraction by lauric acid. The extraction capacity of lauric acid in different solvents decreases in the order: cyclohexane > dichloromethane > toluene ≥ chloroform > 1-octanol > MIBK. The positive value of the free energy (ΔG°) indicated that the copper(II) extraction with lauric acid didn't occurred spontaneously.

References

1. Aidi A, Barkat D (2010) Solvent extraction of copper (II) from sulfate medium with N-(2-hydroxybenzylidene) aniline. J Coord Chem 63: 4136-4144.

2. Barkat D, Derriche Z (2001) Synergistic solvent extraction of zinc (II) and cadmium (II) from sulfate medium by a mixture of 1-phenyl-3-methyl-4-benzoylpyrazol-5-one and methyl-isobutyl ketone. Turk J Chem 25: 381-389.

3. Barkat D, Kameche M (2007) Liquid-liquid extraction of copper (II) with Schiff's bases derived from salicylaldehyde. Phys Chem Liq 45: 289 - 293.

4. Barkat D, Kameche M, Tayeb A, Benabdellah T, Derriche Z, et al. (2004) Ionic strength effect on the liquid–liquid extraction of zinc(II) and cadmium(II) from sulphate medium by 1-phenyl-3-methyl-4-benzoylpyrazol-5-one in chloroform. Phys Chem Liq 42: 53-61.

5. Barkat D, Omari M (2005) Extraction of copper (II) with salicylideneaniline from sulphate and perchlorate media. Asian J Chem 17: 202-208.

6. Zhanga X, Li X, Cao H, Zhang Y (2010) Separation of copper, iron (III), zinc and nickel from nitrate solution by solvent extraction using LK-C2. Separation and Purification Technology 70: 306-313.

7. Nakasuka N, Mitsuoka Y, Tanaka M (1974) Extraction of zinc and cadmium with capric acid. J Inorg Nucl Chem 36: 431-438.

8. Parhi PK, Padhan E, Palai AK, Sarangi K, Nathsarma KC, et al. (2011) Separation of Co (II) and Ni (II) from the mixed sulphate/chloride solutionusing NaPC-88A. Desalination 267: 201-208.

9. Adjel F, Barkat D (2011) Synergistic extraction of copper (II) from sulfate medium with capric acid and tri-n-octylphosphine oxide in chloroform. J Coord Chem 64: 574-582.

10. Baba Y, Inoue K (1988) Solvent extraction equilibria of copper (II) by carboxylic acids with high acid dissociation constants. Hydrometallurgy 21: 203-212.

11. Preston JS (1985) Solvent extraction of metals by carboxylic acids. Hydrometallurgy 14: 171- 188.

12. Bartecki A, Apostoluk W (1978) Extraction of copper(II) from ammonium chloride solutions with capric acid. J Inorg Nucl Chem 40: 109-116.

13. Tanaka M, Niinomi T (1965) Extraction of copper (II) with capric acid dissolved in benzene. J Inorg Nucl Chem 27: 431 -437.

14. Elizalde MP, Castresana JM, Alonso MC (1985) Extraction of Ni(II) by n-dodecanoic acid dissolved in toluene. Polyhedron 4: 2097-2101.

15. Apostoluk W, Bartecki A (1985) Extraction of Chromium(III) from sodium chloride solutions by means of carboxylic acids. Hydrometallurgy 15: 191-202.

16. Elizalde MP, Castresana JM, Elorriaga AI (1987) Extraction of Fe(III) by n-dodecanoic acid dissolved in toluene. Polyhedron 6: 2155-2160.

17. Ghanadzadeh H, Abbasnejad S (2011) Separation of copper (II) with solvent extraction using lauric acid diluted in benzene. J Thermodyn Catal 2

18. Cui Y, Yang J, Yang G, Xia G, Nie Y, et al. (2012) Effect of diluents on extraction behaviour of rare earth elements with N,N,N',N'-tetrabutyl-3-oxy-glutaramide from hydrochloric acid. Hydrometallurgy 121-124: 16-21.

19. Rezaei K, Nedjate H (2003) Diluent effect on the distribution ratio and separation factor of Ni(II) in the liquid–liquid extraction from aqueous acidic solutions using dibutyldithiophosphoric acid. Hydrometallurgy 68: 11-21.

20. Van de Voorde I, Pinoy L, Courtijn E, Verpoort F (2005) Influence of acetate ions and the role of the diluents on the extraction of copper (II), nickel (II), cobalt (II), magnesium (II) and iron (II, III) with different types of extractants. Hydrometallurgy 78

21. Yamada H, Tanaka M (1976) Solvent effects on extraction of copper with capric acid. J Inorg Nucl Chem 38: 1501-1504.

22. Chowdhury M, Sanyal S K (1994) Diluent effect on extraction of tellurium (IV) and b selenium (IV) by tri-n butyl phosphate. Hydrometallurgy 34: 319-330.

23. El-Nadi Y A (2010) Effect of diluents on the extraction of praseodymium and samarium by Cyanex 923 from acidic nitrate medium. J Rare Earths 28: 215-220.

24. Yamada H, Imai S, Takeuchi E (1983) Solvent effects on the extraction of Gallium (III) with decanoic acid. Bull Chem Soc Jpn 56: 1401-1406.

25. Ghebghoub F, Barkat D (2009) The effect of diluents on extraction of copper(II) with di(2-ethylhexyl) phosphoric acid. J Coord Chem 62: 1449-1456.

26. Schweitzer G K, Morris D K (1969) The partition and dimerization of octanoic acid in several organic-aqueous systems. Anal Chim Acta 45: 65-70.

27. Yamada H, Suzuki S, Tanaka M (1981) Extraction of Cu(II) with decanoic acid into 1-octanol. J Inorg Nucl Chem. 43: 1873-1879.

28. Adjel F (2014) Mixtures of various synergists to capric acid in the extraction of transition metals.

29. Oukebdane K, Didi MA, Azzouz A, Villemin D (2004) Solvent extraction of copper (II) by synergistic mixtures of trioctylphosphine oxide and lauric acid. Scientific Study and Research (1-2): 59-64.

30. Al Zoubi W, Kandil F, Chebani M K (2012) Synthesis of macrocyclic schiff bases based on Pyridine-2,6-Dicarbohydrazide and their use in Metal Cations Extraction. Organic Chem Current Res 1

31. Strasser BO, Popov AI (1985) Influence of solvent properties on the kinetics of complexation of the sodium ion with 18-crown-6. J Am Chem Soc 107: 7921-7924.

32. Yamada H, Kato C (1993) Solvent and steric effects on the extraction of copper (II) with pivalic acid. Talanta 40: 1049-1057.

33. Choppin GR (2004) Complexation of metal ions, in: Rydberg J, Cox M, Musikas, Choppin GR (Eds.) Solvent Extraction Principles and Practice, Marcel Dekker Inc., USA.

Thermodynamic Calculation of Melting Heat

Sobko AA*

Department of Physics and Chemistry of New Materials AM Prokhorov Academy of Engineering Sciences 19 Presnensky Val, Moscow 123557, Russia

Abstract

In the present paper, the method for approximate calculation of phase space volumes proposed in Ref. [1] is extended. Expressions for the melting heat of materials with f.c.c., b.c.c. and h.c.p. lattices are derived. Experimental results are compared to numerical calculations for a number of elements.

Keywords: Thermodynamic; Heat; First-order phase transitions

Calculation of a Phase Volume for Solids

A principally new approach to the calculation of the heat of first-order phase transitions was proposed in Ref. [1]. It is based on an approximate calculation of the phase volumes of a system in different phases: solid, liquid or gaseous and calculation of the change of entropy given as a logarithmic ratio of the volumes of old and new phases $\Delta S = k_s \ln\frac{(V_{ph})_n}{(V_{ph})_o}$. The volumes of phase space for liquid and gaseous phases are also calculated therein:

$$V_{phL} = \frac{(3\pi mRT)^{3N/2}}{(3N/2)!} V_L^N \quad (1)$$

$$V_{phG} = \frac{(3\pi mRT)^{3N/2}}{(3N/2)!} V_G^N \quad (2)$$

Where V_L and V_G are the volumes of liquid and gas, respectively. The expression for the evaporation heat is obtained in Ref. [1] in terms of other thermodynamic parameters of the process. A good agreement of numerical results and experimental data suggests that the method is versatile and can be used for calculating the transition heat of other first-order phase transitions. As it was shown in Ref. [1], the calculation should take into account the work performed against external forces during a phase transition with employment of the free volume. The expression for the phase volume of liquid was obtained in Ref. [1], whereas the calculation of a melting heat requires the phase space volume of solid state. The following model is used to find this volume. Every atom (molecule, ion) of a solid is assumed to vibrate near the equilibrium state, and its energy in a self-consisted periodic field is $e_n(\vec{r},\vec{p})$.

Quadratic expansion of energy into powers of $\vec{p}$ gives:

$$e_n(\vec{r},\vec{p}) = e_n(\vec{r}) + \sum_{i,j,}^{3} \frac{\partial^2 e_n}{\partial p_i \partial p_j} p_i p_j$$

$e_n(\vec{r})$ can be interpreted as an effective potential energy $U_n(\vec{r})$, $\frac{\partial e_n}{\partial p_i}\Big|_{p_i=0} = 0$.

By diagonalizing the quadratic form according to the standard technique, we obtain:

$$\frac{p_x^2}{2m_{xx}} + \frac{p_y^2}{2m_{yy}} + \frac{p_z^2}{2m_{zz}} = e_n - U(\vec{r}) \quad (3)$$

Where m_{xx}, m_{yy}, m_{zz} are components of the effective mass tensor. $1/m_{ij} = \partial^2 e_n / \partial p_i \partial p_j |_{0,0}$. This approach is utilized in solid state physics, for example, in Ref. [2]. Equation (3) is the elliptic equation with the axes $(2m_{xx}(e_n - U_n(\vec{r})))^{\frac{1}{2}}, (2m_{yy}(e_n - U_n(\vec{r})))^{\frac{1}{2}}, (2m_{zz}(e_n - U_n(\vec{r})))^{\frac{1}{2}}$, the corresponding volume of the ellipsoid is

$$\frac{4\pi}{3}(8m_{xx}m_{yy}m_{zz})^{1/2}(e_n - U_n(\vec{r}))^{\frac{3}{2}}$$

Integration over $\vec{p}$ in the phase space gives:

$$V_{ph} = \iint_{3N} \prod_{n} \frac{4\pi}{3}(8m_{xx}m_{yy}m_{zz})^{\frac{1}{2}}(e_n - U_n(\vec{r}))^{\frac{3}{2}} d^3\vec{r}$$

It was assumed in Ref. [1] that near a point of first-order phase-transition, the majority of atoms are in the state with the average kinetic energy. This assumption is proved by a good agreement between numerical results and experimental data on evaporation heat. Consequently, according to the equipartition theorem for kinetic energy [3], one can substitute $3KT/2$ for $(e_n - U_n(\vec{r}))$. Thus, the approximate volume of phase space for solid is given by:

$$V_{PhS} = \left(\frac{4\pi}{3}\left(8m_{xx}m_{yy}m_{zz}(3kT/2)^3\right)^{1/2} V_S\right)^N = \left(\frac{4\pi}{3}\left(m_{xx}m_{yy}m_{zz}(3kT)^3\right)^{1/2} V_S\right)^N, \quad (3)$$

Calculation of the Melting Heat

The volume of phase space for solid is given by (3). In the first part of the present work [1] it was shown that the employment of the free volume (for unit mole, the free volume is $V_f = V - N_A V_a$, where V is the geometrical volume, N_A is the Avogadro constant, and V_a is the volume of atom (molecule, ion)) gives a substantially better agreement between numerical and experimental results. The phase space volume for solid state is expressed as:

$$V_{phS} = \left(\frac{4}{3}\pi\left(m_{xx}m_{yy}m_{zz}(3kT)^{\frac{3}{2}} V_{Sf}\right)^N \quad (4)$$

and the phase space volume for liquid state is, respectively, (1):

$$V_{phL} = \frac{(3\pi mRT)^{3N/2}}{(3N/2)!} V_{Lf}^N$$

Correspondingly, the logarithm of the ratio V_{phL}/V_{phS} is equal to:

$$\ln\left((3\pi mRT)^{3N/2} V_{Lf}^N / (3N/2)!\right) - \ln\left(4\pi\left((m_{xx}m_{yy}m_{zz})(3kT)^3\right)^{1/2} V_{Sf}/3\right)^N$$

By using the Stirling formula $\ln((3N/2)!) = 3N(\ln(3N/2))/2 - 3N/2$ and taking into account R=kNA, we obtain:

$$\Delta S = \frac{R}{2}\left(\ln\frac{\pi e^3}{6} + \ln\frac{m^3}{m_{xx}m_{yy}m_{zz}}\right) + R\ln\frac{V_{Lf}}{V_{Sf}}$$

***Corresponding author:** Sobko AA, Department of Physics and Chemistry of New Materials, AM Prokhorov Academy of Engineering Sciences 19 Presnensky Val, Moscow 123557, Russia, E-mail: ainrf@mail.ru

Thus, the expression for the melting heat has the form:

$$\lambda = \frac{RT}{2}\left(\ln\frac{\pi e^3}{6} + \ln\frac{m^3}{m_{xx}m_{yy}m_{zz}}\right) + RT\ln\left(1+\frac{\Delta V}{V_{Sf}}\right) + P\Delta V \qquad (5)$$

Under normal pressure, the term $P\Delta V$ is on the order of $10^5 * 10^{-6} \sim 10^{-1}$, and the value of the first term is $\sim 10^4$, hence, up to pressures of $\sim 10^{10}$ Pa the term $P\Delta V$ can be neglected. $V_{Lf} = V_{Sf} + \Delta V$. Finally, the expression for the melting heat at pressures below 10^{10} Pa has the form:

$$\lambda = \frac{RT}{2}\left(\ln\frac{\pi e^3}{6} + \ln\frac{m^3}{m_{xx}m_{yy}m_{zz}}\right) + RT\ln\left(1+\frac{\Delta V}{V_{Sf}}\right) \qquad (6)$$

Let us term the dimensionless expression

$$S_0 = \frac{1}{2}\left(\ln\left(\frac{\pi e^3}{6}\right) + \ln\left(\frac{m^3}{m_{xx}m_{yy}m_{zz}}\right)\right) \qquad (7)$$

the structural melting constant. Its first term is determined by a substantial difference between liquid and solid states, and the second one makes allowance for the corrections related to symmetry and forces of interaction between atoms for a particular crystal. Note that the expression $\ln\left(\frac{m^3}{m_{xx}m_{yy}m_{zz}}\right)$ is sensitive to the values of effective masses. Let the effective masses be slightly greater than m, then $m_{xx} = m(1+0.0n), m_{yy} = m(1+0.0n), m_{zz} = m(1+0.0n)$, where n is ~3-7 and we have $\ln\left(\frac{m^3}{(1.0n)^3 m^3}\right) = 3\ln\left(1-\frac{n}{100}\right) = -\frac{3n}{100}$, $\ln\left(\frac{\pi e^3}{6}\right)$=2.353. Taking effective masses into account may reduce the value of the structural melting constant by 10-15% since the effective mass is, as a rule, greater than the mass of a free particle.

Calculation of effective masses is beyond the scope of the present work because it is a hard computational task for a particular material. The effective masses can be found from the analysis of phonon spectra as well, but this is the subject to a separate work. The present work considers only geometric factors in the expressions for effective masses for three different types of lattice: the face-centered lattice (f.c.c.), body-centered lattice (b.c.c.), and hexagonal close-packed lattice (h.c.p.). In calculations of metal ion volumes, the ion radii are used, and in calculations of the volumes of atoms Ne, Ar, Kr, and Xe the corresponding van der Waals radii are used [4,5].

The expression for the melting heat (6) has the form:

$$\lambda = \frac{RT}{2}\left(\ln\frac{\pi e^3}{6} + \ln\frac{m^3}{m_{xx}m_{yy}m_{zz}}\right) + RT\ln\left(1+\frac{\Delta V}{V_S}\right)$$

Since, according to the Clapeyron-Clausius equation

$\frac{dT}{dP} = \frac{\Delta V}{\Delta S}$ and $\Delta S = \frac{\lambda}{T}$ we have $\Delta V = \frac{\lambda}{T}\frac{dT}{dP}$

By substituting this result into (6) we obtain the equation for calculating λ:

$$\lambda = RTS_0 + \ln\left(1+\frac{\lambda}{TV_{Sf}}\frac{dT}{dP}\right), \qquad (8)$$

Where dT/dP is the initial slope of the melting curve [6]. The obtained equation is simply solved numerically by the iterative method and provides the value of the melting heat. The results of calculations by formula (6) and the solutions of equation (8) are given in Tables 1-3. Experimental data for the calculations are taken from Ref. [4,5]. In Tables 1-3 the number of polymorphic phase transitions before melting is marked with asterisks *. Spaces in the tables mean that there is no available data on the jumps of volume in melting.

Melting heat for the f.c.c. lattice

To the f.c.c. lattice in the $\vec{p}$-space, in which the expansion was performed, corresponds the b.c.c. lattice. In view of the symmetry of b.c.c. lattice, we take: m_{xx}, m_{yy}, m_{zz} Then, the melting heat for the f.c.c. lattice takes the form:

$$\lambda = \frac{1}{2}RT\ln\left(\frac{\pi e^3}{6}\right) + RT\ln\left(1+\frac{\Delta V}{V_{Sf}}\right) \qquad (9)$$

The structural melting constant for the f.c.c. lattice is S_0=1.1765.

For low-melting elements with weak interaction between atoms the effective masses are approximately equal to atomic masses, whereas for hard-melting materials the difference can be substantial. Results of calculations of the melting heat for f.c.c. lattices by formula (9) and solutions of equation (8) are given in Table 1.

T is the melting temperature in K, λ_{ex} [J/mol] is the experimental value of the melting heat, V_S*10^5[m³/mol] is the molar volume of the solid state, ΔV*10^5[m³/mol] is the volume jump in melting, r*10^{10}[m] is the radius of atom (ion), V_{Sf}*10^5[m³/mol] is the free volume, (dT/dP)*10^8[K/Pa] is the initial slope of the melting curve, λ_{T1} [J/mol] is the molar melting heat calculated by formula (9), λ_{T2} [J/mol] is the molar melting heat obtained by solving equation (8), δ_1 and δ_2[%] are inaccuracies of λ_{T1} and λ_{T2}, respectively.

El-t	T	λ_{ex}	V_S	ΔV	r_a	V_{Sf}	λ_{T1}	δ_1	dT/dP	λ_{T2}	δ_2
Ne	24,48	324	1,397	0,22	1,60	0,365	335,4	-3,5	13,16	317	2,0
Ar	83,87	1210	2,412	0,35	1,91	0,656	1118	7,6	24,89	1102	8,9
Kr	116,6	1640	2,968	0,45	1,98	1,012	1497	8,7	30,13	1444	11,9
Xe	161,3	2290	3,709	0,56	2,18	1,098	2129	7,0	38,9	2082	9,1
Pb	600,7	4770	1,826	0,07	1,32	1,246	6146	-28,8	7,73	6182	-30,0
Al	933,5	10670	1,000	0,07	0,57	0,953	9676	9,3	6,41	9648	9,6
Ar	1235	11300	1,027	0,06	0,89	0,849	12775	-13,1	4,9	12664	-12,1
Au	1338	12700	1,019	0,05	0,91	0,829	13732	-8,1	6,2	13914	-9,6
Cu	1357	13000	0,709	0,04	0,72	0,615	13978	-7,5	4,69	14129	-8,7
Ac	1370	14200	2,260	...	1,18	1,846	...	...	19,4	14603	-2,8
Ni	1726	17600	0,659	0,05	0,78	0,539	18146	-3,1	3,7	17858	-1,5
Pd	1825	17200	0,885	...	0,64	0,819	...	...	6,4	19030	-10,6
Pt	2045	19700	0,910	0,06	0,70	0,824	21188	-7,6	6,2	21275	-8,0
Rh	2239	22600	0,829	...	0,86	0,669	...	...	5,9	23540	-4,2
Ir	2683	26400	0,857	...	0,89	0,679	...	...	6,2	28280	-7,1

Table 1: Calculation results of melting heat for f.c.c. lattices.

El-t	T	λ_{ex}	V_S	ΔV	r_a	V_{Sf}	λ_{T1}	δ_1	dT/dP	λ_{T2}	δ_2
Cs	301,5	2090	7,096	0,169	1,65	5,964	2150	-2,8	25,9	2156	-3,2
Rb	312,2	2200	5,579	0,193	1,49	4,745	2257	-2,6	21,1	2234	-1,6
K	336,8	2400	4,536	0,117	1,33	3,943	2405	-0,2	17,7	2411	-0,5
Na	371	2640	2,368	0,064	0,98	2,131	2650	-0,4	8,77	2648	-0,3
Li	453,7	3130	1,300	0,032	0,78	1,180	3230	-3,2	3,41	3205	-2,4
Pb	600,7	4770	1,826	0,070	1,32	1,246	4416	7,4	7,73	4363	8,5
Ba	983	7530	3,821	0,030	1,43	3,084	6859	8,9	4,3	6990	8,7
Sr**	1042	8000	3,450	0,177	1,27	2,934	7694	3,8	16,5	7531	5,9
Eu	1095	9200	2,898	0,140	1,12	2,544	8040	12,6	12,8	7875	14,4
Yb*	1097	7700	2,484	0,100	1,07	2,175	7976	-3,6	17,0	8075	-4,9
La**	1194	10040	2,260	0,012	1,14	1,887	8298	17,3	2,5	8325	17,1
Pr*	1204	8200	2,080	...	1,06	1,780	...	...	2,2	8389	-2,3
Nd*	1294	10680	2,059	0,019	0,99	1,814	9037	15,4	2,6	9031	15,4
Sm*	1350	10900	2,000	0,072	1,20	1,564	9816	9,9	6,0	9612	11,8
U**	1405	9200	1,256	0,029	0,97	1,026	10016	-8,9	2,8	9920	-7,8
Mn***	1517	12070	0,738	0,042	0,91	0,548	11392	5,6	3,3	11000	8,8
Be*	1551	13000	0,488	0,056	0,35	0,477	12128	6,7	5,0	11675	10,2
Gd*	1586	10200	1,990	0,041	0,97	1,760	11243	-10,2	5,7	11237	-10,2
Tb*	1629	10800	1,931	0,061	0,89	1,753	11699	-8,3	7,4	11636	-7,7
Dy*	1685	10900	1,900	0,093	0,89	1,722	12358	-13,4	12,0	12316	-13,0
Ho*	1747	14160	1,875	0,140	0,95	1,659	13226	6,6	14,3	12947	8,5
Y*	1795	11390	1,989	...	1,06	1,689	...	...	16,5	13432	-17,9
Er*	1802	17200	1,844	0,166	0,93	1,641	13872	19,3	15,0	13413	22,0
Fe**	1808	13800	0,709	0,030	0,83	0,565	13248	4,0	3,0	13033	5,5
Sc*	1814	15900	1,504	...	0,83	1,360	...	...	15,1	13726	13,7
Ti*	1933	15000	1,055	...	0,90	0,871	...	...	15,4	15454	-3,0
Th*	2023	15600	1,980	...	1,11	1,635	...	...	18,6	15342	1,6
Cr	2110	21000	0,723	...	0,84	0,574	...	...	15,9	18336	12,7
Zr*	2125	20000	1,402	...	1,09	1,076	...	...	16,3	16635	16,8
V	2160	23100	0,834	...	0,88	0,662	...	...	15,7	18158	21,4
Hf*	2503	20935	1,341	...	0,84	1,192	...	...	17,3	19489	6,9
Nb	2741	27200	1,084	...	0,77	0,969	...	...	16,2	21739	20,1
Mo	2890	27600	0,939	0,010	0,92	0,743	20254	26,6	0,96	20146	27,0
Ta	3269	31400	1,087	0,056	0,77	0,972	24069	23,3	5,4	23613	24,8
W	3680	35200	0,953	0,125	0,77	0,838	29634	15,8	7,8	27430	22,1

Table 2: Calculation results of melting heat for b.c.c. lattices.

El-t	T	λ_{ex}	V_S	ΔV	r_a	V_{Sf}	λ_{T1}	δ_1	dT/dP	λ_{T2}	δ_2
Cd	594,1	6110	1,300	0,04	1,14	0,927	5516	9,7	5,3	5565	8,9
Zn	692,7	6670	0,917	0,04	0,83	0,773	6479	2,9	4,8	6516	2,3
Mg	992	9040	1,398	0,04	0,78	1,278	9116	-0,8	7,5	9303	-2,9
Ca**	1112	9330	2,586	0,13	1,06	2,286	10445	-11,9	14,9	10485	-12,4
Pm	1441	12600	2,010	...	1,06	1,710	...	...	31,3	14956	-18,7
Co*	1768	15200	0,662	0,01	0,82	0,523	15014	1,2	3,5	15578	-2,5
Tm	1818	18400	1,812	0,13	0,87	1,646	17389	5,5	12,0	17251	6,2
Lu	1936	19200	1,778	0,06	0,85	1,623	17879	6,9	15,9	18753	2,3
Tc	2445	23810	0,860	...	0,95	0,644	...	...	5,5	23441	1,5
Ru	2583	23700	0,814	...	0,77	0,699	...	...	6,1	24802	-4,6
Os	3327	29300	0,843	...	0,89	0,665	...	...	6,5	32220	-10,0
Re	3453	33100	0,886	0,03	0,72	0,792	31913	3,6	3,2	31898	3,6

Table 3: Calculation results of melting heat for h.c.p. lattices.

Experimental values of the melting heat and the line of a simple regression are shown in Figure 1. The slope of the regression line is 1.1670, which well coincides (with a good accuracy of ~ 0.85%) with the structural melting constant of the f.c.c. lattice S_0=1.1765.

Melting heat for the b.c.c. lattice

To the b.c.c. lattice in the $\vec{p}$-space corresponds the f.c.c. lattice. Hence, taking into account geometrical factors we take

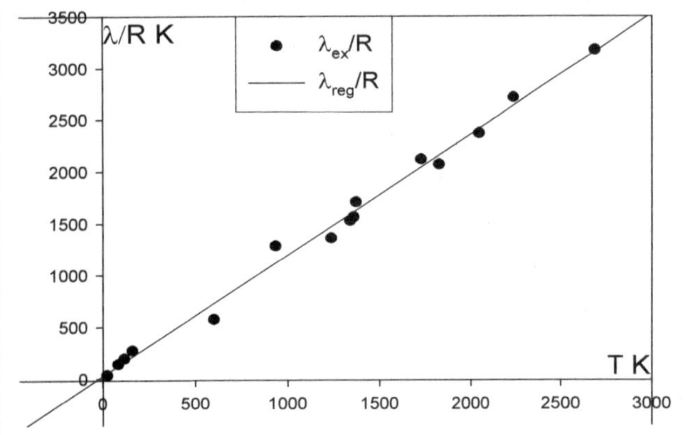

Figure 1: Experimental values of the reduced melting heat for elements with the f.c.c. lattice and the line of simple regression.

$m_{xx} = m_{yy} = m, m_{zz} = m\sqrt{2}$. Then the expression for the melting heat in the case of b.c.c. lattice takes the form

$$\lambda = \frac{1}{2}RT\left(\ln\left(\frac{\pi e^3}{6}\right) + \ln\left(\frac{m^3}{\sqrt{2}m^3}\right)\right) + RT\ln\left(1+\frac{\Delta V}{V_{Sf}}\right)$$

or

$$\lambda = \frac{1}{2}RT\ln\left(\frac{\pi e^3}{6\sqrt{2}}\right) + RT\ln\left(1+\frac{\Delta V}{V_{Sf}}\right) \qquad (10)$$

The structural melting constant for the b.c.c. lattice is S_0=1.0032.

Results of calculations of the melting heat for b.c.c. lattices by formula (10) are given in Table 2 along with the results of solving equation (8).

T is the melting temperature in K, λ_{ex}[J/mol] is the experimental value of the melting heat, V_S*10^5[m^3/mol] is the molar volume of the solid state, ΔV*10^5[m^3/mol] is the volume jump in melting, r*10^{10}[m] is the radius of atom (ion), V_{Sf}*10^5[m^3/mol] is the free volume, (dT/dP)*10^8[K/Pa] is the initial slope of melting curve, λ_{T1}[J/mol] is the molar melting heat calculated by formula (10), λ_{T2}[J/mol] is the molar melting heat obtained by solving equation (8), δ_1 and δ_2[%] are the inaccuracies of λ_{T1} and λ_{T2}, respectively.

Experimental values of the melting heat and the line of simple regression are presented in Figure 2 for the most low-melt elements. First seven metals were chosen with the lowest melting points for which the most relevant experimental data on melting temperature and melting heat are available. The slope of the regression line is 0.967 and matches the structural melting constant of the b.c.c. lattice S_0=1.0032 with a good accuracy of ~ 3.6%.

Melting heat for the h.c.p. lattice

h.c.p lattice transforms into itself in the $\vec{p}$ - space. The effective masses are, respectively, $m_{xx} = m_{yy} = m, m_{zz} = 2\frac{a}{c}m$. Thus, the expression for the melting heat for h.c.p. lattice has the form

$$\lambda = \frac{1}{2}RT\left(\ln\left(\frac{\pi e^3}{6}\right) + \ln\left(\frac{c}{2a}\right)\right) + RT\ln\left(1+\frac{\Delta V}{V_{Sf}}\right), \qquad (11)$$

where a and c are parameters of the h.c.p. lattice. Since the ratio *c/a* for the elements does not differ much from the ideal case $\frac{c}{a} = \sqrt{8/3}$, we may replace real values of the ratio *a/c* in the expression for S_0 with the ideal value and obtain the relationship for the melting heat for the h.c.p. lattice:

$$\lambda = \frac{1}{2}RT\ln\left(\frac{\pi e^3}{3\sqrt{6}}\right) + RT\ln\left(1+\frac{\Delta V}{V_{Sf}}\right) \qquad (12)$$

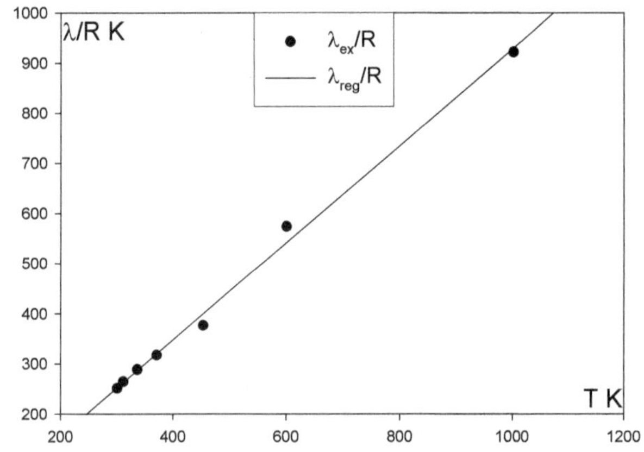

Figure 2: Experimental values of melting heat and the line of simple regression for elements with b.c.c. lattice and the lowest melting points.

The structural melting constant for the h.c.p. lattice is S_0=1.075.

Results of calculations by formula (12) are given in Table 3. The results of solving equation (8) for h.c.p. lattices are also presented in the table.

T is the melting temperature in K, λ_{ex} [J/mol] is the experimental value of the melting heat, V_S*10^5[m^3/mol] is the molar volume of the solid state, ΔV*10^5 [m^3/mol] is the volume jump in melting, r*10^{10}[m] is the radius of atom (ion), V_{Sf}*10^5 [m^3/mol] is the free volume, (dT/dP)*10^8 [K/Pa] is the initial slope of the melting curve, λ_{T1} [J/mol] is the molar melting heat calculated by formula (12), λ_{T2} [J/mol] is the molar melting heat obtained by solving equation (8), δ_1 and δ_2 [%] are the inaccuracies of λ_{T1} and λ_{T2}, respectively.

Experimental values of the melting heat are presented in Figure 3 along with the line of a simple regression. The slope of the regression line is 1.1186 and matches the structural melting constant of h.c.p. lattice S_0=1.0075 with a good accuracy of ~ 4%.

Analysis and Conclusions

As one can see from Tables 1-3, the proposed model gives a satisfactory description of the melting heat for the elements considered. In Ref. [7], experimental data on temperatures and melting heat values for eight well characterized elements are given. As one can see from Table 4, even for these elements the experimental results noticeably differ.

T_{min} and T_{max} are the minimal and maximal values of experimental melting temperature, λ_{min} and λ_{max} are the lowest and highest values of experimental melting heat. The same trend holds true for the values given in various handbooks. The results only coincide if data are taken from a single source. Consequently, it would not be correct to hope for a good agreement between experimental data and the numerical results calculated by formulae (9-10, 12).

The substantial difference between the experimental value of the melting heat for lead and the corresponding value calculated for the f.c.c. lattice along with the satisfactory agreement with the value of the melting heat calculated for the b.c.c. lattice suggest that a polymorphic transition from the f.c.c. lattice to b.c.c. lattice occurs near the melting point with the phase transition heat equal to $\lambda \approx 1700-1800$ J/mol. The phase transition heat is calculated in the frameworks of the proposed model.

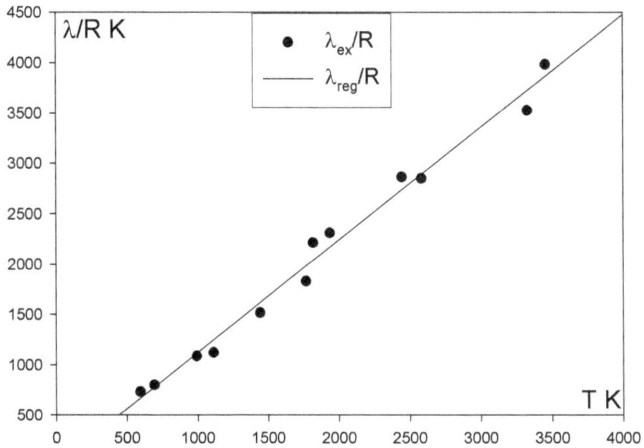

Figure 3: Experimental values of melting heat for elements with the h.c.p. lattice and the line of simple regression.

El-t	T_{min}	T_{max}	λ_{min}	λ_{max}	El-t	T_{min}	T_{max}	λ_{min}	λ_{max}
Al	930	933.74	7928.93	11300.9	Nb	2523	2773	26756.9	26849.8
Cu	1356	1357.8	7968.67	13453.3	Mo	2853	2901	20051.5	41254.2
Ni	1721	1744	15984.2	18663.4	Ta	3053	3275	24699.4	32027.8
Ti	1878	2093	15474.3	20923.6	W	3523	3683	33782.4	54971.1

Table 4: Spread of experimental values on temperatures and melting heat.

Thus, one can assert that the molar melting heat at pressures below 10^{10} Pa is given by the expression

$$\lambda = RTS_0 + RT\ln\left(1 + \frac{\Delta V}{V_{Sf}}\right)$$

or by a solution of the equation

$$\lambda = RTS_0 + \ln\left(1 + \frac{\lambda}{TV_{Sf}}\frac{dT}{dP}\right),$$

Where S_0=1.1765 for the f.c.c. lattice, S_0=1.0032 for the b.c.c. lattice and S_0=1.075 for the h.c.p. lattice. The rest values have been defined earlier.

References

1. Sobko AA (2016) Thermodynamic calculation of a heat of first-order phase transitions. Part I. J Thermodyn Catal 7.
2. JM Ziman (1960) Electrons and Phonons. The Theory of Transport Phenomena in Solids. Oxford at the Clarendon Press, UK.
3. Huang K (1963) Statistical mechanics. 2nd edn. John Wiley & Sons, New York.
4. Elementov S (1976) Properties of Elements. 1st edn. Samsonov GV, Moscow M (eds.).
5. Emsley J (1991) The elements. 2nd edn. Clarendon Press, Oxford, UK.
6. Tonkov EU (1979) Fazovye diagrammy elementov pri vysikom davlenii (Phase diagrams of the elements under high pressure), Handbook, Main Editorial Office of Physico-Mathematical Literature.
7. Stolovich NN, Minitskaya NS (1975) Temperaturnye zavisimosti svoistsv nekotorykh metallov (Tempereture dependencies of the properties for certain metals).

Structure, Potentiometric and Thermodynamic Studies of N-Acryloyl-4-amino salicylic acid and Its Metal Complexes in Monomeric and Polymeric Forms

El-Bindary AA[1*], Ghoneim MM [2], Diab MA [1], El-Sonbati AZ [1] and Serag LS [1]

[1]Chemistry Department, Faculty of Science, Damietta University, Damietta 34517, Egypt

[2]Chemistry Department, Faculty of Science, Tanta University, Tanta, Egypt

Abstract

N-Acryloyl-4-aminosalicylic acid (AAS) was synthesized and characterized using different spectroscopic techniques. The geometrical structures of the ligand are carried out by HF method with 3-21G basis set. The proton-ligand dissociation constants of AAS and its metal stability constants with (Mn^{2+}, Co^{2+}, Ni^{2+} and Cu^{2+}) have been determined potentiometrically in monomeric and polymeric forms using 2,2'-azobisisobutyronitrile as initiator. The potentiometric studies were carried out in 0.1 M (KCl) and 20% (by volume) ethanol-water mixture. The effect of temperature was studied at (298, 308 and 318 K) and the corresponding thermodynamic parameters (ΔG, ΔH and ΔS) were derived and discussed. The dissociation process is non-spontaneous, endothermic and entropically unfavorable. The formation of the metal complexes has been found to be spontaneous, endothermic and entropically favorable.

Keywords: *N*-Acryloyl-4-aminosalicylic acid; Molecular structure; Potentiometry;Thermodynamics

Introduction

Polymer complexes have been given a great deal of attention in recent years. The formation of chelates by polymers has widely been used for speciation [1], concentration, sepration [2] and recoveration [3] of metal ions. Stability constants are key parameters for the investigation of equilibria in solution and are very important in many fields such as industerial chemistry [4], enviromental studies [5], as well as medicinal [6] and analytical chemistry. Potentiometeric method is used to determine the average number of ligands coordinated with metal ion and, further, calculation the stability constants of polymer-metal complexes [7-10].

Salicylic acid and its derivatives are also biologically important ligands, a well known and widely used derivative, aspirin, reduces the risk of many diseases associated with ageing and is used in the treatment of rheumatic fever, pain, the preventation of thrombosis in vascular system analgesic, antipyretic, anti-flammatory bowel diseases (IBD) and anti-tuberculosis drug [11,12].

As a part of our continuous work reporting on the determination of dissociation and stability constants of some organic compounds and their metal complexes by potentiometric techniques [13-16], we report here the proton-ligand dissociation constants of *N*-Acryloyl-4-aminosalicylic acid (AAS) and its metal stability constants with (Mn^{2+}, Co^{2+}, Ni^{2+} and Cu^{2+}) in monomeric and polymeric forms. Furthermore, the corresponding thermodynamic functions are evaluated and discussed. The geometrical structures of the ligand are carried out by HF method with 3-21G basis set.

Experimental

Materials

All the compounds and solvents used were purchased from Aldrich or Sigma and used as received without further purification. Acryloyl chloride was used without further purification. It was stored below -18°C in a tightly glass-stoppered flask. 2,2'-Azobisisobutyronitrile (AIBN) was used as initiator for all polymerizations. It was purified by dissolving it in hot ethanol and left to cool. The pure material was being collected by filtration and then dried [17].

Preparation of N-Acryloyl-4-aminosalicylic acid monomer

N-Acryloyl-4-aminosalicylic acid monomer was prepared previously [13] by the reaction of equimolar amounts of acryloyl chloride and 4-aminosalicylic acid in dry benzene until the evolution of hydrogen chloride ceased forming a gray powder of AAS monomer (Figure 1).

Potentiometeric studies

N-Acryloyl-4-aminosalicylic acid solution (0.001 M) was prepared by dissolving an accurate weight of the solid in ethanol (AnalaR). Metal ion solutions (0.0001 M) were prepared from AnalaR metal chlorides in bidistilled water and standardized with EDTA [18]. Solutions of 0.001 M (HCl) and 1 M (KCl) were also prepared in bidistilled water. A carbonate-free sodium hydroxide solution in a 20% (by volume) ethanol-water mixture was used as titrant and standardized against oxalic acid (AnalaR).

The apparatus, general conditions and methods of calculation were the same as in previous work [13-16]. The following mixtures (i)-(iii) were prepared and titrated potentiometrically at 298 K against standard 0.002 M (NaOH) in a 20% (by volume) ethanol-water mixture:

***Corresponding author:** Chemistry Department, Faculty of Science, Damietta University, Damietta 34517, Egypt
E-mail: abindary@yahoo.com

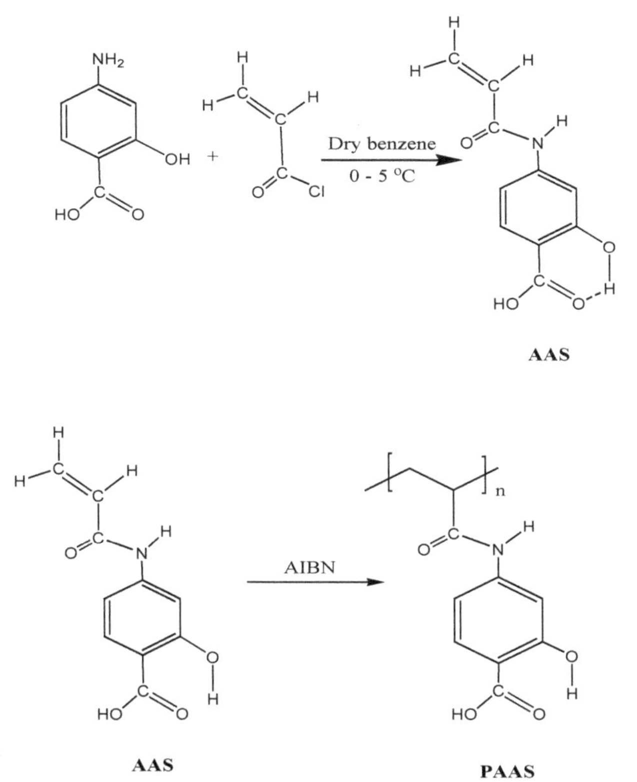

Figure 1: The formation mechanism of N-acryloyl-4-aminosalicylic acid in monomeric (AAS) and polymeric (PAAS) forms.

i) 5 cm^3 0.001 M (HCl)+5 cm^3 1 M (KCl)+10 cm^3 ethanol.

ii) 5 cm^3 0.001 M (HCl)+5 cm^3 1 M (KCl)+5 cm^3 0.00l M ligand + 5 cm^3 ethanol.

iii) 5 cm^3 0.001 M (HCl)+5 cm^3 1 M (KCl)+5 cm^3 0.001 M ligand + 10 cm^3 0.0001 M metal chloride + 5 cm^3 ethanol.

For each mixture, the volume was made up to 50 cm^3 with bidistilled water before the titration. The titrations also were carried out in the presence of 5 ml of AIBN (0.001 M) as initiator for the polymerization step (Figure 1). These titrations were repeated for temperatures of 308 and 318 K. All titrations have been carried out between pH 3.0 and 11.0.

Measurments

Spectroscopic data were obtained using the following instruments: FT-IR spectra (KBr discs, 4000-400 cm^{-1}) by Jasco-4100 spectrophotometer; the ^{1}H NMR spectrum by Bruker WP 300 MHz using DMSO-d_6 as a solvent containing TMS as the internal standard. The pH measurements were carried out using VWR Scientific Instruments Model 8000 pH-meter accurate to ± 0.01 units. The pH- meter readings in the non–aqueous medium were corrected [19]. The electrode system was calibrated according to the method of Irving et al. [20]. Titrations were performed in a double walled glass cell in an inert atmosphere (nitrogen) at ionic strength of 0.1 M KCl. Potentiometric measurements were carried out at different temperature. The temperature was controlled to within ± 0.05 K by circulating thermostated water (Neslab 2 RTE 220) through the outer jacket of the vessel.

The molecular structures of the investigated compound were optimized by HF method with 3-21G basis set. The molecules were built with the Perkin Elmer ChemBio Draw and optimized using Perkin Elmer ChemBio 3D software [21,22]. Quantum chemical parameters such as the highest occupied molecular orbital energy (E_{HOMO}), the lowest unoccupied molecular orbital energy (E_{LUMO}) and HOMO–LUMO energy gap (ΔE) for the investigated molecules are calculated.

Results and Discussion

IR spectra of N-Acryloyl-4-aminosalicylic acid

Examination of IR spectrum of the N-Acryloyl-4-aminosalicylic acid (AAS) shows a broad band appears at ~3458 cm^{-1} corresponding to amide ν_{NH} stretching vibration followed by medium broad band appears at 3006 and 2889 cm^{-1} which assigned to OH of phenolic, carboxylic OH, respectively [17]. This broadening is an evidence for the presence of weak hydrogen bond between OH and carboxylic groups [23]. IR spectrum of AAS exhibits a bands at 1879 and 1612 cm^{-1} is assigned to the antisymmetric streatching vibration of carboxylic and amidic carbonyl group, respectively. The bands at 1616, 1569 and 1415 cm^{-1} are assigned to the (C-H), ν(C=C) and ν(C-C) bands, respectively [24]. The bands in the region 1236-1106 cm^{-1} are due to the C-H in plane deformation, while the out-of-plane deformation vibration between 727-962 cm^{-1} and the C-C out-of-plan deformation at 500 cm^{-1} are assigned.

^{1}H-NMR spectra

The ^{1}H NMR spectrum of AAS revealed six resonances at 10.80, 7.57, 7.28, 3.20, 2.53 and 2.10 ppm relative to TMS which may be assigned to an acid group COOH, CONH, aromatic protons, OH phenolic, vinylic -CH- and -CH_2 protons, respectively [25]. The peak at (10.80/3.20/7.57) ppm, which is due to the exchangeable hydrogen-bonded carboxyl/hydroxyl/amine (COOH/OH/NH) proton, respectively which disappears upon exchange with D_2O and can be associated with the COOH/OH/NH protons involved in intramolecular hydrogen bonding. The ^{1}H NMR spectrum of the ASS monomer showed the expected peaks and pattern of the vinyl group (CH_2=CH) δ 6.82 ppm (dd, J = 17, 11 Hz) for the vinyl CH proton and proton δ 6.32 ppm (AM part of AMX system dd, J = 17,1 Hz) for the vinyl CH_2 protons, respectively. These peaks disappeared on polymerization while a triplet at δ 2.53 ppm (t, J = 7 Hz) and a doublet at 2.1 0 ppm (d, J = 7 Hz) appeared, indicating that the polymerization of AAS monomer occurs on the vinyl group [26]. It is worth noting that the rest of the proton spectrum of the monomer and polymer remain almost without change.

Molecular structure

The molecular structures of the N-Acryloyl-4-aminosalicylic acid (AAS) were optimized by HF method with 3-21G basis set. The molecules were built with the Perkin Elmer ChemBioDraw and optimized using Perkin Elmer ChemBio3D software. The calculated molecular structures for AAS are shown in Figure 2. Selected geometric parameters bond lengths and bond angles of AAS are tabulated in Table 1. Molecular structures (HOMO & LUMO) for AAS and its dimeric form are presented in Figure 3.

The HOMO–LUMO energy gap, ΔE, which is an important stability index, is applied to develop theoretical models for explaining the structure and conformation barriers in many molecular systems. The smaller is the value of ΔE, the more is the reactivity of the compound has [21,22,27]. The calculated quantum chemical parameters are given in Table 2.

Additional parameters such as separation energies, ΔE, absolute

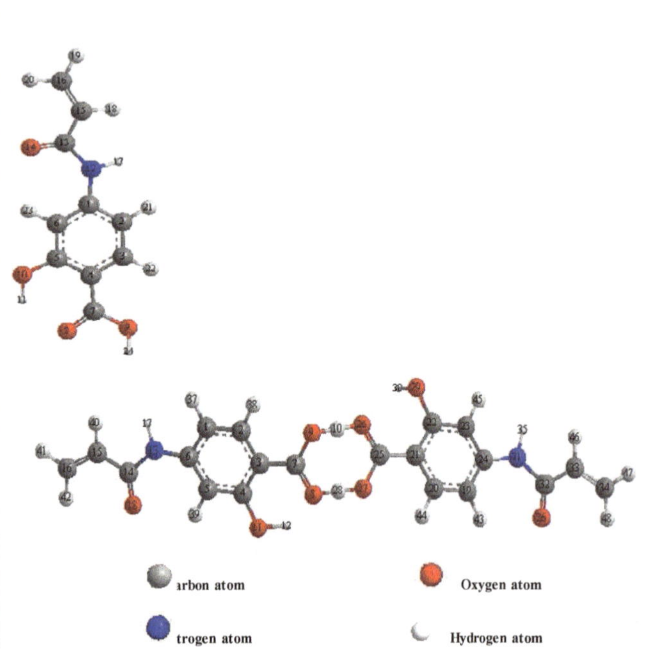

Figure 2: Molecular structure with atomic numbering of AAS in monomeric and dimeric forms.

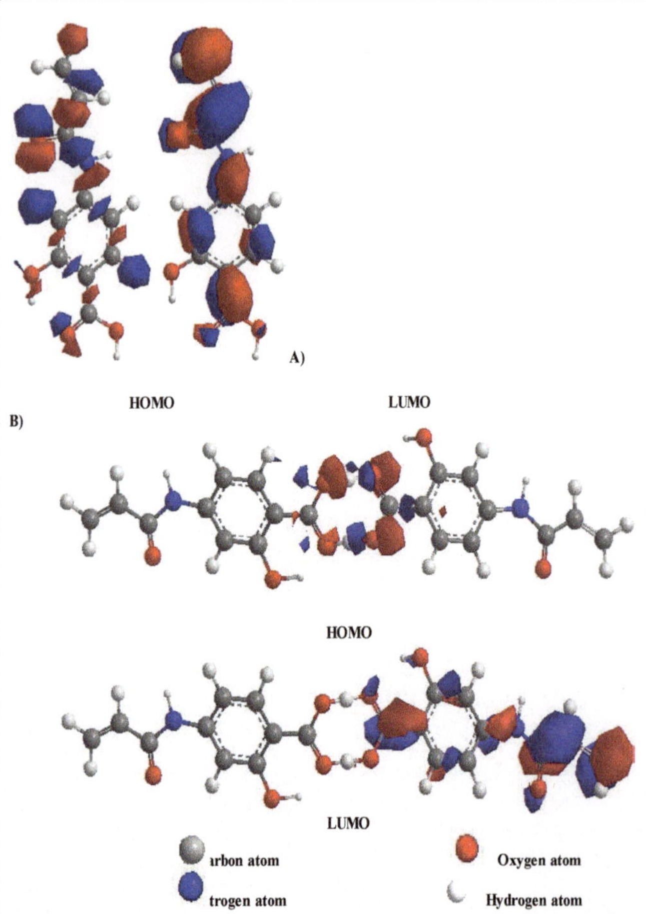

Figure 3: Molecular structures (HOMO & LUMO) for AAS (A) and its dimer (B).

electronegativities, χ, chemical potentials, Pi, absolute hardness, η, absolute softness, σ, global electrophilicity, ω, global softness, S, and additional electronic charge, ΔN_{max}, have been calculated according to the following equations (1-8) [21,22,28]:

$$\Delta E = E_{LUMO} - E_{HOMO} \quad (1)$$

$$\chi = \frac{-\left(E_{HOMO} + E_{LUMO}\right)}{2} \quad (2)$$

$$\eta = \frac{E_{LUMO} - E_{HOMO}}{2} \quad (3)$$

$$\sigma = {}^{1}\!/_{\eta} \quad (4)$$

$$Pi = -\chi \quad (5)$$

$$S = \frac{1}{2\eta} \quad (6)$$

$$\omega = {}^{Pi^2}\!/_{2\eta} \quad (7)$$

$$\Delta N_{max} = {}^{-Pi}\!/_{\eta} \quad (8)$$

In the ring part, the CC bond lengths of the benzene rings are observed in the range 1.339 -1.35 Å. The CO bond lengths in the carboxylic acid group conform to the average values are tabulated for an aromatic carboxylic acid in which C=O is 1.229 Å and C-O is 1.349 Å. In AAS, C5-C4-C7 angle is smaller than C3-C4-C7 because of interaction between the carboxyl acid (COOH) and hydroxyl (OH) group (Figure 4A). These results are in agreement with literatures [23,29-31].

The torsional angles C5-C4-C7-O9 and C3-C4-C7-O8 are 178.8° and 179.3°, respectively. The tilt angles are calculated 180°. The dihedral angles are nearly the same among the all conformers. The geometric dimer structure of AAS is also calculated (Figure 2). Stabilization energy lowest value of dimer showed higher stability than AAS. Interamolecular hydrogen bonds can be responsible for the geometry and the stability of a predominant conformation; the formation of hydrogen bonding between two molecules via a hydroxyl group and -COOH cause the structure of the dimer to be the most stable conformer (Figure 4B). The intermolecular hydrogen bonds are almost linear (the O–H…O angle equals 179.4°) and their length is 2.063 Å. The intramolecular hydrogen bonds between the hydroxyl groups and the oxygen atoms of the carbonyl groups are strongly bent (the O–H…O angle equals 164.7°) and the O…O distance is 2.056 Å [32].

Potentiometric studies

The interaction of a metal with an electron donor atom of a ligand (H_2L) is usually followed by the release of H^+. Alkaline potentiometric titrations are based on the detection of the protons released upon complexation. The main advantage of this technique, compared to other methods is that from the titration curves it is possible to follow complexation continuously as a function of pH and to detect exactly at which pH complexation takes place. Furthermore, it is possible to calculate the dissociation constants and the stability constants of its

Theoretical			
Bond lengths (Å)			
O(9)-H(24)	0.971	N(12)-C(13)	1.365
C(6)-H(23)	1.098	O(10)-H(11)	1.0
C(3)-H(22)	1.102	C(5)-O(10)	1.379
C(2)-H(21)	1.104	C(7)-O(9)	1.349
O(8)-H(11)	1.007	C(7)-O(8)	1.229
N(12)-C(1)	1.35	C(4)-C(7)	1.366
C(16)-H(20)	1.1	C(6)-C(1)	1.346
C(16)-H(19)	1.102	C(5)-C(6)	1.345
C(15)-H(18)	1.105	C(4)-C(5)	1.35
N(12)-H(17)	1.01	C(3)-C(4)	1.339
C(15)-C(16)	1.342	C(2)-C(3)	1.342
C(13)-C(15)	1.36	C(1)-C(2)	1.348
C(13)-O(14)	1.206		
Bond angles (°)			
H(20)-C(16)-C(15)	122.966	O(9)-C(7)-O(8)	119.333
C(15)-C(13)-O(14)	124.552	O(9)-C(7)-C(4)	125.455
C(15)-C(13)-N(12)	111.542	O(8)-C(7)-C(4)	115.212
O(14)-C(13)-N(12)	123.905	C(7)-C(4)-C(5)	114.09
C(1)-N(12)-H(17)	109.793	C(7)-C(4)-C(3)	124.218
C(1)-N(12)-C(13)	133.388	N(12)-C(1)-C(6)	126.721
H(17)-N(12)-C(13)	116.819	N(12)-C(1)-C(2)	115.026
Torsional angles (°)			
C(5)-C(4)-C(7)-O(9)	178.807		
C(3)-C(4)-C(7)-O(8)	179.382		
Intermolecular H- bond (Dimer)			
Bond lengths(Å)		***Bond angles (°)***	
O(9)-H(10)	1.010	O(26)-H(10)-O(9)	176.73
H(10)-O(26)	1.075	O(8)-H(28)-O(27)	179.427
O(27)-H(28)	0.991	O(11)-H(12)-O(8)	164.782
H(28)-O(8)	1.072	O(29)-H(39)-O(26)	164.113
O(11)-H(12)	0.995		
H(12)-O(8)	1.061		
O(29)-H(39)	1.003		
O(26)-H(39)	1.073		

Table 1: Selected geometric parameters for N-Acryloyl-4-aminosalicylic acid (AAS) and its dimer form.

complexes from the potentiometric titration

The following equilibria were used for the determination of the pK_a values of H_2L (AAS) (eqs. 9 and 10) and the metal stability constants (eqs. 11 and 12):

$$H_2L = HL^- + H^+ \qquad (9)$$

$$HL^- = L^{2-} + H^+ \qquad (10)$$

$$H_2L + M^{2+} = ML + 2H^+ \qquad (11)$$

$$H_2L + ML = ML_2^{2-} + 2H^+ \qquad (12)$$

Figures 5,6 shows a typical titration curve of the free acid in the absence and presence of compound AAS and its metal ion complexes. It can be seen that for the same volume of NaOH added, the compound titration curves show a lower pH value than the titration curve of free acid. From these titration curves, the average number of protons associated with N-Acryloyl-4-aminosalicylic acid molecule, $\bar{n}_A$, in monomeric (AAS) and polymeric (PAAS) forms were determined at different pH values applying the following Eq. 13:

$$\bar{n}_A = Y \pm \frac{(V_1 - V_2)(N^o + E^o)}{(V^o - V_1)TC_L^o} \qquad (13)$$

where Y is the number of available protons in AAS (Y=2) and V_1 and V_2 are the volumes of alkali required to reach the same pH on the titration curve of hydrochloric acid and reagent, respectively, V^o is the initial volume (50 cm^3) of the mixture, TC^o_L is the total concentration of the reagent, N° is the normality of the sodium hydroxide solution and E° is the initial concentration of the free acid. The titration curves ($\bar{n}_A$ vs pH) for the proton-ligand systems were constructed and found to extend between 0 and 2 on the $\bar{n}_A$ scale (Figure 7). This means that AAS and PAAS have two dissociable protons (the hydrogen ion of the

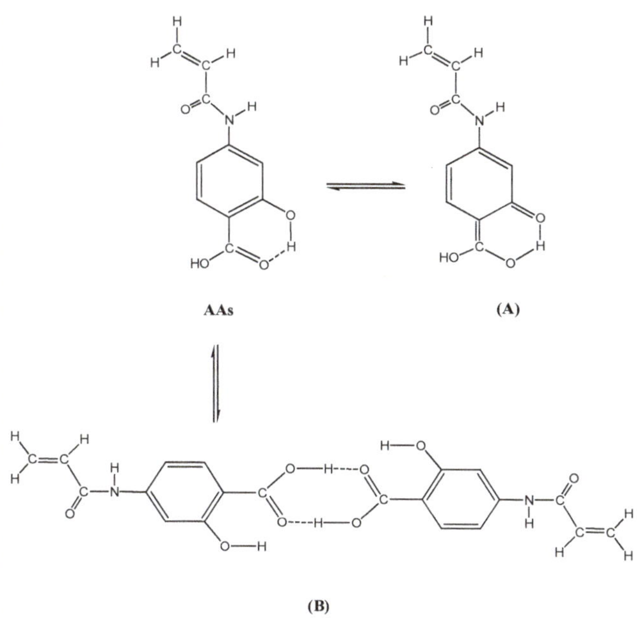

Figure 4: Intra- (A) and inter-molecular (B) hydrogen bonding of N-acryloyl-4-amino salicylic acid (AAS).

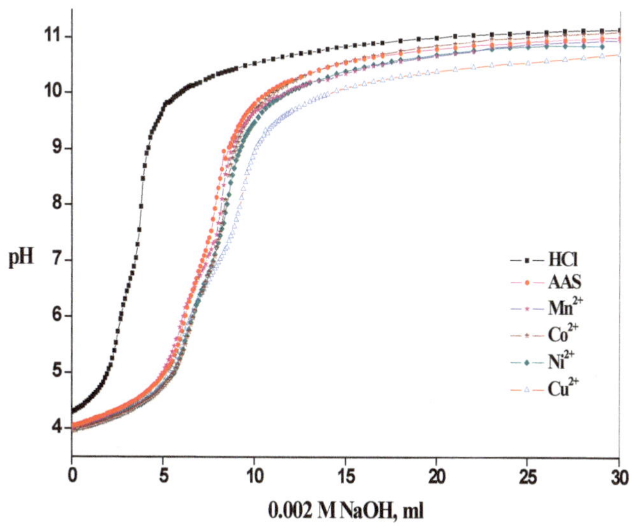

Figure 5. Potentiometric titration curves of AAS and its metal complexes.

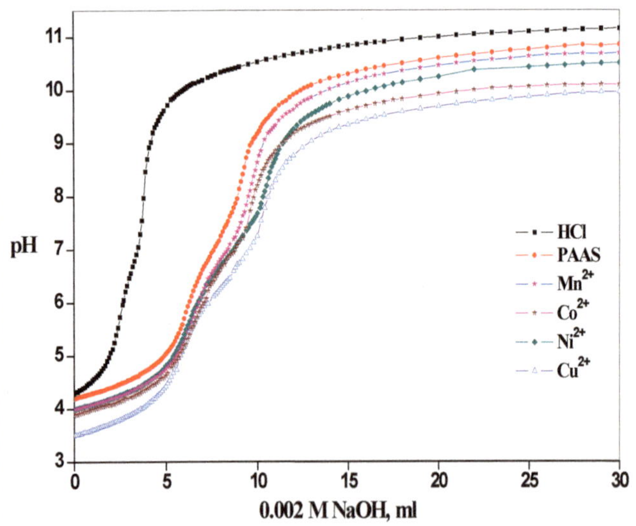

Figure 6. Potentiometric titration curves of PAAS and its metal complexes.

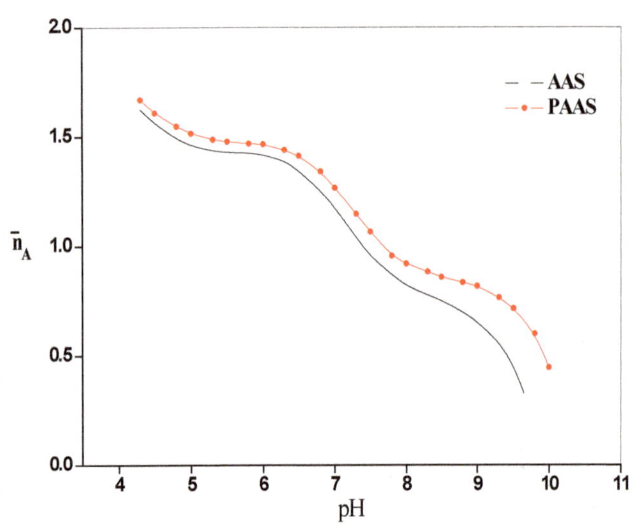

Figure 7. The relation between $\bar{n}_A$ vs. pH for ASS and PAAS.

-OH phenolic moiety, pK_1^H and the COOH group, pK_2^H). Different computional methods [33] were applied to evaluate the dissociation constants. Three replicate titrations were performed; the average values obtained are listed in Table 3. The PAAS has a lower acidic character (higher pK^H values) than AAS. This is quite reasonable because the presence of the vinyl group (H_2C=CH) in monomeric form will decrease the electron density, whereby weaker O-H bond is formed. The absence of vinyl group in polymeric form will lead to the opposite effect (i.e., retard the removal of the ligand proton and hence increase the basicity of PAAS).

The formation curves for the metal complexes were obtained by plotting the average number of the ligands AAS and PAAS attached per metal ions ($\bar{n}$) versus the free ligand exponent (pL), according to Irving and Rossotti [34]. The average number of the reagent molecules attached per metal ion, $\bar{n}$, and free ligand exponent, pL, can be calculated using the Eqs. 14 and 15:

$$\bar{n} = \frac{(V_3 - V_2)(N^o + E^o)}{(V^o - V_2).\bar{n}_A.TC_M^o} \quad (14)$$

and

$$pL = \log_{10} \frac{\sum_{n=o}^{n=J} \beta_n^H \left(\frac{1}{[H^+]} \right)^n}{TC_L^o - \bar{n}.TC_M^o} . \frac{V^o + V_3}{V^o} \frac{n!}{r!(n-r)!} \quad (15)$$

where TC^o_M is the total concentration of the metal ion present in the solution, β^H_n is the overall proton-reagent stability constant. V_1, V_2 and V_3 are the volumes of alkali required to reach the same pH on the titration curves of hydrochloric acid, organic ligand and complex, respectively. These curves were analyzed and the successive stability constants were determined using different computional methods [35,36]. The values of the stability constants (log K_1 and log K_2) are given in Table 4.

The following general remarks can be made:

i) The maximum value of $\bar{n}$ was ~ 2 indicating the formation of 1:1 and 1:2 (metal:ligand) complexes only [37].

ii) The metal ion solution used in the present study was very dilute (2 x 10^{-5} M), hence there was no possibility of formation of metal hydroxide and polynuclear complexes [38,39].

iii) The metal titration curves were displaced to the right-hand side of the ligand titration curves along the volume axis, indicating proton release upon complex formation of the metal ion with the ligand. The large decrease in pH for the metal titration curves relative to ligand titration curves point to the formation of strong metal complexes [40,41].

iv) For all the complexes, the stability constants of PAAS are higher than AAS. This is quite reasonable because the ligand in polymeric forms are better complexing agent [15].

v) For the same ligand (AAS and PAAS) at constant temperature, the stability of the chelates increases in the order Mn^{2+}, Co^{2+}, Ni^{2+}, Cu^{2+} [42 - 44]. This order largely reflects that the stability of Cu^{2+} complexes are considerably larger as compared to other metals of the 3d series. Under the influence of both the polarizing ability of the metal ion [45] and the ligand field [46], Cu^{2+} will receive some extra stabilization due to tetragonal distortion of octahedral symmetry in its complexes. The greater stability of Cu^{2+} complexes is produced by the well known *Jahn–Teller* effect [47].

Effect of temperature

The dissociation constants (pK^H) for ASS and PAAS as well as the stability constants of its complexes with Mn^{2+}, Co^{2+}, Ni^{2+} and Cu^{2+} have been evaluated at (298, 308 and 318) K, and are given in Tables 3, 4. The enthalpy change (ΔH) for the dissociation and complexation process was calculated from the slope of the plot pK^H or log K vs. $^1/_T$ using the graphical reperesentation of the van't Hoff Eqs. 16 and 17:

$\Delta G = -2.303\, RT \log K = \Delta H - T\, \Delta S$ (16)

or

$$\log K = \left(\frac{-\Delta H}{2.303R} \right) \left(\frac{1}{T} \right) + \frac{\Delta S}{2.303R} \quad (17)$$

From the ΔG and ΔH values one can deduce the S using the well known relationships 16 and 18:

$\Delta S = (\Delta H - \Delta G) / T$ (18)

Comp.	E_{HOMO} (a.u)	E_{LUMO} (a.u)	ΔE (a.u)	χ (a.u)	η (a.u)	σ $(a.u)^{-1}$	Pi (a.u)	S $(a.u)^{-1}$	ω (a.u)	ΔN_{max}
AAS	-0.348	-0.170	0.178	0.259	0.089	11.223	-0.259	5.6117	0.3755	2.903
Dimer	-0.347	-0.174	0.173	0.260	0.086	11.510	-0.260	5.7554	0.3907	2.999

Table 2 : Stabilization energy of AAS and its dimer, HOMO, LUMO and other additional parameters.

Comp.	Temp. K	Dissociation constant		Free energy change kJ mol^{-1}		Enthalpy change kJ mol^{-1}		Entropy change J mol^{-1} K^{-1}	
		pK_1^H	pK_2^H	ΔG_1	ΔG_2	ΔH_1	ΔH_2	$-\Delta S_1$	$-\Delta S_2$
AAS	298	9.41	4.71	53.69	26.87	32.63	23.58	70.68	9.91
	308	9.24	4.54	54.49	26.77			70.98	9.89
	318	9.05	4.37	55.10	26.60			70.67	9.90
PAAS	298	9.89	5.00	56.43	29.44	29.88	24.50	89.48	13.52
	308	9.72	4.86	57.32	29.36			90.25	13.51
	318	9.64	4.73	58.69	29.28			87.94	13.53

Table 3: Thermodynamic functions for the dissociation of AAS and PAAS in 20% (by volume) ethanol-water mixture in the presence of 0.1 M KCl at different temperatures.

Comp.	M^{n+}	298 K log K	298 K log K	308 K log K	308 K log K	318 K log K	318 K log K
AAS	Mn^{2+}	5.00[1]	4.20[2]	5.18[1]	4.37[2]	5.37[1]	4.56[2]
	Co^{2+}	5.18	4.37	5.35	4.53	5.47	4.71
	Ni^{2+}	5.24	4.43	5.41	4.58	5.52	4.75
	Cu^{2+}	5.44	4.61	5.62	4.79	5.70	4.90
PAAS	Mn^{2+}	7.44	6.41	7.67	6.63	7.89	6.82
	Co^{2+}	7.63	6.59	7.85	6.78	8.06	6.98
	Ni^{2+}	7.72	6.65	7.91	6.86	8.13	7.07
	Cu^{2+}	7.95	6.87	8.14	7.07	8.35	7.26

Table 4: Stepwise stability constants for ML and ML_2 complexes of AAS and PAAS in 20% (by volume) ethanol-water mixture in the presence of 0.1 M KCl at different temperatures.

where $R = 8.314$ J K^{-1} mol^{-1} is the gas constant, K is the dissociation constant for the ligand or the stability constant of the complex, and T is absolute temperature.

All thermodynamic parameters of the dissociation process of ASS and PAAS are recorded in Table 3. From these results the following conclusions can be made:

i) The pK^H values decrease with increasing temperature, i.e., the acidity of the ligand increase.

ii) A positive value of ΔH indicates that the process is endothermic.

iii) A large positive value of ΔG indicates that the dissociation process is not spontaneous [45].

iv) A negative value of ΔS is obtained due to the increased order as a result of the solvation process.

All the thermodynamic parameters of the stepwise stability constants of AAS and PAAS complexes are recorded in Table 5.

It is known that the divalent metal ions exist in solution as octahedrally hydrated species [36] and the obtained values of ΔH and ΔS can then be considered as the sum of two contributions: (a) release of H_2O molecules, and (b) metal-ligand bond formation. Examination of these values shows that:

i) The stepwise stability constants (log K_1 and log K_2) for AAS and PAAS complexes increases with increasing temperature [48].

ii) The negative value of ΔG for the complexation process of AAS and PAAS suggests the spontaneous nature of such processes [49,50].

iii) The ΔH values of AAS and PAAS are positive, meaning that these processes are endothermic and favourable at higher temperature.

iv) The ΔS values for the complexes of AAS and PAAS are positive, confirming that the complex formation is entropically favourable [14].

Conclusion

N-Acryloyl-4-aminosalicylic acid (AAS) has been synthesized and characterized using spectroscopic techniques (IR and ^{1}H NMR). The geometrical structures of these ligands are carried out by HF method with 3-21G basis set. The geometric dimer structure more stable than AAS. The protonation constants (pK_1 and pK_2) of the AAS and PAAS were determined by Irving-Rossetti pH titration technique. Also metal-ligand stability constants of their complexes with metal ions (Mn^{2+}, Co^{2+}, Ni^{2+} and Cu^{2+}) have been determined potentiometrically. It appears that PAAS are better complexing agent with metal ions compaier d to AAS with the order: Mn(II) < Co(II) < Ni(II) < Cu(II) . The corresponding thermodynamic parameters (ΔG, ΔH and ΔS) were derived and discussed. The dissociation process is non-spontaneous,

Comp.	M^{n+}	T/K	free energy change (kJ mol^{-1})		Enthalpy change (kJ mol^{-1})		Entropy change (J mol^{-1} K^{-1})	
			$-\Delta G_1$	$-\Delta G_2$	ΔH_1	ΔH_2	ΔS_1	ΔS_2
	Mn^{2+}	298	28.52	23.96	33.55	32.63	208.31	189.91
AAS		308	30.54	25.77			208.11	189.61
		318	32.69	27.76			208.32	189.92
	Co^{2+}	298	29.55	24.93	26.35	30.82	187.60	187.09
		308	31.55	26.71			187.98	186.80
		318	33.30	28.67			187.59	187.10
	Ni^{2+}	298	29.89	25.27	25.45	29.00	185.73	182.13
		308	31.90	27.00			186.21	181.84
		318	33.61	28.92			185.72	182.14
	Cu^{2+}	298	31.03	26.30	31.63	26.37	210.30	176.75
		308	33.14	28.24			210.31	177.33
		318	34.70	29.83			208.60	176.74
PAAS	Mn^{2+}	298	42.45	36.57	40.82	37.21	279.43	247.59
		308	45.23	39.09			279.39	247.75
		318	48.04	41.52			279.43	247.59
	Co^{2+}	298	43.53	37.60	39.01	35.36	276.99	244.83
		308	46.29	39.98			276.96	244.62
		318	49.07	42.49			276.99	244.84
	Ni^{2+}	298	44.04	37.94	35.36	30.01	266.47	228.03
		308	46.64	40.45			272.07	228.78
		318	49.50	43.04			272.49	229.74
	Cu^{2+}	298	45.36	39.19	36.26	35.38	273.89	250.26
		308	48.00	41.69			273.58	250.24
		318	50.84	44.20			273.90	250.26

Table 5: Thermodynamic functions for ML and ML_2 complexes of AAS and PAAS in 20% (by volume) ethanol-water mixture and 0.1 M KCl.

endothermic and entropically unfavorable. The formation of the metal complexes has been found to be spontaneous, endothermic and entropically favorable.

References

1. Nurchi VM, Crespo-Alonso M, Toso L, Lachowicz JI, Crisponi G (2013) Iron^{III} and aluminium^{III} complexes with substituted salicylaldehydes and salicylic acids. J Inorg Biochem 128: 174-182.
2. Rivas BL, Pereira ED, Moreno-Villoslada I (2003) Water-soluble polymer–metal ion interactions. Prog Polym Sci 28: 173-208.
3. Ghoneim MM, El-Sonbati AZ, Diab MA, El-Bindary AA, Serag LS (2014) Supramolecular Assembly on coordination of azopolymer complexes: A review. Polym-Plast Technol Eng
4. Elbagerma MA, Azimi G, Edwards HGM, Alatjtal AI, Scowen IJ (2010) In stiu monitoring of pH titration by ramman spectroscopy. Spectrochim Acta Part A 75: 1403-1410.
5. Lin J, Sahakian DC, Morais SMD, Xu JJ, Polzer RJ et al.(2003) The role of absorption, distibution, metabolism, excretion and toxicity in drug discovery. Curr Top Med Chem 3: 1125-1154.
6. Acar N, Tulun T (2001) Interaction of polymer-small molecule complex with cupric(II) ions in aqueous ethanol solution. Europ polym J 37: 1599-1605.
7. Rhazia M, Desbrièresb J, Tolaimatea A, Rinaudob M, Votterod P (2002) Influence of the nature of the metal ions on the complexation with chitosan: Application to the treatment of liquid waste. Europ polym J 38: 1523-1530.
8. Morlay C, Cromer M, Mouginot Y, Vittori O (1998) Potentiometric study of Cu(II) and Ni(II) complexation with two high molecular weight poly(acrylic acids). Talanta 45: 1177-1188.
9. Morlay C, Cromer M, Mouginot Y, Vittori O (1999) Potentiometric study of Cd(II) and Pb(II) complexation with two high molecular weight poly(acrylic acids); comparison with Cu(II) and Ni(II). Talanta 48: 1159-1166.
10. Mouginot Y, Morlay C, Cromer M, Vittori O (2003) Potentiometric study of copper(II) and nickel(II) complexation by a cross-linked poly(acrylic acid) gel. Anal Chim Acta 407: 337-345.
11. Zhao ZB, Zheng HX, Wei YG, Liu J (2007) Synthesis of azo derivatives of 4-aminosalicylic acid. Chin Chem Lett 18: 639-642.
12. Sheng SF, Zheng HX, Liu J, Zhao ZB (2008) Synthesis of phenol-class azo derivatives of 4-aminosalicylic acid. Chin Chem Lett 19: 419-422.
13. Ghoneim MM, El-Ghamaz NA, El-Sonbati AZ, Diab MA, El-Bindary AA et al. (2015) Optical and thermal properties of azo derivatives of salicylic acid thin films. Spectrochim Acta A 137: 1039-1049
14. Issa RM, El-Sonbati AZ, El-Bindary AA, Kera HM (2002) Polymer complexes XXXIV. Potentiometric and thermodynamic studies of monomeric and polymeric complexes containing 2-acrylamidosulphadiazine. Europ Polym J 38: 561-566.
15. El-Sonbati AZ, El-Bindary AA, El-Deeb NA (2002) Polymer complexes XXXII. Potentiometric and thermodynamic studies of cinnamaldehyde anthranilic acid and its metal complexes in monomeric and polymeric forms. React Funct Polym 50: 131-138.
16. Issa RM, El-Sonbati AZ, El-Bindary AA, Kera HM (2003) Polymer complexes XLIV. Potentiometric and thermodynamic studies of monomeric and polymeric complexes containing 2-acrylamido-2-methylpropanesulfonic acid. J Inorg Organom Polym 13: 269-283.
17. Khairou KS, Diab MA (1994) Thermal degradation of poly(acryloyl chloride) and copolymers of acryloyl chloride with metyl acrylate. Polym Deg Stab 43: 329-333.
18. Jeffery GH, Bassett J, Mendham J, Deney RC (1989) Vogel's Textbook of Quantitative Chemical Analysis, 5th Edition. Longman, London.
19. Bates RG, Paabo M, Robinson RA (1963) Interpretation of pH measurements in alcohol-water solvents. J Phys Chem 67: 1833-1838.
20. Irving HM, Miles MG, Pettit LD (1967) A study of some problems in determining the stoicheiometric proton dissociation constants of complexes by potentiometric titrations using a glass electrode. Anal Chim Acta 38: 475-488.

21. El-Ghamaz NA, Ghoneim MM, El-Sonbati AZ, Diab MA, El-Bindary AA (2014) Synthesis and optical properties studies of antipyrine derivatives thin films. J Saudi Chem Soc

22. El-Bindary AA, El-Sonbati AZ, Diab MA, El-Katori EE, Seyam HA (2014) Potentiometric and thermodynamic studies of some Schiff-base derivatives of 4-aminoantipyrine and their metal complexes. Int J Adv Res 2: 493-502.

23. Takahashi M, Ishikawa Y, Ito H (2012) The dispersion correction and weak-hydrogen-bond network in low-frequency vibration of solid-state salicylic acid. Chem Phys Lett 531: 98-104.

24. Diab MA (1994) Thermal stability of poly homopolymer and copolymers of acryloyl chloride with methyl acrylate. Acta Polym 41: 731-738.

25. El-Sonbati AZ, Belal AAM, El-Wakeel SI, Hussien MA (2004) Stereochemistry of new nitrogen containing heterocyclic compounds X. Supramolecular structures and stereochemical versatility of polymeric complexes. Spectrochim Acta A 60: 965-972.

26. El-Sonbati AZ, El-Bindary AA, Diab MA (2003) Polymer complexes XXXX. Supramolecular assembly on coordination models of mixed-valence-ligand poly [1-acrylamido-2-(2-pyridyl)ethane] complexes. Spectrochim Acta A 59: 443-454.

27. El-Ghamaz NA, El-Sonbati AZ, Diab MA, El-Bindary AA, Awad MK (2014) Dielectrical, conduction mechanism and thermal properties of rhodanine azodyes. Mater Sci Semicond Process 19: 150-162.

28. El-Sonbati AZ, Diab MA, El-Bindary AA, Morgan SM (2014) Supramolecular spectroscopic and thermal studies of azodye complexes. Spectrochim Acta A 127: 310-328.

29. Kivelson D, Niemen R (1961) ESR studies on the bonding in copper complexes. J Chem Phys 35: 149-155.

30. Karabacak M, Kurt M (2009) The spectroscopic (FT-IR and FT-Raman) and theoretical studies of 5-bromo-salicylic acid. J Mol Struct 919: 215-222.

31. Karabacak M, Kose E, Kurt M (2010) FT-Raman, FT-IR spectra and DFT calculations on monomeric and dimeric structures of 5-fluoro- and 5-chloro-salicylic acid. J Raman Spectroscopy 41: 1085-1097.

32. Boczar M, Boda L, Wòjcik MJ (2006) Theoretical modeling of infrared spectra of hydrogen bonded crystals of salicylic acid. Spectrochim Acta A 64: 757-760.

33. Irving H, Rossotti HS (1954) The calculation of formation curves of metal complexes from pH titration curves in mixed solvents. J Chem Soc 2904-2910.

34. Irving H, Rossotti HS (1953) Methods for computing successive stability constants from experimental formation curves. J Chem Soc 3397-3405.

35. Rossotti FJC, Rossotti HS (1955) Graphical methods for determining equilibrium constants. I. Systems of mononuclear complexes. Acta Chem Scand 9: 1166-1176.

36. Beck MT, Nagybal I (1990) Chemistry of Complex Equilibrium, Wiley, New York.

37. Boraei AAA, Mohamed NFA (2002) Equilibrium studies of ternary systems involving divalent transition metal ions, aliphatic acids and triazoles. J Chem Eng Data 47: 987-991.

38. Sanyal P, Sengupta GP (1990) Potentiometric studies of complex-formation of some trivalent rare-earths with p,p'-bromosulphonosalicylidene. Anil J Ind Chem Soc 67: 342-346.

39. Sridhar S, Kulanthaipandi P, Thillaiarasu P, Thanikachalam V, Manikandan G (2009) Protonating and chelating efficiencies of some biologically important thiocarbonohydrazides in 60 % (v/v) ethanol-water systems by potentiometric and spectrophotometric methods. J Chem 4: 133-140.

40. Athawale VD, Lele V (1996) Stability constants and thermodynamic parameters of complexes of lanthanide ions and (±)-norvaline. J Chem Eng Data 41: 1015-1019.

41. Athawale VD, Nerkar SS (2000) Stability constants of complexes of divalent and rare earth metals with substituted salicynals. Monatsh Chem 131: 267-276.

42. Tirkistani FAA, El-Bindary AA (2005) Potentiometric and thermodynamic studies of rhodanineazosulfonylazide and its metal complexes. Bull Electrochem 21: 265-269.

43. Ibañez GA, Escandar GM (1998) Complexation of cobalt(II), nickel(II) and zinc(II) ions with mono and binucleating azo compounds: A potentiometric and spectroscopic study in aqueous solution. Polyhedron 17: 4433-4441.

44. Malik WU, Tuli GD, Madan RD (1984) Selected Topics in Inorganic Chemistry, 3rd Edition, Chand S. & Company LTD, New Delhi.

45. Harlly FR, Burgess RM, Alcock RM (1980) Solution Equilibria, p. 257. Ellis Harwood, Chichester.

46. Orgel LE (1966) An introduction to transition metal chemistry ligand field theory, p. 55. Methuen, London.

47. Bebot-Bringaud A, Dange C, Fauconnier N, Gerard C (1999) NMR, potentiometric and spectrophotometric studies of phytic acid ionization and complexation properties toward Co^{2+}, Ni^{2+}, Cu^{2+}, Zn^{2+} and Cd^{2+}. J Inorg Biochem 75: 71-78.

48. Al-Sarawy AA, El-Bindary AA, El-Sonbati AZ, Omar TY (2005) Potentiometric and thermodynamic studies of 3-(4-methoxyphenyl)-5-azorhodanine derivatives and their metal complexes with some transition metals XIV. Chem Pap 59: 261-266.

49. Mubarak AT, El-Sonbati AZ, El-Bindary AA (2004) Potentiometric and conductometric studies on the complexes of some transition metals with rhodanine azosulfonamide derivatives XI. Chem Pap 58: 320-323.

50. Bhesaniya KD, Baluja S (2014) potentiometric determination of dissociation constant and thermodynamic parameters of dissociation process of some newly synthesized pyrimidine derivatives in MeOH/DMF-water medium at different temperatures. J Mol Liq 190: 190-195.

The Reaction Mechanisms of H_2S Decomposition into Hydrogen and Sulfur: Application of Classical and Biological Thermodynamics

Startsev AN*

Pensioner, Novosibirsk 630090, Russia

Abstract

Four possible pathways of H_2S decomposition into hydrogen and elemental sulfur are considered. In the thermal reversible process, H_2S dissociation results in the formation of diatomic both hydrogen and sulfur in the singlet state according to the rule of spin conservation:

$$H_2S \xleftrightarrow{\Delta \dot{O} > 1000°\tilde{N}} H_2 + {}^1\!/_2\,{}_1S_{2(gas)} \quad \textbf{(i)}$$

On the surface of sulfide catalysts, irreversible H_2S splitting proceeds at low temperature through the stage of disulfane, H_2S_2 formation as a key surface intermediate, followed by its decomposition due to release of hydrogen into the gas phase and recombination of the adsorbed singlet sulfur into cyclooctasulfur:

$$2H_2S \xleftrightarrow{sulfide\ catalyst\ 25°C} \{H_2S_2\}_{ads} + H_2 \rightarrow H_2 + {}^1\!/_4S_{8(ads)} \quad \textbf{(ii)}$$

On metal catalysts, irreversible H_2S decomposition occurs at low temperature through the stage of H_2S dissociation into the adsorbed atomic surface species resulted in the formation of both diatomic reaction products in the ground electronic state -the singlet hydrogen and the triplet diatomic sulfur:

$$H_2S \xrightarrow{Metal\ catalyst\ 25°C} H_2 + {}^1\!/_2\,{}_3S_{2(gas)} \quad \textbf{(iii)}$$

The reactions (**ii**) and (**iii**) are considered to be realized under the principles of biological thermodynamics, in the absence of catalyst both reactions are thermodynamically prohibited in the gas phase. The mechanism of H_2S assimilation by sulfur bacteria is suggested to occur in the processes of chemosynthesis resulted in the formation of colorless sulfur globules and activated hydrogen:

$$H_2S \xrightarrow{sulfur\ bacteria,\ ambient\ conditions} 2\,H_2{}^* + \{{}_3S_2\}_{(globules)} \quad \textbf{(iv)}$$

Similar to the bacterial process, the hydrogen production from H_2S is realized with the efficiency of 99.6% on metal catalysts immersed into the liquid which is capable well-dissolving H_2S.

Keywords: H_2S decomposition; Hydrogen production; Classical and biological thermodynamics; Reaction mechanism; Diatomic sulfur; Triplet and singlet sulfur; Sulfide and metal catalysts; Sulfur bacteria

Introduction

Hydrogen sulfide is one of the most toxic substances produced as an obligatory and unavoidable by-product in the mining and processing industries in the total annual amount of tens and hundreds of millions of tons. The reserves of hydrogen sulfide in the entrails and water reservoirs of the Earth are estimated by the billions of tons, its contents in the explored deposits of natural gas may be more than 50%. At the same time, it is the "useless" matter, which did not find any commercial application. Therefore, H_2S must be removed from waste gases and waters to the level of sanitary norms. Processes of hydrogen sulfide utilization are implemented worldwide by the Claus method, developed in the 19th century; as a result, the final products of its disposal are water and solid sulfur.

H_2S, on the other hand, is a simple molecule which has a structure similar to that of water, however, unlike water, the H_2S molecule is much less stable and is decomposed into elements even at temperatures as low as 500°C [1]. This means that hydrogen can be released from H_2S by an array of techniques including simple heating, irradiation, electrolysis, photolysis etc. [2,3]. Several of these processes have been demonstrated on a significant scale, but none of them has yet achieved commercial success because of high cost of hydrogen produced [4,5]. The goal of this article aims to consider the processes of H_2S decomposition from two opposite standpoints. For high-energy, reversible processes, when the system is continuously supplied with energy in the form of heat, the decomposition of hydrogen sulfide is subject to the laws of classical equilibrium thermodynamics for isolated systems, which are permeable neither to heat nor matter. For processes at room temperature, an external source of energy is absent, so the decomposition of hydrogen sulfide is examined from the standpoint of biological thermodynamics, when exchanges of both matter and energy take place in an open system [6].

Thermal Processes of H_2S Decomposition

Thermal dissociation of H_2S is a well researched route for the hydrogen production:

$$2\,H_2S \xrightleftharpoons{\Delta \dot{O} > 1000°\tilde{N}} 2\,H_2 +_1 S_{2(gas)} \quad (1)$$

Since conversion is very low, only 15% at 1000 $_o$C and 1 atm., it is beneficial to increase the reaction temperature. Rapid quenching the reaction products is important to maximize H_2 yields. To shift the reaction equilibrium to the right, various ways of intermediate

***Corresponding author:** Startsev AN, Pensioner, Novosibirsk 630090, Russia
E-mail: anatolii.startsev@gmail.com

removing the reaction products have been presented [2,3]. Reversibility of the reaction (1) means that process runs equally well forwards and backwards and is always in a near equilibrium state. In accordance with the rule of spin conservation, diatomic sulfur formed in the reaction (1) is in the metastable singlet state which does not convert into the ground triplet state even at high temperature and in the presence of metal catalysts [7-9]:

$$_{1}S_{2\,(gas)} \neq {}_{3}S_{2\,(gas)} \quad (2)$$

As is well known, diatomic sulfur is isoelectronic to oxygen [10,11], therefore the S_2 ground state is a triplet $X\,{}_3\Sigma_{-g}$ and the first exited state is a singlet $a_1\Delta_g$ which is above the ground state by ~4400 cm^{-1} (Table 1). However, unlike to singlet oxygen which is spontaneously converted into the ground triplet state [12], the diatomic singlet sulfur prefers to aggregate from the gas phase into the solid S_8 structure. Another high-energy processes of H_2S decomposition reviewed in [2,3] resulted in the formation of solid sulfur as well.

Application of Biological Thermodynamics to the Catalytic Systems

Biological thermodynamics is the quantitative study of the energy transductions that occur in and between living organisms, structures, and cells and of the nature and function of the chemical processes underlying these transductions. All biological organisms require energy to survive. No matter what the type of living species, all living organisms must capture, transduce, store, and use energy to live. In application of thermodynamics to biological systems functioning at ambient conditions, the following peculiarities of organization of living systems are accepted [6]:

- biological systems are open to flows of matter and energy;
- processes in living systems are irreversible and far from equilibrium;
- biological systems are heterogeneous, structured, and individual phases may have a small ultimate number of molecules.

All these features distinguish biological systems from the isolated and close to equilibrium systems in the classical thermodynamics. It is well recognized that for an adequate description of the properties of living systems it is necessary to apply the thermodynamics of irreversible processes which deals with the processes in time. A fundamental concept in classical thermodynamics is the notion of equilibrium state. In thermodynamics of irreversible processes an important concept is the notion of a steady state of the system [6]. Unlike thermodynamic equilibrium, the steady state is characterized by a constant flow of substances in and removing the reaction products out of the system, a continuous expenditure of free energy, which maintains the constancy of the concentrations in the system and the constancy of the thermodynamic parameters (including the internal energy and entropy). It is important that an open biological system can exist only due to the inflow of energy from outside and outflow of energy in the environment [6]. Catalytic reactions, which run at ambient conditions (room temperature and normal pressure) and do not obtain energy from the surroundings (solar energy is excluded, we consider only dark processes), seem to be referred to the open systems which are permeable to heat and matter flows [6]. Actually, H_2S is a stable molecule at ambient conditions and exists for a long period of time in the absence of action of external forces. However, on passing H_2S over metal catalysts at room temperature, two reaction products are observed which are hydrogen and gaseous sulfur [7-9]. Hence, the reaction is likely to occur at the expense of the molecule free energy. The energy stored in H_2S can be released through the molecule dissociation on the catalyst surface.

H_2S Decomposition on Sulfide Catalysts

The initial step of any heterogeneous catalytic reaction is an adsorption of substrate molecules on the surface of solid catalyst. The active component of the sulfide hydrodesulfurization (HDS) catalysts consists of a single slab of MoS_2 with Ni or Co atoms being located in its edge plane in the center of a square pyramid made of sulfur atoms (Figure 1). The active sites for H_2S adsorption are Co(III) or Ni(IV) atoms in the d_6 electronic configuration (Figures 2 and 3) [13-15]. The essential element of the active component is hydrogen occluded into the MoS_2 matrix (Figures 1 and 3). The occluded hydrogen appeared in the active component during the catalyst sulfiding as a result of its oxidative addition to the Ni (Co) atoms. It localizes in the center of "empty" trigonal prism of a MoS_2 matrix made from six sulfur atoms under Ni (Co) atom at the distance of 1.5 Å (Figure 1). The active metal ions are oxidized by the occluded hydrogen to Co(III) and Ni(IV) and acquired d_6 electron configuration (Figure 3) [13-15]. Two unoccupied orbitals of the active metal atom, $3d_{z2}$ and $3d_{xz}$ interact effectively with the n_σ and n_π orbitals of the H_2S molecule (Figure 3). In the absence of occluded hydrogen, H_2S molecule is not adsorbed by a Co(II) or Ni(III) atom with d_7 electron configuration, because the $3d_{z2}$ orbital is occupied. The molecular H_2S adsorption on the d_6 Co(III) ion is an exothermic process (ΔH_{298}=-12.7 kcal/mol) and occurs spontaneously (ΔG_{298}=-0.8 kcal/mol) [16,17], while H_2S adsorption on the d_7 Co(II) ion is hardly possible (ΔH_{298}=-2.2 kcal/mol, ΔG_{298}=+6.6 kcal/mol).

The second step -dissociative chemisorption of two H_2S molecules on two adjacent cobalt atoms (Figures 2 and 4), occurs spontaneously (ΔH_{298}=-28.7 kcal/mol, ΔG_{298}=-17.4 kcal/mol) on the d_6 Co(III) ions as well, and does not possible for the d_7 Co(II) ions (ΔH_{298}>0, ΔG_{298}>0) [16,17]. It is significant that, in the adsorbed state, two H_2S molecules interact to one another, as is indicated by their coming together within a short distance of 2.18 Å of each other (Figure 4). As a result, a new sulfur–sulfur chemical bond is formed in the surface intermediate, and this process is accompanied by the release of a hydrogen molecule into the gas phase and by the formation of adsorbed disulfane H_2S_2 (Figures 4 and 5).

It is very important, that formation of disulfane, H_2S_2 from H_2S in the gas phase in the absence of catalyst is thermodynamically prohibited:

$$2\,H_2S \neq H_2S_2 + H_2 \quad (3)$$

($\Delta_r H_{o298}$=+13.6 kcal/mol, $\Delta_r S_{o298}$=-3.9 cal/mol. K, $\Delta_r G_{o298}$=+14.1 kcal/mol, initial parameters for calculations are taken from [18]). Disulfane was also established in the gas phase during the high temperature metal-catalyzed reaction of H_2S decomposition [19].

The next step of surface reaction -decomposition of adsorbed disulfane into molecular hydrogen and adsorbed disulfur, is a strongly endothermic process and requires large energy consumption ΔH_{298}=25.5 kcal/mol. There is, however, the possibility of terminating the catalytic cycle, which is the recombination (oligomerization) of molecular sulfur S_2 into cyclooctasulfur S_8 on the catalyst surface without it's desorption into the gas phase. This coupled reaction is the highly exothermic process and occurs spontaneously (ΔH_{298}=-78.4 kcal/mol, ΔS_{298}=-116.0 cal/mol. K, ΔG_{298}=-43.6 kcal/mol [16,17]). Therefore, the overall catalytic reaction:

$$2\,H_2S_{(gas)} \xrightarrow{\text{sulfide catalyst}} 2\,H_{2(gas)} + \tfrac{1}{4}\,S_{8(ads)} \quad (4)$$

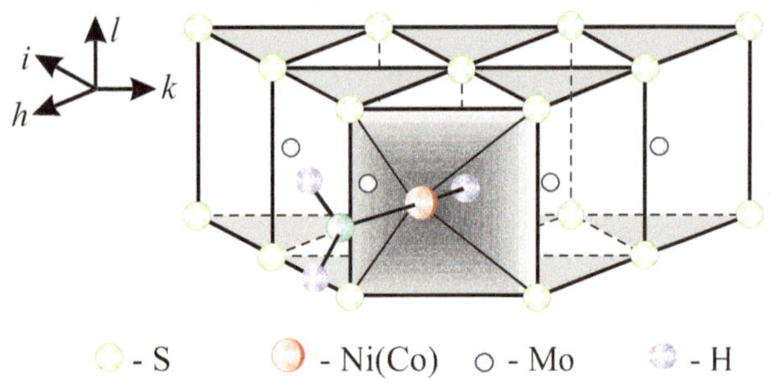

Figure 1: Structure of the active component of the sulfide HDS catalysts with the occluded hydrogen atom and adsorbed H_2S molecule.

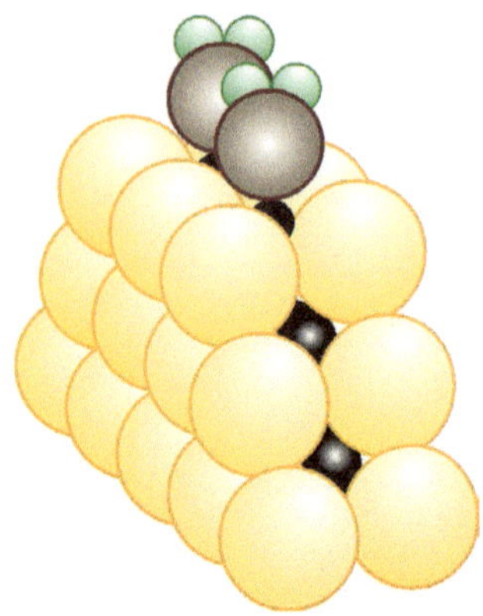

Figure 2: A fragment of the active component of the sulfide HDS catalyst with two molecules of H_2S adsorbed on two adjacent atoms of Ni (Co).

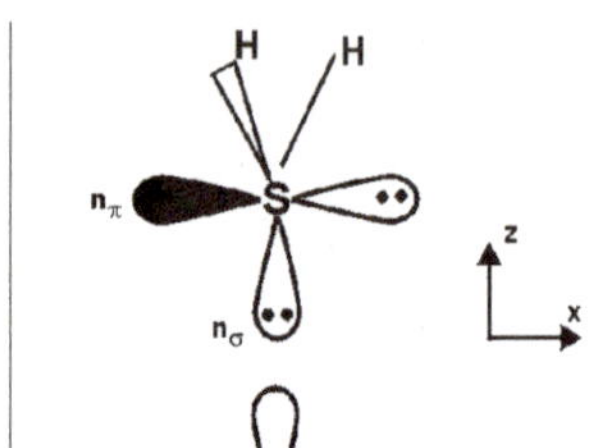

Figure 3: Scheme of interaction of n_σ and n_π orbitals of a hydrogen sulfide molecule with $3d_{z2}$ and $3d_{xy}$ orbitals of the Co(III) atoms in the d^6 electronic configuration [15].

runs also spontaneously at room temperature (ΔH_{298}=-23.0 kcal/mol, ΔS_{298}=-53.2 cal/mol. K, ΔG_{298}=-7.2 kcal/mol [16,17]) (Figure 5). Removal of adsorbed sulfur from the catalyst surface can be realized by immersing solid catalyst into liquid solvent (see next sections).

H_2S Decomposition on Metal Catalysts

An unexpected, unpredictable phenomenon was observed when hydrogen sulfide was passed through metal catalysts at room temperature: along with hydrogen, diatomic gaseous sulfur was discovered as a reaction product [7-9,20].

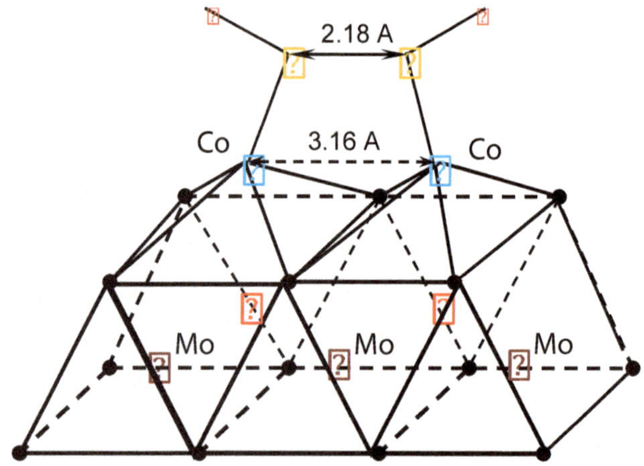

Figure 4: The key surface intermediate $\{H_2S_2\}^{ads}$ resulted from H_2S dissociation on the surface of sulfide catalysts after H_2 molecule being released in the gas phase [16,17].

Atoms of the active component of the sulfide HDS catalysts:

• - sulfur, • - molybdenum, • - cobalt, • - occluded hydrogen, hydride ion H^-.

Atoms of the adsorbed species:

• sulfur, • hydrogen ion, proton H^+.

$$2\,H_2S \xrightarrow{metal\ catalyst,\ 25^\circ C} 2\,H_2 +_3 S_{2(gas)} \quad (5)$$

This irreversible gasphase reacion occurs on the platinum catalyst at ambient conditions under steady state regime with the effiviency of ~15%. Let's consider the reaction (5) in the light of biological thermodynamics. Adsorption of H_2S molecule on the surface of any metals is an exothermic process. In the case of Pt(111) [21], the energy of molecular H_2S adsorption is E_{ads}=-21.0 kcal/mol. The minimum energy pathway of the adsorbed H_2S is then considered for its dissociation in two steps [21]:

$$H_2S_{(ad)} \rightarrow SH_{(ad)} + H_{(ad)} \quad (6)$$

$$SH_{(ad)} \rightarrow S_{(ad)} + H_{(ad)} \quad (7)$$

The first dissociation step (6) occurs via the reaction barrier of E_a=1.6 kcal/mol with the reaction energy ΔE_{rxn}=-19.6 kcal/mol, while for the second step (7) these values are E_a=0.7 kcal/mol and ΔE_{rxn}=-18.2 kcal/mol. Hence, the total reaction energy of H_2S dissociation on Pt(111) into the adsorbed S and H atoms is $\Delta E_{diss.total}$=-58.8 kcal/mol or $\Delta E_{diss.total}$=-117.6 kcal/mol for two H_2S molecules of the reaction (5). It is very important, that H_2S dissociation is a facile process which is kinetically and thermodynamically favorable [21]. On the other hand, from the point of view of thermo-chemistry, the bond dissociation energy D_o of HS -H bond in H_2S molecule is 92 kcal/mol [18]. In the reaction (5) four HS -H bonds should be cleaved, what is ΣD_o=368 kcal/mol, while two H -H bonds (ΣD_o=2 x 104=208 kcal/mol) and one S -S bond D_o=102.5 kcal/mol [18] are formed. From this it follows that the total energy of bond dissociation of initial substance is over ΔD_o=57.5 kcal/mol than that of reaction products, therefore H_2S molecule is stable and does not convert into H_2 and S_2 at room temperature. However in the adsorbed state the required energy for H_2S splitting can be invested from the surface dissociation energy of H_2S adsorption. The resulted energy surplus ($\Delta E_{diss.total}$ +ΔD_o)=-117.6+57.5=-60.1 kcal/mol seems to be consumed for desorption of the reaction (5) products into the gas phase. Actually, the adsorption energy of S_2 species on

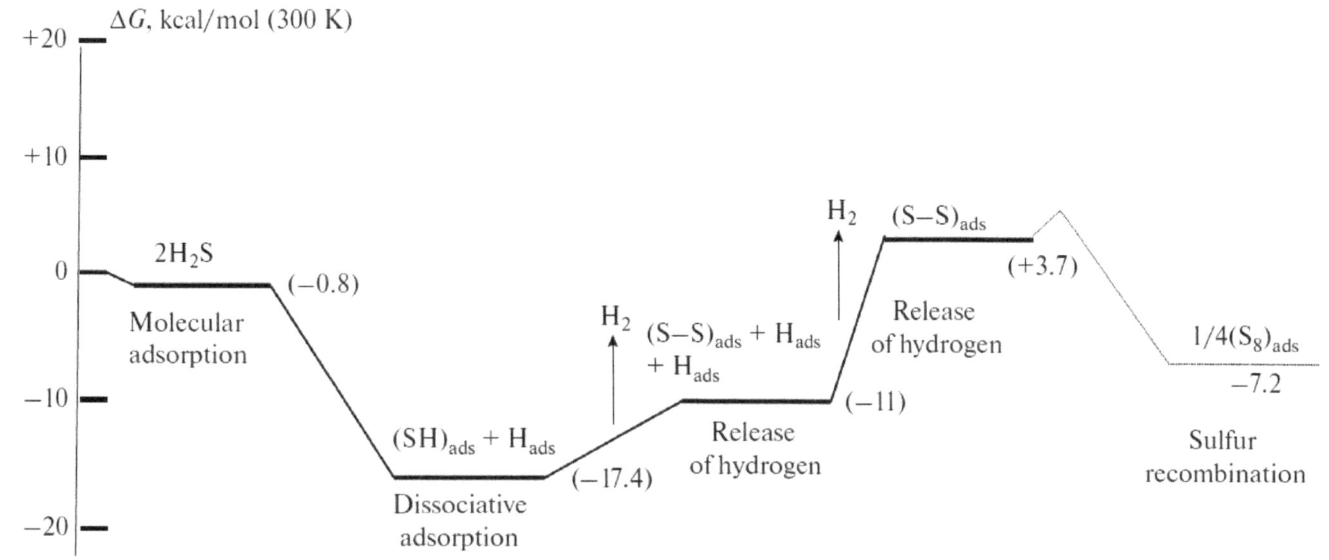

Figure 5: The DFT calculated Gibbs energy profile of H_2S decomposition on the Co(III) ions with the d^6- electronic configuration entering the composition of the active component of the sulfide HDS catalysts [16,17].

Pt(111) is ~45 kcal/mol [22], while molecular hydrogen desorbs from the surface even at 230 K [23]. Undoubtedly, this rough approximation of energy transduction from H_2S dissociation into the surface catalytic reaction should be carefully investigated with both theoretical and experimental methods similar to that made for sulfide catalyst (see above). However, this approach seems to offer new opportunity to understand mechanisms of the low temperature catalytic reactions by means of biological thermodynamics on the atomic scale.

H_2S Assimilation with Sulfur Bacteria

As distinct from solid sulfur, the diatomic sulfur formed in the reaction (5) is well soluble in water, over 5 g of sulfur per liter of H_2O [24]. The pH value and refractive index of the colorless sulfur solution are the same as those of the initial water, and the Raman spectrum of the solution exhibits only bands characteristic of water. It is typical of solutions of diatomic gases that do not interact with water (N_2, O_2, H_2, etc.). When the reaction (5) is carried out on metal catalyst immersed into water, white sulfur globules were obtained from the saturated solutions (Figure 6) [24]. The size of the white spherical globules is up to 10 μm, while the smallest globules are practically transparent and colorless. The elemental analysis of the single globule obtained with the energy dispersive spectroscopy evidences in the only element -sulfur [24]. Electron diffraction on the separate sulfur globule indicates a hexagonal structure of the sulfur crystallites with the interplanar spacing of 0.45, 0.29, and 0.15 nm. More than 30 allotropic modifications of solid sulfur are currently known [25,26] with changes in color from red to black, but no white hexagonal sulfur was found in the database. The only single absorption band at 880 cm_{-1} was observed in the Raman spectra of saturated water solution over precipitate of white globules, while the globules themselves are irreversible transferred into the solid S_8 under the laser beam of a Raman spectrometer. Taking into account the hydrophilic character of white globules, we suggest that this substance is obviously composed of S_2 condensed phase with sulfur in the ground triplet state.

Surprisingly, the white sulfur globules obtained from H_2S on metal catalysts under layer of water bear a great resemblance to colorless globules produced with sulfur bacteria in aqueous medium containing H_2S [27-34]. The role of autotrophic (self-feeding) bacteria in the biosphere was first recognized by Winogradsky at the end of 19 century on the example of sulfur bacteria used hydrogen sulfide as an energy source in the chemosynthesis [27]. Bacteria are the simplest living cells which exhibit all the essential features of a living organism. The ability to harness energy from various sources and to channel it into growth and biological work is a fundamental property of these organisms. The bacteria carry out a remarkable variety of energy transductions, conversions of one form of energy to another [6].

There is a wide range of different types of colorless sulfur bacteria with very diverse properties and with equally diverse environmental habitats. The colorless sulfur bacteria are found in marine and freshwater sediments, soils, and waste water treatment systems. They are growing in aqueous media from acid to alkaline pH values, at temperature ranging from +4 to +95°C, in both aerobic and completely anaerobic conditions [30,32-34]. Many of the properties of "biological sulfur" S_0 produced with sulfur bacteria do not match the properties of any known chemical allotrope of elemental sulfur [25,26]. S_0 forms transparent droplets in diameter of up to 1 μm which are deposited inside or outside of the bacteria cells. S_0 is hydrophilic and of white or pale-yellow color. The buoyant density of S_0 has been determined as 1.22 g/cm_3 [35] in contrast to density of liquid sulfur 1.80 g/cm_3 and solid S_8 2.07 g/cm_3. Therefore, the question was raised of whether S_0 is a new form of elemental sulfur or rather sulfur-rich compound [28].

To answer this question, several models of known sulfur compounds were critically evaluated and it was concluded that S_0 most likely consists of long-chain polythionates [28]. However, the purposeful searching experiments have shown complete absence of polythionates and a new model was proposed where S_0 globules were composed of spherical sulfur droplets, each covered by a monomolecular layer of an amphiphilic compound [36]. Later [32,37] some other models were considered. S_0 was also found in the phototrophic organisms under aerobic conditions [38].

It is generally accepted that sulfur bacteria oxidize H_2S along

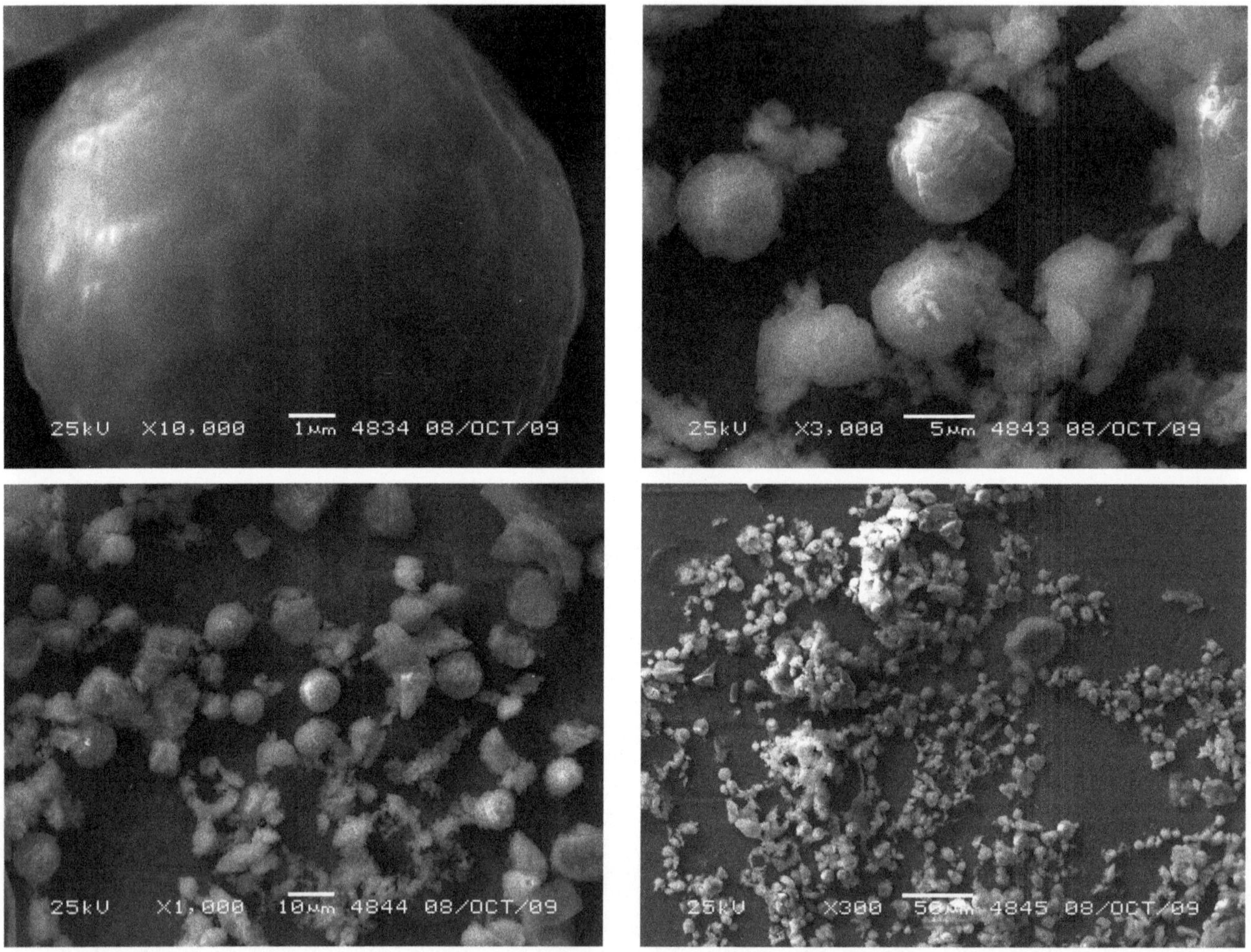

Figure 6: The scanning electron microscope images of white globular sulfur obtained from H_2S on the metal catalyst immersed into water [20,24].

a pathway that includes the formation of zero-valent sulfur as a characteristic intermediate [32,34]:

$$2\ H_2S+O_2 \rightarrow 2\ \{S_0\}+2\ H_2O \tag{8}$$

As believed, the energy from this reaction is used to reduce carbon dioxide to create carbohydrates. However, following this reaction (8), hydrogen is irreversibly bound into water and can't participate in the chemosynthesis. At the same time, it is well recognized that hydrogen is the most important inorganic energy carrier in anaerobic pathways in nature. H_2 can be detected in all environments including waters and the atmosphere. Hydrogen does not accumulate but is rapidly recycled where it is produced, and remains at very low concentrations in aquatic environments. H_2 is utilized as an energy source by many anaerobic bacteria including sulfate reducers [39]. Therefore, it is not clear how H_2 is produced in oxidic waters from H_2S if it follows to the reaction (8).

The astonishing resemblance between the hydrophilic globules isolated by us from saturated aqueous solutions upon hydrogen sulfide decomposition (Figure 6) and the globules of biological sulfur produced by colorless sulfur bacteria [27-34,38] suggests that hydrogen sulfide assimilation by these bacteria can be also resulted in the formation of diatomic sulfur in the ground triplet state via the reaction:

$$2\ H_2S \xrightarrow{\text{sulfur bacteria}} 2\ H_2{}^* + \{{}_3S_2\}_{(globules)} \tag{9}$$

where $H_2{}^*$ is an activated *in situ* hydrogen participating in chemosynthesis:

$$H_2{}^* + CO_2 \xrightarrow{\text{sulfur bacteria}} \{HCOOH\} \tag{10}$$

Obviously, the reaction (9) takes place as a result of the nucleation of diatomic sulfur as a condensed hydrophilic S_2 phase. Since the spherical shape and hydrophilic nature of the globules produced by sulfur bacteria is independent of their nature and habitat [27-34,38], reaction (9) seems to occur both under anaerobic conditions and in the presence of oxygen. Doubtless this hypothesis should be comprehensively verified as well.

H_2S Decomposition on Metal Catalysts Under Layer of Solvents

Similar to the biological processes of H_2S assimilation with sulfur bacteria in aqueous medium, the efficiency of H_2S decomposition can be essentially improved (compared to the gas phase process) when solid catalyst is being immersed into a solvent capable of readily dissolving both hydrogen sulfide and the final products of the reaction (5) [20,40].

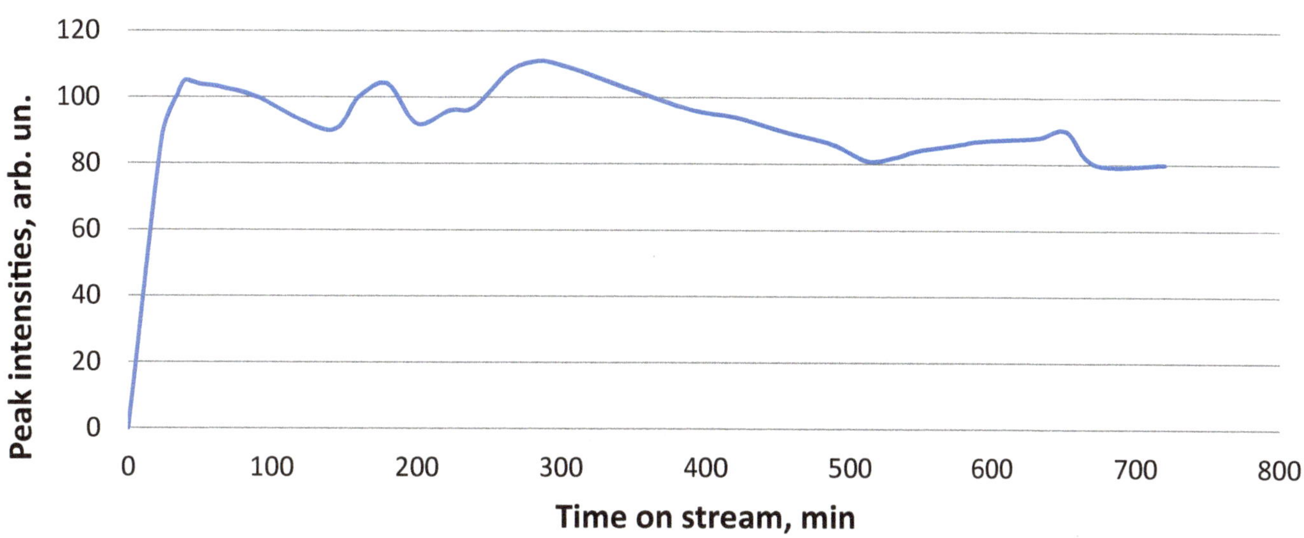

Figure 7: Evolution of hydrogen in the process of H_2S decomposition at room temperature on the stainless steel chip immersed into the aqueous 5% MEA solution [40].

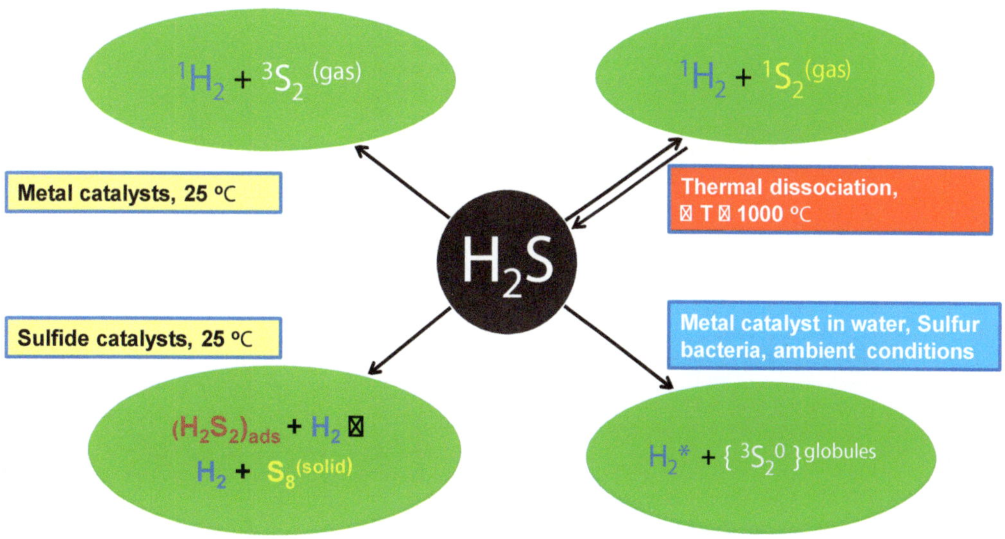

Figure 8: The possible pathways of H_2S decomposition into hydrogen and elemental sulfur. The catalytic pathways are considered from the standpoints of biological thermodynamics.

The use of liquids significantly increases the volumetric and surface concentration of hydrogen sulfide compared with the gas phase. This, in turn, leads to a substantial increase in the overall rate of chemical transformation due to the increasing coverage of the catalyst surface by adsorbed molecules of hydrogen sulfide. Moreover, the reaction product -diatomic sulfur, accumulates in the solvent thereby freeing the catalyst surface for adsorption of new molecules of hydrogen sulfide.

As is well known, purification of any wastes from hydrogen sulfide begins with its extraction from gas streams in absorbers with liquid solvent to the required sanitary standards. Concentrated solutions of the absorbers are thereupon directed to the regeneration stage where H_2S is recovered by solvent boiling and is utilized mainly with the Claus method or is used to obtain sulfuric acid. Instead of this high temperature procedure of H_2S utilization, the concentrated with H_2S solutions could be passed through the metal catalyst at ambient conditions to provide the reaction (5) and to get hydrogen as the target product, while sulfur is accumulated in solution.

Indeed, the aqueous monoethanolamine (MEA) solution, a well-known and widespread hydrogen sulfide absorber, is turned out one of the most suitable solvent for the reaction (5) realization. Decomposition of H_2S occurs at room temperature with hydrogen evolution to the gas phase (Figure 7) and sulfur accumulation in solution with efficiency towards H_2S utilization of 97.9% (Table 2) [40]. Efficiency of H_2S decomposition can be improved to 99.6% in the periodical regime (Table 3) [20], when, after solution saturation with H_2S, the reaction is allowed to be completed in a closed system. No doubt, the possibilities of this process to achieve close to 100% efficiency of H_2S decomposition is practically unlimited taking into account a wide range of choice of catalysts, solvents, regimes, etc.

State	T_e* O_2	T_e* S_2
$c\,^1\Sigma^-_u$	33 057	20 203
$b\,^1\Sigma^+_g$	13 195	7 981
$a\,^1\Delta_g$	7 923	4 395
$X\,^3\Sigma^-_g$	0	0

Table 1: Comparison of excitation energy (in cm^{-1}) of the three most reliably studied low-lying molecular orbitals of the S_2 molecule with a molecule of oxygen O_2. T_e - energy of electronic term (in cm^{-1}) [11].

Solution		5 % MEA	Na_2CO_3 [Na] = 0.84 %
Volume, ml		200	100
Fed H_2S , mmol		110.6	53.4
H_2S decomposed, mmol (g)		108.3 (3.69)	42.5 (1.33)
H_2S conversion, %		**97.9**	**79.6**
Sulfur content in solution*,	% mass	1.62	1.30
	g	3.24	1.30

* - X-ray fluorescent analysis data after removal of non reacted H_2S from solution with Ar flow

Table 2: H_2S decomposition over stainless steel chips placed into aqueous solutions. Catalyst mass is 5 g. Reaction temperature is ambient, Ar flow is 10 ml/min, H_2S flow is 3 ml/min.

Adsorption No	Time on stream, min	H_2S, mmol		Conversion, %
		Fed	Non-reacted	
1	130	10.45	traces	
2	111	8.91	traces	
3	120	9.64	traces	
4	270	21.7	traces	
Σ	631	50.7	0.2	**99.6**

Table 3: The periodical regime of hydrogen sulfide decomposition at room temperature over stainless steel chips placed in aqueous 5% MEA solution. Argon flow rate, 6 ml/min; H_2S flow rate, 1.8 ml/min; catalyst weight, 2 g; solution volume, 150 ml [20]. After saturation with H_2S, the reactor inlet and outlet were closed and overnight later the H_2S feeding starts again.

Analysis of technologies of H_2S disposal has shown that the method proposed can be adapted to the existing industrial technologies of H_2S utilization. The implementation of this method on an industrial scale actually means the gradual replacement of the metal- and energy-intensive Claus processes for a new mode of operation at ambient conditions to produce hydrogen -a recognized energy resource of the future and a valuable chemical reagent. As liquid absorbers, it is proposed to use the known commercially available solvents of hydrogen sulfide with well proven technology of their operation.

Conclusion

"Useless" and hazardous hydrogen sulfide, which in nature is inexhaustible and renewable, plays a very important role in chemosynthesis as an energy and hydrogen source. Due to the sulfur bacteria, H_2S is easily refined into the vitally important substances under ambient conditions in the natural springs, so it is not accumulated in the atmosphere. However it creates a lot of problems in the refinery of petroleum, natural gas and coal, requiring the energy- and metal-intensive processes of H_2S utilization. Therefore, the catalytic reaction of low-temperature decomposition of H_2S into hydrogen and sulfur seems to offer new opportunity to use the biologically-similar process at the industrial scale. The stumbling block of our researches was to explain experimental data in the framework of classical thermodynamics, because it is hardly possible to understand formation of hydrogen and gaseous diatomic sulfur in the ground triplet state from H_2S at room temperature. In terms of biological thermodynamics, the role of solid catalysts becomes clear as an energy transducer from the exothermic processes of molecule adsorption to the surface chemical reactions, which are impossible to be done in the gas phase. Therefore, there exist two regions of the reaction of H_2S decomposition (Figure 8). At high temperature this reaction follows to the laws of the classical equilibrium thermodynamics and results in formation of both diatomic reaction products in the singlet state. At the ambient conditions, this reaction occurs either on the surface of solid catalysts or in the bacteria cells. However, the final reaction products are different depending on the catalyst and the reaction conditions. Of the special interest is the diatomic sulfur in the ground triplet state $X\,_3\Sigma_{-g}$, which can be obtained in the gas phase only (?) on the surface of metal catalysts via H_2S decomposition [7-9,20]. The standard thermodynamical reference data [18] are concerned with the metastable single state $a\,_1\Delta_g$ of this molecule (Table 1). Following the analogy with oxygen, the standard enthalpy of the $_3S_2$ molecule formation should be equal to zero as well. However, the generally accepted thermodynamical reference point is the orthorhombic solid sulfur. Therefore, to solve this conflicting problem, it should be necessary to compare the full energy of the S_8 molecule with that of the diatomic triplet sulfur. We made attempts to calculate full energy of cyclooctasulfur by the available computational methods, but unsuccessfully. So, the problem is open for discussion.

References

1. Kaloidas VE, Papayannakos NG (1987) Hydrogen production from the decomposition of hydrogen sulfide. Equilibrium studies on the system $H_2S/H_2/S_i$, (i=1,...,8) in the gas phase. Int J Hydrogen Energy 12: 403-409.
2. Zaman J, Chakma A (1995) Production of hydrogen and sulfur from hydrogen sulfide. Fuel Proc Technol 41: 159-198.
3. Luinstra EA (1995) Hydrogen from H_2S: Technologies and Economics. Sulfotech Research. Calgary, Alberta, Canada, pp: 1-122.
4. Fletcher EF, Noring JE, Murray JP (1984) Hydrogen sulfide as a source of hydrogen. Int J Hydrogen Energy 9: 587-593.
5. Cox BG, Clarke PF, Pruden BB (1998) Economics of thermal dissociation of H_2S to produce hydrogen. Int J Hydrogen Energy 23: 531-544.
6. Haynie DT (2008) Biological Thermodynamics. 2nd edn. Cambridge University Press, The Edinburgh Building, Cambridge CB2 8RU, UK.
7. Startsev AN, Kruglyakova OV (2013) Diatomic Gaseous Sulfur obtained at Low Temperature Catalytic Decomposition of Hydrogen Sulfide. J Chem Chem Eng 7: 1007-1013.
8. Startsev AN, Kruglyakova OV, Ruzankin SP, Bulgakov NN, Chesalov YA, et al. (2014) Peculiarity of Low Temperature Catalytic Decomposition of Hydrogen Sulfide. Russ J Phys Chem 88: 943-956.
9. Startsev AN, Bulgakov NN, Ruzankin SP, Kruglyakova OV, Paukshtis EA (2015) The reaction thermodynamics of hydrogen sulfide decomposition into hydrogen and diatomic sulfur. Journal of Sulfur Chemistry 36: 234-239.
10. Swope WC, Lee YP, Schafer HF (1979) Diatomic sulfur: Low lying bound molecular electronic states of S_2. J Chem Phys 70: 947-953.
11. Smirnov BM, Yatsenko AS (1997) Dimers. Novosibirsk: "Nauka" (in Russian).
12. Schweitzer C, Schmidt R (2003) Physical Mechanisms of Generation and Deactivation of Singlet Oxygen. Chem Rev 103: 1685-1758.
13. Startsev AN (2009) Concept of acid-base catalysis by metal sulfides. Catalysis Today 144: 350-357.
14. Zakharov II, Startsev AN, Zhidomirov GM, Parmon VN (1999) Oxidative addition of dihydrogen as the key step of the active center formation in the HDS sulfide bimetallic catalysts: ab initio MO/MP2 study. J Mol Catal A: Chem 137: 101-111.

15. Zakharov II, Startsev AN (2000) An ab Initio Molecular Orbital Study of the Hydrogen Sorbed Site in Co/MoS_2 Catalysts. J Phys Chem B104: 9025-9028.

16. Startsev AN, Zakharov II, Voroshina OV, Pashigreva AV, Parmon VN (2004) Low-Temperature Decomposition of Hydrogen Sulfide under the Conditions of Conjugate Chemisorption and Catalysis. Doklady Physical Chemistry 399: 283-286.

17. Zakharov II, Startsev AN, Voroshina OV, Pashigreva AV, Chashkova NA, et al. (2006) The Molecular Mechanism of Low-Temperature Decomposition of Hydrogen Sulfide under Conjugated Chemisorption–Catalysis Conditions. Russian Journal of Physical Chemistry 80: 1403-1410.

18. Benson SW (1978) Thermochemistry and kinetics of sulfur-containing molecules and radicals. Chem Rev 78: 23-35.

19. Raymont MED (1975) Make hydrogen from hydrogen sulfide. Hydrocarb Proc 1: 139-142.

20. Startsev AN (2016) Low-Temperature Catalytic Decomposition of Hydrogen Sulfide into Hydrogen and Diatomic Gaseous Sulfur. Kinet Catal 57: 511-522.

21. Alfonso DR (2008) First-principles studies of H_2S adsorption and dissociation on metal surfaces. Surf Sci 602: 2758-2768.

22. Rodriguez JA, Hrbek J, Kuhn M, Jirsak T, Chaturvedi S, et al. (2000) Interaction of sulfur with Pt(111) and Sn/Pt(111): Effects of coverage and metal–metal bonding on reactivity toward sulfur. J Chem Phys 113: 11284-11292.

23. Koestner RJ, Salmeron M, Kollin EB, Gland JL (1986) Adsorption and surface reactions of H_2S on clean and S-covered Pt(111). Surf Sci 172: 668-690.

24. Startsev AN, Kruglyakova OV, Chesalov YA, Kravtsov EA, Serkova AN, et al. (2015) Aqueous Solutions of Sulfur Produced via Low Temperature Hydrogen Sulfide Catalytic Decomposition. Russ J Phys Chem A 89: 19-23.

25. Meyer B (1976) Elemental sulfur. Chem Rev 76: 367-387.

26. Steudel R, Eckert B (2003) Solid Sulfur Allotropes. Top Curr Chem 230: 1-79.

27. Schlegel HG, Bowien B (1989) Autotrophic Bacteria. Madison, WI. Science Tech Publishers.

28. Steudel R (1989) On the Nature of the "Elemental Sulfur" (S^0) Produced by Sufur-oxidizing Bacteria – a Model for S0 Globules. In: Autotrophic Bacteria. Schlegel HG, Bowien B (eds.), Madison, WI. Science Tech Publ, pp: 289-303.

29. Janssen A, de Keizer A, van Aelst A, Fokkink R, Yangling H, et al. (1996) Surface characteristics and aggregation of microbiologically produced sulfur particles in relation to the process conditions. Colloids Surf. B Biointerfaces 6: 115-129.

30. Janssen AJH, Lettinga G, de Keizer A (1999) Removal of hydrogen sulfide from wastewater and waste gases by biological conversion to elemental sulfur. Colloidal and interfacial aspects of biologically produced sulfur particles. Colloids Surf A 151: 389-397.

31. Pasteris JD, Freeman JJ, Goffredi SK, Buck KR (2001) Raman spectroscopic and laser scanning confocal microscopic analysis of sulfur in living sulfur-precipitating marine bacteria. Chem Geol 180: 3-18.

32. Kleinjan WE, de Keizer A, Janssen AJH (2003) Biologically Produced Sulfur. Top Curr Chem 230: 167-188.

33. Sorokin DY, Banciu H, Robertson LA, Kuenen JG (2006) Haloalkaliphilic Sulfur-Oxidizing Bacteria. Prokaryotes 2: 969-984.

34. Robertson LA, Kuenen JG (2006) The Colorless Sulfur Bacteria. Prokaryotes 2: 985-1011.

35. Guerrero R, Mas J, Pedros-Alio C (1984) Buoyant density changes due to intracellular content of sulfur in Chromatium warmingii and Chromatium vinosum. Arch Microbiol 137: 350-356.

36. Steudel R, Holdt G, Visscher PT, van Gemerden H (1990) Search for polythionates in culture of Chromayium vinosum after sulfide incubation. Arch Microbiol 153: 432-437.

37. Pickering J, George GN, Yu EY, Brune DC, Tuschak Ch, et al. (2001) Analysis of Sulfur Biochemistry of Sulfur Bacteria Using X-ray Absorption Spectroscopy. Biochemistry 40: 8138-8145.

38. Dahl CH, Friedrich C, Kletzin A (2008) Sulfur Oxidation in Prokaryotes. In: Encyclopedia of Life Sciences (ELS). John Wiley & Sons Ltd., Chichester.

39. Jorgensen B (1989) Biogeochemistry of Chemoautotrophic Bacteria. In: Autotrophic Bacteria. Schlegel HG, Bowien B (eds.), Madison, WI. Science Tech Publishers, pp: 117-145.

40. Startsev AN, Kruglyakova OV, Chesalov YA, Paukshtis EA, Avdeev VI, et al. (2016) Low temperature catalytic decomposition of hydrogen sulfide on metal catalysts under layer of solvent. J Sulf Chem 37: 229-240.

Thermodynamic Calculation of a Heat of First-Order Phase Transitions

Sobko AA*

Department of Physics and Chemistry of New Materials, AM Prokhorov Academy of Engineering Sciences, 19 Presnensky Val, Moscow 123557, Russia

Abstract

A transition heat is the most important characteristics of first-order phase transitions. Black [1] was first who discovered in 1762 that in the transfer of water to vapor, some quantity of heat is absorbed, which he termed the latent evaporation heat. In spite of more than the two-hundred-year period of the heat transfer concept existence there are no analytical expressions relating the transition heat with other parameters of phase transitions. For example, the fundamental "Physics Encyclopedia", articles devoted to the transition heat, evaporation heat, and so on, comprises no formulae but only tables of experimental data. One can also mention monographs [2-9] which have no relationships except for the conventional definition of the transition heat λ=TΔS. Hence, obtaining the relationships between the transition heat and other parameters of first-order phase transitions will be a substantial contribution into the theory of first-order phase transitions.

Keywords: Thermodynamic; Heat; First-order phase transitions

Experimental Methods

Calculation of the evaporation heat for liquid gases and metals

General expressions for the transition heat of first-order phase transitions: The conventional expression for a transition heat $\lambda = T\Delta S$ has two substantial drawbacks. First, in some phase transitions, for example, in evaporation, not only entropy changes but a system does the work, which can only be supplied by an external source of heat. Second, the transition heat is expressed in terms of the entropy variation ΔS, which cannot be measured experimentally. In the preset work, the transition heat is defined as

$$\lambda = T\Delta S + A$$

Where T is the transition temperature in K^0, ΔS is the change of system entropy, A is the work that the system does. Entropy is calculated from the general definition [3]

$$S = k\ln\frac{V_{ph}}{(2\pi\hbar)^s},$$

Where K is the Boltzmann factor, V_{ph} is the volume of phase space occupied by the system, and S is the number of degrees of freedom. Thus, we obtain:

$$S = k\ln\frac{V_{ph}}{(2\pi\hbar)^s},$$

where $(V_{ph})_o$ and $(V_{ph})_n$ are the volumes of the old and new phases, respectively. The general expression for the heat for a phase transition has the form:

$$\lambda = kT\ln\frac{(V_{ph})_n}{(V_{ph})_o} + A \qquad (1)$$

The volumes of phase spaces and the expressions for the work are specified for each particular phase transition. In the present work, all calculations are performed for one mole of substance; hence, all extensive values refer to one mole.

Approximate calculation of the phase space volumes for liquid and gaseous states

For liquid and gaseous states, the energy of system has the form:

$$\frac{p_{x1}^2}{2m_1} + \frac{p_{y1}^2}{2m_1} + \frac{p_{1z}^2}{2m_1} + \cdots + \frac{p_{zN}^2}{2m_N} + U(\vec{r}_1,\ldots,\vec{r}_N) = E \cdot 1 \le i \le 3, 1 \le n \le N_A \cdot \qquad (2)$$

Since we consider one mole of a single-component substance, all masses are equal and $N = N_A$ is the Avogadro number. The volume of the phase space is:

$$V_{ph} = \int_{6N_A} \!\!..\!\! \int \prod_n^{6N_A} d^3\vec{p}_n d^3\vec{r}_n$$

Equation (2) can be rewritten in the form:

$$\sum_{in} p_{in}^2 = 2m\left(E - U(\vec{r}_1,..,\vec{r}_{N_A})\right),$$

which is the equation of a 3N-dimensional sphere in $\vec{p}$-space; hence, the 3N-dimensional integral over pulses $\vec{p}$ is equal to the volume of this sphere of the radius $\left(2m\left(E - U(\vec{r}_1,...,\vec{r}_{N_A})\right)\right)^{\frac{1}{2}}$, and the expression for the volume of phase space takes the form:

$$\iint_{3N} \frac{\pi^{3N/2}}{(3N/2)!}\left(2m\left(E - U(\vec{r}_1,...\vec{r}_{N_A})\right)\right)^{3N_A/2} \prod_n^{3N_A} d^3\vec{r} \cdot$$

In the result of integration the dimensionality of the integral has changed from 6 to 3; however, one cannot take the rest integral without further assumptions. For calculating the 3N-dimenstional integral, let us consider the behavior of the function of kinetic energy distribution near a point of a first-order phase transition. Not specifying exactly the distribution function one may assert that it has a bell shape with a maximum that shifts with temperature to right. Near the point of phase transition such an evolution of the distribution function is impossible because the temperature of the system does not change and energy incoming still continues. Hence, the only way for the distribution function to vary is its narrowing and, in the limit, it transfers to the δ-function. In the latter case, the most probable and the average values of the kinetic energy will coincide. It is not a rigorous proof of the distribution function narrowing; however, it suggests a principal

***Corresponding author:** Sobko AA, Department of Physics and Chemistry of New Materials, AM Prokhorov Academy of Engineering Sciences, 19 Presnensky Val, Moscow 123557, Russia, E-mail: ainrf@mail.ru

assumption of the present work, which is confirmed by a satisfactory agreement between experimental data and calculation results.

Note: Near the first-order phase transition the most of atoms (molecules, ions) are in the state with the average kinetic energy.

Since (E-U($\mathbf{r}_1$,..$\mathbf{r}_n$)) is the kinetic energy of the system, according to the theory about equal distribution of a kinetic energy over degrees of freedom [4] one can substitute it for the average value $3N_AKT/2=3RT/2$, where $R=KN_A$ is the universal gas constant:

$$V_{ph} = \int_{3N}^{..}\int \frac{\pi^{3N/2}}{(3N/2)!}\left(\frac{3}{2}2mRT\right)^{3N/2}\prod_n d^3r_n = \frac{\pi^{3N/2}}{(3N/2)!}(3mRT)^{3N/2}V^N$$

Thus, the volume of a phase space for liquid and gas is expressed as

$$V_{phL} = \frac{(3\pi mRT)^{3N/2}}{(3N/2)!}V_L^N \quad (3)$$

$$V_{phG} = \frac{(3\pi mRT)^{3N/2}}{(3N/2)!}V_G^N \quad (4)$$

Where V_L and V_G are the volumes of liquid and gas, respectively.

Calculation of the evaporation heat for liquid gases

By using the general expression (1) for a transition heat one can find the expression for the evaporation heat on the saturation curve. Since the volumes of phase spaces for liquid and gas are known (3), (4), the change of entropy in evaporation has the form:

$$\Delta S = k\ln\left(\frac{V_G}{V_L}\right)^N = kN\ln\left(\frac{V_G}{V_L}\right) = R\ln\left(1+\frac{\Delta V}{V_L}\right),$$

Where ΔV is the volume jump, V_L is the volume of liquid, V_G is the gas volume.

The work on volume expansion is A_1=PΔV. In the transit liquid-gas, in addition to the work on volume expansion, also the work against surface tension forces is done $A_2=\sigma FN_1$, where σ is the surface tension coefficient, F is the area of liquid surface, $N_1=V_a/Fd$ is the number of mono-molecular layers, and V_a is the volume occupied by atoms (molecules, ions), d is the thickness of a mono-molecular layer (in the present work it is αr, where r is the radius of atom (molecule, ion), α=1.717 is the packing factor).

Thus, we have $A_2=\sigma FN_l=\sigma FV/Fd=\sigma V_\alpha/d$.

The expression for the evaporation heat on the saturation curve has the form

$$\lambda = RT\ln\left(1+\frac{\Delta V}{V_{Lf}}\right)+P\Delta V+\frac{\sigma V_a}{\alpha r_a}, \quad (5)$$

Where R is the universal gas constant; T, P are the temperature and pressure on the saturation curve, respectively; ΔV is the jump of volume in the process of evaporation; V_L is the volume of liquid; V_α is the volume occupied by atoms (molecules, ions); r_α is the effective atomic (molecular, ion) radius; and α=1.817 is the sphere packing factor. All extensive values refer to one mole of substance.

The expression for the evaporation heat λ comprises the volume occupied by atoms V_α and the volume occupied by liquid V_L. The question arises which volume one should employ – the geometrical (experimental) volume or the free volume $V_{Lf}=V_{Lf}-N_AV_0$, where V_0 is the volume occupied by atom (molecule, ion). Here, the evaporation heat is calculated by using as the geometrical liquid volume V_L, so and the free volume of liquid V_{Lf}.

Experimental data on the saturation curve are taken from Ref. [10], the radii of atoms and ions are taken from Ref. [11,12]. The radii of two-atomic molecules are taken as half the distance between nuclei centers [13] plus the Van der Waals radii [12]. The results are given in Table 1 and Figure 1.

TK^0 is the evaporation temperature, $P*10^{-5}$ [Pa] is the pressure, V_L*10^5 [m³/mole] is the molar volume of liquid, $\Delta V*10^5$ [m³/mole] is the jump of volume in evaporation, $\sigma*10^3$ [N/m] is the surface tension coefficient, $r*10^{10}$ [m] is the radius of atom (molecule, ion), λ_{ex} [J/mole] is the experimental value of molar evaporation heat, λ_{T1} [J/mole] is the molar evaporation heat calculated by using the geometrical volume, λ_{T2} [J/mole] is the molar evaporation heat by using the free volume, δ_1 and δ_2 [%] are inaccuracies of λ_{T1} and λ_{T2}, respectively [14].

The calculated values of evaporation heat in Table 1 are only presented for the evaporation lines, for which the experimental values of surface tension coefficient have been found. For hydrogen, data on a surface tension coefficient along the entire evaporation curve are known; the corresponding experimental and calculated values of evaporation heat are plotted in Figure 1. A small difference between experimental and calculated values at low temperatures is related with the fact that the calculation of a phase volume should make allowance for quantum effects.

From Table 1 and Figure 1 one can see that the calculations of the evaporation heat performed by the obtained formula well agree (within several percent) with experimental results. Hence, the assumption, that the most of atoms (molecules, ions) near a point of a first-order phase transition are in the state with the average kinetic energy, is valid. Also valid is the assumption that in a calculation of the evaporation heat one should take into account the work done by the system. The employment of the free volume in calculations also gives a better agreement between experimental and calculated results. Thus, one can assert that the molar evaporation heat λ on the saturation curve is described by the expression:

$$\lambda = RT\ln\left(1+\frac{\Delta V}{V_{Lf}}\right)+P\Delta V+\frac{\sigma V_a}{\alpha r_a}, \quad (6)$$

where all the values have been defined above.

Specific features of calculating the evaporation heat for liquid metals

As one can see from the results presented in Table 1, calculations of evaporation heat should be performed with the "free volume". For

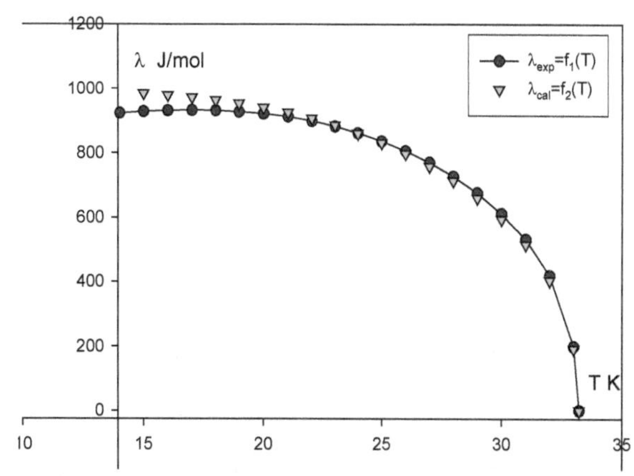

Figure 1: Experimental and calculated evaporation heat versus temperature.

determining the "free volume" of liquid metals one should know the radii of ions. Handbooks [11,12] comprise two metal ion radii: radii for ions M^{+1}, M^{+2} and so on, and for metal. Since it is not clear which radius should be used in the calculations of the free volume, the latter was calculated by using as the metal radius so and the ion radius. The results with the employment of the ion radius are given in Table 2, and the results based on metal radius are presented in Table 3.

T[K] is the evaporation temperature, $P*10^{-5}$ [Pa] is the pressure, V_L*10^5 [m³/mole] is the molar volume of liquid, $\Delta V*10^5$ [m³/mole] is the volume jump in evaporation, $\sigma*10^3$ [J/m²] is the surface tension coefficient, r_m*10^{10} [m] is the ion radius in Table 2 and the metal radius in Table 3, V_m*10^5 [m³/mole] is the volume occupied by ions, $V_{Lf}*10^5$ [m³/mole] is the "free volume" of liquid, λ_{ex} [J/mole] is experimental evaporation heat, λ_T [J/mole] is calculated value of evaporation heat, and δ % is the calculation error λ_T.

From Table 2 one can see that the employment of ion radii in calculations of the free volumes of liquid gives a good agreement with experimental data; however, the presence of free electrons in liquid and screening of ions make one to assume that more realistic are metal radii. Results of calculations with metal radii are given in Table 3. One can see that the calculated values of the evaporation heat in this case are systematically greater than the experimental values. This can be explained by the fact that in evaporation of metals, in addition to endothermic processes, there are also exothermic processes.

Recombination of ions and electrons with the origin of neutral atoms occurs for all metals at the interface liquid-gas (vapor). In this case, the energy is released and completely or partially participates in the evaporation process. In addition, alkali metal atoms in the gaseous state form two-atomic molecules [10] with an energy release. Thus, $\Delta\lambda=\lambda_T-\lambda_{ex}$ is the evaporation heat received by a system due to the exothermic processes considered above. The energy released due to generation of two-atomic molecules is proportional to the fraction of two-atomic molecules in gas, and the energy released in the ion recombination is constant and independent of thermodynamic parameters.

Hence,

$$\Delta\lambda=\beta C+Q_0, \qquad (7)$$

Where C is the part of two-atomic molecules in gas, β is the energy, released in the process of producing 0.5 mole of two-atomic molecules, Q_0 is the heat released in recombination of one mole of metal ions.

Calculation results for the evaporation heat, experimental values of evaporation heat in a wide temperature range on the evaporation curve for all alkali metals and mercury, and data on the part of two-atomic molecules in gas for alkali metals are given in Table 4.

Basing on these data, dependences $\Delta\lambda=f(C)$ for alkali metals were plotted in Figure 2. One can see that the assumption about linear dependence of $\Delta\lambda$ on C is confirmed with a high accuracy. Moreover, one can assert that the energies released in generating two-atomic molecules Na_2, K_2, Rb_2, Cs_2 are similar, and the energy released in generating Li_2 is substantially higher. The values of Q_0 are, respectively, Q_0^{Li}=4510 J/mole, Q_0^{Na}=3510 J/mole, Q_0^{K}=2510 J/mole, Q_0^{Rb}=2030 J/mole, and Q_0^{CS}=1220 J/mole and well correlate with the ionization energy for these metals. This does not mean that the linear dependence will still exist at very high temperatures close to the critical temperature, because λ_{ex} and λ_T tend to zero as the temperature approaches the critical value; hence, $\Delta\lambda$ also tends to zero. The mechanism of λ_{ex} reduction at high temperatures is most probably related with the fact that an exothermic process of dissociation of two-atomic molecules starts. Unfortunately, there are no experimental data for calculating λ_T and no information about the behavior of λ_{ex} and the part of two-atomic molecules in gas at high temperatures close to the critical temperature.

In Figure 3 one can see a dependence of $\Delta\lambda$ on temperature for mercury. A linear extrapolation of $\Delta\lambda=f(T)$ turns $\Delta\lambda$ to zero at a point close to the value of T_c.

Conclusion

We may assert that suggested expression (6), which only comprises measurable parameters, allows one to calculate the evaporation heat of liquid gases with a high accuracy taking into account considerations of Sect. 4, and the evaporation heat of liquid metals.

Su-s	T	P	V_L	ΔV	σ	r	λ_{ex}	λ_{T1}	δ_1	λ_{T2}	δ_2
Ne	25	0.51	1.63	394	5.50	1.60	1790	1549	13.5	1757	1.80
Ar	90	1.34	2.90	533	10.53	1.92	6307	5185	17.8	5895	6.53
Kr	150	6.56	3.87	164	10.00	1.98	7886	6356	19.4	7220	8.44
Xe	200	5.22	4.87	284	12.00	2.18	11327	9100	19.7	10362	8.52
H_2	30	8.08	3.67	15.5	0.33	1.44	612	547	10.5	595	2.83
N_2	90	3.60	3.75	182	6.16	2.09	5057	3968	21.5	4668	7.69
O_2	100	2.55	2.95	303	10.70	2.00	6490	5256	19.0	6209	4.32
F_2	95	2.78	2.65	261	10.70	2.06	6775	5024	25.9	6426	5.16
Cl_2	201	0.13	4.26	14570	33.00	2.47	21934	18408	16.1	22110	-0.80
CH_4	105	0.56	3.70	1521	15.80	2.30	8390	7333	12.6	8872	-5.74

T K⁰ is the evaporation temperature, $P*10^{-5}$ [Pa] is the pressure, V_L*10^5 [m³/mole] is the molar volume of liquid, $\Delta V*10^5$ [m³/mole] is the jump of volume in evaporation, $\sigma*10^3$ [N/m] is the surface tension coefficient, $r*10^{10}$ [m] is the radius of atom (molecule, ion), λ_{ex} [J/mole] is the experimental value of molar evaporation heat, λ_{T1} [J/mole] is the molar evaporation heat calculated by using the geometrical volume, λ_{T2} [J/mole] is the molar evaporation heat by using the free volume, δ_1 and δ_2 [%] are inaccuracies of λ_{T1} and λ_{T2}, respectively.

Table 1: Calculation results of evaporation heat.

El-s	T	P	V_L	ΔV	σ	r_i	V_i	V_{Lf}	λ_{ex}	λ_T	δ
Li	1600	0,91	1,73	13058	273	0,78	0,12	1,61	134720	133995	0,5
Na	1150	0,96	3,09	8760	120	0,99	0,24	2,84	89590	86899	3,0
K	1000	0,73	5,82	10516	65	1,33	0,59	5,23	75994	72581	4,5
Rb	950	0,92	7,25	7893	50,7	1,49	0,83	6,42	69742	65089	6,7
Cs	800	0,20	8,56	31264	47,5	1,69	1,22	7,34	65126	63799	2,0
Hg	613	0,75	1,57	6820	384	1,12	0,35	1,22	59275	56159	5,3

T[K] is the evaporation temperature, $P*10^{-5}$ [Pa] is the pressure, $VL*10^5$ [m^3/mole] is the molar volume of liquid, $\Delta V*10^5$ [m^3/mole] is the volume jump in evaporation, $\sigma*10^3$ [J/m^2] is the surface tension coefficient, $rm*10^{10}$ [m] is the ion radius in Table 2 and the metal radius in Table 3, $Vm*10^5$ [m^3/mole] is the volume occupied by ions, $VLf*10^5$ [m^3/mole] is the "free volume" of liquid, λex [J/mole] is experimental evaporation heat, λT [J/mole] is calculated value of evaporation heat, and δ % is the calculation error λT.

Table 2: Calculation results for the evaporation heat on the saturation curve for metals by using the ion radii.

El-s	T	P	V_L	ΔV	σ	r_m	V_m	V_{Lf}	λ_{ex}	λ_T	δ
Li	1600	0,91	1,73	13058	273	1,57	0,97	0,75	134720	151512	-12,5
Na	1150	0,96	3,09	8760	120	1,89	1,70	1,39	89590	98320	-9,7
K	1000	0,73	5,82	10516	65	2,36	3,31	2,51	75994	82311	-8,3
Rb	950	0,92	7,25	7893	50,7	2,53	4,08	3,17	69742	73767	-5,8
Cs	800	0,20	8,56	31264	47,5	2,74	5,18	3,37	65126	72209	-10,9
Hg	613	0,75	1,57	6820	384	1,60	1,03	0,54	59275	67675	-14,2

T[K] is the evaporation temperature, $P*10^{-5}$ [Pa] is the pressure, V_L*10^5[m^3/mole] is the molar volume of liquid, $\Delta V*10^5$ [m^3/mole] is the volume jump in evaporation, $\sigma*10^3$ [J/m^2] is the surface tension coefficient, r_m*10^{10}[m] is the ion radius in Table 2 and the metal radius in Table 3, V_m*10^5 [m^3/mole] is the volume occupied by ions, $V_{Lf}*10^5$[m^3/mole] is the "free volume" of liquid, λ_{ex}[J/mole] is experimental evaporation heat, λ_T[J/mole] is calculated value of evaporation heat, and % is the calculation error λ_T.

Table 3: Calculation results for the evaporation heat on the saturation curve for metals by using the metal radii.

El-s	T	P	V_L	ΔV	σ	r_m	V_m	V_{Lf}	λ_{ex}	λ_T	C	Δλ
	1400	0,18	1,64	64895	265,5	1,55	0,939	0,701	139563	154081	8,67	14518
	1500	0,43	1,69	29223	251,5	1,55	0,939	0,751	137114	153172	10,55	16058
Li	1600	0,91	1,73	14622	237,5	1,55	0,939	0,791	134719	152305	12,60	17586
	1700	1,77	1,77	7983	223,5	1,55	0,939	0,831	132394	151553	13,57	19159
	1800	3,19	1,82	4691	209,5	1,55	0,939	0,881	129487	150689	15,05	21202
	900	0,05	2,87	145285	145,3	1,89	1,702	1,168	95107	102618	6,37	7511
	1000	0,20	2,95	41885	135,3	1,89	1,702	1,248	92849	102068	8,58	9219
	1100	0,60	3,05	15340	125,3	1,89	1,702	1,348	90628	101145	10,79	10517
Na	1200	1,50	3,15	6636	115,3	1,89	1,702	1,448	88481	100064	12,89	11583
	1300	3,22	3,25	3360	105,3	1,89	1,702	1,548	86469	99339	14,74	12870
	1400	6,26	3,37	1859	95,3	1,89	1,702	1,668	84434	98268	16,58	13834
	1500	11,01	3,49	1129	85,3	1,89	1,702	1,788	82547	97294	18,12	14747
	800	0,06	5,43	108807	78,75	2,36	3,313	2,117	79866	85079	4,05	5213
	900	0,24	5,62	30668	72,15	2,36	3,313	2,307	77942	84272	5,98	6330
	1000	0,73	5,82	11355	65,55	2,36	3,313	2,507	75944	83606	7,97	7662
K	1100	1,86	6,04	4904	58,95	2,36	3,313	2,727	73899	82454	9,88	8455
	1200	3,91	6,28	2703	52,35	2,36	3,313	2,967	71823	82814	11,61	10991
	1300	7,30	6,54	1475	45,75	2,36	3,313	3,227	69739	80698	13,10	10959
	1400	12,44	6,81	899	39,15	2,36	3,313	3,497	67674	78990	14,31	11316

El-s	T	P	V_L	ΔV	σ	r_m	V_m	V_{Lf}	λ_{ex}	λ_T	C	Δλ
Rb	700	0,03	6,61	183451	65,0	2,53	4,082	2,528	72632	76715	3,44	4083
	800	0,16	6,85	42015	59,35	2,53	4,082	2,768	70718	76303	5,46	5585
	900	0,55	7,11	13667	53,55	2,53	4,082	3,028	68777	75483	7,61	6706
	1000	1,47	7,39	5661	47,75	2,53	4,082	3,308	66862	74680	9,70	7818
	1100	3,30	7,70	2768	41,95	2,53	4,082	3,618	64931	73783	11,62	8852
	1200	6,47	8,03	1535	36,15	2,53	4,082	3,948	62999	72816	13,26	9817
	1300	11,43	8,40	937	30,35	2,53	4,082	4,318	61075	71729	14,62	10655
Cs	700	0,04	8,26	132419	50,3	2,68	4,851	3,409	69366	72071	3,03	2705
	800	0,20	8,56	32798	47,5	2,68	4,851	3,709	67731	71982	4,91	4251
	900	0,66	8,89	11300	42,7	2,68	4,851	4,039	66016	71321	6,95	5305
	1000	1,69	9,24	4843	37,9	2,68	4,851	4,389	64209	70410	8,98	6201
	1100	3,63	9,65	2478	33,1	2,68	4,851	4,799	62361	69605	10,86	7244
	1200	6,79	10,1	1440	28,3	2,68	4,851	5,249	60487	68785	12,05	8298
	1300	11,41	10,7	924	23,5	2,68	4,851	5,849	58600	67779	13,85	9179
Hg	373	0,0004	1,50	8285191	452	1,55	0,939	0,561	60848	70421		9573
	423	0,0038	1,52	931431	439	1,55	0,939	0,581	60518	69241		8723
	473	0,0232	1,53	169895	429	1,55	0,939	0,591	60194	68471		8277
	523	0,0996	1,54	43651	416	1,55	0,939	0,601	59872	67662		7790
	573	0,3302	1,56	14417	402	1,55	0,939	0,621	59546	66801		7255
	623	0,8990	1,57	5748	378	1,55	0,939	0,631	59206	65695		6489

T [K] is the evaporation temperature, $P*10^{-5}$ [Pa] is the pressure, V_L*10^5 [m³/mole] is the molar volume of liquid phase, $\Delta V*10^5$ [m³/mole] is the volume jump, $\sigma*10^3$ [J/m²] is the surface tension coefficient, r_m*10^{10} [m] is the metal radius for ion, V_m*10^5 [m³/mole] is the volume occupied by ions, $V_{Lf}*10^5$ [m³/mole] is the "free volume" of liquid, λ_{ex} [J/mole] is the experimental value of evaporation heat, λ_T [J/mole] is the calculated value of evaporation heat, C[%] is the concentration of two-atomic molecules in a gas of alkali metals, $\Delta\lambda=\lambda_{ex}-\lambda_T$ [J/mole].

Table 4: Calculation of Δλ.

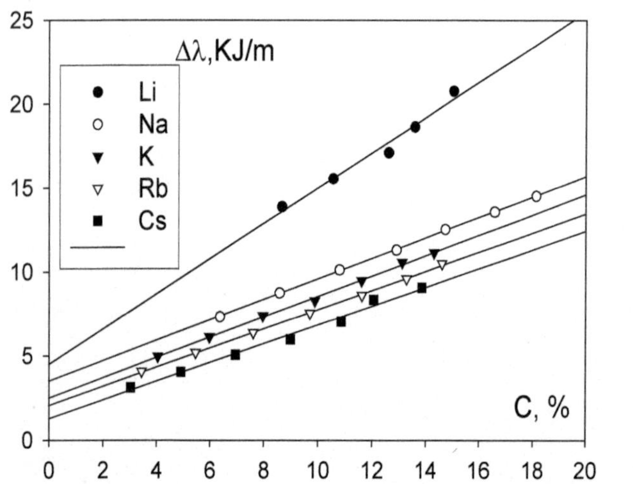

Figure 2: Dependence of $\Delta\lambda=\lambda_{ex}-\lambda_T$ on the concentration of two-atomic molecules in gas for alkali metals.

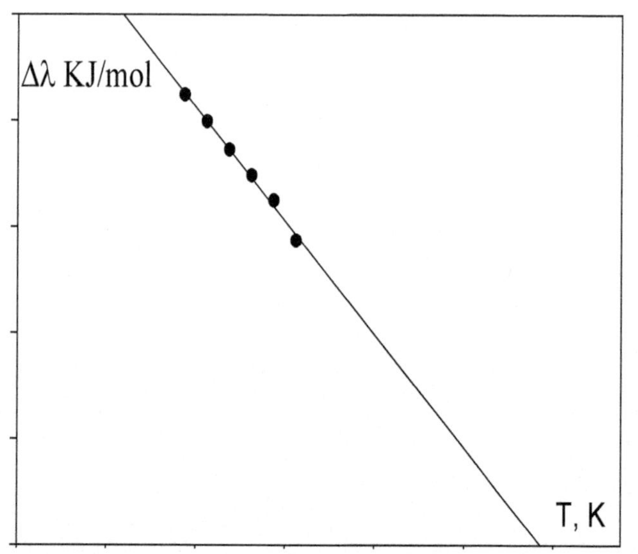

Figure 3: Dependence of $\Delta\lambda=\lambda_{ex}-\lambda_T$ on temperature for mercury.

References

1. Black J (1803) Lectures on the elements of chemistry: delivered in the University of Edinburgh. 1st edn.
2. Fizicheskaya Entsiklopedia (1998) Physics Encyclopedia. Prokhorov AM (ed.), Moscow: Bol'shaya Rossiiskaya Entsiklopediya.
3. Landau LD, Lifshits EM, Statisticheskaya F (1980) (Moscow: Nauka 1976). Landau LD and Lifshitz EM (eds) Statistical Physics, Part 1, NY: Pergamon.
4. Huang K (1963) Statistical mechanics. 2nd edn. John Wiley & Sons, Inc., New York, London KK.
5. Mayer JE, Mayer MG (1977) Statistical mechanics. A Wiley-Interscience Publication, John Wiley & Sons, New York.
6. Isihara A (1971) Statistical physics. Academic Press, New York.
7. Kubo R (1968) Thermodynamics. North-Holland Publishing Company, Amsterdam, John Wiley & Sons, Inc., New York.
8. Kubo R (1965) Statistical mechanics. North-Holland Publishing Company, Amsterdam.
9. Kittel CH (1977) Statisticheskaya termodinamika (Moscow: Nauka).
10. Vargaftik NB (1972) Spravochnik po teplovym svoistvam gazov i zhidkostei (Handbook on Thermal Properties of Gases and Liquids) (Moscow: Nauka).
11. Svoistva Elementov (1976) Properties of Elements. Editor Samsonov GV, Moscow: Metallurgiya.
12. Emsley J (1991) The elements. Clarendon Press Oxford.
13. Kratkii spravochnik fiziko-khimicheskikh velichin (1983) Brief Handbook on Physical-Chemical Values. Revdel AA, Ponomareva AM (eds.) Leningrad: Khimiya.
14. Fizicheskiye Velichiny (1991) Physics Values. Grigor'ev IS, Meilikhov EZ (eds.) Moscow: Energoatomizdat.

Spectro-Potentiality and Thermodynamic Studies on Schiff Base Hydrazone Derivatives Complexes with Some Transition Elements

El Batouti M[1]*, El Mossalamya EH[2], Nawmosyc G[3]

[1]*Chemistry Department, Faculty of Science, Alexandria University, Egypt*
[2]*Chemistry Department, Faculty of Science, Benha University, Egypt*
[3]*Chemistry Department, Faculty of Science and Arts, Al Qassim University, Saudi Arabia*

Abstract

A new hydration Schiff base consisting of 7-chloro-4-(o-hydroxy benzilidenehydrazo) quinoline, with metal complexes of some divalent metal ions, Mn(II), Fe(III), Co(II), Ni(II), Cu(II), Zn(II), and Cd(II) complexes were studies. Spectrophotometrically obeyed Beer's law up to certain values, 6.6×10^3, $8.0 \times 10^3 dm^3 mol^{-1} cm^{-1}$, respectively, indicating that the ligand HL could be used to detect Co(II) and Ni(II) in such concentrations, the study was performed also for Fe (III)-complexes. Conductance and thermodynamically used molar ratio method to determine the stoichiometry of the complexes obtained confirmed the formation of 1:1 and 1:2, M:L complexes. The standard thermodynamic parameters viz. ΔG, ΔH and ΔS, for the proton-legend and the stepwise metal-ligand complexes eluated were analyzed in terms of the electrostatic (eel) and non-electrostatic (crates, c) components. ΔHc was found to be linearly correlated with the acceptor number of the metal ion (ANM), and ΔH el was linearly correlated with the ionic radii of the metal ion. The calculated values for ΔG°, ΔH° and ΔS° concluded that the complexion process processed spontaneously.

Keywords: Hydrazone; Spectro-potentiality; Transition metal complexes; Thermodynamic parameters.

Introduction

Hydration Schiff base compounds containing quinoline moiety are well known in natural materials, and show interesting biological and antiviral activities [1-3]. Many derivatives of hydration compounds form colored complexes with different metal ions and can be used as analytical reagents for their determinations [4-9]. The coordination compounds of aroylhydrazones have been reported to act as enzyme inhibitors [10] and are useful due to their pharmacological applications [11-13]. The study of Schiff base hydrazones has been growing interest because of their antimicrobial, anti-tuberculosis, and anti-tumor activity [1-12]. The azomethine (C=N) linkage in Schiff bases imports in elucidating the mechanism of transmission and reexamination reactions in biological system [14]. The importance of metals and transition elements are formed organic compound due to its presence in many of the enzymes associated with mineral particles of large ring molecules such as hem [15]. Effect of the nature of the center metal ion and the medium, composition of fixed factors on the stability of the complexes formed Hydrazones of Schiff bases with ions transition elements were studied by using, spectral studies of solutions, and reduced the pollution of both heavy metals or organic compounds unwanted. Thermodynamic parameters are computed and analyzed in order to investigate the bond character between the metal ligand. The magnetic moments and electrical conductance of the complexes were also determined.

Experimental

Reagents and materials

All chemicals used were for anal, grade and the solutions of metal nitrates of, Mn (II), Fe (III), Co (II), Ni (II), Cu (II), Zn (II), and Cd (II) ions were prepared in carbonate free double distilled water and they were stabilized using EDTA titrations [16]. The HL ligand was prepared as described previously [17]. The structure of the ligand was elucidated by IR, Mass, H-NMR and electronic (UV-VIS) spectroscopy, as well as micro analytical analyses [17]. The results have been previously published [17] and weighed quantity of the ligand was dissolved in 75% (v/v) ethanol-water medium. Ethanol-water was freshly refluxed and distilled over magnesium powder and iodine [16].

Synthesis of the organic ligand

7-chloro-4-(o-hydroxy benzilidenehydrazo) quinoline, HL the ligand (HL) was synthesized as follows. 7-chloro-4-hydrazinoquinoline; 2 g (1 mmol) was dissolved in 10 ml absolute ethanol. To this solution salicylaldehyde, 13 ml (1.1 mmol) was added. The reaction mixture was refluxed for 2 h. After cooling, the formed yellow precipitate was collected, filtered, and finally washed with 5-10 ml absolute ethanol and purified by recrystallization from ethanol (80% m mp 235°C) (Figure 1).

Measurement of electrical conductivity

Electrical conductance conducted measurements using a WTW. D 8120 Weilheim LF. 42.

Spectral measurements

Spectral measurements were conducted, on a Shimadzu Model UV-probe.

Potentiometric measurements

A Hanna pH-meter model 302 digital with conventional pH-electrode assembly was used for pH measurements at 10, 20, 30 and

***Corresponding author:** El Batouti M, Chemistry Department, Faculty of Science, Alexandria University, Alexandria, Egypt
E-mail: mervette_b@yahoo.com

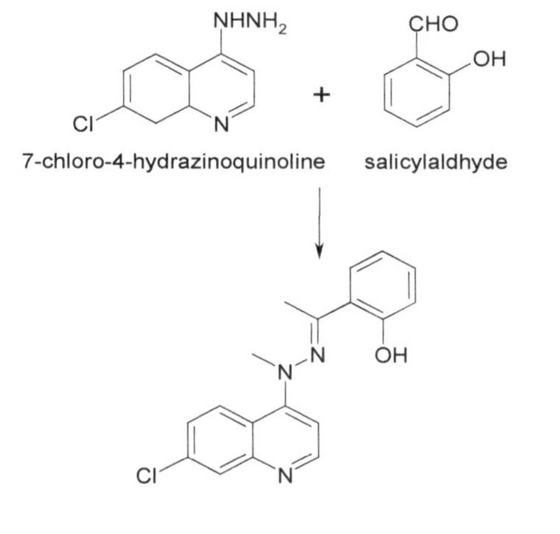

Figure 1: 7 - chloro-4-(o-hydroxybezilidenehydrzo)quinoline.

40°C. The ionic strength of the medium was kept virtually constant at 0.05 mol.dm^{-3} with KNO_3 as background electrolyte. The temperature was maintained constant by use of double-jacket cells with water circulated from a constant-temperature bath. Purified nitrogen gas was bubbled through the solution before and during the titrations multiple titrations were carried out for each system and the pH meter readings were recorded [6-9]. Electronic absorption spectra of the free ligand and its complexes in 75% (v/v) ethanol-water were recorded using a Shimadzu model UV-probe spectrophotometer. The composition of the complexes in solutions was determined by the molar ratio method [18,19].

Result and Discussion

Thermodynamic parameters

Thermodynamic parameters, (Table 1) are calculated from the formation constants in a mixture of 75% ethanol-water, at different temperatures as follows:

1. Free energy change $\Delta G°$

$-\Delta G° = 2.2303T \log K$ (1)

Vant-Hoff isotherm $\Delta H°$

$d\log K/dt = \Delta H°/4.57T$ (2)

By solving (1) and (2) we obtained

$\log K = -\Delta H°/4.47T + \text{constant}$

By plotting the relation between log K vs. 1/T we obtained a straight line its slope is equal to $=\Delta H°/4.57$ T and $\Delta H°$ is in cal.mol^{-1}.

The change in entropy $\Delta S°$ is calculated from Gibbs_Helmhoholtz equation:

$\Delta G° = \Delta H° - T\Delta S°$

Thermodynamic parameters provide valuable information about the nature of forming complexes, the degree of covalent bonding, and the change in hydration. The thermodynamic parameters results and the stability constants of the metal ions at a temperature 30°C in 75% (v/v) prove that the stability constants of complexes ligand log K_1 would be in the following order:

Ni (II)> Co (II)> Mn (II)> Cd (II)> Zn

The Fe (III) ion was excluded because it must be compared with Triple charges ions. If we compare {Zn (II), Cd (II)} ions and {Co (II), Ni (II)} ions, we find that their positions are consistent with previous publication research [21-24]. Figure 2 illustrates the form of a linear relationship between the dissociation constant of ligand and static stability of the complexes log K_1. This linear relationship reflects the similarity of the ionic nature of the reaction between the metal and ligand [21].

We can also use functions to distinguish between thermodynamic parameters formed inside the outer sphere and inner sphere complexes. According to Choppin [25] reflects the function's thermodynamic parameters and any type of these complexes may be as follows:

Therefore, the results confirm thermodynamic parameters which are internal complexes. Also, the table shows that the complexion process decreased with the increase of temperature, indicating that the temperature at least is the favorite for the complexion process and leading to negative values for $\Delta H°$ which means that the interaction between the metal ion and ligand molecule reaction is exothermic.

According to Nancollas et al. [26], the values of $\Delta H°$ reflect the change in the number and strength of the bonds that are formed by breaking during the interaction between the metal ion and ligand. The composition of the complex and hence the values of $\Delta H°$ is directly related to the type of interdependence between the metal ion (M^{n+}) and ligand molecules, as well as with the structure of complex formed. If we take into account the fact that the areas of crystal-fields produced by the consistency of oxygen centers O-coordinating sites are similar to the consistency of molecules [27], including water, $\Delta H°$ conclude that the effect is not affected by moving the replacement of water molecules. Therefore, the negative value of $\Delta H°$ is due to N-coordinating sites in the ligand molecule as described in the following structure.

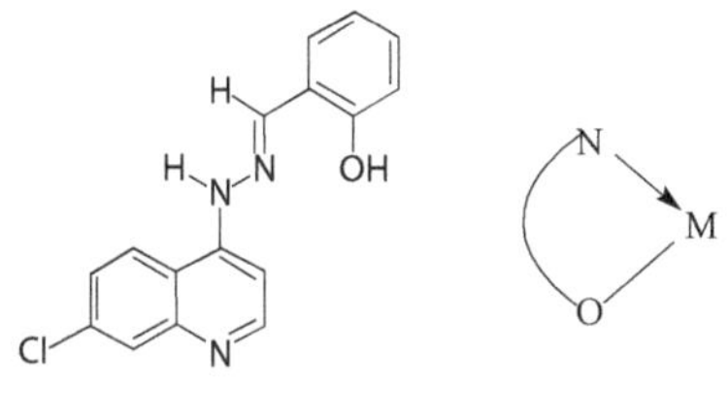

The table shows the values of $\Delta H°1 > -\Delta H°2$ of the ligand complexes with ions Co (II), Ni (II), Fe (III), and Mn (II). This demonstrates that there is a change in the properties of X-rays of the ligand, change in the dentate character. In other words, the decline in values -H°2 indicated to change donor centers in ligand of tridentate (ONN) donor in 1:1 complexes to bidentate (ON) donor complexes to 1:2 (M:L) to

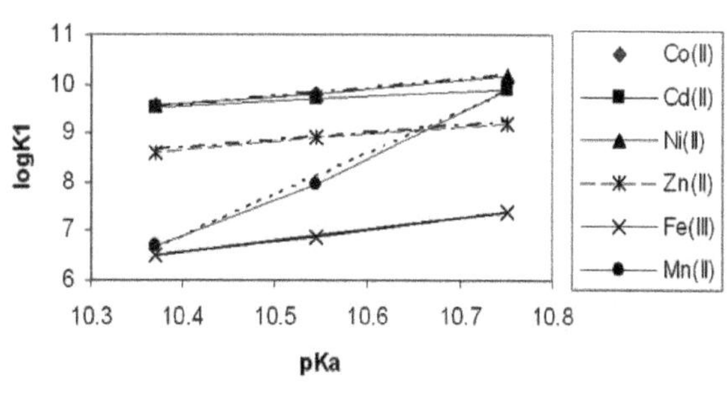

Figure 2: Relationship between the values of logK1 and pKa at 30°C in 75% (v/v) ethanol-water.

Complex (HL)	t°C						K. cal. mol^{-1}				K. cal. mol^{-1}	
	30		40		50		$-\Delta G^\circ_{30}$		$-\Delta H^\circ_{30}$		$-\Delta S^\circ_{30}$	
	logk1	logk2	logk1	logk2	logk1	logk2	$-\Delta G^\circ_1$	$-\Delta G^\circ_2$	$-\Delta H^\circ_1$	$-\Delta H^\circ_2$	ΔS°_1	ΔS°_2
Co(II)	10.09	10.56	9.79	10.4	9.55	10.2	13.9903	14.6419	0.120863	0.0803863	0.0457736	0.0480579
Cd(II)	9.88	10.99	9.69	10.6	9.52	10.1	13.6991	15.2381	0.080528	0.198704	0.0449457	0.0496351
Ni(II)	10.18	10.57	9.8	10.32	9.55	10.11	14.115	14.6558	0.141149	0.10293	0.0461185	0.0480296
Zn(II)	9.21	9.69	8.93	9.31	8.61	9	12.7701	13.4356	0.134038	0.15442	0.0417032	0.0438324
Fe(III)	7.41	8.4	6.9	8.1	6.5	7.9	10.2743	11.647	0.203699	0.112015	0.0332363	0.0380692
Mn(II)	9.91	10.61	7.96	10.47	6.7	10.3	13.7407	14.7113	0.719272	0.0692309	0.0429749	0.0483235
pka	10.751	--	10.544	--	10.371	--	14.9068	--	0.0850294	--	0.0489166	--

ΔH° and ΔS	ΔG°	Type of complex
Large positive	Large negative	inner sphere complex
Small positive	Small negative	outer sphere complex

Table 1: Thermodynamics functions and stepwise formation constants of 1:1 and 1:2 metal complexes at 30, 40, and 50°C in 75% ethanol-water.

Complex (HL)	t°C						K. cal. mol^{-1}				K. cal. mol^{-1}	
	30		40		50		$-\Delta G^\circ_{30}$		$-\Delta H^\circ_{30}$		$-\Delta S^\circ_{30}$	
	logk1	logk2	logk1	logk2	logk1	logk2	$-\Delta G^\circ_1$	$-\Delta G^\circ_2$	$-\Delta H^\circ_1$	$-\Delta H^\circ_2$	ΔS°_1	ΔS°_2
Co(II)	10.09	10.56	9.79	10.4	9.55	10.2	13.9903	14.6419	0.120863	0.0803863	0.0457736	0.0480579
Cd(II)	9.88	10.99	9.69	10.6	9.52	10.1	13.6991	15.2381	0.080528	0.198704	0.0449457	0.0496351
Ni(II)	10.18	10.57	9.8	10.32	9.55	10.11	14.115	14.6558	0.141149	0.10293	0.0461185	0.0480296
Zn(II)	9.21	9.69	8.93	9.31	8.61	9	12.7701	13.4356	0.134038	0.15442	0.0417032	0.0438324
Fe(III)	7.41	8.4	6.9	8.1	6.5	7.9	10.2743	11.647	0.203699	0.112015	0.0332363	0.0380692
Mn(II)	9.91	10.61	7.96	10.47	6.7	10.3	13.7407	14.7113	0.719272	0.0692309	0.0429749	0.0483235
pka	10.751	--	10.544	--	10.371	--	14.9068	--	0.0850294	--	0.0489166	--

Table 2: Thermodynamics functions and stepwise formation constants of 1:1 and 1:2 metal complexes at 30, 40, and 50°C in 75% ethanol-water.

overcome the steric hindrance in the complexes 1:1 (M: L). Which is identical to the published Evans et al. [28], where he obtained the same results. On the contrary, illustrates Table 2 that the values of Ho2>ΔHo1 for complexes with ions of Cd (II) and Zn (II) and takes this as a characteristic sign of the effect of Trans influence [29] of the second coordination.

In contrast, high negative values for ΔG° indicates that the complexity is spontaneously, and that the positive values of ΔS° for all complexes are consistent with the hypothesis that during the process complexity remove a large number of water molecules with the possibility of changes in the number of coordination [29]. This shows that the change in entropy reflects the formation of complexes. Figure 3 shows the relationship between the change in ΔG°_1, H°_1, ΔS°_1 and the number of electrons 3d and found that relationship is between metal ions and the number of electrons 3d, which indicates that the change is a valuable thermodynamic parameters probably due to the change in the electron affinity to ions metals. In order to get information on the nature of the bonds in the complexes, (Table 2) would be advisable to analyze the values of ΔG°, ΔH°, ΔS° into two parts: Part a, the part that temperature-dependent ΔX°_{el}, electrostatic (el) interaction due to reaction of dipole or ion with long-range electrostatic forces of the powers of the electrostatic solvent medium.

Part b, which does not depend on temperature ΔX°_{non} non-electrostatic or also called Kratk ΔX°_c which is due to the molecule or ion is produced from a short term or short range of covalent bonds with the surrounding environment and hence the,

$$\Delta X^\circ_{non} = \Delta X^\circ_{el} + \Delta X^\circ_c$$

If we look at the degree of covalent bonding, we find that the ion metal has a softener donor [30] and Figure 4 shows the relationship

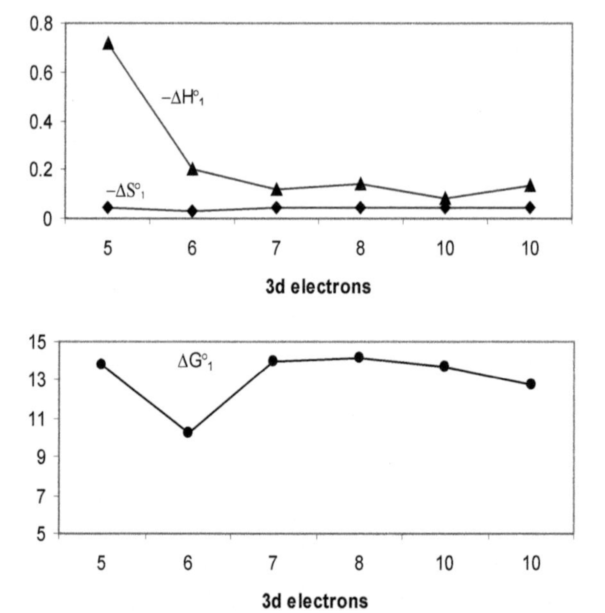

Figure 3: Relationship between the thermodynamics functions, (ΔG°_1, ΔH°_1, K. cal. mol^{-1}) and ΔS°_1 (cal. mol^{-1} k^{-1}) and the 3d electrons.

between ΔH°_{non} and quantity $E_n^{\#}$ described in Klopman [30] as a measure of hardness and a softness, where we note that the metal ion soft features with negative values of high-value $E_n^{\#}$ and vice versa (Table 3).

As is clear from the Figure 4 that there is a linear relationship

Complex (HL)	G° (K. cal. Mol^{-1})				ΔH° (K. cal. Mol^{-1})				S° (K. cal. Mol^{-1})	
	-ΔG°el		-ΔG°non		-ΔH°non		ΔH°el		ΔS°el	
	Log K1	Log K2	Log K1	Log K2	Log K1	Log K2	Log K1	Log K2	Log K1	Log K2
Co(II)	11.3959	11.8801	2.5944	2.76177	5.01251	5.17988	4.89164	5.09949	53.7543	56.0383
Cd(II)	11.2204	12.2145	2.4787	3.0236	4.89681	5.44171	4.81629	5.24301	52.9262	57.6155
Ni(II)	11.4689	11.8741	2.64602	2.78171	5.06413	5.19982	4.92298	5.09689	54.0987	56.0097
Zn(II)	10.533	10.9843	2.23716	2.45127	4.65526	4.86937	4.52123	4.71497	49.6838	51.8128
Fe(III)	8.73795	9.76253	1.53631	1.88442	3.98	4.30252	3.75072	4.19052	41.2168	46.0497
Mn(II)	10.8026	11.9365	2.93815	2.77479	5.35626	5.19289	4.63697	5.12367	50.9557	56.304

Table 3: Electrostatic (el) and non-electrostatic thermodynamic parameters for the reaction of metal ions with HL ligand in 75 % a(v/v) ethanol-water at 30°C.

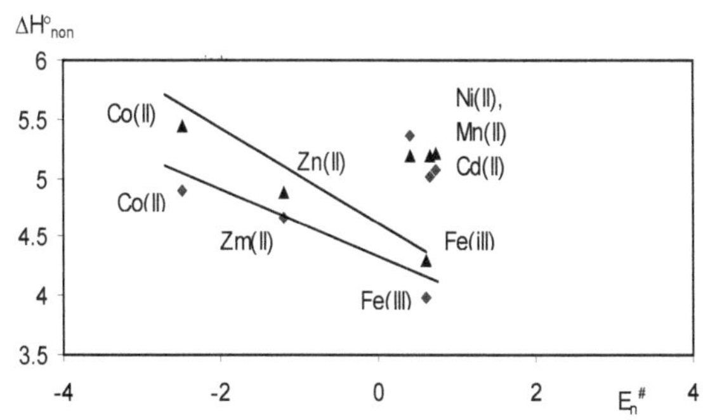

Figure 4: Relationship of hardness and softness, En# (eV), of metal ions and non-electrostatic heat change, ΔH°_{non}, of the formation of the complex.

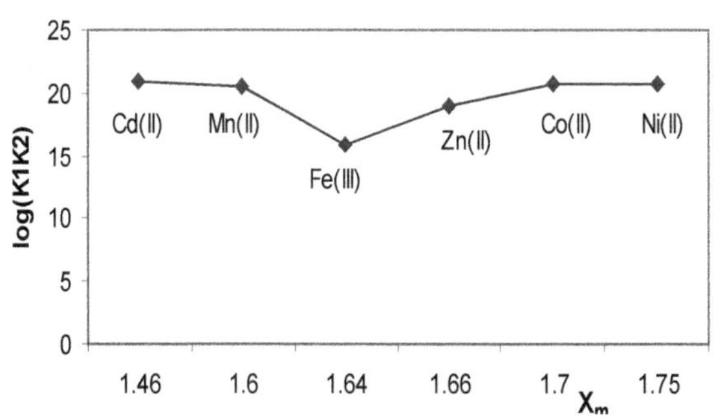

Figure 5a: Relationship of the X_m of the metal ions with log(K_1 K_2).

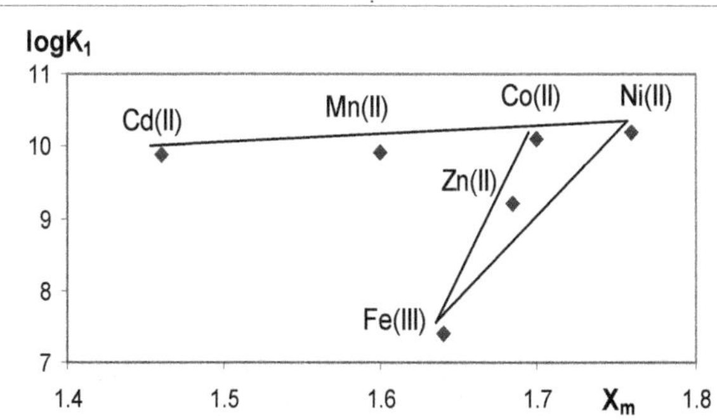

Figure 5b: Relationship between the electronegativity of the metal ionsand second ionization potential.

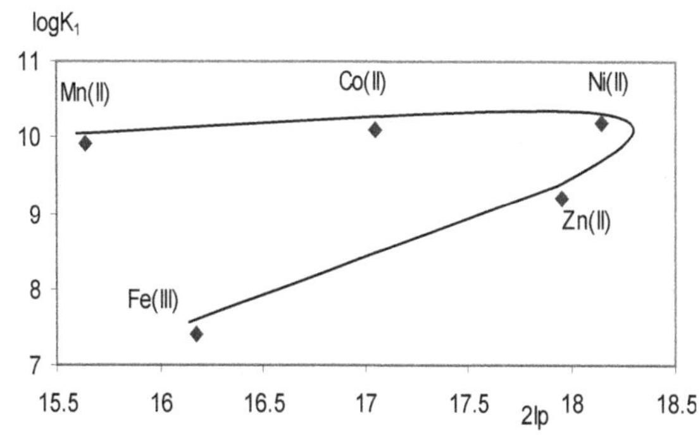

Figure 6: Relationship and logK_1 with 2Ip including Fe (III).

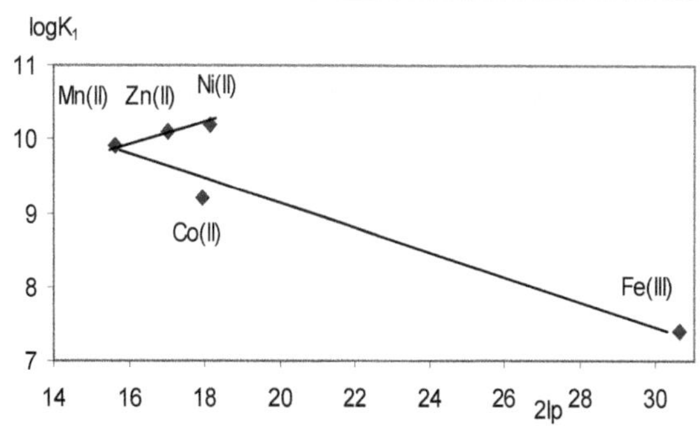

Figure 7: Relationship and logK_1 with 2Ip including 3Ip of Fe (III).

between ΔH°_{non} and the number of $E_n^{\#}$ ΔH°_{non} in the sense that increases as the increased softness of the softer metal ion. According to many authors [31-33], the change in electrostatic enthalpy ΔH°_{non} arise from changes that occur in the area of changes in the ligand field (LSF) associated with the composition of the complex. The following forms (Figures 5a, 5b, 6, 7 and 8) and after excluding the ion Fe (III), show the relationship between each of the electronegativity (X_m) and second ionization potential (2Ip) for ions metals with a fixed configuration log K_1 as well as with a fixed configuration log (K_1K_2) and note that the constants of the configuration log (K_1K_2) of the complexes are growing linearly with the increase electronegativity (X_m) or increase the ionization potential of the metal ion bilateral charge (2Ip).

This reflects the stability of the complexes are getting an increase of X_m and hence the link between the metal and ligand increase in the properties of covalent and confirmed by the linear relationship between ΔH_{non} and X_m (Figure 9).

On the other hand, the property of the bond in covalent complexes can infer from the linear relationship (Figure 10) between ΔH_{non} and ANm acceptor number of the cations [34].

Acceptor number (AN) known as a quantitative measure of Lewis

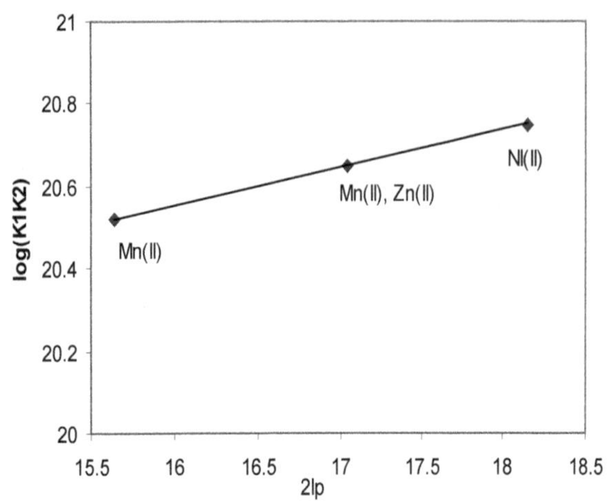

Figure 8: Relationship and log (K_1K_2) with 2Ip of divalent cations excluding Fe(III).

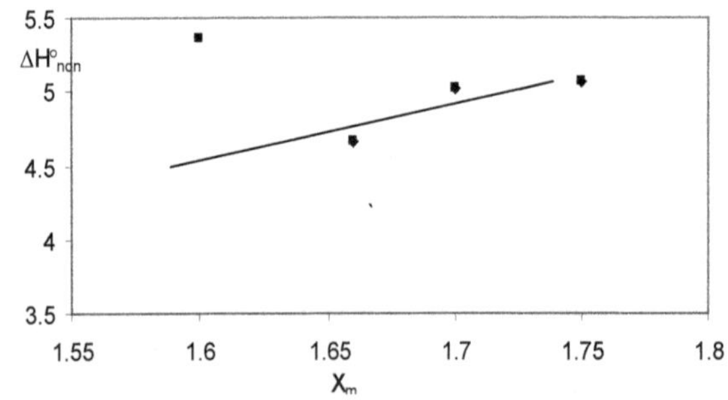

Figure 9: Relationship of the Xm of the metal ions with ΔH_{non}.

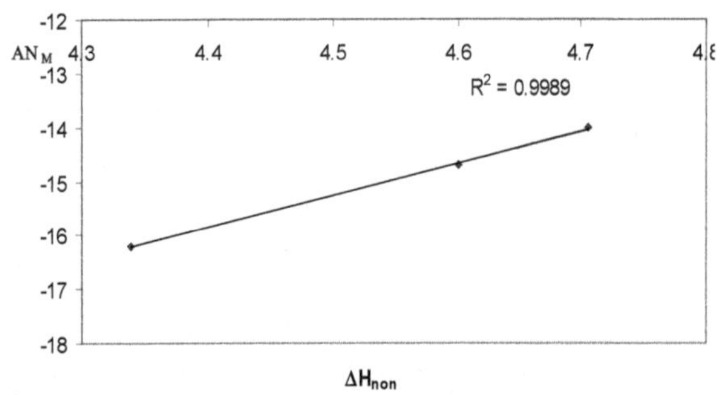

Figure 10: Relationship of acceptor number of metal ions, AN_M, with ΔH°_{non}.

acidity according to the ability to form a covalent bond with a pair of electrons. We can summarize the above, we found that the ΔH°_{non} associated with a positive relation with the donor acceptor number while ΔH°_{el} and ΔS°_{el} linked to a positive relation with half radii of metal ion (Figures 11 and 12).

Conductometric titration

In order to get information on the composition of complexes we have had adjustments electrical conductivity. It is known that this measurement shows that the measured amount reflect and directly proportional to the concentration of one or more of the ions. When the interaction between metal ions and ligand we get calibration curve, which consists of two or more lines transmit at the equivalent point (the point that determine the structure of the complex) [35].

Figure 13 showed conductometric titration curve of the ligand, HL, with metal ions, in which electrical conductivity increases continuously whenever we add ligand to the metal ion and continue to increase until we reach the equivalence point for the titration. It is also clear that this behavior was attributed to the replacement of some metal ions with some of complex molecules. When you add the following drop of ligand we note that there is a slight increase in the values of electrical conductivity, which reflects that the added ligand remains

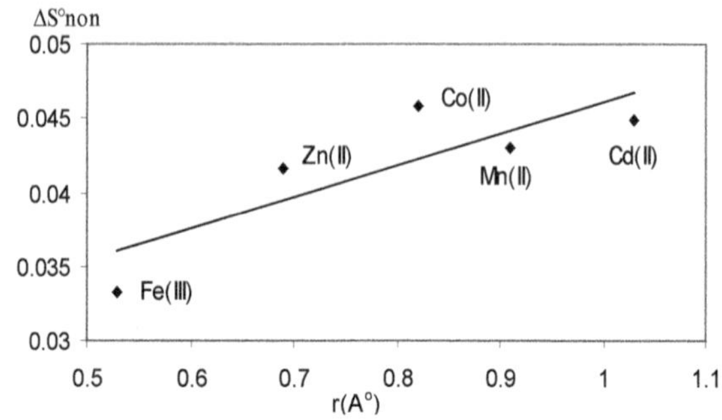

Figure 11: Relationship of ΔS°_{non} with ionic radii of metal ions.

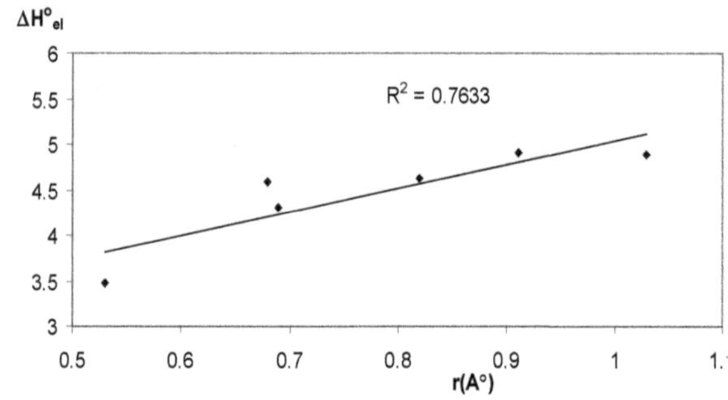

Figure 12: Relationship of ΔH°_{el} with ionic radii of metal ions.

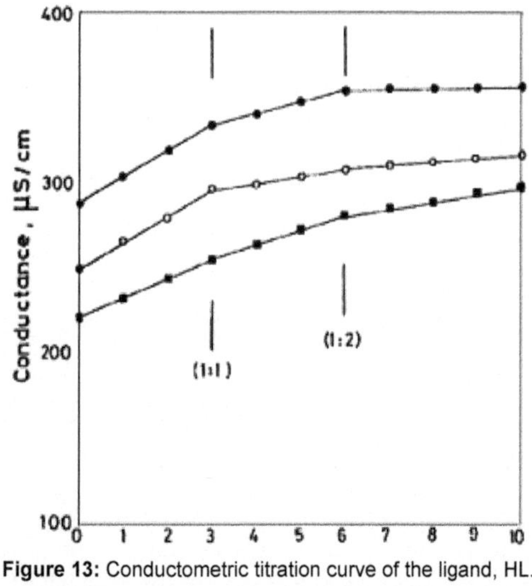

Figure 13: Conductometric titration curve of the ligand, HL.

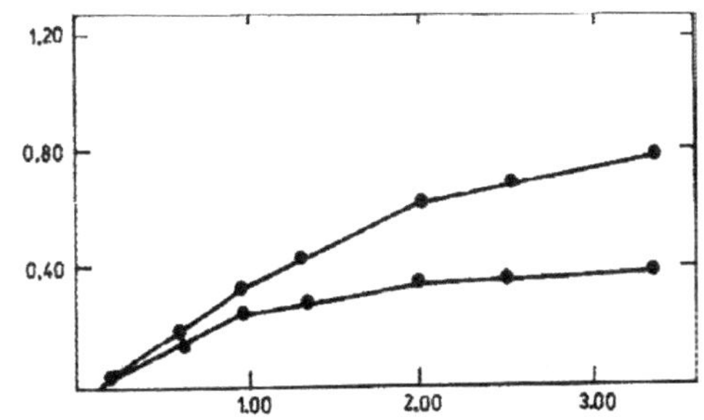

Figure 14: Mole ratio method for Co (II)-ligand (a) and Ni (II)-ligand (b).

without reaction and that increase could be due to the accumulation of ligand molecules. Titration curve and electrical conductivity appear to be complex double transmittance which reflects the percentage of M:2L and M:L, and corresponds with the results of potential titrations. Titration electrical conductivity has been made under the same conditions of potential titration.

Spectrophotometric measurements

The spectra of organic molecules are usually affected by the nature of the solvent and observed the influence of peaks intensity and position of the absorption top λ_m. The influence depends on several factors, including, for example, the physical properties of solvent such as dielectric constant and the determination of the dipole moment and refractive index as well as the ability of the solvent on the interaction through the formation of hydrogen bonds.

Determination of stoichiometry of the formed complex

The method used to determine the percentage molar ratio to form the complex.

$$M^{n+} + nHL \leftrightarrow ML_n + nH^+ ....1$$

$$\frac{[ML\ [H\ ]}{[M^n\ ][HL]^n}$$

1. Job's method [35-39] derived from equation (1), the interaction between the metal ion M^{n+} and organic ligand.

2. Molar ratio method [40-42], where the metal ion concentration was constant while changing the concentration of [ligand]/metal ion] and when we plot the values of absorption of these solutions against molarity ratio we get the straight lines intersect each two of them at a certain rate. Contrary to the method that we can make ligand concentration fixed and change the metal ion concentration was measured [metal ion]/[ligand] absorption of solutions at required pH.

3. Slope-ratio method in this way prepares two series of solutions. Save in the first series ion concentration element constant while changing the ligand concentration. In the second series saves the ligand concentration ion concentration fixed and changes the element. The pH value of the two methods was constant and measured the absorption using both the contents of the standard solution except the ion element.

We used the Molarity ratio method to determine the Molarity ratio of the complexes Co (II)-ligand, and Ni (II)-ligand as follows the preparation of a series of solutions, size 10 ml consisting of 0.2 ml of a solution of concentration [2-10 M] of the ions Co (II), Ni (II) and Mn (II). Then added 0.05, 0.1, 0.2, 0.3, 0.4, 0.5, 0.6, 0.7 and 0.8 ml of a solution focus ligand [2-10 M] added buffer (3 ml) and then to complete the 10 ml of 70% (v/v) Ethanol-water, when ionic strength m=0.5 M and then measured the spectrum of the figure turns out to be ML1 and ML2 (Figure 14). We will discuss the spectra of the complexes Co (II)-ligand (as an example to the rest of the complexes) as a function of ligand concentration and analysis of the results at wavelengths different that the ions Co (II) and Ni (II) are two entities essential and the two main complex species are 1:1 and 2:1 metal to the ligand.

n (CL)/n (CM)

The results of this method agreed with the results of measuring the pH adjustments previously discussed, where the Co (II), Ni (II), as an example to the rest from complexes, ML and ML_2 overlapped with ligand HL.

Validity of Beer's law

Using a constant concentration of ligand HL (1×10^{-3} M) and change the concentration of ions Co(II), Ni(II) and Mn(II) in a mixture of ethanol-water percentage of 75% (v/v) and when the ionic strength (=0.05 using potassium nitrate and ligand reference when λ=500 nm and found that they all follow the Beer law of standard deviation 0.002, 0.006 and 0.03 respectively during the concentration of metal ion complexes with Co (II) of 1×10^{-5}.

WTWD-812 Weilneium-conductivity meter, model LBR, fitted with a cell model LTA100 to 4×10^{-4} M With Ni (II) and the values of molar absorptivity (ε) at 500 nm for the complexes of Co (II), Ni (II), Mn (II) are respectively 6.6 × 103, 8.0 × 103 and 1.4 × 103 $dm^3mol^{-1}cm^{-1}$, indicating that HL can be used as a sensitive factor in the specific analysis to determine the copper (II) and nickel (II) form (Figure 15).

The pH metric titrations in 75% (v/v) ethanol-water of the free ligand and its mixtures with the metal ions were carried out as described previously [6-9]. Using the equations of Irving and Rossoti [20] the caves were used to calculate the formation constants of the following equilibria:

$$M^{2+} + HL \rightleftharpoons ML^+ + H^-$$

and

$$ML^+ + HL \rightleftharpoons ML_2 + H^+$$

The experimental reading was used to calculate the values of $\overline{\eta}$ and PL. From which the stability constant log K_1 and log K_2 were calculated with the help of the following equations given by Irving and Rossoti [20].

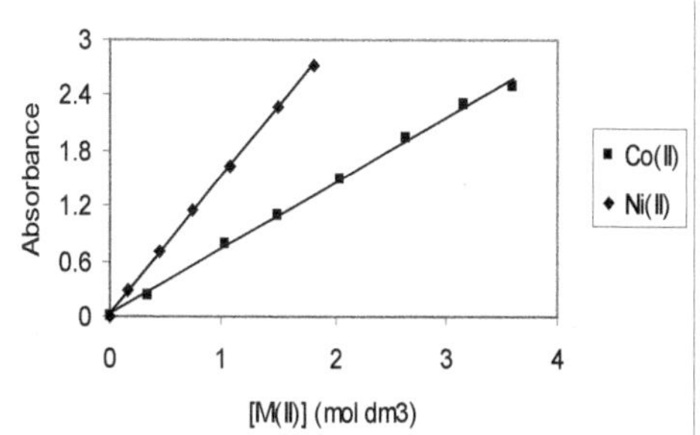

Figure 15: Validity of Beer's law, a) Co (II)-ligand at pH=8; b) Ni(II)-ligand at pH=8.

$$\log\frac{\bar{n}}{1-\bar{n}}=\log K_1+PL$$

$$\log\frac{\bar{n}-1}{2-\bar{n}}=\log K_2+PL$$

The metal-ligand stability constant was obtained using a linear plot method by plotting $\log\frac{\bar{n}}{1-\bar{n}}$ or $\log\frac{(\bar{n}-1)}{2-\bar{n}}$ against. Where $\bar{n}$ is defined by Irving and Rossetti [20] as the average number of the reagent molecules attached per metal ions. It can be obtained from the following relations:

$$\bar{n}=\frac{(V_3-V_2)(N^0+E^0)}{(V^0+V_2)\bar{n}_A TC_M^0}$$

And

$$PL=\log_{10}\frac{\sum_{n=0}^{n=1}\beta_n^H\left(\frac{1}{anti\log pH}\right)V^0+V_2}{TC_L^0-\bar{n}TC_M^0}$$

Where V_2 and V_2 are the volumes of alkali required to reach the same pH value, V_0 is the total volume of titrating mixture, N^0 and E^0 are the concentration of free acid, and pL, is the free ligand exponent. The experimental reading was used to calculate the values of $\bar{n}$ and from which the stability constant log K1 and log K2 were calculated with the help of the following equations given by Irving and Rossetti [20].

The variations of pKH, log K1 and log K2 vs 1/T gave straight lines which enable us to calculate the thermodynamic parameters ΔGo, ΔHo, and ΔSo of the complex formation by the Vant-Hoff equation and other relationships [21,22]. Conduct metric titrations were carried out at room temperature by titrating 30 ml metal ion solution (1×10^{-3} mol dm^{-3}) with the ligand solution (1×10^{-2} mol dm^{-3}) as tyrant using.

Conclusion

The present work describes the complex formation equilibria was investigated to ascertain the composition and stability constants of the complexes. The effect of ligand properties on the stability of the complexes was investigated. After tracing the complexation properties of discrete ligands towards different metals, it is also easier to correlate the data to the origin matter. As phenolic acids are a significant part of the chemical structure of substances it can be postulated, from the data obtained, that they can have an important role in their complication properties, mainly due to the presence of a hydroxyl group. To note that the discrete ligands used in previous studies only possess a phenolic function, usually in an ortho-position.

From the results obtained one can propose this type of structure corresponds to suitable discrete legends. Owning to their natural origin, phenolic acids could also be proposed, as the stability constants were found to be rather high. The protonation constants (pK1 and pK2) were determined by Irving-Rossetti pH titration technique. Also metal-ligand stability constants of their complexes with metal ions (Mn^{2+}, Co^{2+}, Ni^{2+} and Cu^{2+} Zn^{2+}, Cd^{2+} and Fe^{3+}) have been determined potentiometrically. It appears that PAAS are a better complexion agent with metal ions compared with the other: Mn (II)<Co (II)<Ni (II)<Cu (II)<Zn (II)<Cd (II)<Fe (III). The corresponding thermodynamic parameters (ΔG, ΔH and ΔS) were derived and discussed the dissociation process is non-spontaneous.

References

1. Heilman PW, Geilman RD, Scozzie AJ, Wayner RJ, Gullo MJ et al. (1980) Synthesis and anti-inflammatory evaluation of 3-methylthio-1,2,4-triazines, 3-alkoxy-1,2,4-triazines, and 3-aryloxy-1,2,4-triazines. J Pharm Sci 69: 282-287.
2. Schilt AA, Dunbar WE (1969) Talanta 16: 519.
3. Borges F, Guimarães C, José Lima LFC, Pintoc I, Reis S (2005) Potentiometric studies on the complexation of copper(II) by phenolic acids as discrete ligand models of humic substances. Talanta 66: 670-673.
4. DOĞAN A, KILIÇ E (2005) Potentiometric Studies on the Protonation Constants and Solvation of Some a-Amino Acid Benzyl- and t-Butyl- Esters in Ethanol-Water Mixtures. Turk J Chem 29: 41-48.
5. El-Bindary AA, Ghoneim MM, Diab MA, El-Sonbati AZ Serag LS (2014) Structure, Potentiometric and Thermodynamic Studies of N-Acryloyl-4-amino salicylic acid and Its Metal Complexes in Monomeric and Polymeric Forms. J Thermodyn Catal 5:2.
6. Ramadan AAT, El -Behairy MA, Ismail AII, Mahmoud M (1994) Synthesis and properties of iron(II) and iron(III) complexes with 3-(α,γ-dicarboxy-n-propylidene-hydrazino)-5,6-diphenyl-1,2,4-triazine (DCPT). Monatsh Chem 125: 1171-1182.
7. Ramadan AAT, Rahman RMA, Seda MH (1992) Studies on complexes of Cu(II), Ni(II), Co(II) and Ln(III) with 3-(α-benzoyl) benzylidenehydrazino-5, 6-diphenyl-1,2,4-triazine. Asian J Chem 4: 569-578.
8. Ramadan AAT, Rahman RMA, El-Behairy MA, Ismail AI, Mahmoud M (1999) The thermodynamics of complexation of transition and lanthanide ions by 3-(α-carboxymethylaminobenzyli-denehydrazino)-5,6-diphenyl-1,2,4 triazine (HipHt). Thermochim Acta 222: 291-303.
9. patel RN, singh N, Shrivastava RP, shukla KK, Singh PK (2002) Potentiometric and spectrometric study: Copper(II), nickel(II) and zinc(II) complexes with potentially tridentate and monodentate ligands. Proc Indian Acad Sci (Chem Sci) 114:115-124.
10. Dilworth JR (1976) The coordination chemistry of substituted hydrazines. Coord Chem Rev 21: 29-62.
11. Mubarak AT, Al-Shihri AS, Nassef HM, El-Bindary AA (2010) Potentiometric and Thermodynamic Studies of Vanillin and Its Metal Complexes. J Chem Eng Data 55: 5539-5542.
12. Nurchi VM, Crespo-Alonso M, Toso L, Lachowicz JI, Crisponi G (2013) IronIII and aluminiumIII complexes with substituted salicylaldehydes and salicylic acids. J Inorg Biochem 128: 174-182.
13. Rivas BL, Pereira ED, Moreno-Villoslada I (2003) Water-soluble polymer-metal ion interactions. Prog Polym Sci 28: 173-208.
14. Lau KY, Mayr A, Cheung KK (1999) Synthesis of transition metal isocyanide complexes containing hydrogen bonding sites in peripheral locations .Inorg Chim Acta 285: 223-232
15. Mansour AK, Eid MM, Khalil NSAM (2003) Synthesis and reactions of some new heterocyclic carbohydrazides and related compounds as potential anticancer agents. Molecules 8: 744-755.
16. Vogel AI (1978) Quantitative Inorganic Analysis. Longman, London.
17. El-begery M, El-Twigry H (2007) Synthesis, magnetic, spectral, and antimicrobial studies of Cu(II), Ni(II) Co(II), Fe(III), and UO_2(II) complexes of a new Schiff base hydrazone derived from 7-chloro-4-hydrazinoquinoline. Spectrochimica Acta Part A: Molecular and Biomolecular Spectroscopy 66: 28-36.
18. Ghoneim MM, El-Sonbati AZ, Diab MA, El-Bindary AA, Serag LS (2014) Supramolecular Assembly on coordination of azopolymer complexes: A review. Polym-Plast Technol Eng 54: 100-117.
19. Elbagerma MA, Azimi G, Edwards HGM, Alatjtal AI, Scowen IJ (2010) In situ monitoring of pH titration by ramman spectroscopy. Spectrochim Acta Part A 75: 1403-1410.
20. Irving HM, Rossotti HS (1954) The calculation of formation curves of metal complexes from pH titration curves in mixed solvents. J Chem Soc 2904-2910.

21. Ramadan AAT, El Shetary BA, Abdel-Moez MS, Seleim HS (1993) Acta Chim Hung 130: 25.

22. Kole N, Chaudhury KN AK (1981) J Inorg Nucl Chem 43: 2471.

23. GergelyA, Kiss T (1976) J Inorg Nucl Chem 39:109.

24. Lin J, Sahakian DC, Morais SMD, Xu JJ, Polzer RJ et al.(2003) The role of absorption, distibution, metabolism, excretion and toxicity in drug discovery. Curr Top Med Chem 3: 1125-1154.

25. Mouginot Y, Morlay C, Cromer M, Vittori O (2003) Potentiometric study of copper(II) and nickel(II) complexation by a cross-linked poly(acrylic acid) gel. Anal Chim Acta 407: 337-345.

26. Degischer C, Nancollas GH (1970) Thermodynamics of ion association. Part XX. Interpretation of the enthalpy changes. J Chem Soc A 1125 -1128.

27. Nancollas GH (1966) Interaction on Electrolytic Solutions Elsevier Amsterdam.

28. Rowlands CC, Morgan PH (1979) J Coord Chem 9: 19.

29. Linert W, Taha A (1993) J Coord Chem 29: 265.

30. Klopman G (1968) Chemical reactivity and the concept of charge- and frontier-controlled reactions. J Am Chem Soc 90: 233-234.

31. Paoletti P, Vacca A (1964) Thermochemical studies. Part XIII. Heats and entropies of reaction of tetraethylenepentamine with protons and bivalent transition-metal ions. J Chem Soc 5051-5057.

32. Ciapolini M, Paoletti P, Sacconic L (1960) Thermochemical studies. Part III. Heats and entropies of reaction of transition metal ions with ethylenediamine. J Chem Soc 4553-4561.

33. Jorgenson CK (1956) Comparitive crystal studies II . Niclle (II) and Copper(II) complex with polydenate ligands and the behavior of the residual places for co-ordination. Ada Chem Scand 10: 887-910.

34. Linert W, Jameson RF, Bauer G, Taha (1997) Estimation of the acceptor numbers of cations by means of an acid-base indicator A. J Coord Chem 42: 211-229.

35. Kealy D Principle and Practice of Analytical Chemistry (1995) Blackie Academic and Professional, London 261-371.

36. Job B (1928) Anna Chim 9:133.

37. Lin J, Sahakian DC, Morais SMD, Xu JJ, Polzer RJ et al.(2003) The role of absorption, distibution, metabolism, excretion and toxicity in drug discovery. Curr Top Med Chem 3: 1125-1154.

38. Mouginot Y, Morlay C, Cromer M, Vittori O (2003) Potentiometric study of copper(II) and nickel(II) complexation by a cross-linked poly(acrylic acid) gel. Anal Chim Acta 407: 337-345.

39. Vosburgh WC, Cooper GR. (1941) Complex Ions. I. The Identification of Complex Ions in Solution by Spectrophotometric Measurements. J Am Chem Soc 63: 437-442.

40. Zhao ZB, Zheng HX, Wei YG, Liu J (2007) Synthesis of azo derivatives of aminosalicylic acid. Chin Chem Lett 18: 639-642.

41. Sheng SF, Zheng HX, Liu J, Zhao ZB (2008) Synthesis of phenol-class azo derivatives of 4-aminosalicylic acid. Chin Chem Lett 19: 419-422.

42. Ghoneim MM, El-Ghamaz NA, El-Sonbati AZ, Diab MA, El-Bindary AA et al. (2015) Optical and thermal properties of azo derivatives of salicylic acid thin films. Spectrochim Acta A 137: 1039-1049.

Thermodynamic Modeling of Surface Tension of Aqueous Electrolyte Solution by Competitive Adsorption Model

Mohamad Javad Kamali[1*], Zakarya Kamali[2] and Gholamhossein Vatankhah[1]

[1]Department of Chemical Engineering, Islamic Azad University, Bushehr branch, Bushehr, Iran
[2]National Iranian Gas Company (NIGC), Fajr-e Jam Gas Company, Bushehr, Iran

Abstract

Thermodynamic modeling of surface tension of different electrolyte systems in presence of gas phase is studied. Using the solid-liquid equilibrium, Langmuir gas-solid adsorption and ENRTL activity coefficient model, the surface tension of electrolyte solutions are calculated. The new model has two adjustable parameters which could be determined by fitting the experimental surface tension of binary aqueous electrolyte solution in single temperature. Then the values of surface tension for other temperature in binary and ternary system of aqueous electrolyte solution are predicted. The average absolute deviations for calculation of surface tension of binary and mixed electrolyte systems by new model are 1.98 and 1.70%, respectively.

Keywords: Surface tension; Aqueous solution; Langmuir adsorption; E-NRTL activity model

Introduction

For studying the aqueous electrolyte solution in porous media, distillation, extraction process and liquid-liquid dispersion, the calculation of surface tension of aqueous solutions is required [1]. So, different equations have been developed to calculate surface tension of mineral salts. Ariyama [2], Lorenz [3], Young and Grinstead [4], and Glejm and Shelomov [5] formulated useful equations as group contribution method for calculation of surface tension of some limited binary electrolyte-water systems. Oka [6] proposed an equation for calculation of surface tension based on the concentration of solution, electronic charge, dielectric constant of water, ionic charge and Avogadro's constant. Later Hovarth [7] developed this equation by introducing the ionic strength and degree of dissociation into the Oka's model [6]. Onsager and Samaras [8], obtained a relation based on the temperature, dielectric constant of water and concentration of solution for calculation of surface tension of electrolyte solution. Schmutzer [9],considered the osmotic coefficient as an important factor for calculation of surface tension of electrolyte solution. Adding a proportional factor of anion concentration to the surface tension of water, the surface tension of aqueous electrolyte solution was determined by Abramazon and Gaukhberg [10]. This parameter was considered as a function of the inverse of square of ionic radius and anion charge. Li et al. [1] developed a new model for calculation of surface tension of single and mixed electrolyte solution. In this model, the surface layer is considered as a distinct phase where the electrolytes could be entered into it from another phases. The surface tension was obtained using the proportion of molality of salt in surface layer to liquid bulk phase. While this model had satisfactory results in low concentration of electrolytes, in high concentration the calculated surface tensions were not in good agreement with experimental data. Yu et al. [11] combined Li et al. [1] model with modified mean spherical approximation (MSA) as osmotic coefficient model. The results showed that the calculated surface tensions in highly concentrated regions were improved. Furthermore, the Langmuir gas-solid adsorption model was used at equilibrium condition for calculation of surface tension of mineral salts by Li and Lu [12]. The results indicated the satisfactory agreement with experimental data. Sadeghiet al. [13] used the combination of MSA model [14] with the Ghotbi and Vera [15] and the Mansoori et al. [16] equations of state, for correlation of the surface tension of single aqueous solution. Also the surface tensions of different mixed aqueous solutions were predicted by this approach. The results indicate the satisfactory agreement between calculated and experimental data [13]. In this paper, a new model for calculation of surface tension of the electrolyte systems is developed using the Langmuir adsorption equation and E-NRTL [17] model. The adjustable parameters of this model are obtained by experimental data of surface tension at single temperature. Then the model is verified by prediction of surface tension of 65 binary electrolyte-water systems and 17 ternary electrolyte systems.

Thermodynamic Modeling

For calculation of surface tension of electrolyte system, the aqueous electrolyte solution-vapor system is supposed as three different phases: bulk vapor phase, surface phase and bulk liquid phase (Figure 1). The surface phase is considered as distinct layer for adsorption of electrolyte from liquid phase. In this system, the chemical potential of water in liquid bulk phase and surface would be defined as below [1],

$$\mu_w^L = \mu_w^{L_0} + RT\ln a_w^L \tag{1}$$

$$\mu_w^S = \mu_w^{S_0} + RT\ln a_w^S - \bar{A}_w\,\sigma_{sol} \tag{2}$$

Where μ, a, $\bar{A}$ and σ_{sol} represent the chemical potential, activity, partial molar area and surface tension of solution. The subscripts w, L, S, L_0 and S_0 refer to water, liquid phase, surface phase, reference state of liquid phase and reference state of surface phase, respectively.

At equilibrium condition, the chemical potential of water in surface and liquid phase are equal. So we have,

***Corresponding author:** Mohamad Javad Kamali, Department of Chemical Engineering, Islamic Azad University, Bushehr branch, Bushehr, Iran
E-mail: kamali.mj@gmail.com

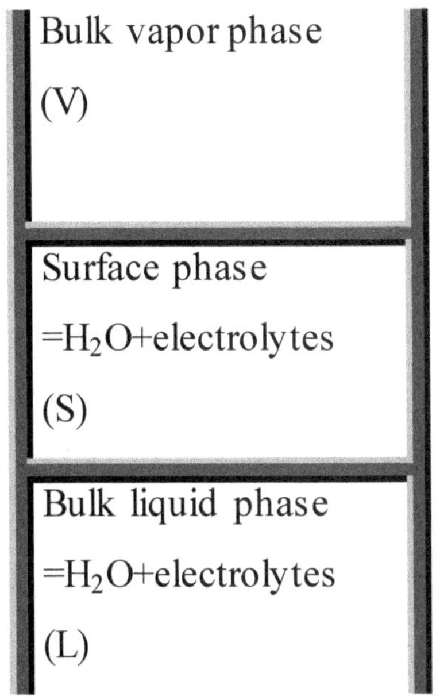

Figure 1: Different phases in aqueous electrolyte solution-vapor system [1].

$$\mu_w^{S0} - \mu_w^{L0} + RT\ln a_w^S - RT\ln a_w^L = \bar{A}_w \sigma_{sol} \tag{3}$$

Rewriting above equation for pure water, it yields the following equation,

$$\sigma_w A_w = \mu_w^{S0} - \mu_w^{L0} \tag{4}$$

It would be worth to noting that the activity of pure water is unity and the partial molar area for pure water is equal to the molar area.

Substituting the above equation into the equation (3), it results to below equation,

$$\sigma_{sol} = \sigma_w + \frac{RT}{A_w}\ln\frac{a_w^S}{a_w^L} \tag{5}$$

It is assumed that in the above equation the partial molar area and molar area of water are equal. The molar area of water is calculated by ref. [18],

$$\bar{A}_w = A_w = (V_w)^{2/3}(N_A)^{1/3} \tag{6}$$

where V_w and N_A are molar volume of water and Avogadro number.

Using the osmotic definition (ϕ) instead of activity of water the equation (5) converts to the following equation,

$$\sigma_{sol} = \sigma_w + \frac{vRT}{55.51A_w}\left(m^L\varphi^L - m^S\varphi^S\right) \tag{7}$$

Where m is molality of electrolyte in aqueous solution and v is stoichiometric coefficient. For calculation of the molality of electrolyte in surface phase (m^S), the fraction of adsorbed electrolyte (θ) on the interface between vapor and liquid phases could be related to the excess area (Γ) as [12],

$$\theta = \frac{\Gamma}{\Gamma^0} \tag{8}$$

Where superscript 0 means saturated condition.On the other hand, the excess area is defined as moles of electrolyte in surface per area of surface [19]. So,

$$\theta = \frac{n^S / A_s}{n^0 / A_s} = \frac{n^S}{n^0} = \frac{m^S}{m^0} \tag{9}$$

Where in the above equation, n is mole number and superscript S indicates the surface phase. The saturated molality of surface phase is considered as fraction of molality of liquid bulk phase, or

$$m^0 = K^0 m^L \tag{10}$$

where L is liquid bulk phase and K^0 is a constant.

Using the Langmuir gas-solid adsorption [20] and equality of adsorption and desorption rate for an electrolyte in surface phase [21], the adsorbed fraction of electrolyte on the surface phase is obtained as [12],

$$\theta = \frac{K\,a}{1+K\,a} \tag{11}$$

Combining equations 7-9, the molality of electrolyte in surface phase would be obtained as,

$$m^S = \frac{K^* a}{1+K\,a} m^L \tag{12}$$

Where $K^*=K^0K$.

So, substituting equation (12) into the equation (7), the surface tension of pure aqueous electrolyte solution is calculated. It is mentioned that, the two adjustable parameters $K^*$and K are obtained from fitting the experimental surface tension data to the calculated values (equation 7) at single temperature.

For mixed electrolyte solution the surface tension would be calculated by following equation,

$$\sigma_{sol} = \sigma_w + \frac{RT}{55.51A_w}\left(\varphi^L \sum_i v_i m_i^L - \varphi^S \sum_i v_i m_i^S\right) \tag{13}$$

Assuming competitive adsorption between electrolytes in surface phase [12], the molality of surface phase is,

$$m_i^S = \frac{K_i^* a_i}{1+\sum_j K_j a_j} m_i^L \tag{14}$$

Results and Discussion

For studying the surface tension of electrolyte solution, 65 binary electrolyte systems and 17 ternary electrolyte systems are selected. The surface tension of these electrolyte solutions are obtained by competitive adsorption model (Eq. 7 or 13). In this model, for calculation of osmotic and activity coefficients, E-NRTL [17] model is used. Using the regression of model with experimental surface tension of binary electrolyte-water system, the adjustable parameters of competitive adsorption model (K^*, K)are optimized. The objective function is defined as following,

$$AAD = \frac{100}{N_p}\sum_{i=1}^{N_p}\frac{\left|\sigma_{sol}^{cal} - \sigma_{sol}^{exp}\right|}{\sigma_{sol}^{exp}} \tag{15}$$

Where, AAD, N_p, superscripts cal and exp are average absolute deviation, number of data points, calculated and experimental data of surface tension, respectively. The results for different binary electrolyte system are given in Table 1. As it is shown in Table 1, when the first adjustable parameter (K^*), tends to zero or small positive value or the second adjustable parameter (K), becomes a large number, the molality of electrolyte system in surface phase closes to zero. Moreover, when $K^*$is lower than unity and the K is not moderately large number, the molality of electrolyte in surface phase decreases with respect to the liquid bulk phase. In these cases, which point out to the mineral salts, the rate of adsorption in surface layer would lower than desorption rate and surface tension of electrolyte solution is increased with respect to the surface tension of water. Strong inorganic acids, organic acetates

System	T(°C)	Np	K*	K	AAD%
$AgNO_3$	20	4	6.99×10^{-2}	6.16×10^{14}	0.23
$Al_2(SO_4)_3$	30	12	8.95×10^{-1}	3.93×10^{22}	0.46
$BaCl_2$	30	5	0	4.53×10^{4}	0.61
$CaCl_2$	30	11	6.60×10^{-1}	4.27×10^{4}	0.81
$CdCl_2$	20	5	7.34×10^{-1}	2.19×10^{1}	0.08
$CdSO_4$	20	2	7.47×10^{-1}	3.62×10^{-2}	9.15×10^{-7}
$CoCl_2$	20	2	6.20×10^{-1}	7.51×10^{-5}	1.05×10^{-4}
CsCl	25	15	1.30×10^{-1}	1.65×10^{16}	0.43
CsI	25	11	1.91×10^{-6}	1.91×10^{-6}	0.45
$CuSO_4$	30	3	7.83×10^{-1}	5.64×10^{14}	0.22
HBr	18	2	1.09×10^{0}	1.86×10^{16}	0.07
HCl	20	7	1.05×10^{0}	1.53×10^{1}	0.94
$HClO_4$	25	10	1.06×10^{0}	2.86×10^{14}	2.13
HNO_3	20	7	1.42×10^{0}	6.03×10^{14}	2.38
KBr	20	5	2.36×10^{-1}	6.02×10^{14}	0.21
$KC_2H_3O_2$	30	3	1.14×10^{0}	9.27×10^{-2}	0.40
KCNS	25	12	9.36×10^{-1}	4.71×10^{0}	0.34
KCl	20	10	1.07×10^{-1}	1.96×10^{16}	0.33
K_2CrO_4	30	15	2.49×10^{-1}	1.14×10^{18}	0.35
$K_3Fe(CN)_6$	25	16	1.91×10^{-6}	1.91×10^{-6}	0.67
KI	25	12	1.91×10^{-6}	1.91×10^{-6}	3.17
KNO_3	25	6	1.66×10^{-1}	2.73×10^{15}	0.38
KOH	20	4	3.07×10^{-1}	8.46×10^{15}	0.35
K_2SO_4	25	12	1.91×10^{-6}	1.91×10^{-6}	0.72
LiBr	30	4	6.40×10^{-1}	9.99×10^{-2}	0.34
LiCl	25	7	4.70×10^{-1}	1.31×10^{1}	0.61
LiI	18	2	6.54×10^{-1}	1.31×10^{0}	2.99×10^{-6}
LiOH	20	4	1.28×10^{-1}	5.96×10^{14}	0.40
Li_2SO_4	18	2	4.36×10^{-1}	3.57×10^{-2}	0.15
$MgCl_2$	20	12	7.28×10^{-1}	2.50×10^{-1}	1.13
$MgSO_4$	10	12	8.31×10^{-1}	5.43×10^{-3}	0.75
$MnCl_2$	18	6	6.44×10^{-1}	7.20×10^{-3}	0.23
NH_4Cl	25	6	2.90×10^{-1}	5.68×10^{14}	0.10
NH_4NO_3	20	9	5.20×10^{-1}	6.42×10^{-2}	0.32
$(NH4)_2SO_4$	30	4	5.77×10^{-1}	6.08×10^{0}	0.48
NaBr	20	4	4.85×10^{-1}	1.77×10^{-2}	0.21
$NaCHO_2$	30	12	3.71×10^{-1}	1.86×10^{16}	0.29
$NaC_2H_3O_2$	30	12	1.12×10^{0}	2.04×10^{16}	1.04
$NaC_3H_5O_2$	30	12	1.91×10^{0}	1.27×10^{18}	0.26
$NaC_4H_7O_2$	30	12	3.28×10^{0}	1.94×10^{16}	0.77
NaCl	20	9	2.22×10^{-1}	2.05×10^{16}	0.66
$NaClO_3$	15	2	1.38×10^{0}	4.31×10^{-1}	0.03
$NaClO_4$	25	3	8.61×10^{-1}	2.16×10^{16}	0.17
Na_2CrO_4	30	4	4.23×10^{-1}	1.59×10^{-2}	0.27
NaI	25	8	6.15×10^{-1}	9.07×10^{14}	0.16
NaNO3	20	4	2.59×10^{-1}	1.16×10^{15}	0.21
NaOH	18	6	3.47×10^{-1}	2.46×10^{-3}	0.46
Na_2SO_4	30	3	1.50×10^{-1}	5.63×10^{17}	0.22
$Na_2S_2O_3$	40	4	4.48×10^{-1}	9.00×10^{-3}	0.21
$NiSO_4$	15	2	7.14×10^{-1}	8.60×10^{-2}	1.75×10^{-7}
$Pb(NO_3)_2$	20	3	1.91×10^{-6}	1.91×10^{-6}	3.77
RbCl	25	11	1.94×10^{-1}	9.65×10^{15}	0.36
$SrCl_2$	20	9	5.95×10^{-1}	4.41×10^{-3}	0.49
$Sr(NO_3)_2$	18	8	3.39×10^{-1}	8.14×10^{14}	0.50
UO_2SO_4	30	12	8.70×10^{-1}	1.23×10^{1}	0.32
$Zn(NO_3)_2$	40	5	7.67×10^{-1}	1.27×10^{-1}	0.37
Overall		**404**			**0.55**

Table 1: The optimized values for adjustable parameters (K*, K) in new model. Experimental data are from ref. [10].

and propionates which have high vapor pressure, attend to escape from liquid phase to surface phase and consequently the concentration of these components in surface phase would be greater than liquid phase [11]. For these electrolytes, the first adjustable parameter is higher than unity and so the molality of electrolyte in surface phase becomes greater than the liquid bulk phase and consequently the surface tension of aqueous electrolyte solution increases with respect to the liquid bulk phase. For other temperatures, the value of surface tension of electrolyte aqueous solution is predicted and the results are given in Table 2. Using the competitive adsorption model, the experimental and predicted values of surface tension of aqueous binary electrolyte system are illustrated in Figures 2-8. These figures indicate the agreement between predicted and experimental values of surface tension values for $AgNO_3$, $BaCl_2$,$CaCl_2$, KBr, HNO_3, NaCl and UO_2SO_4 using E-NRTL [17] model. For ternary systems, the surface tension of the aqueous solutions is predicted in vast range of temperatures and molalities.The values of AAD for prediction of surface tension of 16 ternary systems by competitive adsorption model are shown in Table 3. The overall AAD for prediction of surface tension of mixed electrolyte solutions is 1.7%. The predicted values of surface tension of NH_4NO_3-KNO_3-water, KBr-KCl-water and NH_4Cl- $(NH_4)_2SO_4$-water systems by new surface tension model and experimental values are illustrated in Figures 9-11, respectively. The agreement between predicted and experimental values in these figures represents the satisfactory results of the competitive adsorption model.

Conclusion

Based on the Gibbs thermodynamic and distinct area for surface phase a new model for calculation of surface tension of single and mixture of electrolyte is developed. The molality in surface phase is calculated using Langmuir gas-solid adsorption theory for electrolytes. The osmotic coefficient model in the competitive adsorption model is E-NRTL model. The adjustable parameters of the model are obtained by correlating the experimental values of surface tension of binary electrolyte solution in single temperature. For other temperatures and ternary systems, competitive adsorption model could predict the surface tension of aqueous solution. The agreement between experimental and calculated values of the competitive adsorption model could introduce

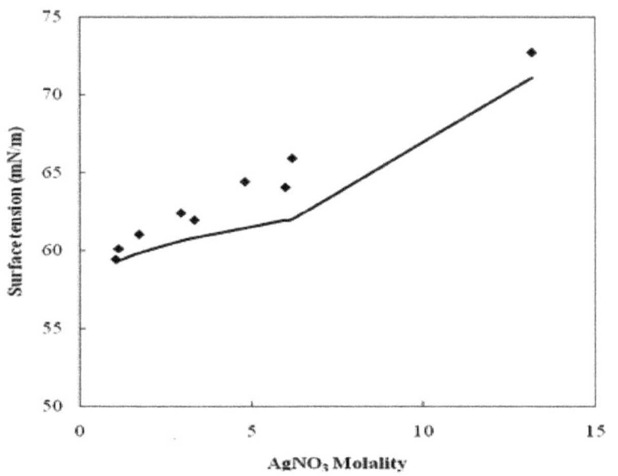

Figure 2: Prediction of surface tension of $AgNO_3$ water binary system at 283.15K using the new model. Experimental data are taken from ref. [10].

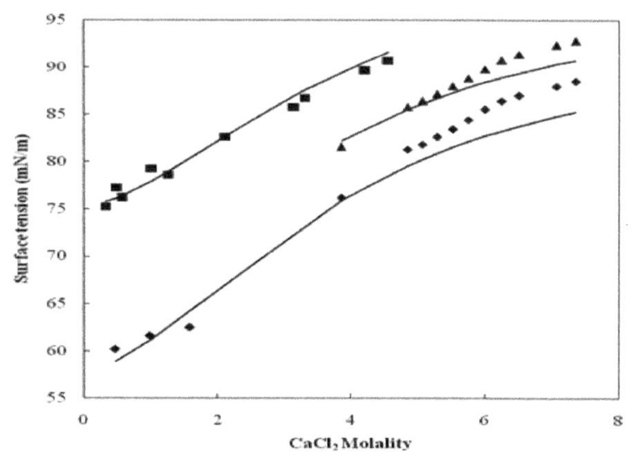

Figure 4: Prediction of surface tension of $CaCl_2$-water binary system using the new model. Experimetal data are taken from ref. [10]. (■: T=283.15K; ▲: T=333.15K; ♦: T=373.15K).

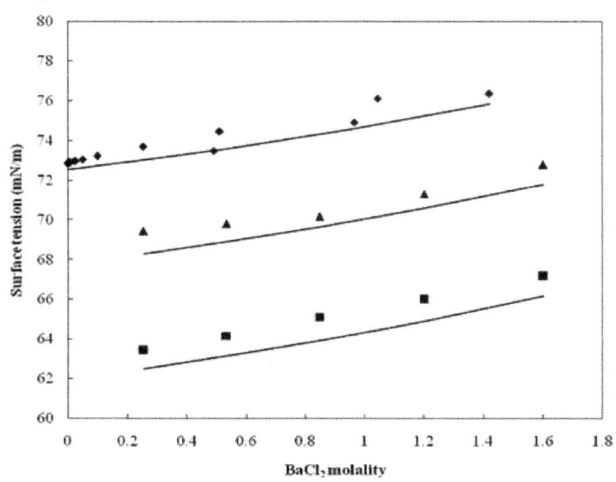

Figure 3: Prediction of surface tension of $BaCl_2$-water binary system using the new model. Experimental data are taken from ref. [10]. (♦: T=298.15K; ▲: T=323.15K; ▪: T=353.15K).

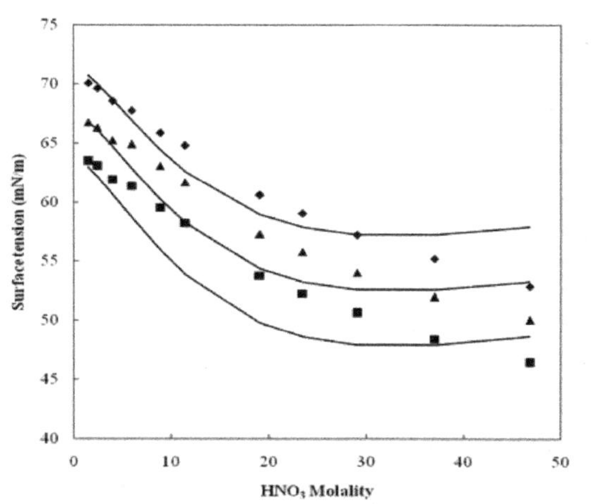

Figure 5: Prediction of surface tension of HNO_3-water binary system using the new model. Experimetal data are taken from ref. [10]. (♦: T=303.15K; ▲: T=323.15K; ■: T=343.15K).

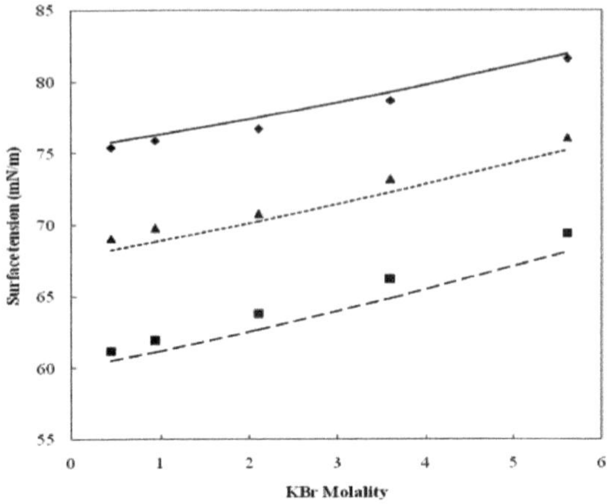

Figure 6: Prediction of surface tension of KBr-water binary system using the new model. Experimetal data are taken from ref. [10]. (♦: T=283.15K; ▲: T=323.15K; ■: T=363.15K).

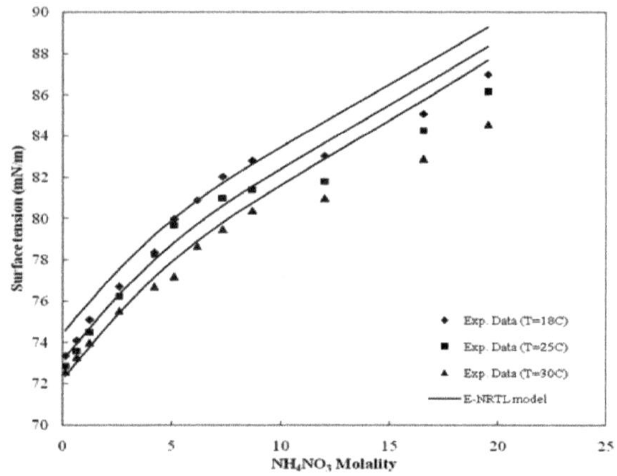

Figure 9: Prediction of surface tension of NH_4NO_3-KNO_3-water ternary system using the competitive adsorption model with non-competitive approach (KNO_3 molality=0.52). Experimental data are taken from ref. [10]. (♦: T=291.15K; ■: T=298.15K; ▲: T=303.15K).

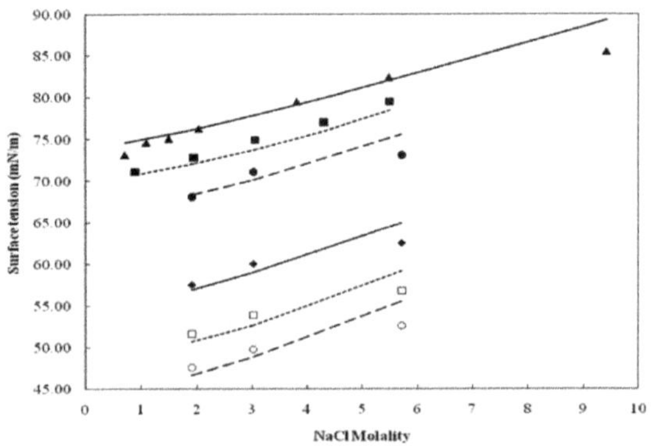

Figure 7: Prediction of surface tension of NaCl-water binary system using the new model. Experimetal data are taken from ref. [10]. (▲: T=291.15K; ■: T=313.15K; ●: T=333.15K; ♦: T=393.15K; □: T=423.15K; O: T=443.15K).

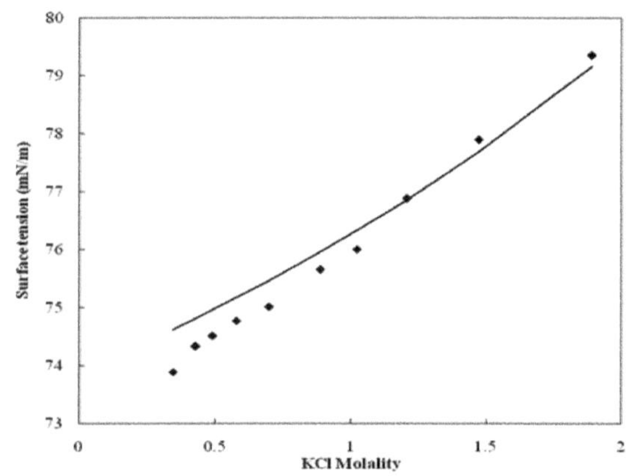

Figure 10: Prediction of surface tension of KBr-KCl-water ternary system using the competitive adsorption model at 291.15K (KBr molality/KCl molality=1). Experimental data are taken from ref. [10].

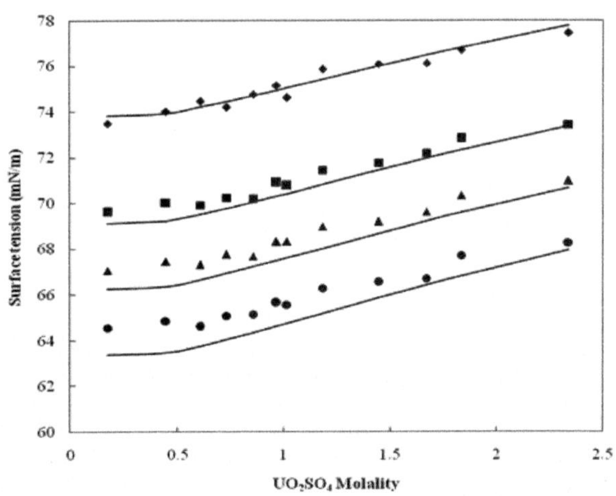

Figure 8: Prediction of surface tension of UO_2SO_4-water binary system using the new model. Experimental data are taken from ref. [10]. (♦: T=293.15K; ■: T=318.15K; ▲: T=333.15K; ●: T=348.15K).

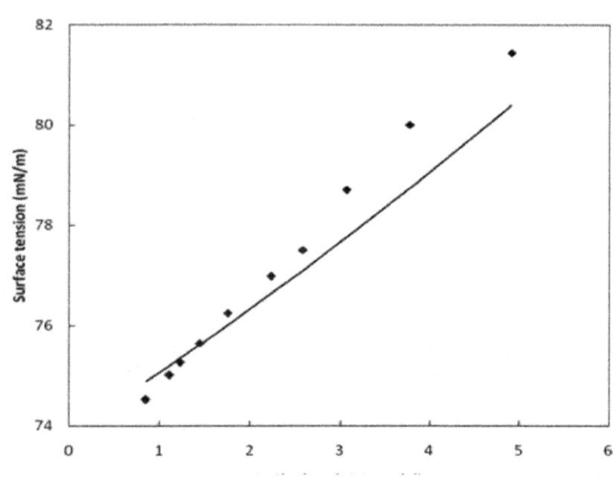

Figure 11: Prediction of surface tension of NH_4Cl-$(NH_4)_2SO_4$-water ternary system using the competitive adsorption model at 291.15K (NH_4Cl molality/ $(NH_4)_2SO_4$ molality=1.5). Experimental data are taken from ref. [10].

System	T(°C)	molality	Np	%AAD
$AgNO_3$	100	1.03-13.16	10	3.81
$BaCl_2$	10-80	0.01-5.96	51	1.10
$CaCl_2$	10-100	0.1-7.37	122	1.38
CsCl	20-30	0.06-8.88	28	0.52
$CuSO_4$	10-80	0.33-1.11	21	2.12
HCl	25-90	0.3-3	18	3.82
$HClO_4$	15-50	0.51-25.92	23	2.56
HNO_3	30-80	1.55-46.86	66	3.71
KBr	10-90	0.44-5.6	40	1.15
$KC_2H_3O_2$	0-80	0.5-23.78	36	4.82
KCl	25-80	0.71-5.16	56	1.03
$K_3Fe(CN)_6$	12.35-20.8	0.29-0.62	4	0.76
KI	20-60	0.01-0.12	55	4.45
KNO_3	18-100	0.1-2.63	27	1.28
KOH	30-95	4.77-13.28	22	4.39
LiBr	10-90	1.28-17.27	48	1.52
LiCl	10-90	1.19-15.73	79	1.97
$MgCl_2$	10-70	0.55-3.5	23	2.90
NH_4Cl	19-60	1.02-7.21	39	1.01
NH_4NO_3	40-95	0.61-29.43	29	1.21
$(NH_4)_2SO_4$	18-95	0.7-5.61	33	2.07
NaBr	10-90	0.0007-6.48	58	3.14
$NaC_2H_3O_2$	0-25	0.50	6	0.94
$NaC_4H_7O_2$	0-50	0.50	10	1.63
NaCl	10-200	0.71-9.42	81	1.73
NaI	20-50	0.33-8.81	32	0.89
$NaNO_3$	18-100	1.02-11.77	42	1.59
NaOH	20-70	0.49-6.25	26	1.30
Na_2SO_4	10-193.4	0.2-1.24	62	2.87
RbCl	20-30	0.11-6.93	22	0.59
$SrCl_2$	10-25	0.48-1.92	7	1.24
UO_2SO_4	20-75	0.18-2.34	48	0.81
$Zn(NO_3)_2$	21	2.62	1	1.10
Overall			**1215**	**1.98**

Table 2: The absolute average deviation percent (%AAD) in prediction of new model for binary systems. Experimental data are from ref. [10].

Ternary system	m_1	m_2	T(°C)	Np	%AAD
$BaCl_2$-HCl	0.45-1.13	0.10	25	3	0.47
$CaCl_2$-HCl	0.37-1.48	0.10	25	4	1.92
LiCl-HCl	0.55-4.88	0.10	25	5	2.23
$SrCl_2$-HCl	0.48-1.92	0.10	25	4	1.13
KBr-KCl	0.35-2.08	0.34-1.89	18	10	0.65
KBr-NaBr	0.44-3.57	0.51-4.2	10.1-90.8	45	2.59
KNO_3-NH_4NO_3	0.52-2.43	0.12-19.54	18-30	143	6.93
KNO_3-$Pb(NO_3)_2$	0-1	0-1	20	33	2.66
KNO_3-$Sr(NO_3)_2$	0.23-1.3	0.19-1.17	18	10	0.92
NH_4Cl-$(NH_4)_2SO_4$	0.51-2.86	0.33-2.04	18	10	0.63
NH_4NO_3-$Pb(NO_3)_2$	0.33	0-0.333	35	6	2.07
$NaNO_3$-$Sr(NO_3)_2$	0.71-3.48	0.25-1.17	18	9	1.24
$NaClO_4$-HCl	0.5-1.34	0.10	25	3	0.70
KNO_3-NH4Cl	0.23-1.3	0.49-2.86	18	10	1.06
NH_4Cl-$Sr(NO_3)_2$	0.51-2.86	0.19-1.17	18	10	0.91
$(NH_4)_2SO_4$-$NaNO_3$	0.33-2.04	0.56-3.47	18	9	1.03
Average				**314**	**1.70**

Table 3: The absolute average deviation percent (%AAD) in prediction of new models for ternary systems. Experimental data are from ref. [10].

this new model as effective one.

References

1. Li ZB, Li YG, Lu JF (1999) Surface Tension Model for Concentrated Electrolyte Aqueous Solutions by the Pitzer Equation. Ind Eng Chem Res 38: 1133-1139.
2. Ariyama K (1937) A theory of surface tension of aqueous solutions of inorganic acids. Bull Chem Soc Japan 12: 109-113.
3. Lorenz PB (1950) The Specific Adsorption Isotherms of Thiocyanate and Hydrogen Ions at the Free Surface of Aqueous Solutions. J Phys Chem 54: 685-690.
4. Young TF, Grinstead SR (1949) The Surface Tensions of Aqueous Sulfuric Acid Solutions. Ann New York Acad Sci 51: 765-780.
5. Glejm VG, Shelomov IK (1957) J Appl Chem USSR 30: 29-35.
6. Oka S (1932) Zur Theorie der Oberflächenspannung beliebiger verdünnter Elektrolyte mit einer Untersuchung über die Struktur der Kapillarschicht. Proc Phys Math Soc Japan 14: 649-664.
7. Horvath AL (1985) Aqueous electrolyte solutions, physical properties, estimation and correlation methods, Halsted press, England.
8. Onsager L, Samaras NNT (1934) The Surface Tension of Debye-Hückel Electrolytes. J Chem Phys 2: 528-536.
9. Schmutzer E (1955) Zur Theorie der Oberflächenspannung von Lösungen. Bemerkungen über den Zusammenhang zwischen radialer verteilungsfunktion und den thermo-dynamischen Funktionen. Z Phys Chem 204: 131-156.
10. Abramzon AA, Gaukhberg RD (1993) Surface tension of aqueous solutions

of inorganic salts. Russian Journal of Applied Chemistry 66: 1428-1438.

11. Yu Y, Gao G, Li Y (2000) Surface tension for aqueous electrolyte solutions by the modified mean spherical approximation. Fluid Phase Equilib 173: 23-28.

12. Li Z, Lu B (2001) Surface tension of aqueous electrolyte solutions at high concentrations - representation and prediction. Chem Eng Sci 56: 2879-2888.

13. Sadeghi M, Taghikhani V, Ghotbi C (2009) Application of the MSA-Based Models in Correlating the Surface Tension for Single and Mixed Electrolyte Solutions. J Chem Thermodynamics 41: 1264–1271.

14. Lee LL (1988) Molecular Thermodynamics of Nonideal Fluids, Butterworth Publishers.

15. Ghotbi C, Vera JH (2001) Extension to Mixtures of Two Robust Hard-Sphere Equations of State Satisfying the Ordered Close-Packed Limit. Can J Chem Eng 79: 678-686.

16. Mansoori GA, Carnahan NF, Starling KE, Leland TW (1971) Equilibrium Thermodynamic Properties of the Mixture of Hard Spheres. J Chem Phys 54: 1523-1525.

17. Chen CC, Zhu Y, King JA, Evans LB (1992) A molecular thermodynamic approach to predict the secondary structure of homopolypeptides in aqueous systems. Biopolymers 32: 1375-1392.

18. Sprow FB, Prausnitz JM (1967) Surface thermodynamics of liquid mixtures. Can J Chem Eng 45: 25-28.

19. Adamson AW, Gast AP (1997) Physical chemistry of surfaces. 3rd edn, Wiley Interscience, New York.

20. Langmuir I (1918) The Adsorption Of Gases On Plane Surfaces Of Glass, Mica And Platinum. J Am Chem Soc 40: 1361-1403.

21. Desnoyer C, Masbernat O, Gourdon C (1997) Predictive Model for the Calculation of Interfacial Tension in Nonideal Electrolytic Systems. J Colloid Interface Sci 191: 22-29.

Thermal Degradation Kinetics of Poly (3-hydroxybutyrate)/Cellulose Nano-crystals based Nanobiocomposite

Prodyut Dhar, Sai Phani Kumar Vangala, Pankaj Tiwari, Amit Kumar and Vimal Katiyar*

Department of Chemical Engineering, Indian Institute of Technology Guwahati, Assam, India

Abstract

Polyhydroxybutyrate (PHB)/ Cellulose nanocrystals (CNCs) nanobiocomposites are prepared by solvent exchange cum solution casting technique at various loading fractions. The effects of CNC loading on dispersion in polymer matrix are studied. Acid hydrolysis of cellulose pulp from bamboo (Bambusabalcooa) yields crystalline rod shaped CNCs having width in the range of 10-20 nm and length being 300-400 nm. Morphological and X-ray diffraction (XRD) studies revealed improved interfacial adhesion of PHB with hydroxyl groups on CNC surface at a threshold loading of 3 wt%. Thermogravimetric analysis (TGA) showed that thermal stability of the nanobiocomposites (PHB/CNC) slightly improved at 3 wt% CNC loading compared to pristine PHB. Further, kinetic analysis of the PHB/CNC nanobiocomposites at different loadings are investigated using isoconversional methods to predict the kinetic triplet. Kinetic parameters predicted from isoconversional methods using Ozawa Flynn Wall (OFW) and Kissinger Akahira Sunose (KAS) models showed that activation energy does not significantly vary with the degree of degradation, revealing that overall degradation follows a single step mechanism. The predicted activation energy values from both OFW and KAS models are in the range of 100-130 kJ/mol The activation energy values are high at higher CNC loadings, showing enhancement in thermal degradation rate due to agglomeration of CNCs. Thermal degradation phenomenon is further studied using Coats Redfern method considering phase boundary controlled models, first order reaction model and power law model. Overall investigation and comparison of the kinetic parameters led to the conclusion that thermal degradation mechanism of PHB/CNC nanobiocomposites followed phase boundary controlled first order reaction models (contracting volume) with random chain scission mechanism.

Keywords: Nanobiocomposite; Polymers; Thermal properties; Cellulose nanocrystals

Introduction

Recent technological advancements and development of novel synthesis processes for fabrication of biodegradable polymers have attracted polymer industries worldwide as a commodity of their interest. These biodegradable polymers have shown potential to substitute conventional petroleum based polymers at industrial scale. Due to their biodegradable and environmental-friendly nature, the use of these biodegradable polymers reduces carbon foot print which subsequently leads to improvement in ecological balance. Biodegradable polymers are usually classified as biopolymers or bio-based polymers depending on their source of origin. The properties of biodegradable polymers need to be critically tailored using different nanofillers and blending techniques to enhance their market applications. Till date several biodegradable polymers have been produced; however, only a few are manufactured at industrial scale such as poly (lactic acid) (PLA) chemically derived from lactic acid precursor, microbially synthesized poly (3-hydroxy butyrate) (PHB), poly (ε-caprolactone) (PCL) from petroleum feedstocks, etc [1,2].

Polyhydroxybutyrate (PHB) is a biodegradable thermoplastic polyester from the family of polyhydroxyalkanoates (PHAs), usually produced from cheap renewable resources by fermentation technique using various microorganisms. However, higher production cost, lower productivity, brittle characteristics and lower thermal stability at processing temperature have limited the wide end applications [3]. Recent studies on incorporation of the nanofillers, usually at lower fractions (1 to 5 wt. %) have shown drastic improvement in the barrier, thermal and mechanical properties of composites compared to pristine polymers [4]. As per literature reports, most widely used nanofillers are nanoclays, metallic nanoparticles, carbon nanotubes etc. The surface characteristics of nanofillers are chemically modified to obtain better dispersion into the polymer matrix. But incorporation of such fillers disturbs the overall biodegradation cycle of biopolymers and problems related to leaching of nanoparticles are observed. Hence, a shift in trend of using non-toxic bionanofillers derived from plant and marine based natural polymers such as cellulose, starch, silk, chitosan etc. have shown improvement in properties while maintaining the biodegradation integrity [5,6]. Biopolymer "cellulose" is abundantly available in the form of lignocellulosic biomass and could be employed as a filler to synthetic biodegradable polymers. Cellulose in its crystalline form usually exist in 20-400 nm range and acts as strong reinforcement agent in polymer through the formation of percolation network due to the strong inter molecular hydrogen bonding [7,8]. Cellulose nanocrystals (CNCs) possess several favorable properties such as nanoscale dimension, high aspect ratio, high specific strength and modulus, high surface area, unique optical properties, etc. [9-11]. These properties make CNC a promising reinforcement nanomaterial with potential application in nanocomposite films, drug delivery, protein immobilization and metallic reaction template.

In present work, hydrophilic CNCs at different loading fractions have been dispersed into hydrophobic PHB through solvent exchange cum solution casting process. Studies carried out by Srithep, et al. [12] for the formation of PHB/CNC nanobiocomposite films through

***Corresponding author:** Vimal Katiyar, Department of Chemical Engineering, Indian Institute of Technology Guwahati, Assam, India
E-mail: vkatiyar@iitg.ac.in

melt compounding showed drastic degradation in molecular weight of PHB. Here we report a cost effective process for the fabrication PHB/CNC composites, with an aim to obtain uniform dispersion of CNCs in the biopolymer which subsequently leads to improvement in the thermal properties. The thermal degradation kinetics of PHB and its CNC composites have been investigated using thermogravimetric analysis (TGA) at various heating rates. Detailed kinetic analysis of the degradation process has been performed using model based isoconversional methods to predict the kinetic parameters, i.e. pre-exponential factor, A and activation energy, E. Further, the effect of surface charged groups on CNCs, which have a catalyzing effect on the degradation of PHB, is discussed in detail.

Experimental

Pretreatment of cellulose fibers

Bamboo pulp, supplied by Nagaon Paper Mill, Hindustan Paper Corporation Limited, India has been used as cellulosic biomass source. Pretreatment of the bamboo fibers to remove hemicellulose and lignin content using sodium hydroxide, hydrogen peroxide, (Merck, India) and sodium hypochlorite (Biochem Processing Lab, India) is carried out as per the methodology reported in literature [13].

Synthesis of CNCs

Extracted cellulose from bamboo pulp (1.0 g) is hydrolyzed with 64 wt% sulfuric acid (Merck, India) under vigorous mechanical stirring for 2 hrs at room temperature. Chilled deionized water (Millipore) at 5°C is added to stop the reaction and the suspension is dilute by 10 folds. Further, the suspension is centrifuged and washed with water (several times) to attain pH in the range of 3-4. Homogenization of CNC suspension using IKA-Homogenizer-24 is carried out at ~20000 rpm for 2 min in two cycles. Subsequently, the suspension is dialysed against distilled water to remove excess acid using cellulose acetate dialysis membrane with ~14,000 Da molecular weight cut off (Sigma Aldrich, India). The final suspension is freeze dried to obtain dried powdered cellulose nanocrystals.

Fabrication of PHB/CNC nanocomposites

CNCs have been dispersed into PHB at various loadings through solvent exchange cum solution casting technique which led to improved dispersion without altering the polymer properties. Poly (3-hydroxybutyrate) (purchased from Sigma Aldrich, India) under reflux conditions is dissolved in chloroform at 90°C for 2 hrs. CNCs are solvent exchanged in chloroform through sequential solvent exchange process from water to acetone by centrifugation (10000 rpm) at each step. The uniformly dispersed CNC suspension in chloroform is finally added to PHB melt solution under strong mechanical stirring at 60°C for 20 min. PHB/CNC films at different loadings (1-5 wt%) are solution casted on Teflon plates and dried in fume hood under ambient conditions.

Characterization methods

Wide angle X-ray diffraction (XRD), (Bruker D8 advance, Germany) equipped with Cu Kα generator (λ=0.154 nm) as X-ray source , operating (40 kV, 40 mA) at a scan rate of 0.05° per 0.5 s and scattering intensity measured 2° from 1° to 40° is used to examine the crystal structure of the PHB/CNC nanocomposites. Thermogravimetric Analysis (TGA) (Netzsch) for nanocomposites (~5 mg) is performed under nitrogen atmosphere (at flow rate of 250 mL/min) from 25°C to 500°C at different heating rates. Morphology studies of PHB/CNC nanocomposites and CNC are carried out using field emission scanning electron microscope (FESEM) (Sigma, Zeiss) and atomic force microscopy (AFM) (Agilent, Model 5500 series). WSxM 5.0 Develop 6.5 software (Nanotec Electronica S.L.) has been used for processing the AFM images. For calculation of diameter of CNCs an arbitrary area (consisting of a dense CNC region) is selected on AFM micrograph and corresponding 3D image is generated and average height profile calculated. However, the average length distribution of CNCs is calculated from the AFM topography image.

Kinetic modeling of TGA data: theoretical background

Different kinetic models are proposed to understand the polymer degradation mechanism through prediction of the kinetic parameters based on the data obtained from TGA curves [14].

The kinetic parameters (i.e., A, n and E) can be calculated from TGA data by using the basic rate equation:

$$d\alpha/dt=k(T)\ f(\alpha) \quad (1)$$

where α represents the extent of reaction, which is determined from the TGA data (fractional mass loss), t is time, k(T) represents the temperature dependent rate constant expressed by an Arrhenius type expression and f(α) denotes the particular reaction model, which determines the dependence of the reaction rate on the extent of reaction.

The temperature dependence of the rate constant can be expressed by Arrhenius equation

$$k=A\exp((-E)/RT) \quad (2)$$

where A and E are pre-exponential factor and activation energy, respectively.

Equation (1) can be written as

$$d\alpha/dt= A\exp((-E)/RT)f(\alpha) \quad (3)$$

In order to determine the kinetic triplet (A, E and f(α)), various methods have been developed. They are classified as isoconversional and model fitting methods. In this study two isoconversional models (i.e. Ozawa Flynn Wall and Kissinger Akahira Sunose) and two model fitting methods (i.e. Kissinger and Coats-Redfern) are considered for determining the kinetics of thermal degradation and thermal stability of pristine PHB and PHB/CNC nanocomposites.

Ozawa Flynn Wall (OFW) Model

OFW model is an integral isoconversional method derived by using Doyle's approximation using multiple heating rates TGA data [15]. The model expression is given by

$$\mathrm{Log}\ \beta=\log AE/g(\alpha)R-2.315-0.4567\ E/RT \quad (4)$$

The plot of log β vs. 1/T gives a straight line whose slope is equal to −0.4567E/RT from which activation energy can be calculated. Pre-exponential factor is calculated from the intercept of the resulting straight line by assuming a reaction model.

Kissinger akahira sunose (KAS) model

KAS is an isoconversional model based on the numerical approximations of the Arrhenius integral over a wide range of thermal history [14]. The model expression is written as

$$\ln(\beta/T^{2})=\ln(AR/(Eg(\alpha)))-E/RT \quad (5)$$

Plot of $\ln(\beta/T^2)$ vs. $1/T$ yields a straight line. The values of E and A can be calculated from slope and intercept for a particular reaction model.

Kissinger model (KGR)

This is a maximum rate method and applicable only to the multiple heating rate TGA or DTG data [16]. The temperatures (Tm or Tp) at which the rates reach maximum are used to predict the single values of E and A. The expression for this model is

$$\ln(\beta/T_p^2) = -(E/(RT_p)) + \ln(AR/E) \qquad (6)$$

Coats-Redfern model

This method is most widely used for analysis of single heating rate TGA data. The equation for this method is

$$\ln(g(\alpha)/T^2) = \ln[(AR/\beta E)(1-(2RT)/E)] - E/RT \qquad (7)$$

Various solid state reaction models have been reported by several authors previously [17]. Insertion of various expressions shown in Table 1 for $g(\alpha)$ yields different set of E and A values.

Results and Discussions

PHB/CNC composite: morphology and crystalline structure

Hydrolysis of cellulose extracted from the pretreated bamboo pulp with sulphuric acid yields defect free CNCs of typical dimensions ranging from 300-400 nm in length and 10-20 nm in width, with an average aspect ratio ~30. Figure 1a and b illustrate the AFM images of the fabricated CNCs and their corresponding height profile for a selected region (as shown with blue line). The height pattern shows that the diameter of the fabricated CNCs varied within a range of 10-20 nm.

The CNCs are hydrophilic nanoparticles, and hence tend to agglomerate when directly dispersed into hydrophobic polymers such as PHB. Therefore, solvent exchange of freeze dried CNCs is carried out from the easily dispersible solvent water to chloroform in which PHB is readily soluble. This process led to the formation of the stable dispersion of the CNC solution in chloroform, which is also supported from the zeta potential values.

The solvent exchanged CNCs are dispersed into PHB at various loadings and fabricated into films, whose transparency and morphology are studied in detail. CNC loadings above 2 wt% show agglomerations due to poor dispersion and could be detected by visual examination as small white dots on the films. It has been further supported from the transparency values which drastically decreased from 58% to 49%, as the loading was increased to 2 wt%.

Morphological behavior of PHB/CNC nanobiocomposites is further studied by FESEM to understand dispersion and interaction of CNCs with PHB. Two types of orientation of rod like CNCs are found in PHB matrix from thorough investigation of the micrographs. In the first type, CNCs are vertically embedded into PHB matrix, which are seen as white dots/circles, giving the cross sectional view of CNCs. In the second type, CNCs are randomly aligned on the PHB surface. Figure 2a, shows homogenous dispersion of CNCs (at 1 wt% loading) on smooth polymer surface, where majority of nanofillers are embedded into polymer surface, providing more contact area of adhesion. This suggests better adhesion between hydrophilic CNCs and hydrophobic polymer. However, at higher loading fractions (~5 wt.%) agglomeration of the CNCs is observed as shown in Figure 2b.

XRD patterns for pure PHB film and PHB/CNC nanocomposites at various loadings are studied as shown in Figure 3. XRD studies showed broadening of peak at 13.5° and shift to lower 2θ angles, suggesting possible intercalation of CNCs. The hydrogen bonding interaction between the functional groups on the surface of the nanofiller and

Reaction model	Notation	f(α)	g(α)
First order	F1	$(1-\alpha)$	$-\ln(1-\alpha)$
Second order	F2	$(1-\alpha)^2$	$\alpha(1-\alpha)^{-1}$
Phase boundary controlled (contracting area)	R2	$2(1-\alpha)^{1/2}$	$[1-(1-\alpha)^{1/2}]$
Phase boundary controlled (contracting volume)	R3	$3(1-\alpha)^{2/3}$	$[1-(1-\alpha)^{1/3}]$
Power law model	P4	$2/3\ \alpha^{-1/2}$	$\alpha^{3/2}$

Table 1: Various reaction model expressions for f(α) and g(α).

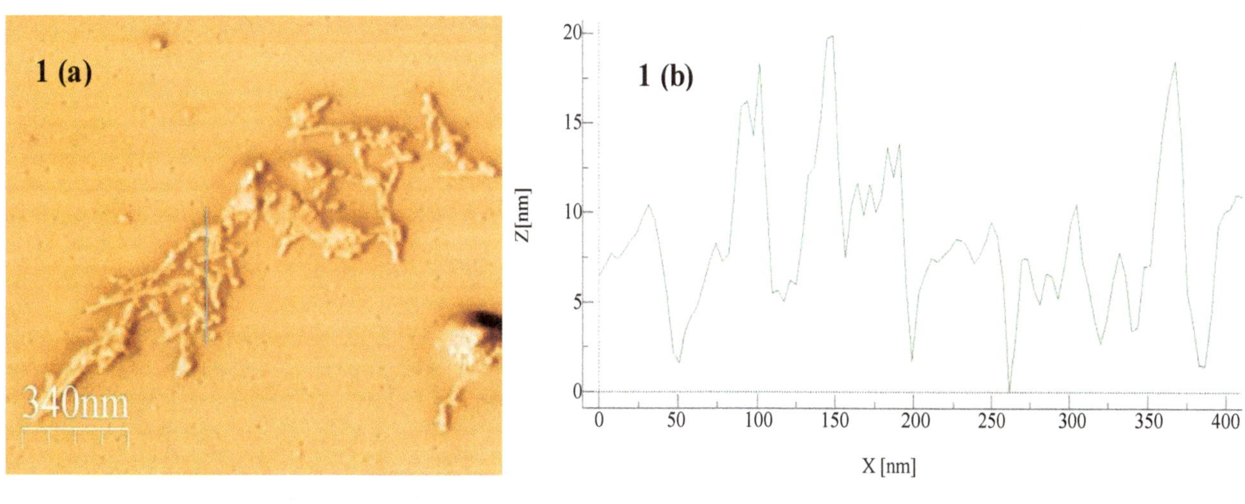

Figure 1: (a) AFM images of the Cellulose Nanocrystals; (b) Height profile of the selected region in AFM image (shown in blue line).

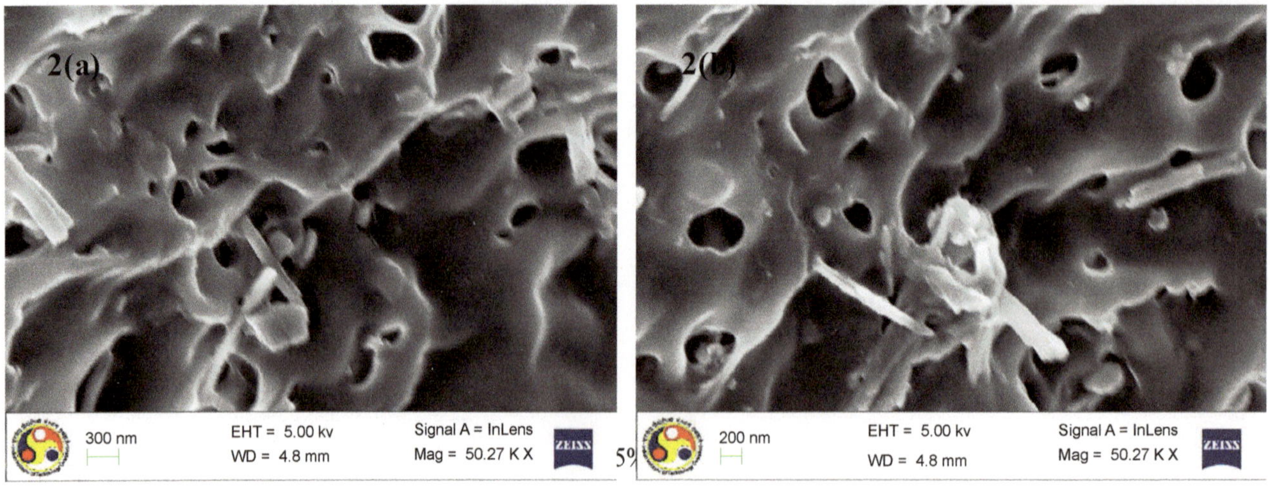

Figure 2: FESEM of PHB film with (a) 1% CNC and (b) 5% CNC.

polymer matrix leads to shifting of the XRD peak at θ=13.5° towards lower angles. The presence of peaks at 2θ=16.5° is a prominent indication of PHB and CNC crystallites in the composites. Moreover, it is found that there is an increase in intensity of cellulose diffraction peaks at 2θ=16.5° and 34.5°, corresponding to (110) and (004) planes on increasing CNCs content to 5 wt%,

PHB/CNC composites: thermal properties

Thermal degradation behavior and thermal stability of pure PHB and PHB/CNC nanocomposites have been determined from single as well as multiple heating rates TGA data shown in Figure 4a and b. It is observed from Figure 4a that the thermal degradation of PHB and its nanocomposites follows the single step mass loss. PHB nanocomposite with 3% CNC loading shows improved thermal stability compared to the pure PHB. However, with increased CNC loadings (~upto 5 wt%), onset degradation temperature decreased from 246°C to 232°C. This shows that CNCs act has a synergistic effect on degradation of the PHB polymer. The enhanced biodegradation might be due to the presence of pendant sulphate and hydroxyl groups on CNC surface which act as nucleation sites for the chain scission of PHB at higher temperature. Thus, biofillers such as CNCs help in enhancing thermal degradation rate of the polymer which is advantageous from environmental point of view. However, rate of degradation (αp) decreases with increase in CNC loadings upto ~3 wt%, reaching a minimum αp value of approximately 24.2 wt%. This is due to the improved compatibility between the PHB/CNCs at lower CNC loadings which led to better hydrogen bonded interaction between nanofiller and polymer subsequently suppressing the polymer degradation rate. However, αp increases at higher CNC loadings (~5wt. %) and the nanocomposite behaves as neat PHB. It could be due to the formation of isolated CNC aggregates in PHB. The residual mass at 500 °C increased with CNC loadings.

Figure 4b shows multiple heating rate TGA curves for PHB/CNC (1%) nanocomposite at 5, 10, 15 and 20°C/min. Like pristine PHB, PHB/CNC (1%) nanocomposite also follows the same degradation behavior with single stage mass loss. The onset (start and end) degradation temperatures increase with increasing heating rate for nanocomposites. Similar TGA curves are obtained for PHB/CNC (5%) nanocomposite at various heating rates.

Estimation of effective activation energy

Evaluation of kinetic parameters from isoconversional kinetic models requires the data obtained from multiple heating rate TGA experiments. Thermal degradation of most of the thermoplastic aliphatic polyesters occurs via random chain scission mechanism. Likewise PHB also degrades by random chain scission via β-elimination of crotonic acid and crotonoate oligomers [18]. First order reaction model is considered in KAS and OFW models for the determination of kinetic parameters using multiple heating rates data. Figure 5a and b shows the plot of log β vs. 1/T from OFW model for (a) PHB/CNC (1%) and (b) PHB/CNC (5%) nanocomposites at various extents of conversions with four linear intervals of heating rates. The resulting straight lines in Figure 5a are shifted to left hand side from α = 0.1–0.9 with increase in slope up to α=0.6 and slight decrease for α=0.7 and 0.8. The slopes of the straight lines are gradually reduced from α=0.1–0.9 in case of PHB/CNC (5%) as shown in Figure 5b. Kinetic parameters calculated from OFW model for PHB/CNC (1 and 5%) nanocomposites are tabulated in Table 2.

It is observed from Table 2, that 'E' and 'A' values increased for nanocomposite containing 1% CNC till α=0.6 and then reduced suddenly. The best fit (R2=1) is obtained at α=0.8 but the 'E' and 'A' values are lower. In case of PHB/CNC (5%) nanocomposite, the 'E' and 'A' values changes rapidly with extent of conversion. The change in activation energy and pre-exponential factor values are notable at initial stages of degradation (i.e. α=0.1–0.2 and α=0.3–0.4).

The plots obtained by KAS model for PHB/CNC (1%) and PHB/CNC (5%) nanocomposites at multiple heating rates are shown in Figure 6 a and b respectively. The kinetic parameters calculated from KAS model for PHB/CNC (1 and 5%) nanocomposites are listed in Table 3.

The kinetic parameters (i.e. E and A) calculated from KAS model are in good agreement with the values evaluated from OFW model. The variation of 'E' with extent of conversion indicates that the occurrence of complex reactions involving different mechanisms rather than degradation by simple mechanisms. The 'E' and 'A' values determined from Kissinger model (maximum rate method) are 105.6 kJ/mole and 2.3379×1010 min^{-1} respectively for PHB/CNC (1%) and 108.7 kJ/mole and 1.2927×1011 min^{-1} respectively for PHB/CNC (5%) nanocomposite.

Coats Redfern method is applied to determine kinetic parameters and the mechanism of thermal degradation using several reaction

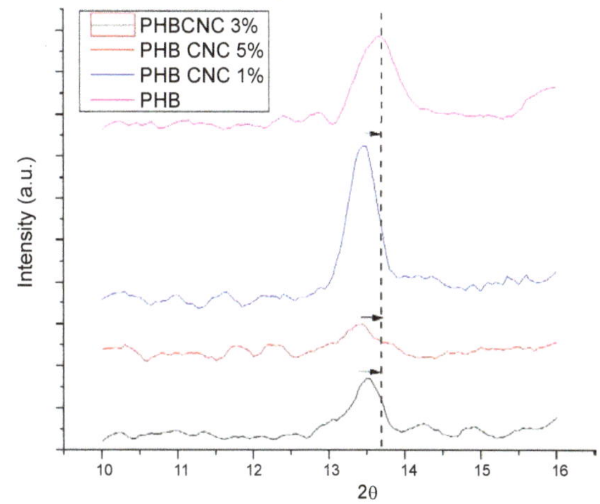

Figure 3: XRD comparison of dspacing and Dhkl of PHB film with different CNC loaded films.

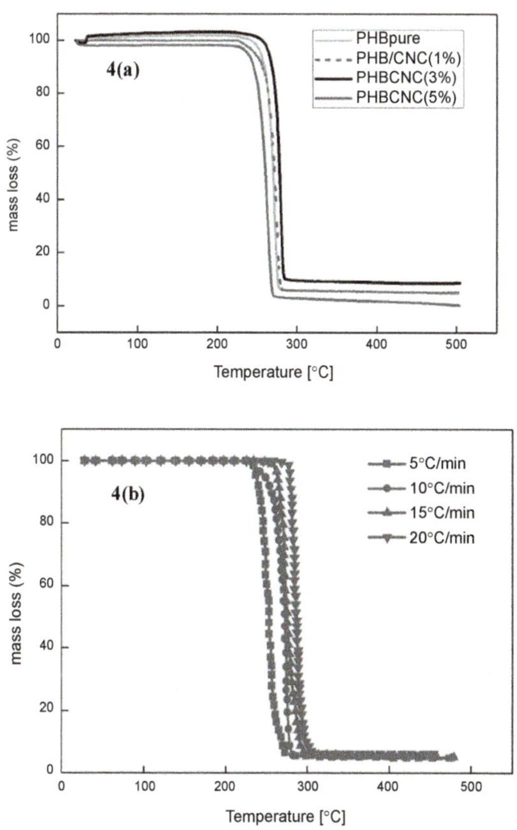

Figure 4: (a) TGA curves of pure PHB and PHB/CNC with various amounts of CNC at 10°C/min. (b) TGA curves of PHB/CNC (1%) nano-composites at multiple heating rates.

models indicated in Table 4 for single heating rate TGA curves at 10°C/min, as shown in Figure 6. Five reaction models have been used in Coats Redfern method in order to determine the best suitable reaction model which represents the experimental data effectively. Figure 7 shows Coats Redfern plot, ln(g(α)/T^2) vs. 1/T for pure PHB obtained from TGA curve at a heating rate of 1°C/min.

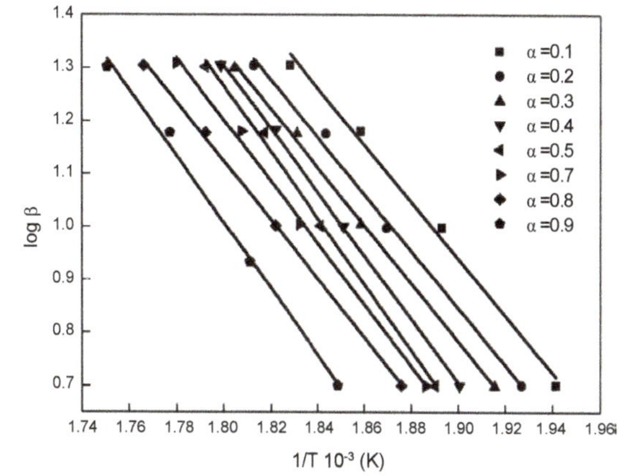

Figure 5: OFW plots of (a) PHB/CNC (1%) and (b) PHBCNC (5%) nanocomposites at different conversions obtained from multiple heating rate TGA curves.

For all five models the trend is same but the slopes and intercepts are different. The 'E' and 'A' values are found from the slope and intercept of each straight line. The values of activation energies evaluated for pristine PHB and for bionanocomposites from the Coats Redfern method are summarized in Table 4.Various reaction models yield different 'E' values. Phase boundary controlled models (i.e. contracting area and contracting volume) yield lower activation energy values while the regression coefficients are high. The 'E' values evaluated from first order reaction are intermediate between second order and phase boundary controlled models. For pure PHB as well as for nanobiocomposites, first order reaction model represents best fits (R2~1). The values of 'E' calculated from second order reaction model are quite high compared to other reaction models. Power law model presents most relevant values of 'E' which are also close to values reported in earlier investigations [18]. Selection of a single model from the different models considered is very difficult and each model has its own limitation. At low and high percentages of CNC, thermal stability of nanobiocomposite is less than the pristine PHB while at 3% of CNC thermal stability is increased due to better interaction between the polymer matrix and the filler and also due to good dispersion of nanofiller. At higher percentages of nanofiller, agglomeration of nanofiller in polymer matrix occurs, which subsequently leads to enhanced degradation. This is revealed by kinetics from isoconversional and model fitting analyses. Overall, thermal degradation of pure PHB and PHB/CNC nanocomposites containing various percentages of nanocomposites follows first order reaction model with random chain scission mechanism and phase boundary controlled reaction models (contracting volume).

Conclusion

Fabricated PHB/CNC nanobiocomposites showed better dispersion at a threshold CNC loading of ~3 wt%, as suggested from the morphological orientation of the CNC in FESEM micrographs. XRD studies showed shift of the PHB peak to lower theta angles, representing improved interaction of the hydrophilic CNCs with hydrophobic PHB. Thermal degradation kinetics study of PHB/CNC nanobiocomposites revealed that thermal stability slightly improved upto CNC loading of ~3 wt%. Isoconversional analysis of thermal properties and evaluation of kinetic parameters using OFW and KAS models didn't show significant change in activation energies. This could be attributed to the degradation mechanism which follows single step degradation and

Extent of conversion (α)	PHB/CNC (1%)			PHB/CNC (5%)		
	E	A	R^2	E	A	R^2
0.1	96.9	2.1385×10^{8}	0.9910	127	6.2565×10^{11}	0.9949
0.2	97.4	4.1162×10^{8}	0.9915	116	6.917×10^{10}	0.9716
0.3	100.8	1.208×10^{9}	0.9831	114.3	5.8834×10^{10}	0.9959
0.4	109.2	1.023×10^{10}	0.9667	103.5	5.839×10^{9}	0.9925
0.5	113.4	3.1451×10^{10}	0.9681	112	4.9713×10^{10}	0.9914
0.6	114	5.1431×10^{10}	0.9785	105.1	1.0919×10^{10}	0.9922
0.7	107	1.3969×10^{10}	0.9906	101.5	5.7666×10^{9}	0.9912
0.8	101	3.5711×10^{9}	1	106.9	2.4427×10^{10}	0.9949
0.9	111.2	3.80×10^{10}	0.9829	103.8	1.4496×10^{10}	0.9928

Table 2: Kinetic parameters with regression coefficients from OFW model for PHB/CNC (1 and 5%) nanocomposites.

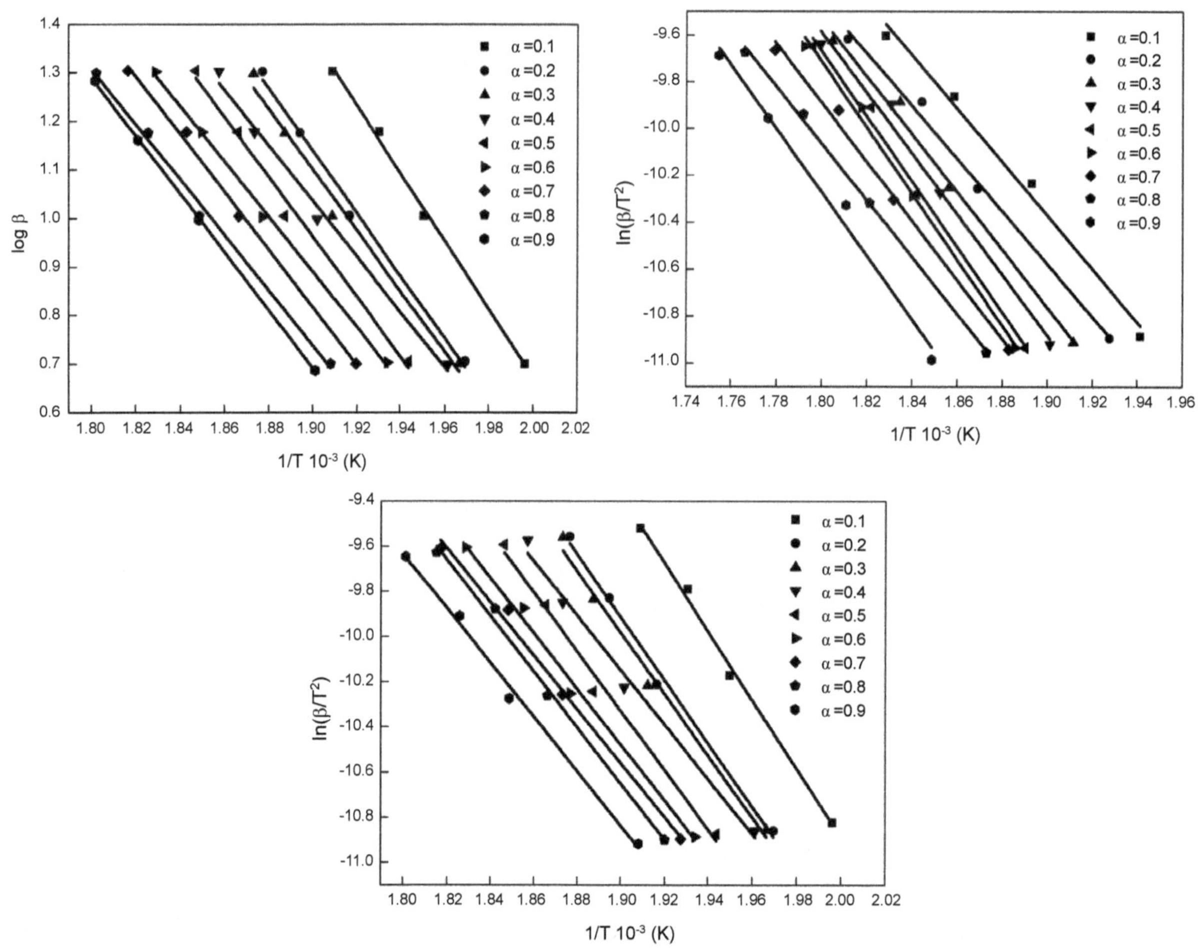

Figure 6: KAS plots of (a) PHB/CNC (1%) and (b) PHB/CNC (5%) nanocomposites at different conversions obtained from multiple heating rate TGA data.

Extent of conversion (α)	PHB/CNC (1%)			PHB/CNC (5%)		
	E	A	R2	E	A	R2
0.1	93.1	6.5025×10^{7}	0.9890	125.2	3.6072×10^{11}	0.9941
0.2	93.6	1.2454×10^{8}	0.9897	115.3	4.1902×10^{10}	0.9825
0.3	101	9.6962×10^{8}	0.9818	111.5	2.6229×10^{10}	0.9953
0.4	105.9	3.8482×10^{9}	0.9610	100.1	2.0946×10^{9}	0.9912
0.5	110.3	1.2687×10^{10}	0.9629	109.2	2.0797×10^{10}	0.99
0.6	118	1.0058×10^{11}	0.9761	101.7	3.9245×10^{9}	0.9908
0.7	104.4	5.0134×10^{9}	0.9889	97.8	1.9049×10^{9}	0.9894
0.8	100.7	2.4592×10^{9}	1	103.5	8.9542×10^{9}	0.9940
0.9	111	2.979×10^{10}	0.9838	100.3	4.9376×10^{9}	0.9915

Table 3: Kinetic parameters with regression coefficients from KAS model for PHB/CNC (1 and 5%) nanocomposites.

Reaction model	PHB pure	PHB/CNC (1%)	PHB/CNC (3%)	PHB/CNC (5%)
	E	E	E	E
First order (F1)	108	83.3	118	83.2
Second order (F2)	152	121	174.6	117.6
Phase boundary controlled (R2)	91	70.2	104	71
Phase boundary controlled (R3)	96.4	74.3	110	76.2
Power Law (P4)	117.6	91.4	133	93.3

Table 4: Activation energy values evaluated from Coats Redfern method using five reaction models for pure PHB and PHB nanocomposites.

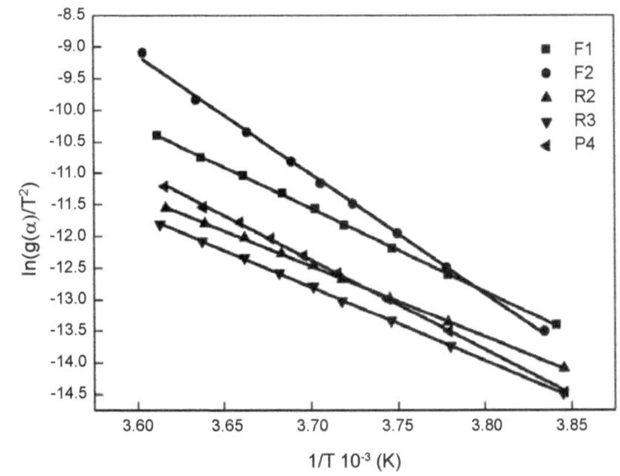

Figure 7: Coats Redfern plots using five reaction models for pure PHB at 10°C/min

remains unchanged for neat PHB and all other nanobiocomposites. On further evaluation of the kinetic parameters with different models, it is concluded that PHB/CNC nanobiocomposites follows first order reaction and phase boundary controlled reaction models (contracting volume) with random chain scission mechanism.

Acknowledgement

Authors would like to express their gratitude to Central Instruments Facilities (CIF) and Department of Chemical and Petrochemical (DCPC, GoI) funded Centre of Excellence for Sustainable Polymers (CoE-SusPol), Indian Institute of Technology, Guwahati, India for providing the research and analytical facilities.

References

1. Chen GQ (2009) A microbial polyhydroxyalkanoates (PHA) based bio- and materials industry. Chem Soc Rev 38: 2434-2446.
2. Lunt J (1998) Large-scale production, properties and commercial applications of polylactic acid polymers. Polym Degrad Stab 59: 145–152.
3. Doi Y, Kanesawa Y, Kunioka M, Saito T (1990) Biodegradation of microbial copolyesters: poly(3-hydroxybutyrate-co-3-hydroxyvalerate) and poly(3-hydroxybutyrate-co-4-hydroxybutyrate). Macromolecules 23: 26–31.
4. Bordes P, Pollet E, Avérous L (2009) Nano-biocomposites: Biodegradable polyester/nanoclay systems. Prog Polym Sci 34: 125–155.
5. Ten E, Turtle J, Bahr D, Jiang L, Wolcott M (2010) Thermal and mechanical properties of poly(3-hydroxybutyrate-co-3-hydroxyvalerate)/cellulose nanowhiskers composites. Polymer 51: 2652–2660.
6. Svagan AJ, Åkesson A, Cárdenas M, Bulut S, Knudsen JC, et al. (2012) Transparent films based on PLA and montmorillonite with tunable oxygen barrier properties. Biomacromolecules 13: 397-405.
7. Umesh Bhardwaj (2013) Evaluation of discrete kinetics during pyrolysis of lignocellulosic biomass and its extracted pulp (ENFL606).
8. Wu D, Wu L, Wu L, Zhang M (2006) Rheology and thermal stability of polylactide/clay nanocomposites. Polym. Degrad Stab 91: 3149–3155.
9. Habibi Y, Lucia LA, Rojas OJ (2010) Cellulose nanocrystals: chemistry, self-assembly, and applications. Chem Rev 110: 3479-3500.
10. Ten E, Bahr DF, Li B, Jiang L, Wolcott MP (2012) Effects of Cellulose Nanowhiskers on Mechanical, Dielectric, and Rheological Properties of Poly(3-hydroxybutyrate-co-3-hydroxyvalerate)/Cellulose Nanowhisker Composites. Ind Eng Chem Res 51: 2941–2951.
11. Xu X, Liu F, Jiang L, Zhu JY, Haagenson D, Wiesenborn DP (2013) Cellulose Nanocrystals vs. Cellulose Nanofibrils: A Comparative Study on Their Microstructures and Effects as Polymer Reinforcing Agents. ACS Appl Mater Interfaces 5: 2999–3009.
12. Srithep Y, Ellingham T, Peng J, Sabo R, Clemons C, Turng LS, et al. (2013) Melt compounding of poly (3-hydroxybutyrate-co-3-hydroxyvalerate)/nanofibrillated cellulose nanocomposites. Polym Degrad Stab 98: 1439–1449.
13. Yu M, Yang R, Huang L, Cao X, Yang F, et al. (2012) Preparation and characterization of bamboo nanocrystalline cellulose. BioResources 7: 1802–1812.
14. Vyazovkin S, Sbirrazzuoli N (2006) Isoconversional Kinetic Analysis of Thermally Stimulated Processes in Polymers. Macromol Rapid Commun 27: 1515–1532.
15. Carrasco F, Pérez-Maqueda LA, Sánchez-Jiménez PE, Perejón A, Santana OO, et al. (2013) Enhanced general analytical equation for the kinetics of the thermal degradation of poly(lactic acid) driven by random scission. Polym Test 32: 937–945.
16. Cho YS, Shim MJ, Kim SW (1998) Thermal degradation kinetics of PE by the Kissinger equation. Mater Chem Phys 52: 94–97.
17. Ebrahimi-Kahrizsangi R, Abbasi MH (2008) Evaluation of reliability of Coats-Redfern method for kinetic analysis of non-isothermal TGA. Trans Nonferrous Met Soc China 18: 217–221.
18. Ariffin H, Nishida H, Shirai Y, Hassan MA (2008) Determination of multiple thermal degradation mechanisms of poly(3-hydroxybutyrate). Polym Degrad Stab 93:1433–1439.

Thermodynamic Models for Determination of the Solubility of N-Acetylglycine in (Methanol+Acetonitrile) Binary Solvent Mixtures

Wangdan Zhao[1], Wenge Yang[1*], Qirun Guo[1], Shimin Fan[1], Jianfeng Hao[1] and Yonghong Hu[2]

[1]*School of Pharmaceutical Sciences, Nanjing Tech University, No. 30, South Puzhu Road, Nanjing 211816, PR China*

[2]*State Key Laboratory of Materials-Oriented Chemical Engineering, Nanjing Tech University, No.200, North Zhongshan Road, Nanjing 210009, PR China*

Abstract

N-Acetylglycine is an important organic chemical and pharmaceutical intermediate. The solubility of N-Acetylglycine in methanol+Acetonitrile was measured in the temperature ranging from 278.15 K to 333.15 K. In the processing of the experimental data, we use three kinds of equations. There are the Combined Nearly Ideal Binary Solvent/Redlich–Kister (CNIBS/R–K) equations and the simplified Jouyban–Acree equation. All equations can be well applied. The CNIBS/R–K equations are relatively better. We used the van't Hoff equation to determine the thermodynamic properties of N-Acetylglycine further more, including the entropy, enthalpy, and Gibbs energy.

Keywords: N-Acetylglycine; Solubility-temperature dependence; Pharmaceutical intermediate; Correlation; Enthalpy; Jouyban–Acree equation; CNIBS/R–K equation

Introduction

N-Acetylglycine (Figure 1, CAS: 543-24-8, MW=117.10) is a long needle crystal and its crystal structure has been reported in the literature [1,2]. N-Acetylglycine is important organic chemical intermediate for many chemical products, such as glycine, aspartame and other amino acids. It is also important pharmaceutical intermediate for Acetyl glutamine that can be used for the patient to promote department of neurosurgery coma after surgery. N-Acetylglycine was widely used in pharmaceutical [3], pesticide [4], chemical industry [5] and other fields. N-Acetylglycine is prepared from paraformaldehyde; acetamide, carbon monoxide with a catalyst system, then gained with concentrating and filtrating, and can be purified by crystallization (Figure 2). Therefore, to know the relationship between N- Acetylglycine solubility with temperature and solvent composition is very important for the preparation and purification of products. As far as we know, it has not been reported about the solubility data of N-Acetylglycine in (methanol+Acetonitrile) mixtures. So we provided the helpful and valuable information on the crystallization of N-Acetylglycine. We chose methanol and acetonitrile as the binary solvent mixtures system in this paper. The reason to make such a choice is that the solubility of N-Acetylglycine in methanol is better than acetonitrile in previous study. In this paper, the solubility of N-Acetylglycine in methanol+Acetonitrile mixtures was determined between 278.15 K and 333.15 K at standard atmospheric pressure (101.325 kPa) by the dynamic method. The experimental data measured was correlated with two forms of Redlich–Kister (CNIBS/R–K) equation and the simplified Jouyban–Acree equation. The thermodynamic properties of the solution process, including the Gibbs energy, enthalpy, and entropy, were calculated by the van't Hoff analysis.

Experimental

Materials and apparatus

All solvents and N-Acetylglycine (Aladdin Chemistry Co., Ltd) used for experiments were of analytical reagent grade and their mass fraction purities were higher than 99%. More details about the purity of reagents are listed in Table 1. Analytical balance (model: BSA224S) was provided from Sartorius Scientific Instruments (Beijing) CO., Ltd. with an uncertainty of ± 0.0002 g. The smart Acetonitrile-circulator thermostatic bath (model: DC-2006) was provided by Ningbo Scientz Biotechnology Co. Ltd. with an uncertainty of ± 0.1 K.

Methods

The solubility of N-Acetylglycine in methanol+Acetonitrile mixtures was determined from 278.15 K to 333.15 K at atmospheric pressure by the dynamic method [6-8]. Solvents and excess N-Acetylglycine were added into glass test tubes with stoppers. The glass test tubes were maintained at a double layer jacket glass equilibrium cell which was full of solution. The temperature of the inner was controlled by circulating which was from a smart Acetonitrile-circulator thermostatic bath. Stirring the mixture continuously uesed a magnetic stirrer for twentyfour hours until it reaching saturation andequilibrium. Then stop stirring to let the solution settle over nights to make sure the suspended solid precipitated. Next, took 1 mL the clear upper saturated solution by a pipette gun and transferred it into a 10 mL beaker with a cover that had been weighed up beforehand

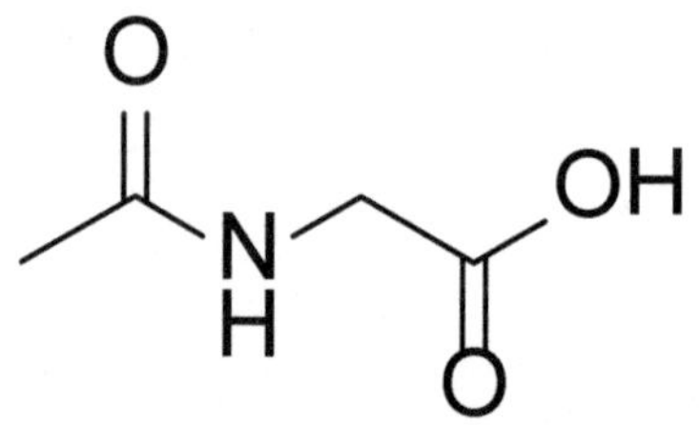

Figure 1: Chemical structure of N-Acetylglycine.

***Corresponding author:** Wenge Yang, School of Pharmaceutical Sciences, Nanjing Tech University, No. 30, South Puzhu Road, Nanjing 211816, PR China, E-mail: wengeyang11@163.com

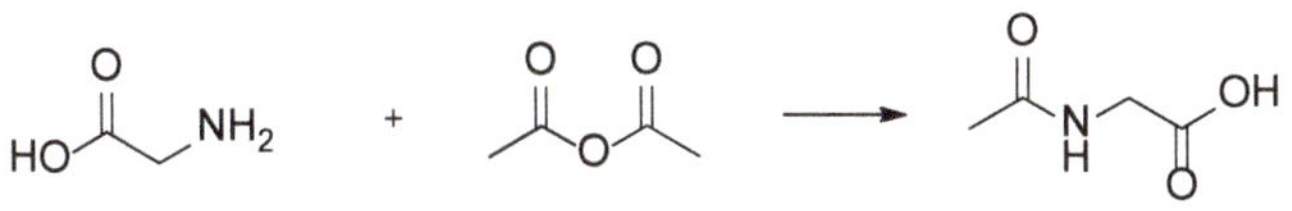

Figure 2: The synthetic route of N-Acetylglycine.

Compound	Provenance	Mass fraction purity
N-Acetylglycine	Aladdin Chemistry	0.990
Methanol	Aladdin Chemistry	0.996
Acetonitrile	Aladdin Chemistry	0.991

Table 1: Provenance and purity of the materials used.

and then measured the total weight rapidly. Then the beaker was put into a dryer and the mass was recorded repeatedly until keeping the constant weight. Each experiment was repeated at least three times and we used the average value to calculate the mole fraction solubility at last. The mole fraction solubility of N-Acetylglycine (x_I) in methanol+Acetonitrile is calculated by equation (1). The composition of solvent mixtures (x_{II}) is defined using equation (2).

$$x_I = \frac{m_I/M_I}{m_I/M_I + m_{II}/M_{II} + m_{III}/M_{III}} \tag{1}$$

$$x_{II} = \frac{m_{II}/M_{II}}{m_{II}/M_{II} + m_{III}/M_{III}}, \quad x_{III} = \frac{m_{III}/M_{III}}{m_{II}/M_{II} + m_{III}/M_{III}} \tag{2}$$

where m_I, m_{II}, m_{III} represent the mass of the solute, methanol, Acetonitrile, and M_I, M_{II}, M_{III} represent the molecular weight of the solute, methanol, Acetonitrile, respectively.

Results and Discussion

Solubility data

The solubility data (x_I) of N-Acetylglycine in methanol+Acetonitrile mixtures with the temperature ranging from 278.15 to 333.15 K are presented in Table 2. For comparison with each of the experimental points, experimental solubility data of N-Acetylglycine in methanol+Acetonitrile mixtures in the temperature range from 278.15 to 333.15 K are presented in Figures 3 and 4. From Table 2 and Figure 3, it can be found that the solubility of N-Acetylglycine in methanol+Acetonitrile mixtures is a function of temperature and solvent composition. More specifically, the solubility of N-Acetylglycine in Methanol+Acetonitrile mixtures increases with the rise of temperature, while at a constant temperature increases with increasing methanol content of the mixed solvent, By these properties, methanol could be used as effective benign solvent in the crystallization process.

Data correction

The relationship of the experimental isothermal mole fraction solubility and binary solvent compositions is described by the Combined Nearly Ideal Binary Solvent /Redlich-Kister (CNIBS/R-K) model [9-11]. The model is presented in equation (3).

$$\ln x_I = x_{II}\ln(x_I)_{II} + x_{III}\ln(x_I)_{III} + x_{II}x_{III}\sum_{i=0}^{N} S_i (x_{II} - x_{III})^i, \tag{3}$$

Where x_{II} and x_{III} represent the initial mole fraction composition of the binary solvent when the solute was not added. S_i is the model constant and the value of N can be 0, 1, 2 and 3. $(x_I)_i$ represents the saturated mole fraction solubility of the solute(N-Acetylglycine) in pure solvent i. When N=2 and substitution of $(1-x_{II})$ for x_{III}, then equation (3) can be converted into equation (4).

$$\ln x_I = \ln(x_I)_{III} + (\ln(x_I)_{II} - \ln(x_I)_{III} + (S_0 - S_1 + S_2)x_{II} + (-S_0 + 3S_1 - 5S_2)x_{II}^2 + (-2S_1 + 8S_2)x_{II}^3 + (-4S_2)x_{II}^4, \tag{4}$$

Which can be further simplified as the following equation:

$$\ln x_I = C_0 + C_1 x_{II} + C_2 x_{II}^2 + C_3 x_{II}^3 + C_4 x_{II}^4, \tag{5}$$

where C_0, C_1, C_2, C_3, C_4 are model constants and we can get these constants by least-squares analysis [12].

Equation (5) is just variant one of CNIBS/R–K model. When N=2 and x_{III} is relaced by $(1-x_{II})$, equation (3) can rearrangements result in equation (6).

This is variant two of CNIBS/R–K model [10]. The parameters S_i would be obtained by regressing

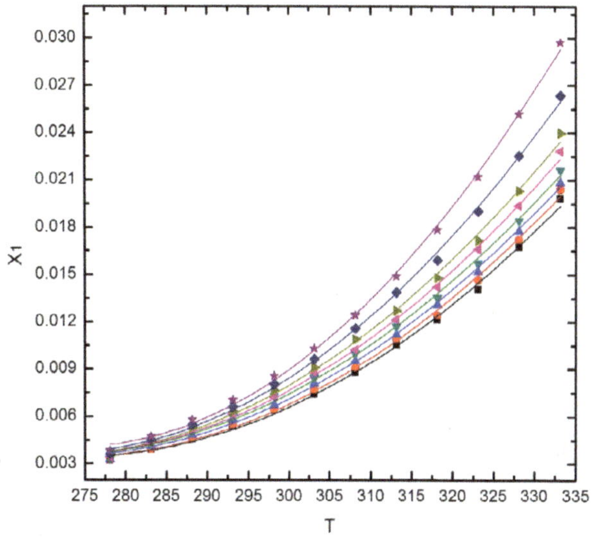

Figure 3: Solubility of N-Acetylglycine in Methanol+Acetonitrile binary solvent mixtures: ■, x_{II}=0; •, x_{II}=9.027%; ▲, x_{II}=19.419%; ▼, x_{II}=31.510%; ◄, x_{II}=45.754%; ►, x_{II}=62.782%; ◆, x_{II}=83.500%; ★, x_{II}=1.

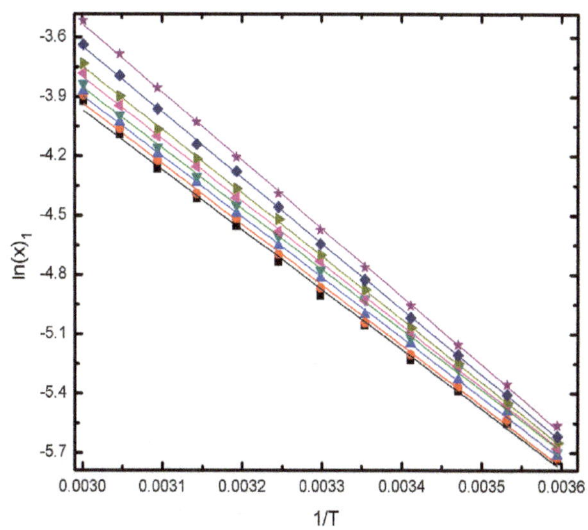

Figure 4: Mole fraction solubility ($\ln(x_I)$) of N-Acetylglycine in the binary solvent against $1/T$ with a straight line to correlate the data: ■, x_{II}=0; •, x_{II}=9.027%; ▲, x_{II}=19.419%; ▼, x_{II}=31.510%; ◄, x_{II}=45.754%; ►, x_{II}=62.782%; ◆, x_{II}=83.500%; ★, x_{II}=1.

x_{II}	$100x_I$	$100(x_I - x_I^{cal})/x_I$(Eq. 5)	$100(x_I - x_I^{cal})/x_I$ (Eq. 6)	$100(x_I - x_I^{cal})/x_I$ (Eq. 9)
		T=278.15 K		
0.000	0.322	-0.086	0.000	-1.266
0.090	0.326	0.168	0.221	-0.042
0.194	0.330	-0.013	-0.057	1.249
0.315	0.336	-0.088	-0.091	0.098
0.458	0.343	-0.091	-0.074	-0.140
0.628	0.352	0.197	0.194	0.097
0.835	0.364	-0.116	-0.119	-0.032
1.000	0.384	0.029	0.000	0.004
		T=283.15 K		
0.000	0.389	0.474	0.000	-2.558
0.090	0.395	-1.098	-1.172	-0.974
0.194	0.412	0.531	0.468	3.340
0.315	0.421	0.264	0.402	0.003
0.458	0.427	0.049	0.051	-0.005
0.628	0.431	-0.498	-0.604	0.003
0.835	0.449	0.346	0.417	-0.002
1.000	0.473	-0.091	0.000	0.000
		T=288.15 K		
0.000	0.457	0.472	0.000	-2.649
0.090	0.465	-1.093	-1.023	-0.975
0.194	0.484	0.530	-0.569	3.421
0.315	0.513	0.265	1.501	0.015
0.458	0.521	0.039	-0.110	-0.021
0.628	0.527	-0.483	-1.012	0.016
0.835	0.549	0.338	0.707	-0.005
1.000	0.579	-0.089	0.000	0.001
		T=293.15 K		
0.000	0.536	0.487	0.000	-3.817
0.090	0.550	-1.097	-1.309	-1.004
0.194	0.582	0.495	0.577	4.428
0.315	0.601	0.309	0.394	0.122
0.458	0.616	-0.015	-0.039	-0.173
0.628	0.630	-0.418	-0.490	0.121
0.835	0.663	0.298	0.353	-0.039
1.000	0.705	-0.077	0.000	0.005
		T=298.15 K		
0.000	0.638	0.438	0.000	-3.746
0.090	0.646	-0.952	-1.406	-0.875
0.194	0.673	0.342	-0.872	4.238
0.315	0.728	0.399	2.863	0.172
0.458	0.724	-0.072	-1.743	-0.244
0.628	0.764	-0.373	-0.069	0.171
0.835	0.803	0.275	0.375	-0.055
1.000	0.856	-0.072	0.000	0.007
		T=203.15 K		
0.000	0.741	0.481	0.000	-4.786
0.090	0.769	-1.075	-0.440	-1.026
0.194	0.809	0.515	0.206	5.251
0.315	0.843	0.218	-0.099	0.131
0.458	0.883	0.048	0.420	-0.182
0.628	0.909	-0.410	-0.503	0.126
0.835	0.964	0.274	0.251	-0.040
1.000	1.034	-0.069	0.000	0.005
		T=308.15 K		
0.000	0.876	0.501	0.000	-4.998
0.090	0.910	-1.100	-0.543	-1.082
0.194	0.954	0.511	0.023	5.489
0.315	1.002	0.182	0.876	0.071

0.458	1.025	0.155	-0.982	-0.098
0.628	1.090	-0.506	0.443	0.068
0.835	1.159	0.314	-0.098	-0.021
1.000	1.245	-0.078	0.000	0.004
T=313.15 K				
0.000	1.052	0.490	0.000	-5.047
0.090	1.094	-1.117	0.208	-1.040
0.194	1.122	0.533	-0.621	5.473
0.315	1.172	0.231	0.416	0.141
0.458	1.216	0.056	0.260	-0.195
0.628	1.273	-0.416	-0.456	0.134
0.835	1.389	0.275	0.220	-0.042
1.000	1.492	-0.069	0.000	0.005
T=318.15 K				
0.000	1.212	0.024	0.000	-2.710
0.090	1.241	-0.053	-0.757	-0.281
0.194	1.311	0.186	0.790	2.975
0.315	1.353	-0.553	-0.433	-0.500
0.458	1.423	0.760	0.249	0.678
0.628	1.480	-0.531	-0.197	-0.453
0.835	1.592	0.190	0.097	0.134
1.000	1.782	-0.036	0.000	-0.015
T=323.15 K				
0.000	1.384	-0.335	0.000	-3.300
0.090	1.406	0.716	0.192	0.505
0.194	1.465	-0.127	-0.128	2.700
0.315	1.517	-0.707	-0.592	-0.297
0.458	1.568	0.517	1.164	0.403
0.628	1.662	-0.013	-0.895	-0.266
0.835	1.717	-0.089	0.326	0.079
1.000	1.904	0.026	0.000	-0.008
T=328.15 K				
0.000	1.673	0.033	0.000	-3.006
0.090	1.723	0.052	0.043	-0.024
0.194	1.776	-0.181	-0.178	2.962
0.315	1.843	-0.164	-0.156	-0.365
0.458	1.941	0.660	0.652	0.495
0.628	2.032	-0.601	-0.599	-0.327
0.835	2.254	0.240	0.240	0.095
1.000	2.516	-0.048	0.000	-0.010
T=333.15 K				
0.000	1.982	-0.042	0.000	-2.843
0.090	2.043	0.108	0.693	-0.110
0.194	2.085	0.063	-0.345	2.782
0.315	2.163	-0.501	-0.434	0.044
0.458	2.283	0.677	0.514	-0.060
0.628	2.399	-0.432	-0.164	0.040
0.835	2.636	0.142	-0.001	-0.011
1.000	2.975	-0.026	0.000	0.001

[a]Standard uncertainties u are $u(T)$=0.1K, $u(x_I)$=0.0002, u(p)=0.02

Table 2: Mole fraction solubility (x_I) of N-Acetylglycine in Methanol+Acetonitrile binary solvent mixtures at the temperature range from (278.15 to 333.15) K under atmospheric pressure[a].

$$\{\ln x_I - (1-x_{II})\ln(x_I)_{III} - x_{II}\ln(x_I)_{II}\} \quad Versus$$

$$\{(1-x_{II})x_{II}[S_0 + S_1(2x_{II}-1) + S_2(2x_{II}-1)^2]\} \qquad (6)$$

The CNIBS/R-K model only can be used to describe the solubility data and to predict solubility data for different concentrations of a mixed solvent at a fixed temperature. To describe the effect of both solvent compositions and temperature on the solubility of N-Acetylglycine, another model called Jouyban-Acree model was adopted [12,13]

$$\ln x_I = x_{II}\ln(x_I)_{II} + x_{III}\ln(x_I)_{III} + x_{II}x_{III}\sum_{i=0}^{N}\frac{J_i(x_{II}-x_{III})^i}{T}, \qquad (7)$$

where J_i is the model constant, T is the absolute temperature, and other symbols denote the same meaning as equation (3). When N=2 and x_{III} is relaced by $(1-x_{II})$ in equation (7), we can obtain equation (8) after rearrangement.

$$\ln x_I = \ln(x_I)_{III} + (\ln(x_I)_{II} - \ln(x_I)_{III})x_{II} + \frac{(J_0 - J_1 + J_2)x_{II}}{T} + \frac{(-J_0 + 3J_1 - 5J_2)x_{II}^2}{T} + \frac{(-2J_1 + 8J_2)x_{II}^3}{T} + \frac{(-4J_2)x_{II}^4}{T}, \quad (8)$$

Equation (8) can also be further simplified as another form:

$$\ln x_I = D_0 + D_1/T + D_2\ln T + D_3 x_{III} + [D_4 x_{III} + D_5 x_{III}^2 + D_6 x_{III}^3 + D_7 x_{III}^4]/T + D_8 x_{III}\ln T \quad (9)$$

where D_0, D_1, D_2, D_3, D_4, D_5, D_6, D_7, D_8 are parameters of this model.

Experimental solubility data of N-Acetylglycine were correlated with equation (5), equation (6) and equation (9). Calculated solubility data (x_I^{cal}) from these models are listed in Table 2.

The mean deviation (MD) [14,15] used as the validity of the models is introduced and calculated by equation (10):

$$MD = 100\frac{\sum \frac{|x_I - x_I^{cal}|}{x_I}}{N}, \quad (10)$$

where N is the number of experimental points, x_I is the experimental solubility and x_I^{cal} is the calculated solubility. The parameter values of equation (5), equation (6) and equation (9) are listed in Tables 3-5 together with the values of MD.

As we can see from Tables 3-5, obtained overall MD value for the variant 1 of CNIBS/R-K equation is 0.384, for the variant 2 of CNIBS/R-K equation is 0.377, and for the Jouyban-Acree equation is 1.042. This result indicated that the experimental data agreed well with the calculated results from the three equations, and the the variant 2 of CNIBS/R-K equation provided the more accurate prediction according to its lower value of the overall MD. The CNIBS/R-K equation can only be used for the prediction of solubility for different concentrations of a mixed solvent at constant temperature, while the Jouyban-Acree equation can be used for predicting solubility at different temperatures and different concentrations.

Thermodynamic analysis of dissolution

We can calculate the standard molar enthalpy of dissolution (ΔH^0_{soln}) from the slope of the $\ln x_I$ versus $1/T$ plot by equation (11). This is the van't Hoff analysis and it is defined as [16-18]:

$$\Delta H^0_{soln} = -R \times \left(\frac{\partial \ln x_I}{\partial(1/T)}\right), \quad (11)$$

The standard molar enthalpy of dissolution (ΔH^0_{soln}) would be valid for the mean temperature (T_{mean}=305.65 K, in the present work) in a certain temperature range. Equation (11) can also be represented as:

$$\Delta H^0_{soln} = -R \times \left(\frac{\partial \ln x_I}{\partial(1/T - 1/T_{mean})}\right), \quad (12)$$

In addition, we can also calculate the standard molar Gibbs energy of dissolution (ΔG^0_{soln}) according to [19]:

$$\Delta G^0_{soln} = -RT_{mean} \times intercept, \quad (13)$$

where the intercept is obtained in plots of $\ln x_I$ versus $(1/T - 1/T_{mean})$.

Then the standard molar entropy of dissolution (ΔS^0_{soln}) is calculated by [20]:

$$\Delta S^0_{soln} = \frac{\Delta H^0_{soln} - \Delta G^0_{soln}}{T_{mean}} \quad (14)$$

The calculation results of the standard Gibbs energy, enthalpy, and entropy are shown in Tables 6 together with ξ_{TS} and ξ_H. The ξ_{TS} and ξ_H is the comparison of the relative contribution to the standard Gibbs energy by enthalpy and entropy in the dissolution process, respectively.

$$\%\xi_H = \frac{|\Delta H^0_{soln}|}{|\Delta H^0_{soln}| + |T\Delta S^0_{soln}|} \times 100, \quad (15)$$

$$\%\xi_{TS} = \frac{|T\Delta S^0_{soln}|}{|\Delta H^0_{soln}| + |T\Delta S^0_{soln}|} \times 100. \quad (16)$$

We can conclude from Table 6 that the standard Gibbs energy and the enthalpy of N-Acetylglycine are positive in the selected binary solvent mixtures, so it means the dissolution process of N-Acetylglycine in Methanol+Acetonitrile is endothermic. Furthermore, the main contributor to the standard molar Gibbs energy of solution is the enthalpy during the dissolution in virtue of values of $\%\xi_H$ are ≥ 64.038 for Methanol+Acetonitrile.

Conclusions

In this work new data were provided for the solubility of

T/K	C_0	C_1	C_2	C_3	C_4	MD
278.15	-5.738	0.080	0.349	-0.665	0.412	0.099
283.15	-5.555	0.387	-0.469	0.129	0.154	0.419
288.15	-5.358	0.396	-0.466	0.107	0.170	0.514
293.15	-5.235	0.529	-0.566	0.138	0.180	0.399
298.15	-5.042	0.507	-0.469	0.001	0.243	0.665
303.15	-4.910	0.632	-0.627	0.128	0.205	0.486
308.15	-4.742	0.648	-0.562	0.004	0.267	0.419
313.15	-4.559	0.663	-0.662	0.158	0.195	0.399
318.15	-4.413	0.256	0.251	-0.612	0.492	0.391
323.15	-4.261	0.296	0.062	-0.362	0.412	0.316
328.15	-4.091	0.327	-0.030	-0.106	0.219	0.248
333.15	-3.896	0.321	-0.239	0.261	0.039	0.249
					Overall	MD=0.384

Table 3: Parameters the variant 1 of CNIBS/R-K equation for N-Acetylglycine in Methanol+Acetonitrile binary solvent mixtures at the temperature range from (278.15 to 333.15) K.

T/K	S_0	S_1	S_2	MD
278.15	-0.073	-0.083	-0.091	0.095
283.15	-0.001	-0.204	-0.067	0.389
288.15	0.085	-0.191	-0.234	0.615
293.15	0.039	-0.228	-0.068	0.395
298.15	0.030	-0.123	-0.187	0.916
303.15	0.059	-0.198	-0.122	0.240
308.15	0.011	-0.152	0.003	0.371
313.15	-0.082	-0.109	0.082	0.273
318.15	-0.094	-0.250	-0.231	0.315
323.15	-0.151	-0.233	-0.020	0.412
328.15	-0.194	-0.164	-0.054	0.234
333.15	-0.214	-0.174	-0.137	0.269
			Overall	MD= 0.377

Table 4: Parameters the variant 2 of CNIBS/R-K equation for N-Acetylglycine in Methanol+Acetonitrile binary solvent mixtures at the temperature range from (278.15 to 333.15) K.

T/K	D_0	D_1	D_2	D_3	D_4	D_5	D_6	D_7	D_8	MD
278.15										0.366
283.15										0.861
288.15										0.888
293.15										1.214
298.15										1.188
303.15										1.443
308.15										1.479
313.15										1.510
318.15										0.968
323.15										0.945
328.15										0.911
333.15										0.736
	-1.281	-61.475	-6.620	-0.070	6.742	11.842	15.437	9.908	1.191	
									Overall	MD=1.042

Table 5: Parameters of the Jouyban-Acree equation for N-Acetylglycine in Methanol+Acetonitrile binary solvent mixtures at the temperature range from (278.15 to 333.15) K.

x/l	$\Delta H^0_{so\ln}$ / KJ·mol^{-1}	$\Delta G^0_{so\ln}$ / KJ·mol^{-1}	$\Delta S^0_{so\ln}$ /J·mol^{-1}·K^{-1}	$\%\xi_H$	$\%\xi_{TS}$
0.000	22.869	12.381	34.944	68.558	31.442
0.090	23.270	12.314	36.503	67.989	32.011
0.194	23.494	12.212	37.587	67.559	32.441
0.315	23.671	12.110	38.516	67.187	32.813
0.458	24.266	12.038	40.739	66.493	33.507
0.628	24.713	11.954	42.509	65.951	34.049
0.835	25.648	11.802	46.128	64.942	35.058
1.000	26.508	11.621	49.596	64.038	35.962

Table 6: Thermodynamic functions relative to solution process of N-Acetylglycine in Methanol+Acetonitrile binary solvent mixtures at mean temperature.

N-Acetylglycine in Methanol+Acetonitrile mixtures at temperature range from 278.15 K to 333.15 K. We can draw the following conclusions: The solubility of N-Acetylglycine in Methanol+Acetonitrile mixtures increases with the rise of temperature, while at a constant temperature increased with increasing methanol content of the mixed solvent. The solubility data could be successfully correlated using the two variants of the combined nearly ideal binary solvent/Redlich–Kister (CNIBS/R–K) equation and the Jouyban–Acree equation, but the variant 2 of CNIBS/R-K equation was more accurate than the other two equations. The thermodynamic properties for the dissolution process including Gibbs energy, enthalpy, and entropy were obtained by the van't Hoff analysis. The enthalpy change and the standard Gibbs energy change are positive proved that the dissolution process of N-Acetylglycine in Methanol+Acetonitrile mixtures was endothermic and contributor to the standard molar Gibbs energy of solution was the enthalpy during the dissolution due to values of $\%\xi H$ are ≥ 64.038. The experimental solubility data and the parameters could be used for optimizing the purification process of N-Acetylglycine in industry.

Acknowledgements

This research work was financially supported by the Science and Technology support program of JiangSu Province (Agriculture) (Grant No. BE2012373, BE2012374), the College Industrialization Project of Jiangsu Province (Grant No. JHB2011-16).

References

1. Karpenter GB, Donohue J (1950) The Crystal Structure of N-Acetylglycine1,2. J Am Chem Soc 72: 2315-2328.
2. Donohue J, Marsh RE (1962) A refinement of the structure of N-acetylglycine. Acta Cryst 15: 941.
3. Asiago VM, Alvarado LZ, Shanaiah N, Gowda GA, Owusu-Sarfo K, et al. (2010) Early detection of recurrent breast cancer using metabolite profiling. Cancer Res 70: 8309-8318.
4. Kim SK, Ubukata M, Isono K (2003) N-Acetylglycine side chain is critical for the antimicrobial activity of xanthostatin. Journal of Microbiology and Biotechnology 13: 998-1000.
5. Harburn JJ, Rath NP, Spilling CD (2005) Efficient synthesis of tyrosine-derived marine sponge metabolites via acylation of amines with a coumarin. J Org Chem 70: 6398-6403.
6. Hu YH, Chen X, Yang WG, Guan J (2009) Measurement and correlation of the solubility of urea l-tartaric acid in aqueous ethanol solution. Thermochim Acta 484: 38-40.
7. Chen ZG, Yang WG, Hu YH, Lei ZY (2011) J Chem Eng Data 56: 2676-2679.
8. Lei ZY, Hu YH, Yang WG, Li L (2011) Solubility of 2-(2, 4, 6-trichlorophenoxy) ethyl bromide in methanol, ethanol, propanol, isopropanol, acetonitrile, n-heptane, and acetone. J Chem Eng Data 56: 2714-2719.
9. Wang JY, Lee LN, Lai HC, Wang SK, Jan IS, et al. (2007) Fluoroquinolone resistance in Mycobacterium tuberculosis isolates: associated genetic mutations and relationship to antimicrobial exposure. J Antimicrob Chemother 59: 860-865.
10. Wang P, Jiang JX, He LT, Li HQ, Li J, et al. (2013) Solubility of isomalt in the water+ethanol solvent system at (288.15, 298.15, 308.15, and 318.15) K. J Chem Eng Data 58: 364-369.
11. Noubigh A, jeribi C, Mgaidi A, Abderrabba M (2012) Solubility of gallic acid in liquid mixtures of (ethanol+ water) from (293.15 to 318.15) K. J Chem Thermodynamics 55: 75-78.
12. Sun H, Li M, Jia JT, Tang FK, Duan EH (2012) Measurement and correlation of the solubility of 2, 6-diaminohexanoic acid hydrochloride in aqueous methanol and aqueous ethanol mixtures. J Chem Eng Data 57: 1463-1467.
13. Hasana M, Sawanta AB, Sawanta RB, Loke PG (2011) Densities, viscosities, speed of sound, and IR spectroscopic studies of binary mixtures of tert-butyl acetate with benzene, methylbenzene, and ethylbenzene at T=(298.15 and 308.15) K. J Chem Thermodynamics 43: 1389-1394.
14. Zhao Y, Wang YI (2013) J Chem Thermodynamics 57: 9-13.
15. Lim J, Jang S, Cho HK, Shin MS, Kim H (2013) Solubility of salicylic acid in pure alcohols at different temperatures. J Chem Thermodynamics 57: 295-300.
16. Wang S, Qin LY, Zhou ZM, Wang JD (2012) Solubility and solution thermodynamics of betaine in different pure solvents and binary mixtures. J Chem Eng Data 57: 2128-2135.
17. Wei DW, Li H, Li YN, Zhu J (2012) Effect of temperature on the solubility of 3-aminopyridine in binary ethanol+ toluene solvent mixtures. J Fluid Phase Equilibria 316: 132-134.
18. Maher A, Croker D, Rasmuson AC, Hodnett BK (2010) Solubility of Form III Piracetam in a Range of Solvents. J Chem Eng Data 55: 5314-5318.
19. Zhou XQ, Fan JS, Li N, Du ZX, Ying HJ, et al. (2012) Solubility of l-phenylalanine in water and different binary mixtures from 288.15 to 318.15 K. Fluid Phase Equilibria. 316: 26-33.
20. Azara VP, Ahmadian S, Martínez F, Acree Jr WE, Jouyban A (2012) Thermodynamic studies of fluphenazine decanoate solubility in PEG 200+ water mixtures. Fluid Phase Equilibria 330: 36-43.

Synthesis of Phase-Pure Zeolite Sodalite from Clear Solution Extracted from Coal Fly Ash

Hums E*

Consulting Environmental Catalysis, PO Box 1848, 91008 Erlangen, Germany

Abstract

The formation of zeolite sodalite (SOD) as a by-product in the synthesis of zeolite X from clear solution extracted from fused South African coal fly ash has recently been studied. However, the present study shows that zeolite X as well as zeolite A should be seen as intermediate structures and that phase-pure sodalite can be obtained by increasing crystallization time and temperature. To investigate the transition from zeolites A and X into SOD, the crystallization was conducted at 70, 80, and 90°C for different times. In addition to structural and chemical analyses by XRD and ICP, respectively, a detailed morphological characterization of the synthesis products by scanning electron microscopy (SEM) was undertaken in order to investigate the effect of the crystallization conditions on different crystal morphologies of zeolite SOD.

Keywords: Coal fly ash; Sodalite; Zeolite; Phase transformation

Introduction

Coal-fired power plants accounted for 41% of the global electricity production in 2006 and are expected to reach 44% in 2030. Apart from activities to replace conventional energy from fossil fuel, coal consumption increased globally for the decade of 2002–2012. A total of 750 million tonnes of fly ash were produced globally in 2012. Around 31 million tonnes of fly ash are produced just in South Africa per year but only about 6% is recycled. With a focus on sustainable resource management, because of the high content of silica and alumina coal fly ash, it is of interest as suitable feedstock in the synthesis of high-value zeolitic material and therefore competing with zeolites made from pure industrial chemicals [1]. Hydroxysodalite $Na_6[AlSiO_4]_6(OH)_2 \cdot 2H_2O$ (SOD) as well as zeolites X, Y (FAU) and zeolite A (LTA) belong to the class of zeolites which have the common characteristic of β-cages (sodalite units) as common building unit. Although it is well-known that sodalite can be ideally formed by direct attachment of sodalite units, there is still a constant interest to study the process of SOD synthesis. Sodalite can be successfully synthesized starting the crystallization either from gels [2], applying hydrothermal reaction conditions to mullite and quartz [3], or transformation of zeolite A [4]. Al-Azmi [5] found that the synthesis temperature and the amount of sodium hydroxide are the most important variables affecting the sodalite formation and the morphology when starting from gels. Furthermore, Huang et al. [6] mentioned that in zeolite A synthesis gel, which was free of structure-directing agents, nanocrystalline sodalite could be obtained when large amounts of ethanol were added. From this point of view, they systematically studied the crystallization of sodalite in the $NaOH–Al_2O_3–SiO_2–H_2O$ system in the presence of ethanol. Structures with different morphologies were found. In particular, clusters of sodalite discs, spherical aggregates of sodalite nanoplates (e.g., thread-ball-like particles), and core-shell/hollow sodalite structures were synthesized by varying the ethanol content and hydrothermal synthesis time. In this context they reported, that zeolite A is more stable in a water solvent system, whereas the presence of ethanol accelerated phase transformation to sodalite. Early works published by Borchert and Keidel [7] and Barrer and White [8] opened the sodalite synthesis via hydrothermal conversion of kaolinite in the presence of sodium hydroxide at low temperatures, whereas Novembre et al. [9] investigated the solid state transformation of metakaolinite to basic sodalite in the presence of NaCl in a so-called high temperature dry process synthesis route at 850°C and ambient pressure. Silica, alumina, and oxides of iron and calcium are the main constituent parts of coal fly ash. Therefore coal fly ash, which is a by-product of coal fired power plants, is an upcoming alternative to other industrial resources used for the synthesis of zeolites [10]. The fly ash which was used in the present study contained more than 70 wt.% $SiO_2+Al_2O_3$ (defined as class F coal fly ash corresponding with ASTM C618) and predominantly contaminated with Fe_2O_3 and low in lime. Although several methods are available to convert fly ash into zeolites, there are still technical and economical obstacles of products and processing at semi-industrial scale to be solved before successful commercialization. Besides the processing costs and low yields, the mobility of heavy and toxic elements in fly ashes in contrast to the synthesis from industrial pure chemicals are of particular relevance [11]. A simple way to obtain zeolites from fly ash is the conventional hydrothermal conversion of a mixture of fly ash and alkaline solution. In this way, fly ash conversion of about 50% into zeolite, especially P and/or hydroxysodalite, was reached as shown by Singer and Berkgaut [12] and Izidoro et al. [13]. But it seems hard from our point of view to remove the insoluble fly ash impurities from the produced zeolitic material. As an alternative to this procedure a mixture of fly ash and sodium hydroxide can be fused at 550°C in order to convert insoluble SiO_2 and Al_2O_3 in the fly ash into soluble sodium silicate and sodium aluminate. Before reacted under hydrothermal conditions the soluble sodium silicate and sodium aluminate can be separated from the insoluble fly ash constituents. The zeolites can subsequently be crystallized from this clear solution extract, which has to be additionally doped with an external solution of aluminate to adjust the Si/Al ratio for the synthesis of selected zeolite types [14]. Crystallization from clear solution yields much lower amounts of zeolite, but the zeolite product exhibits significantly higher purity compared to the conventional hydrothermal fly ash

***Corresponding author:** Hums E, Consulting Environmental Catalysis, PO Box,1848, 91008 Erlangen, Germany
E-mail: humseric@t-online.de

treatment. The high temperature fusion step makes it very energy intensive, therefore the optional use of microwave-assisted heating for the conversion of fly ash should not remain unmentioned to produce zeolites [15,16], although homogenous heating of the reaction mixture is then difficult. More than 60 wt.% of sodalite were formed from coal fly ash in a 3 M NaOH medium in less than 1 hour heating time via microwave-assisted heating [16]. It was stated that synthesis conditions and yield of zeolites obtained are comparable to conventional hydrothermal treatment, but activation time can be drastically reduced using microwave technology. The phase fraction of sodalite containing impurities of zeolite A was determined by XRD measurements and increased with increasing the reaction time from 30 to 45 minutes, suggesting that sodalite is the more stable zeolite phase compared to zeolite A, which is also in agreement with the findings of Maldonado et al. [17]. In most fly ash based zeolite synthesis processes, pure water as solvent for NaOH is used [18-26]. When Musyoka et al. [27] used waste industrial brine solutions instead of pure water, hydroxy sodalite was formed in significant quantity contaminated with the remaining amount of unconverted mullite and hematite from the fly ash feedstock. This is not surprising because both cancrinite and sodalite can be formed in alkaline solution from zeolite A depending on the types of anions [27,28]. The mineral phases observed in a study suggested that Cl^- and NO_2^- direct the nucleation of sodalite, whereas SO_4^{2-} direct the nucleation of cancrinite. It was also mentioned that some anions, such as CO_3^{2-} and NO_3^- can act as templates for both cancrinite and sodalite as discussed by Deng et al. [29]. Subotic et al. [30] worked out details of mechanistic aspects to synthesize hydroxy sodalite by transformation of zeolite A in the presence of concentrated solution of sodium hydroxide. From this point of view our interest was focused on sodalite which we recognized as contaminant when studying the kinetics of zeolite X formation based on in situ ultrasonic data starting from clear solution extracted from alkali fused coal fly ash [31]. The progress of attenuation of ultrasound during synthesis of zeolite X at different temperatures is shown in Figure 1. There, the crystallisation of zeolite X was already finished within 17 h in a temperature range between 80 and 94°C. It was the purpose of the present study to investigate further time- and temperature-dependent effects on zeolite X and SOD formation based on the former synthesis route published in [31] with the aim to examine formation and morphology of phase-pure SOD.

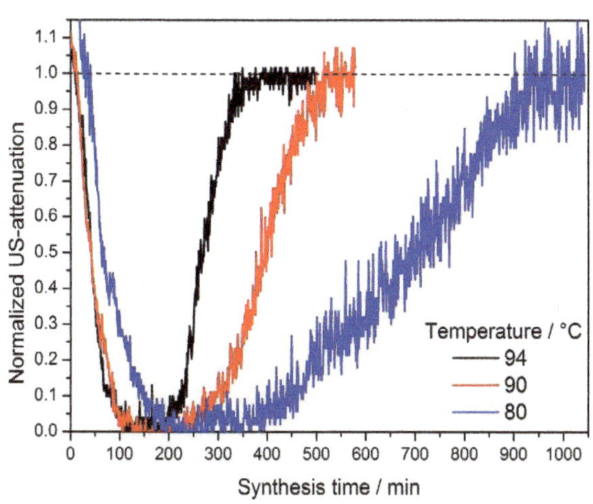

Figure 1: Progress of attenuation of ultrasound during synthesis of zeolite X at different temperatures (Hums et al. [31]).

Experimental

Class F coal fly ash was supplied by ESKOM from a power plant located in the Mpumalanga province of South Africa. The fly ash was stored in sealed containers to preserve its compositional integrity. The mineralogical composition of fly ash, the fused fly ash and the crystallized products were determined by X-ray powder diffraction (XRD) using a Philips X-ray diffractometer with Cu-Kα radiation and graphite monochromator. ICP-OES was used for the chemical characterization after digestion of 0.1 g sample in a solution of 8 ml HF, 2 ml HNO_3, and 2 ml HCl. The morphology of the solid samples was studied using a scanning electron microscope with ultra-high resolution (FESEM, ULTRA 55 Carl Zeiss MST AG). The zeolites were synthesized from clear solution extracted from fused coal fly ash. This was achieved by initially grinding the fly ash with sodium hydroxide in a weight ratio of 1/2 using a ball mill for 20 min to obtain a fine homogenously ground material. The resulting mixture was then fused at 550°C for 2 hours in a muffle furnace to convert insoluble fly ash mineral phases into soluble sodium silicate and sodium aluminate. The solid material was cooled down, crushed, dispersed in demineralized water in a weight ratio of 1/2.5 and stirred for 2 hours. The clear green solution extract obtained was separated from the insoluble material by filtration and subsequent centrifugation and then crystallized without stirring in polypropylene flasks at 70,80 and 90°C for different times. The resulting zeolitic material was isolated by centrifugation, thoroughly washed with demineralized water until the pH of the washing water was below 9. The synthesis products were tried at 75°C overnight before characterization.

Results and Discussion

The composition of raw fly ash was found to be mainly amorphous (62%) with the remaining crystalline phase consisting of quartz (13.5%), mullite (22.3%), hematite (1.6), and lime (0.5%) with a particle size <90 nm. After thermal fusion with sodium hydroxide the main mineral phases were sodium aluminate and sodium silicate; no evidence of quartz and mullite could be found as shown in [32]. Exemplarily for raw fly ash and fused fly ash SEM images are presented in [28]. The majority of the particles of the raw fly ash were observed to be spherical in shape related to the cooling effect of flue gas under power plant conditions. The predominantly spherical particles are transformed into an agglomerated porous mass after fusion. The fly ash used for this study showed insignificantly higher amounts of SiO_2 and Al_2O_3 in comparison with the fly ash used in our previous study [28]. The XRD patterns of the samples obtained after crystallization and the corresponding reference diffractograms for zeolites A (LTA; pdf 00-038-0241), X (FAU; pdf 00-038-0237), P (GIS; pdf 00-040-1464), cancrinite (CAN; pdf 00-046-1332) and sodalite (SOD; pdf 00-037-0476) are shown in Figure 2. It can be seen that only reflections of zeolites A, X and SOD occur in the XRD patterns of samples a-d. Cancrinite, which is frequently reported to crystallize as a accompanying or consecutive phase of SOD [29,33] is not detected. According to the XRD results, hydrothermal treatment at 80°C for 25 hours (sample a) resulted in a mixture of zeolites A, X, and SOD. This is also stated by the corresponding SEM images in Figure 3a which show a mixture of particles with three different morphologies. The octahedral particles represent the typical morphology of zeolite X, zeolite A exhibits the typical cubic morphology and accordingly the thread-ball to armadillo-like clusters can be assigned to sodalite and not to a new morphology of hierarchical zeolite X as published in this context by Musyoka et al. [34-36]. When the crystallization at 80°C was extended from 25 hours (sample a) to 48 hours (sample b) the diffraction peaks of sodalite increased whereas the diffraction peaks of

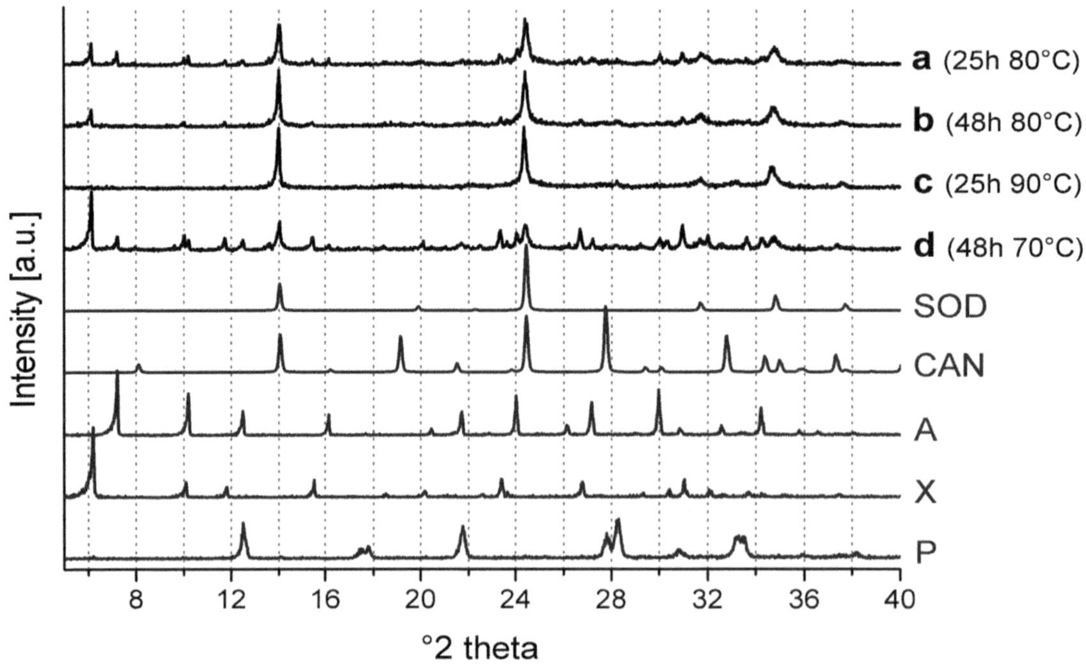

Figure 2: XRD patterns of the synthesis products from clear solution extract after crystallization at different reaction temperature and reaction time: at 80°C after 25 (sample a) and 48 hours (sample b); at 90°C after 25 hours (sample c), at 70°C after 48 hours (sample d) including the references for zeolites A, X, P, CAN, and SOD.

zeolite A disappeared, which is in agreement with the literature [32] and states that a phase transformation from zeolite A to sodalite took place. Also a decrease of the reflections of zeolite X can be observed, which correlates with the SEM images (Figure 3b), where a lower amount of octahedral zeolite X particles can be observed, whereas the amount of thread-ball-like and armadillo-like particles (SOD) increased. Also traces of zeolite A are still visible in the SEM images, but they start to be overgrown with SOD phase, which suggests that the transformation of LTA into SOD structure proceeds from the surface of the particles to the interior. The same mechanism seems to work for the transformation of zeolite X (FAU), because also zeolite X particles are overgrown with SOD phase (Figures 3a and 3b). The increase of the crystallization temperature from 80 to 90°C (sample c) accelerates the crystallization (and transformation) process and the XRD pattern (Figure 2) shows phase-pure SOD already after 25 h, whereas the reflections of zeolites X and A completely disappeared. Accordingly, the SEM images in Figure 3c show solely particles with armadillo-like morphology. However, two different sizes of SOD particles could be found during the SEM analyses of sample c (Figure 4). The large SOD particles seem to stem from the transformation of LTA or FAU zeolite particles, which is also evident from one SEM image where the armadillo-like phase seems to overgrow remaining particles of zeolite A or X (Figure 5c). In contrast, the smaller SOD particles might originate from SOD seeds which started to form and to grow later in the crystallization process. It is noticeable that two different types of SOD morphologies, namely thread-ball like and armadillo-like morphology, were observed depending on the crystallization conditions. The thread-ball-like SOD morphology accompanied the transformed particles of zeolites A and X and thus seemed to grow preferably on the surface of preformed zeolites (which also contain sodalite cages as the secondary building unit). This was also the case for the SOD impurities in zeolite X samples previously reported [30]. In contrast, the SOD particles growing from SOD seeds had armadillo-like morphology. For the phase-pure SOD sample (sample c) it was observed that even the large SOD particles had the coarser armadillo-like morphology, suggesting that the thread-ball like morphology transformed into a more stable (less external surface) armadillo-like morphology. Different types of SOD morphologies were also found in the literature [28] where the morphology was influenced by several factors like NaOH concentration, type of anion, Si/Al ratio, salinity. Finally, a mix of cancrinite and sodalite was obtained by transformation of zeolite A [29,33]. Gualtieri et al. [37] also recognized that zeolite A is replaced by hydroxy-sodalite at higher crystallization temperatures and/or after longer heating periods. They suggested a nucleation of sodalite, because direct solid-state transformation of zeolite A to sodalite seemed impossible due to the large energy required to reform the double-four membered ring units. According to Subotic et al. [29] the transformation should be controlled by a solution-mediated process starting with the dissolution of zeolite A, followed by supersaturation of the synthesis solution and then heterogeneous nucleation as reason for the proposed phase transformation process. In the present study sodalite seems to nucleate close to the phase boundary of zeolite X via Ostwald ripening, proceeding from the surface to the inside of the particles as shown in the SEM images in Figure 5. Results of the elemental composition of the fly ash and the resulting phase-pure sodalite analysed by ICP are summarized in Table 1. It is important to be mentioned that the amount of foreign elements such as Ca, Fe, Ti and Ba in the final zeolite product is significantly reduced compared to the initial fly ash material.

Conclusions

The present study showed that the clear solution extracted from South African type F fly ash exhibited a proper $H_2O/SiO_2/Al_2O_3/NaOH$ ratio for the synthesis of zeolite sodalite without the need to

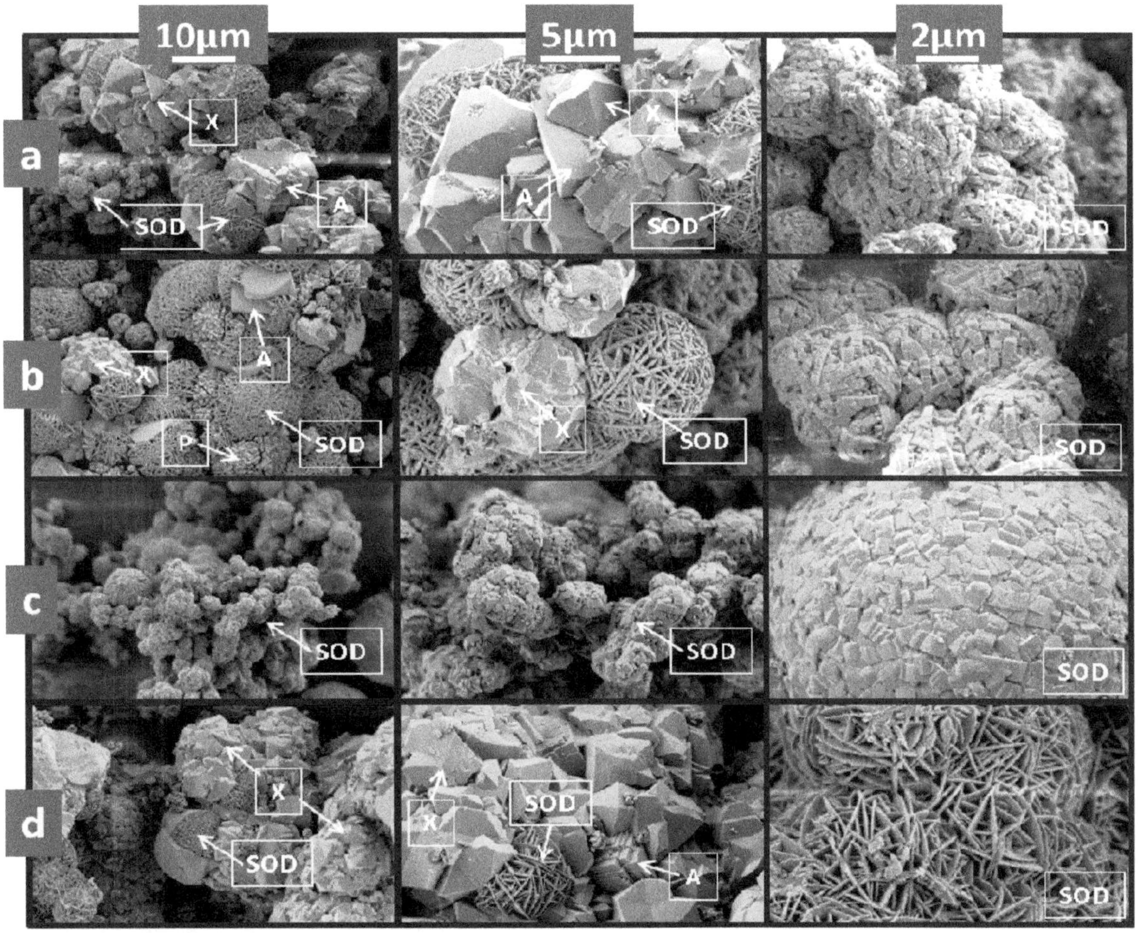

Figure 3: Selected SEM images corresponding to XRD patterns of the products synthesized at 80°C after 25 (sample a) and 48 hours (sample b), at 90°C after 25 hours (sample c) at 70°C after 48 hours (sample d).

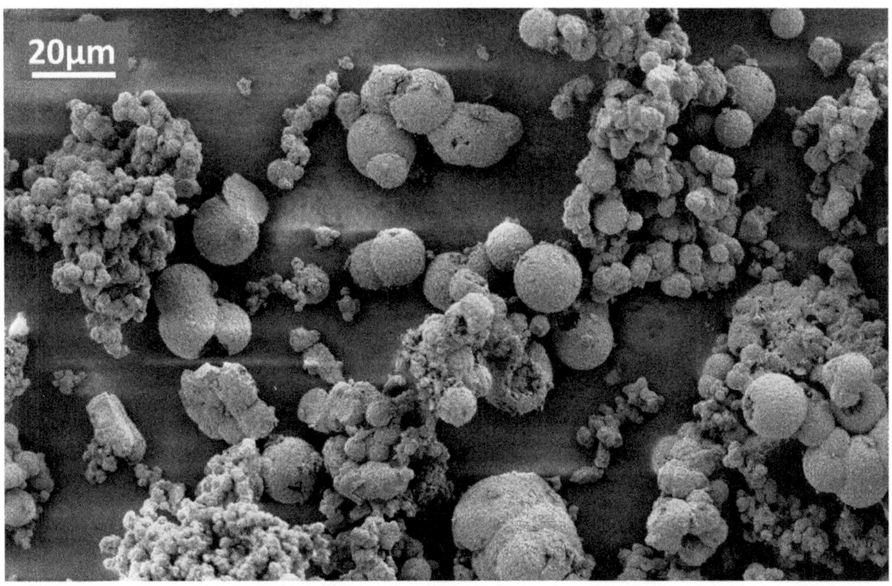

Figure 4: Overview: SEM image of different sizes of SOD.

Figure 5: SEM images of armadillo-like phase overgrowing remaining particles of zeolite A and X showed by SEM images.

Element	Fly ash [wt-%] ICP	After crystallization [wt-%] ICP
Si	26.52	16.20
Al	14.53	11.00
Fe	3.46	0.38
Mn	0.01	0.01
Mg	0.62	0.01
Ca	1.47	0.04
Na	0.03	10.15
K	0.37	<0.22
Ti	0.93	<0.08
P	n.a.	0.09
S	n.a.	n.a.
Ba	0.00	0.00
Ce	n.a.	n.a.
Co	0.01	0.00
Cr	0.02	0.01
Cu	0.00	0.01
Li	0.01	n.a.
Mo	n.a.	<0.14
Nb	n.a.	n.a.
Ni	n.a.	0.01
Pb	n.a.	n.a.
Rb	n.a.	n.a.
Sr	n.a.	n.a.
Th	n.a.	n.a.
U	n.a.	n.a.
V	n.a.	0.00
Y	n.a.	n.a.
Zn	0.01	0.01
Zr	n.a.	n.a.

n.a.=not assigned

Table 1: ICP-OES results of elements of fly ash and synthesized sodalite from clear solution extracted from fused coal fly ash.

add any additional reactants. From XRD and SEM analyses it could be seen that (for the synthesis composition used) the crystallization of SOD, X and A is significantly affected by thermodynamic and kinetic parameters, from which finally zeolite SOD results. On this basis it was possible to suppress the formation of zeolites X and A by raising the crystallization temperature to 90°C. Furthermore, the morphology of the obtained SOD particles was found to vary with the crystallization conditions between thread-ball-like and armadillo-like. Since the zeolites were crystallized from the clear solution extracted from fused fly ash, there was no other solid material after crystallization except zeolite crystals. Also, the content of heavy metals in the product could be clearly reduced due to the lower content of heavy metals in the clear solution extract compared to the content of heavy metals in zeolites which were crystallized directly from fly ash [28]. However, it should be noted that with the procedure applied here, the yield of zeolite in respect to the amount of fly ash used is far away from an economically desirable quantity for the use of fused coal fly ash as industrial feedstock for zeolite syntheses. Nevertheless, in order to increase the yield of zeolite it seems promising to optimize the extraction step of the fused fly ash. Economically promising for a feasible scale-up option to synthesize zeolites is an approach to replace high temperature fusion by a short sonochemical treatment step in order to enrich aluminate and silicate in the clear solution extracted from fly ash. This technique is also profitable to reduce the ageing period and to increase the rate of crystallization [38,39].

Acknowledgements

The author would like to thank A. Inayat from the *University of Erlangen - Nürnberg, Germany* for performing experiments, A. Postatny to assist with ICP analyses and the German Research Foundation (DFG) and Prof. W. Schwieger for funding this study. Thanks are also addressed to L. Petrik from the University of the Western Cape, South Africa for providing the fly ash from ESKOM for this project.

References

1. Hums E, Baser H, Schwieger W (2016) In situ ultrasonic measurements: a powerful tool to control the synthesis of zeolites from coal fly ash. Research on Chemical Intermediates 42: 7513-7532.

2. Hayashi S, Suzuki K, Shin S, Hayamizu K, Yamamoto O (1984) High-resolution solid-state NMR study of the crystallization of hydroxysodalite. Chemical Physics Letters 110: 54-57.

3. Denk G, Menzel W (1971) Uber die Reaktion von Kaolin mit Alkalilaugen. Zeitschrift für anorganische und allgemeine Chemie 382: 209-216.

4. Subotić B, Škrtić D, Šmit I, Sekovanić L (1980) Transformation of zeolite A into hydroxysodalite: I. An approach to the mechanism of transformation and its experimental evaluation. Journal of Crystal Growth. 50: 498-508.

5. Radhi AA. The Effects of Different Synthesis Conditions on Sodalite Crystal Morphology (Doctoral dissertation, Ohio University). Thesis 2001 Ohio University, USA.

6. Huang Y, Yao J, Zhang X, Kong CC, Chen H, et al. (2011) Role of ethanol in

sodalite crystallization in an ethanol–Na2O–Al2O3–SiO2–H2O system. Cryst Eng Comm 13: 4714-4722.

7. Borchert W, Keidel J (1947) Beiträge zur Reaktionsfähigkeit der Silikate bei niedrigen Temperaturen. Heidelberger Beiträge zur Mineralogie und Petrographie 1: 2-16.

8. Barrer RM, White EA (1952) The hydrothermal chemistry of silicates. Part II. Synthetic crystalline sodium aluminosilicates. Journal of the Chemical Society (Resumed) 1952: 1561-1571.

9. Novembre D, Gimeno D, Pasculli A, Di Sabatino B (2010) Synthesis and characterization of sodalite using natural kaolinite: an analytical and mathematical approach to simulate the loss in weight of chlorine during the synthesis process. Fresenius Environmental Bullettin 19: 1109-1117.

10. Ahmaruzzaman M (2010) A review on the utilization of fly ash. Progress in energy and combustion science 36: 327-363.

11. Hums E (2008) Feasibility Study: Synthesis of zeolite from coal fly ash at a commercial scale in the context of fly ash leaching tests (2008 part of ESKOM project: Feasibility of Application of Zeolites made from Fly Ash).

12. Singer A, Berkgaut V (1995) Cation exchange properties of hydrothermally treated coal fly ash. Environmental science & technology 29: 1748-1753.

13. Izidoro JC, Fungaro DA, dos Santos FS, Wang S (2012) Characteristics of Brazilian coal fly ashes and their synthesized zeolites. Fuel Processing Technology 97: 38-44.

14. Musyoka NM, Petrik LF, Hums E, Baser H, Schwieger W (2012) In situ ultrasonic monitoring of zeolite A crystallization from coal fly ash. Catalysis Today 190: 38-46.

15. Yao ZT, Li HY, Xia MS, Ye Y, Zhang L (2009) Hydrothermal synthesis of sodalite from coal fly ash and its property characterization [J]. The Chinese Journal of Nonferrous Metals 19:366-371.

16. Mohamed MH, Lintang HO, Ramli Z (2012) Characterization and Gravimetric Analysis of the Dissolved Quartz in the Conversion of Coal Fly Ash to Sodalite. Malaysian Journal of Analytical Sciences 16: 235-240.

17. Maldonado M, Oleksiak MD, Chinta S, Rimer JD (2013) Controlling crystal polymorphism in organic-free synthesis of Na-zeolites. Journal of the American Chemical Society 135: 2641-2652.

18. Molina A, Poole C (2004) A comparative study using two methods to produce zeolites from fly ash. Minerals Engineering 17: 167-73.

19. Yaping Y, Xiaoqiang Z, Weilan Q, Mingwen W (2008) Synthesis of pure zeolites from supersaturated silicon and aluminum alkali extracts from fused coal fly ash Fuel 87: 1880-1886.

20. Querol X, Moreno N, Umaña JC, Alastuey A, Hernández E, et al. (2002) Synthesis of zeolites from coal fly ash: an overview. International Journal of coal geology 50: 413-423.

21. Querol X, Umaña JC, Plana F, Alastuey A, Lopez-Soler A, et al. (2001) Synthesis of zeolites from fly ash at pilot plant scale. Examples of potential applications. Fuel 80: 857-865.

22. Moriyama R, Takeda S, Onozaki M, Katayama Y, Shiota K, et al. (2005) Large-scale synthesis of artificial zeolite from coal fly ash with a small charge of alkaline solution. Fuel 84: 1455-1461.

23. Moreno N, Querol X, Ayora C, Pereira CF, Janssen-Jurkovicová M (2001) Utilization of zeolites synthesized from coal fly ash for the purification of acid mine waters. Environmental science & technology 35: 3526-3534.

24. Lee DB, Matsue N, Henmi T (2001) Influence of NaOH concentrations dissolved in seawater and hydrothermal temperatures on the synthesis of artificial zeolite from coal fly ash. Clay science 11: 451-463.

25. Takashi H, Kazuharu Y (2008) Preparation of zeolitic adsorbents for environmental improvement from coal fly ash using alkali fusion. Downloaded by [217.229.39.141] at 10:18 21 May 2013 Waste brine used to synthesize hydroxy sodalite 1707 method. Proceedings of the 3rd Japan - Taiwan Joint International Symposium on Environmental Science and Technology. Kitakyushu, Japan.

26. Belviso C, Cavalcante F, Lettino A, Fiore S (2009) Zeolite synthesised from fused coal fly ash at low temperature using seawater for crystallization. Coal Combustion and Gasification Products 1: 7-13.

27. Musyoka NM, Petrik LF, Balfour G, Gitari WM, Hums E (2011) Synthesis of hydroxy sodalite from coal fly ash using waste industrial brine solution. Journal of Environmental Science and Health, Part A 46: 1699-1707.

28. Musyoka NM, Petrik LF, Fatoba OO, Hums E (2013) Synthesis of zeolites from coal fly ash using mine waters. Minerals Engineering 53: 9-15.

29. Deng Y, Harsh JB, Flury M, Young JS, Boyle JS (2006) Mineral formation during simulated leaks of Hanford waste tanks. Applied Geochemistry 21: 1392-1409.

30. Subotić B, Škrtić D, Šmit I, Sekovanić L (1980) Transformation of zeolite A into hydroxysodalite: I. An approach to the mechanism of transformation and its experimental evaluation. Journal of Crystal Growth 50: 498-508.

31. Subotić B, Sekovanić L (1986) Transformation of zeolite A into hydroxysodalite. II: Growth kinetics of hydroxysodalite microcrystals. Journal of crystal growth 75: 561-572.

32. Hums E, Musyoka NM, Baser H, Inayat A, Schwieger W. In-situ ultrasound study of the kinetics of formation of zeolites Na–A and Na–X from coal fly ash. Research on Chemical Intermediates 41: 4311-4326.

33. Hums E, Inayat A, Schwieger W (2016) Presentation, 18th International Zeolite Conference, Rio, Brazil.

34. Reyes CA, Williams C, Alarcón OM (2013) Nucleation and growth process of sodalite and cancrinite from kaolinite-rich clay under low-temperature hydrothermal conditions. Materials Research 16: 424-438.

35. Musyoka NM, Petrik LF, Hums E, Kuhnt A, Schwieger W (2015) Thermal stability studies of zeolites A and X synthesized from South African coal fly ash. Research on Chemical Intermediates 41: 575-582.

36. Musyoka NM, Petrik LF, Hums E, Baser H, Schwieger W (2014) In situ ultrasonic diagnostic of zeolite X crystallization with novel (hierarchical) morphology from coal fly ash. Ultrasonics 54: 537-543.

37. Musyoka NM, Petrik LF, Hums E (2016) Process for the Synthesis of Zeolites. PCT/IB2013/052438.

38. Gualtieri A, Norby P, Artioli G, Hanson J (1997) Kinetics of formation of zeolite Na-A [LTA] from natural kaolinites. Physics and Chemistry of Minerals 24: 191-199.

39. Musyoka NM, Petrik LF, Hums E (2016) Inventors; Eskom Holdings Soc Limited, University of The Western Cape, Assignee. Ultrasonic synthesis of zeolites from fly ash. United States Patent US 9,522,828.

Solvent Extraction of Cadmium Picrate by 18-Crown-6 Ether into Several Less-polar Diluents and Nitrobenzene: Re-evaluation of the Corresponding Overall Extraction Systems

Yoshihiro Kudo[1]*, Shoichi Katsuta[1], Yuuki Ohsawa[2] and Kohei Nozaki[2]

[1]Graduate School of Science, Chiba University, Chiba 263-8522, Japan
[2]Department of Chemistry, Faculty of Science, Chiba University, Chiba 263-8522, Japan

Abstract

TCadmium picrate ($CdPic_2$) in water was extracted at 25°C by 18-crown-6 ether (18C6) into *o*-dichlorobenzene (*o*DCBz), bromobenzene (BBz), dibutylether (DBE), and nitrobenzene (NB). Their extraction constants (K_{ex} and $K_{ex\pm}$) were determined, where K_{ex} and $K_{ex\pm}$ at L=18C6 were defined as $(CdLPic_2)_o/((Cd^{2+})(L)_o(Pic^-)^2)$ and $(CdLPic^+)_o(Pic^-)_o/((Cd^{2+})(L)_o)(Pic^-)^2)$, respectively. The subscript "o" denotes an organic phase, such as *o*DCBz and BBz. The same extraction constants were re-determined for a system with benzene. Also, individual distribution constants ($K_{D,Pic}=(Pic^-)_o/(Pic^-)$) of picrate ion Pic^- into the above o phases were determined with the K_{ex} determination. Properties for the $CdPic_2$ extraction with 18C6 were discussed by using the above constants and those available from the same extraction systems with other diluents. From comparing the experimental log $K_{D,Pic}$ value with the log $K_{D,Pic}^S$ one for only the high-polar NB system, an interfacial potential difference ($\Delta\varphi_{eq}$) at extraction equilibrium was evaluated, where the symbol, log $K_{D,Pic}^S$, shows the log $K_{D,Pic}$ value standardized at Δφeq=0 V and 25°C. In the course of this study, an extraction constant for an HPic extraction into DBE was determined spectrophotometrically at 25°C.

Keywords: Extraction constants; Individual distribution constant; Interfacial equilibrium potential; Solute-solvent interaction; At water/nitrobenzene interface; Cadmium picrate; 18-Crown-6 ether

Introduction

18-Crown-6 ether (18C6) extracts $CdBr_2$, cadmium picrate ($CdPic_2$), or alkaline-earth metal picrates into various diluents [1-3]. In the previous study [1], the authors clarified that the extracted ion-pair complex, $Cd(18C6)Br_2$, very weakly interacts with water molecules and also the diluents ones and they also suggested that another complex $Cd(18C6)Pic_2$ strongly interacts with water molecules. In addition to this, it has been suggested that the latter ion-pair complex has a more polar structure than the former does [1]. However, kinds of the diluents used for the $CdPic_2$ extraction experiments with 18C6 were fewer, compared with kinds of the diluents for the $CdBr_2$ extraction ones.

In the above studies, furthermore, component equilibria which constitute overall extraction equilibria have been considered to be at least 6-10 ones [1,3,4]. For example, the component equilibria are $L \rightleftharpoons L_o$, $CdPic^+ \rightleftharpoons Cd^{2+}+Pic^-$, $H^++Pic^- \rightleftharpoons HPic$, $L+Cd^{2+} \rightleftharpoons CdL^{2+}$, $CdL^{2+}+Pic^- \rightleftharpoons CdLPic^+$, $CdLPic^++Pic^- \rightleftharpoons CdLPic_2$, $CdLPic_2 \rightleftharpoons CdLPic_{2,o}$, $CdLPic_{2,o} \rightleftharpoons CdLPic^+_o+Pic^-_o$, $Pic^- \rightleftharpoons Pic^-_o$, and $CdLPic^+ \rightleftharpoons CdLPic^+_o$ at L=18C6 [1] where the subscript "o" denotes an organic phase. Unfortunately, except for several studies of the M(II) extraction with dibenzo-18C6 into nitrobenzene (NB) [5,6] the authors have not been able to find out studies based on such precise equilibrium analyses. Also, it is very difficult to critically-evaluate equilibrium constants determined in the studies of the NB systems, because the corresponding component equilibrium-constants experimentally-obtained from other methods are few. In the present paper, by employing *o*-dichlorobenzene (*o*DCBz), bromobenzene (BBz), dibutylether (DBE), and NB, we added new data in the $CdPic_2$ extraction by 18C6 and then re-discussed extraction characteristics of 18C6 against $CdPic_2$. Also, extraction constants (K_{ex} and $K_{ex\pm}$) for the same extraction into benzene (Bz) were re-determined, because the data of its Bz system abnormally deviated from the plot [1,4] of log $K_{ex,ip}$ *versus* log $K_{D,L}$. Here, K_{ex}, $K_{ex\pm}$, $K_{ex,ip}$, and $K_{D,L}$ were defined $[MLA_2]_o/([M^{2+}][L]_o[A^-]^2)$, $[MLA^+]_o[A^-]_o/([M^{2+}][L]_o[A^-]^2)$, $[MLA_2]_o/[ML^{2+}][A^-]^2$, and $[L]_o/[L]$ [1,3,4], respectively, with $M^{2+}=Cd^{2+}$ and $A^-=Pic^-$. In the above discussion, an interfacial equilibrium-potential difference ($\Delta\varphi_{eq}$) at the transfer of Cd^{2+}, CdL^{2+}, $CdLPic^+$, and Pic^- from water into the o phase, in particular NB or *o*DCBz, was introduced, as described previously [3]. Furthermore, an extraction constant of picric acid, HPic, into DBE was determined spectrophotometrically at 25°C, in order to precisely analyze extraction data for the Cd(II) extraction by 18C6 into DBE.

Materials and Methods

Chemicals

The preparation method of $CdPic_2{\cdot}nH_2O$ was essentially the same as that reported before [1] this *n* value was determined by a Karl-Fischer titration to be 5.67 ± 0.23 at number (*N*) of run of 3. This finding was in good agreement with that (*n*=5.65) obtained from an EDTA titration of Cd(II) in the aqueous solution which contains $CdPic_2{\cdot}nH_2O$. Also, spectrophotometric analysis of Pic (−I) in the aqueous solution at 355.0 nm showed the molar ratio of Cd(II):Pic(−I)=1:2.1 (see below). Water amount of used 18C6 (>98.0%, Wako) was determined by the Karl-Fischer titration to be 0.062 ± 0.029% at *N*=4. The H(I) amount of an aqueous solution prepared from a commercial HPic (>99.5%, Wako) was determined by an acid-base titration [7]. Other reagents were the same as or similar to those employed previously [1,4].

***Corresponding author:** Yoshihiro Kudo, Graduate School of Science, Chiba University, 1-33, Yayoi-cho, Inage-ku, Chiba 263-8522, Japan
E-mail: iakudo@faculty.chiba-u.jp

Extraction procedures

The experimental procedures for the $CdPic_2$ extraction by 18C6 were essentially the same as those described before [1,4] although the experiments were performed with the prepared $CdPic_2$ for the *o*DCBz (>99.0%, Kanto), DBE (>99.0%, Kanto), NB (>99.5%, Kanto), and Bz (>99.5%, Wako) systems and with mixtures of $CdSO_4$(>99.0%: (3/8) hydrate, Kanto) with HPic for the *o*DCBz and BBz (>98.0%, Kanto) systems. Here, the mixture was prepared by mixing an aqueous $CdSO_4$ solution with a mixture between $Ba(OH)_2$ (>98.0%: 8 hydrate, Wako) and an excess amount of HPic and then that with the precipitate was filtered. Total amounts of Cd(II) back-extracted from the o phases into aqueous solutions of the 0.1 mol dm^{-3} HNO_3 were determined at the wavelength of 228.8 nm by a Hitachi atomic absorption spectrophotometer (type Z-6100) equipped with a Hitachi hollow cathode lamp (Mitorika Co., under the license of Hitachi Ltd.) of Cdin an air-acetylene flame [1]. Total concentrations employed for the experiments were (0.79-1.7) × 10^{-2} mol dm^{-3} for the prepared $CdPic_2$ and 1.1 × 10^{-5}-0.22 for 18C6; (0.79-1.5) × 10^{-2} mol dm^{-3} Cd(II), 0.012-0.030 Pic(−I), and 0.0017 SO_4^{2-} for the mixture and (0.051-5.1) × 10^{-2} for 18C6. Also, the procedures for the HPic extraction into DBE were the same as those reported before [7]. Total amounts of Pic(−I) extracted into DBE were back-extracted into aqueous solutions with 0.1 mol dm^3 NaOH from the DBE phases and then were determined at the wavelength of 355.0 nm spectrophotometrically [7]. Total concentrations of the aqueous HPic solutions employed were in the range of (0.50-3.4) × 10^{-2} mol dm^{-3}.

Data analysis

Data analysis for the extraction experiments was essentially the same as that described before [1,3,4]. The extraction constant parameter, K_{ex}^{mix}, has been introduced, where K_{ex}^{mix} is defined as $([MLA_2]_o+[MLA^+]_o)/([M^{2+}][L]_o[A^-]^2)$ by assuming that $[MLA_2]_o+[MLA^+]_o >> [MA^+]_o+[ML^{2+}]_o+[M^{2+}]_o$ [1,3,4]. Its numerators were determined by AAS measurements (see above) and were expressed as "*Ab*" here. Also, the $[M^{2+}]$, $[L]_o$, and $[A^-]$ values were calculated from the following equations by a successive approximation method:

$$[M^{2+}] = \frac{[M]_t - Ab}{1+K_{MA}[A^-]+(K_{ML}/K_{D,L})\left(1+K_1K_2[A^-]^2\right)[L]_o} \quad (1)$$

$$[L]_o = \frac{[L]_t - Ab}{1+K_{D,L}^{-1}+\left(K_{ML}/K_{D,L}\right)\left(1+K_1K_2[A^-]^2\right)[M^{2+}]} \quad (2)$$

$$[A^-] \approx \frac{\sqrt{b^2-4ac}-b}{2a} \quad (3)$$

With $a=2K_{ML}K_1K_2(M^{2+})(L)_o/K_{D,L}$, (3a)

$b=1+2K_{MA}[M^{2+}](K_{HA}+K_{ex,HA})[H^+]$, (3b)

and $c=2Ab-[A]_t$. (3c)

Here, K_{MA}, K_{ML}, K_1K_2, K_{HA}, and $K_{ex,HA}$ in the above equations denote an ion-pair formation constant (mol^{-1} dm^3 unit) for MA^+ in water, a complex formation one (mol^{-1} dm^3) for ML^{2+} in water, overall ion-pair formation one (mol^{-2} dm^6) between ML^{2+} and $2A^-$ in water, an association one(mol^{-1} dm^3) for an acid, HA, in water, and an extraction one (mol^{-1} dm^3) for HA into the o phase, respectively. These values were either available from a reference [8] or evaluated from those [1,9,10] reported previously. Also, $(j)_t$ shows the total concentration of species with *j*=M(II), L, or A(−I). In the computation, the K_{MA}, K_1K_2, and K_{HA} values were calculated taking account of the ionic strength (*I*) of ionic species in water: $I=(1/2)(4[M^{2+}]+4[ML^{2+}]+[MLA^+]+[MA^+]+[A^-])$ [1,3,4]which was changed into $I=[M^{2+}]+[ML^{2+}]+[A^-]$ fundamentally based on the charge balance equation, $2[M^{2+}]+2[ML^{2+}]+[MLA^+]+[MA^+]=[A^-]$ (see below).

Results and Discussion

On the determination of fundamental extraction data

Compositions of extracted complexes were determined by plotting log $(D/(Pic^-)^2)$ against log $(18C6)_o$ at o=BBz, *o*DCBz, DBE, NB, and Bz [1-4,11]. Here, *D* refers to an experimental distribution ratio of Cd(II) into the o phases and the (Pic^-) and $(18C6)_o$ values were calculated from Equations (3) and (2), respectively. Regression lines of the plots were lines with slope (*a*)=0.81 and intercept (*b*)=3.90 at a correlation coefficient (*R*)=0.976 and *N*=11 for the BBz system, *a*=1.13 and *b*=4.65 at *R*=0.972 and *N*=7 for *o*DCBz with the mixture, *a*=0.76 and *b*=3.39 at *R*=0.944 and *N*=18 for *o*DCBz with the prepared $CdPic_2$, *a*=1.02 and *b*=4.61 at *R*=0.712 and *N*=23 for DBE, *a*=0.55 and *b*=4.80 at *R*=0.992 and *N*=15 for NB, and *a*=0.95 and *b*=4.28 at *R*=0.995 and *N*=15 for Bz. These results indicate that the species composed of M:L:A=1:1:2 are extracted into *o*DCBz from the mixture, DBE, and Bz, where the ratios of Pic(−I)=A were speculated from the charge balance to Cd(II) (see above) [1,3,4,11]. Also, the intercepts *a*,*b* of these systems approximately show their log K_{ex} values, when *a* equals unity [11]. On the other hand, dissociations of the species extracted are suggested for the BBz, NB systems, and *o*DCBz one with the prepared salt. The difference in *a* between the two *o*DCBz systems is caused by that between the experimental log $[18C6]_{oDCBz}$ ranges which were -3.02 to -2.53 for the mixture and -6.73 to -3.09 for the prepared salt. That is, $Cd(18C6)Pic_2$in the *o*DCBz phase dissociates in the lower log [18C6]*o*DCBz range and its ion-pair formation is facilitated in the higher range: the former case causes *a*<*1*, while the latter one does $a \geq 1$. Therefore, further data analyses were performed by assuming the extraction of $CdLPic_2$ into the three diluent-systems [3,4,11]. Next, we determined the K_{ex}, $K_{D,A}$, and $K_{ex\pm}$ values in terms of the following equations:

$$\log K_{ex}^{mix} \approx \log \{K_{ex}+K_{D,A}/([M^{2+}][L]_o[A^-])\} \quad (4)$$

$$\text{and} \approx \log \{K_{ex}+(K_{ex\pm}/[M^{2+}][L]_o[A^-]^2)^{1/2}\} \quad (4a)$$

under the electroneutrality condition of $[MLA^+]_o \approx [A^-]_o$[3,4,11]: see Data Analysis. The non-linear regression analyses of plots [1,3,4,11] of log K_{ex}^{mix} *versus* −log $([M^{2+}][L]_o[A^-])$ {from Equation (4)} and −log $\{([M^{2+}][L]_o)^{1/2}[A^-]\}$ {from Equation (4a)} yielded these values. The latter plots were analyzed here by introducing the K_{ex} values, obtained from the former plots, in Equation (4a). As examples, Figures 1 and 2 show the plots for the extraction of the prepared $CdPic_2$ with 18C6 into *o*DCBz. Lastly, the log $K_{2,org}$, log $K_{D,MLA2}$, and log $K_{ex,ip}$ were evaluated from the following thermodynamic cycles: log $K_{2,org}$=log $(K_{ex}/K_{ex\pm})$, log $K_{D,MLA2}$=log $([MLA_2]_o/[MLA_2])$=log $(K_{ex}K_{D,L}/K_{ML}K_1K_2)$ (see the introduction for the $K_{D,L}$ definition), and log $K_{ex,ip}$=log $(K_1K_2K_{D,MLA2})$=log $(K_{ex}K_{D,L}/K_{ML})$, respectively [3,4]. Table 1 summarizes the thus determined fundamental values.

In Table 1, there were no large differences in log *K* values between the 18C6 extraction from the prepared salt, $CdPic_2·5.7H_2O$, and that from the mixture into *o*DCBz, except for log $K_{D,Pic}$. These facts fundamentally show that the BBz system is comparable with the *o*DCBz, DBE, and NB systems. The difference in log $K_{D,Pic}$ can come from that in I_{oDCBz}, because its value is proportional to the log (I_{org}/I) one in some cases [4] see below for another detailed explanation.

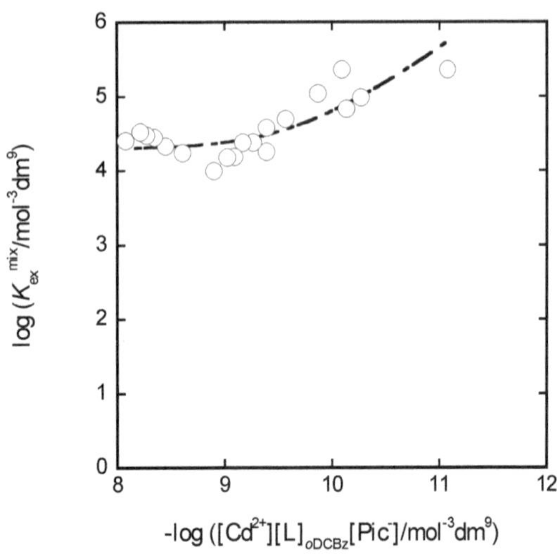

Figure 1: Plot of log K_{ex}^{mix} *versus*–log ([Cd^{2+}][L]$_{oDCBz}$[Pic$^-$])for the Cd(II) extraction with L=18C6 into *o*DCBz.The dotted line shows a regression one based on Equation (4): log K_{ex}^{mix}=log {2.04 × 10^4+(3.22 × 10^{-6}/[Cd^{2+}][L]$_{oDCBz}$[Pic$^-$])} at *R*=0.838.

Diluent[a]	log K_{ex}	log $K_{ex\pm}$	log $K_{D,A}$	log $K_{2,org}$ (I_{org})[b]
*o*DCBz	4.31 ± 0.07 4.21 ± 0.07[e]	-3.03 ± 0.24 -3.26 ± 0.10[e]	-5.49 ± 0.13 -4.24 ± 0.34[e]	7.34 ± 0.25 (2.6 × 10^{-8}) 7.46 ± 0.12[e] (1.6 × 10^{-7})
BBz[e]	4.38 ± 0.08	-2.67 ± 0.37	-4.64 ± 0.20	7.05 ± 0.37 (7.7 × 10^{-8})
DBE	4.25 ± 0.36	-3.26 ± 0.31	-5.07 ± 0.45	7.51 ± 0.47 (3.2 × 10^{-8})
NB	6.14 ± 0.06	2.25 ± 0.14	-2.35 ± 0.05	1.89 ± 0.15 3.3 × 10^{-5})
Diluent[a]	log $K_{D,MLA2}$	log $K_{ex,ip}$	log K_1K_2[c] (I)[b]	log $K_{D,L}$[d]
*o*DCBz	-3.73, -3.74[e]	3.23, 3.13[e]	6.96 (0.012), 6.86[e] (0.025)	-1.13
BBz[e]	-3.55	3.31	6.86 (0.026)	-1.12
DBE	-5.51	1.53	7.04 (0.0058)	-2.77
NB	-1.79	5.19	6.98 (0.011)	-1.00

[a]Abbreviations of the diluents are as follows: *o*-dichlorobenzene (*o*DCBz); bromobenzene (BBz); di butyl ether (DBE); nitrobenzene (NB). [b]Average ionic strength values for the o and w phases: mol dm^{-3} unit. The I_{org} values were calculated from I_{org}≈ (1/2) ([MLA$^+$]$_o$+[A$^-$]$_o$)=[A$^-$]$_o$=[MLA$^+$]$_o$ {>>(1/2)(4[M^{2+}]$_o$+4[ML^{2+}]$_o$+[MA$^+$]$_o$)}. [c]K_1K_2 is defined as [CdLPic$_2$]/[CdL^{2+}][Pic$^-$]^2 and was calculated from the K_1K_2 value at I→ 0 reported before [10]. [d][14]. [e]Values obtained from extraction experiments which were performed by using mixtures of Cd^{2+} with an excess amount of HPic in the presence of SO_4^{2} in the w phase at pH=2.00-2.90.

Table 1: Fundamental data for the CdPic$_2$ extraction by 18C6 into *o*DCBz, BBz, DBE, and NB at 25°C.

The log K_{ex} values were in the order DBE ≤ *o*DCBz ≤ BBz<NB (Table 1). Similar tendencies were observed in the log $K_{ex\pm}$, log $K_{D,Pic}$, and log $K_{D,CdLPic2}$ values. The log $K_{2,org}$ values were in the order DBE ≥ oDCBz ≥ BBz>NB. Although there are differences in I_{org} and amounts of water saturated into the diluents, these three orders seem to fundamentally reflect the diluent's polarity except for the $K_{D,Pic}$ order; the authors were not able to find the data of amount of water into DBE in references (H_2O)$_{NB}$=0.178 mol dm^{-3} at 25°C [12] mole fractions of water in the o phases were 0.0025 for o=*o*DCBz, 0.0021 for BBz, and 0.0149 for NB [13]. The log $K_{D,MLA2}$ order was DBE<*o*DCBz ≤ BBz<NB. This order is strongly reflected into the log $K_{ex,ip}$ one, because the average log K_1K_2 values are in the range of 6.86-7.04 (see Table 1); that is, differences in K_1K_2 among the diluents employed are very small.

Plot of log $K_{ex,ip}$ versus log $K_{D,L}$

Figure 3 shows there-plot of log $K_{ex,ip}$ *versus* log $K_{D,L}$ for the various diluent systems [1,4,14] at L=18C6. A point of the NB extraction system seemed to largely deviate from other points. Its regression line in Figure 3 was log $K_{ex,ip}$=(0.80 ± 0.13) log $K_{D,L}$+(4.13 ± 0.18) at *R*=0.894 and *N*=12 except for the NB system, where the two sets of data were employed for the *o*DCBz system in the calculation. In CdLX$_2$ crystals at L=18C6 with X=Cl, Br, and I [15,16] and at B18C6 with Cl and Br [17] the two X(−I) bind directly to the central Cd(II) and their positions are perpendicular to the mean plane composed of the donor oxygen atoms in L. These complexes have the hexagonal bi pyramidal geometry. One can easily suppose its structure is kept in o phases with less polarities, except for the NB phase; see the caption in Figure 1 about the o phases. From comparing the log $K_{ex,ip}$ *versus* log $K_{D,L}$ plot of the Cd(18C6)Pic$_2$ system with that of Cd(18C6)Br$_2$ one (log $K_{ex,ip}$=1.16log $K_{D,18C6}$+5.27 at *R*=0.993 and *N*=11 [1,4]), we can immediately see that the plots of the both systems largely differ from each other at the *R* and slope values. The *R* value of the Pic$^-$ system suggests that the extracted ion-pair complex Cd(18C6)Pic$_2$ has the higher polar structure, compared with Cd(18C6)Br$_2$, and also its slope (=V_{MLA2}/V_L) suggests that the complex with Pic(−I) has relatively a compact shape [1,4]. Here, *V* denotes a molar volume (cm^3 mol^{-1}) of the species corresponding to the subscript, MLA$_2$ or L. The $V_{CdLPic2}$ value was estimated to be 1.7 × 10^2 cm^3 mol^{-1}from V_{18C6}=214 cm^3 mol^{-1} [14]. This value is in accord with that (=about 200) reported before [1].

Since the intercept is expressed as log K_1K_2+[a term based on interactions between solutes and solvent molecules] [1,14], we can immediately evaluate this latter term from the average log K_1K_2 value among the extraction systems. This interaction term was calculated to be -2.7, when log K_1K_2=6.86 was used. This negative sign suggests a

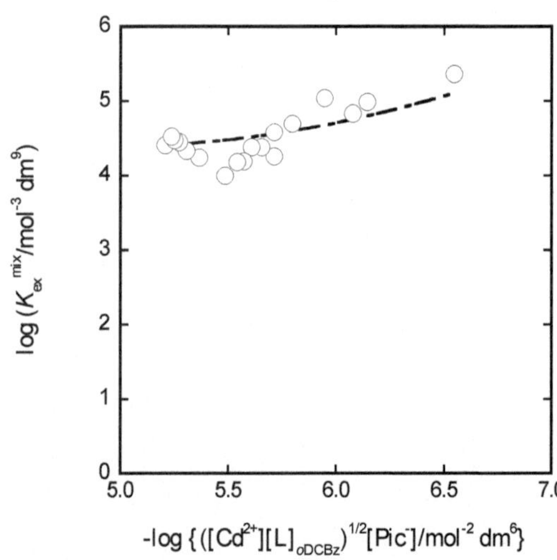

Figure 2: Plot of log K_{ex}^{mix} *versus*–log {([Cd^{2+}][L]$_{oDCBz}$)$^{1/2}$[Pic$^-$]}for the Cd(II) extraction with L=18C6 into *o*DCBz. The dotted line shows a regression one based on Eq. (4a): log K_{ex}^{mix}=log {2.04 × 10^4+(9.40 × 10^{-4}/[Cd^{2+}][L]$_{oDCBz}$[Pic$^-$]^2)$^{1/2}$} at *R*=0.761.

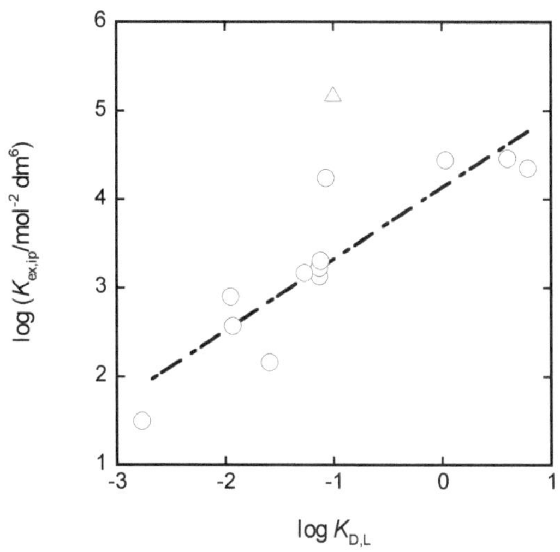

Figure 3: Plot of log $K_{ex,ip}$ *versus* log $K_{D,L}$ at L=18C6. The circles show points of the oDCBz, BBz, DBE, Bz, 1-chlorobutane [1], chloroform [1], 1,2-dichloroethane [1], dichloromethane [1], chlorobenzene [1], toluene [1], and *m*-xylene [1] extraction systems andthedotted line shows a regression one based on log $K_{ex,ip}=(V_{MLA2}/V_L)$log $K_{D,L}$+log K_1K_2+[the constant term] (see text). The triangle shows the point of the NB extraction system.

strong interaction between water molecules and the ion-pair complex Cd(18C6)Pic_2 and/or a weak one between diluent molecules and the complex (see below), compared with the interaction of 18C6 with both the solvent molecules. This suggestion is in agreement with the finding speculated from the *R* value [1]. The same discussion can be satisfied for the plot of log $K_{D,Cd(18C6)Pic2}$ *versus* log $K_{D,18C6}$ [1].

For the $CdPic_2$ extraction by 18C6 into Bz, we easily obtained from the analysis described in the above section log K_{ex}=4.39 ± 0.04 {or 4.2± 0.1 from Equation (4a)}, log $K_{D,Pic}$=-5.1$_8$ ± 0.1$_1$ and log $K_{ex\pm}$=-2.9$_7$ ± 0.1$_8$ {or -2.3$_2$ ± 0.2$_0$ from Equation (4a) without the fixed K_{ex}}. These values also gave log $K_{2,Bz}$ (=log K_{ex}–log $K_{ex\pm}$) =7.4 ± 0.2 at I_{Bz}=8.4 × 10^{-8} mol dm^{-3} and log $K_{D,CdLPic2}$=–3.75{= log K_{ex}-log K_{CdL}-log K_1K_2(average)+log $K_{D,L}$=4.39 +0.05 – 6.92 (at *I*=0.019 mol dm^{-3})-1.27, see above} [1,10,14] at L=18C6. The thus determined log K_{ex} value was used for are-calculationof log $K_{ex,ip}$: log $K_{ex,ip}$ becomes 3.17 (=log K_{ex}+log $K_{D,L}$-log K_{CdL}). In this study, the authors correct the log K_{ex} value from 1.98 reported in ref.1 to 4.39 and also do the logarithmic separation factor [4] log ($K_{ex,Pb}/K_{ex,Cd}$), from 9.73 in ref. 4 to 7.32; the revised log ($K_{ex\pm,Pb}/K_{ex\pm,Cd}$) value became 7.1. By there-determination of K_{ex}, the authors will eliminate a problem on the large deviation [1] of the Bz system from the plot of log $K_{ex,ip}$ versus log $K_{D,18C6}$.

Determination of equilibrium constants for the HPic extraction into DBE

Figure 4 shows a plot of log D_A versus pH for the HPic extraction into DBE. A non-linear regression analysis of this plot gave the log $K_{ex,HPic}$ value from its intercept: $K_{ex,HA}=[HA]_o/[H^+][A^-]$. Here, the relation of

logD_A=log $K_{ex,HA}$-pH-log (1+$K_{HA}10^{-pH}$) (5) in which the K_{HA} value averaged in the experimental *I* range (=0.00043-0.022 mol dm^{-3} at HA=HPic) was introduced, was employed for the analysis [7]. The symbol D_A refers to the distribution ratio of species with A^-. The line seems to be somewhat higher than the plots in the lower range of pH. This deviation may be due to that of the averaged K_{HPic}. The thus obtained value was log $K_{ex,HPic}$=1.06 ± 0.01 at 25°C with the regression line at *R*=0.997. This value was used for the K_{ex} calculation {see Equations (3) and (3b)}. Also, using the average log K_{HPic} value (=0.55 ± 0.03) [8], we evaluated log $K_{D,HPic}$ to be 0.51 ± 0.03. This log $K_{D,HPic}$ value was the smallest of those (=0.89-1.97) [7] reported previously by the authors on the HPic extraction into various diluents.

On the interfacial equilibrium-potential difference at extraction

According to our previous papers [3,18], a difference between log $K_{D,Pic}^S$ and log $K_{D,Pic}$ for the NB system shows the presence of an interfacial potential difference,$\Delta\varphi_{eq}$, at extraction equilibrium. From the relation [1,3,18]

$$\Delta\varphi_{eq}=\Delta\varphi_A^{0\prime}-(2.303RT/F)\log K_{D,A} \qquad (6)$$

for the univalent anion A^-, we can easily calculate the $\Delta\varphi_{eq}$ value. Here, the symbol, $\Delta\varphi_A^{0\prime}$, refers to the formal potential standardized at $\Delta\varphi_{eq}$=0 V based on the $Ph_4As^+BPh_4^-$ assumption [19] and is expressed as the function log $K_{D,A}$(= $F\Delta\varphi_A^{0\prime}/2.303RT$)=$\Delta\varphi_A^{0\prime}$/0.05916 (defined as log $K_{D,A}^S$) at 25°C [1,3,18,20]. Also, *F*, *R*, and *T* show usual meanings. From $\Delta\varphi_{Pic}^{0\prime}$=0.030 V [21] and the log $K_{D,Pic}$ value of the NB extraction system in Table 1, the $\Delta\varphi_{eq}$ value was calculated to be 0.14 V. At the same time, this value indicates that log $K_{D,Cd}$, log $K_{D,CdL}$, and log $K_{D,CdLPic}$ values are expressed as $2(\Delta\varphi_{eq}-\Delta\varphi_{Cd}^{0\prime})/0.05916$, $2(\Delta\varphi_{eq}-\Delta\varphi_{CdL}^{0\prime})/0.05916$,and $(\Delta\varphi_{eq}-\Delta\varphi_{CdLPic}^{0\prime})/0.05916$ at $\Delta\varphi_{eq}$=0.14 V, respectively [3,18]: "+2" of the former two terms refer to the formal charges of Cd^{2+} and CdL^{2+}(see Equation (7a)).The same is essentially true of the other extraction systems, although their $\varphi_{Pic}^{0\prime}$ values have not been determined in many cases. For the 1,2-dichloroethane and dichloromethane systems, the s values are available from references [21,22]. The difference between log $K_{D,Pic}$ values (see Table 1) in the *o*DCBz extraction system can be explained in terms of those between their charge balance equations, namely the difference between the $\Delta\varphi_{eq}$ values and is also relevant to the difference between the I_{oDCBz} values (see above). The experimental condition with the extraction of the prepared $CdPic_2$ by L yields the charge balance equation of

$$2[Cd^{2+}]_o+2[CdL^{2+}]_o+[CdLPic^+]_o=[Pic^-]_o \qquad (7)$$

(or actually $[CdLPic^+]_o\approx[Pic^-]_o$, see above) in the o phase, while the other condition with the extraction of the mixture can yield the equation of $2[Cd^{2+}]_o+2[CdL^{2+}]_o+[CdLPic^+]_o+[H^+]_o+[HL^+]_o=[Pic^-]_o$ (8)

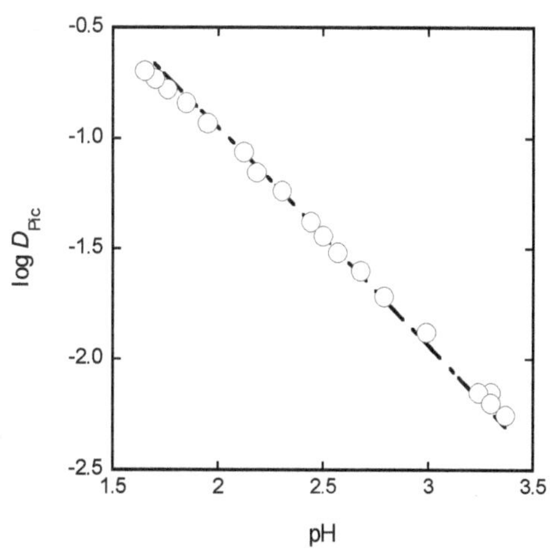

Figure 4: Plot of log D_{Pic} *versus* pH for the HPic extraction into DBE. The dotted line shows a regression one based on Equation (5) (see text).

(or similarly $[\mathrm{CdLPic^+}]_o+[\mathrm{H^+}]_o+[\mathrm{HL^+}]_o \approx [\mathrm{Pic^-}]_o$). Probably, the $[\mathrm{H^+}]_o+[\mathrm{HL^+}]_o$ term in Equation (8) corresponds to an increase in I_o for the extraction of the mixture with L (see Table 1).

Applying Equiation (6) to Equation (7) and the term in Equation (8), we can rewrite them as

$$2[\mathrm{Cd^{2+}}]\exp\{2F(\Delta\varphi_{eq}-\Delta\varphi_{Cd}^{0\prime})/RT\}+2[\mathrm{CdL^{2+}}]\exp\{2F(\Delta\varphi_{eq}-\varphi_{CdL}^{0\prime})/RT\} +[\mathrm{CdLPic^+}]_o=[\mathrm{Pic^-}]\exp\{-F(\Delta\varphi_{eq}\Delta\varphi_{Pic}^{0\prime})/RT\} \quad (7a)$$

And $[\mathrm{H^+}]\exp\{F(\Delta\varphi_{eq}-\Delta\varphi_{H}^{0\prime})/RT\}+[\mathrm{HL^+}]\exp\{F(\Delta\varphi_{eq}-\varphi_{HL}^{0\prime})/RT\}$ (8b)

respectively [18,23]. The symbol $\Delta\varphi_j^{0\prime}$ denotes the standard formal potential for the species j {= Cd(II), CdL(II), CdLPic(I), Pic(−I), H(I), and HL(I)}. Defining $x=\exp(F\Delta\varphi_{eq}/RT)$ in Equation (7a) or Equation (8a) that Equation (8b) was added to the left-hand side of Equation (7a), then we can easily obtain the cubic equations, $ax^3+bx+c=0$, solve them for x, and also for $\Delta\varphi_{eq}$ [18,23]. For example, $a=2[\mathrm{Cd^{2+}}]\exp(-2F\varphi_{Cd}^{0\prime}/RT)+2[\mathrm{CdL^{2+}}]\exp(-2F\varphi_{CdL}^{0\prime}/RT)$, $b=[\mathrm{CdLPic^+}]_o$(see below for the NB system), and $c=-[\mathrm{Pic^-}]\exp(F\Delta\varphi_{Pic}^{0\prime}/RT)$ in the case of Equation (7a).

One can easily see that this $\Delta\varphi_{eq}$ value is commontoall processes of $\mathrm{Cd^{2+}} \rightleftharpoons \mathrm{Cd^{2+}}_o$, $\mathrm{CdL^{2+}} \rightleftharpoons \mathrm{CdL^{2+}}_o$, $\mathrm{CdLPic^+} \rightleftharpoons \mathrm{CdLPic^+}_o$, and $\mathrm{Pic^-} \rightleftharpoons \mathrm{Pic^-}_o$ or to those of $\mathrm{Cd^{2+}} \rightleftharpoons \mathrm{Cd^{2+}}_o$,$\mathrm{CdL^{2+}} \rightleftharpoons \mathrm{CdL^{2+}}_o$, $\mathrm{CdLPic^+} \rightleftharpoons \mathrm{CdLPic^+}_o$, $\mathrm{H^+} \rightleftharpoons \mathrm{H^+}_o$, $\mathrm{HL^+} \rightleftharpoons \mathrm{HL^+}_o$, and $\mathrm{Pic^-} \rightleftharpoons \mathrm{Pic^-}_o$[18] at o=*o*DCBz. Thus, these $\Delta\varphi_{eq}$ values solved by the above charge-balance equations are equivalent to those determined from log $K_{D,Pic}$ [18] and accordingly both the $\Delta\varphi_{eq}$ values can reflect materials employed in extraction experiments. Therefore, the upper log $K_{D,Pic}$ value (= -5.49) in Table 1 is different from the lower value (=-4.2$_4$) and the increase in the lower value from Equation (7) can be caused by Equation (8b). Unfortunately, we were not able to find out the basic $\Delta\varphi_j^{0\prime}$ data {j=Cd(II), Pic(−I), or H(I)} in references with respect to the *o*DCBz system.

Estimate of ion-transfer formal potential of CdLPic$^+$ at the water/NB interface

On the basis of the above discussion, the $\Delta\varphi_{CdLPic}^{0\prime}$ value at the water/NB interface was evaluated at L=18C6. The $[\mathrm{CdLPic^+}]_{NB}$ values were evaluated from the relation $[\mathrm{MLA^+}]_o=(K_{ex}^{mix}-K_{ex})[\mathrm{M^{2+}}][\mathrm{L}]_o[\mathrm{A^-}]^2$ (>0) and then the [CdLPic$^+$] ones were calculated from $[\mathrm{MLA^+}]=K_1[\mathrm{ML^{2+}}][\mathrm{A^-}]=(K_1K_{ML}/K_{D,L})[\mathrm{M^{2+}}][\mathrm{L}]_o[\mathrm{A^-}]$. Here, K_1 was evaluated from the relation $\log K_1\{=\log K_1^0-\log(y_{II+}y_-/y_{I+})\}\approx \log K_1^0-\log y_{II+}$ with $y_-\approx y_{I+}$. The symbols, K_1^0,y_{II+}, y_-, and y_{I+}, denote K_1 at $I\rightarrow 0$, an activity coefficient of $\mathrm{ML^{2+}}$, that of $\mathrm{A^-}$, and that of $\mathrm{MLA^+}$ in water at 25°C, respectively [10,24]; $K_1^0=3.4\times 10^4$ mol^{-1} dm^3. From the above calculation, -0.4_3(± 0.15 at N=8) was obtainedas the average log $K_{D,CdLPic}$ value. Hence, using the equation $(\Delta\varphi_{eq}-\Delta\varphi_{CdLPic}^{0\prime})/0.05916=\log K_{D,CdLPic}$ and $\Delta\varphi_{eq}=0.14$ V (see above), we immediately was able to calculate $\Delta\varphi_{CdLPic}^{0\prime}$ to be 0.17 V at 25°C. This value is much larger than that ($\Delta\varphi_{Pic}^{0\prime}=0.003$ V) [21] of Pic$^-$ in the ion transfer at the water/NB interface. This fact suggests the stronger interaction of CdLPic$^+$ than Pic$^-$ with water molecules, as described above.

Conclusion

By adding the log $K_{ex,ip}$ values for the three diluent systems except for the NB one, the plot of log $K_{ex,ip}$ *versus* log $K_{D,L}$ for the present $\mathrm{CdPic_2}$-18C6 extraction system was re-analyzed in the number of data comparable with the plot of the $\mathrm{CdBr_2}$-18C6 extraction one. Although the extraction data for the Bz system were revised, the result for the interaction between $\mathrm{Cd(18C6)^{2+}(Pic^-)_2}$ and water molecules, obtained from the former plot, was in agreement with the previously-reported result [1]. Also, the presence of $\Delta\varphi_{eq}$ in the Cd(II) extraction system was shown from the $K_{D,Pic}$ determination, as well as that in the M(I) extraction ones with L. Moreover, the authors were able to estimate the ion-transfer formal potential of Cd(18C6)Pic$^+$ at the water/NB interface.

References

1. KudoY, Horiuchi N, KatsutaS, TakedaY (2013) Extraction of cadmium bromide and picrate by 18-crown-6 ether into various less-polar diluents: Analysis of overall extraction equilibria based on their component equilibria with formation of their ion pairs in water. J Mol Liq 177: 257-266.
2. Takeda Y, Tanaka A (1986) Thermodynamic study on solvent extraction of alkali metal picrates with 18-crown-6 into $\mathrm{CHCl_3}$. Bull Chem Soc Jpn 59: 733-735.
3. Kudo Y, Takeuchi T (2014) On the interfacial potential differences for the extraction of alkaline-earth metal picrates by 18-crown-6 ether derivatives into nitrobenzene. J Thermodyn Catal 5: 6.
4. Kudo Y, Takahashi Y, Numako C, Katsuta S (2014) Extraction of lead picrate by 18-crown-6 ether into various diluents: Examples of sub-analysis of overall extraction equilibrium based on component equilibria. J Mol Liq 194: 121-129.
5. VaňuraP, Makrlík E (1998) Extraction of ^{133}Ba with bis[undecahydro-7,8-undecaborato(2−)]cobaltate(1−) in the presence of dibenzo-18-crown-6. Collect Czech Chem Commun 63:338-346.
6. Nový P, Vaňura P, Makrlík E (1998) Extraction of ^{85}Sr by the nitrobenzene solution of bis-1,2-dicarbollylcobaltate in the presence of dibenzo-18-crown-6. J Radioanal Nucl Chem 231: 65-68.
7. Kudo Y, Takahashi Y, Katsuta S (2013) Distribution of picric acid into various diluents. Journal of Chemistry 2013.
8. Kortüm G, Vogel W, Andrussow K (1961) Dissociation Constants of Organic Acids an Aqueous Solution, IUPAC: Section of Analytical Chemistry Commission on Electrochemical Data, Butterworths, London, 454.
9. Kudo Y, Takeuchi S, Kobayashi Y, Katsuta S, Takeda Y, et al. (2007) Potentiometric determination of ion-pair formation constants for cadmium, calcium salts, and cadmium-18-crown-6 ether derivative complexes with a sulfate ion in water. J Chem Eng Data 52: 1747-1752.
10. Kudo Y, Kobayashi Y, Katsuta S, Takeda Y (2009) Ion-pair formation constants of cadmium-18-crown-6 ether complex ions with picrate and chloride ions in water and distribution of its picrate ion-pair complex into benzene. J Mol Liq 146: 60-65.
11. Kudo Y, Takahashi Y, Katsuta S, Takeda Y (2013) Extraction of several divalent metal picrates by 18-crown-6 ether derivatives into benzene: A refinement of methods for analyzing extraction equilibria. Int J Org Chem 5: 90-102.
12. Iwachido T, Minami M, Kimura M, Sadakane A, Kawasaki M, et al. (1980) The coextraction of water into alkali and alkaline earth metal salts of 2,2',4,4'6,6'-hexanitrodinitroamine in the presence of several crown ethers and cryptands. Bull Chem Soc Jpn 53: 703-708.
13. IwachidoT, MinamiM, NaitoH, TôeiK (1982)The extractability of potassium as picrates (KPic and K(L)Pic, L: 18-crown-6) and its relationship to the properties of solvents. Bull Chem Soc Jpn 55: 2378-2382.
14. Takeda Y (2002) Extraction of alkali metal picrates with 18-crown-6, benzo-18-crown-6, and dibenzo-18-crown-6into various organic solvents. Elucidation of fundamental equilibria governing the extraction-ability and -selectivity. Analyt Chem 51: 515-525.
15. Paige CR, Richardson MF (1983) Complexes with macrocyclic ligands. II. The crystal structure of $\mathrm{HgCl_2}$·18-crown-6 and $\mathrm{CdCl_2}$·18-crown-6, and the FAB mass spectrum of $\mathrm{HgCl_2}$·18-crown-6. Canadian Journal of Chemistry 62: 332-335.
16. Hazell A (1988) Complexes of cadmium(II) bromide and Iodide with 18-crown-6 ether. Acta Cryst C44: 88-92.
17. Pickardt J, Wiese S, Wischlinski P (1997) Reactions of cadmium and mercury(II) halides and pseudohalides with crown ethers: crystal structures of complexes of benzo-18-crown-6 with $\mathrm{CdX_2}$ and $\mathrm{HgX_2}$ (X = Cl, Br) and of dibenzo-18-crown-6 with $\mathrm{Hg(CN)_2}$. Z Naturforsch C 52b: 1296-1300.

18. Kudo Y, Katsuta S (2015) On an expression of extraction constants without interfacial equilibrium-potential differences for the extraction of univalent and divalent metal picrates by crown ethers into 1,2-dichloroethane and nitrobenzene. American Journal of Analytical Chemistry 6: 350-363.

19. Rais J (1971) Individual extraction constants of univalent ions in the system water-nitrobenzene. Collect Czech Chem Commun 36: 3253-3262.

20. Kudo Y, Ogihara M, Katsuta S, Takeda Y (2014) An electrochemical understanding of extraction of silver picrate by benzo-3*m*-crown-*m* ethers (*m* = 5, 6) into 1,2-dichloroethane and dichloromethane. American Journal of Analytical Chemistry 5: 433-444.

21. Kudo Y, Harashima K, Hiyoshi K, Takagi J, Katsuta S, et al. (2011) Extraction of some univalent salts into 1,2-dichloroethane and nitrobenzene: Analysis of overall extraction equilibrium based on elucidating ion-pair formation and evaluation of standard potentials for ion transfer at the interfaces between their diluents and water. Anal Sci 27: 913-919.

22. Danil de Namor AF, Traboulssi R, Salazar FF, Dianderasde Acosta V, Fernández de Vizcardo Y, et al. (1989) Transfer and partition free energies of 1:1 electrolytes in the water-dichlorometane solvent system at 298.15 K. J Chem Soc Faraday Trans 185: 2705-2712.

23. Takeda Y, Ezaki T, Kudo Y, Matsuda H (1995) Distribution study on electroneutral and protonated amino acids between water and nitrobenzene. Determination of the standard ion-transfer potentials of protonated amino acids. Bull Chem Soc Jpn 68: 787-790.

24. Kudo Y (2013) Potentiometric Determination of Ion-Pair Formation Constants of Crown Ether-Complex Ions with some Pairing Anions in Water Using Commercial Ion-Selective Electrodes in Electrochemistry, Khalid MAA, InTech, Rijeka, Croatia.

Thermodynamic Models: Application to the Brines of Chotts in Algerian North-Eastern Sahara

Abdelaziz B* and Messaoud H

Laboratoire Géologie du Sahara, Université de Kasdi Merbah, BP 511, Ouargla 30000, Algérie

Abstract

The Brines of chotts in Algerian Northeastern Sahara are highly concentrated solutions of mineral salts with a salinity that far exceeds that of the sea water (120 to 440 g/l). The first interpretation of data allows to specify the chemical facies of these brines form their chemical compositions. They have mainly a very marked sodic chlorinated character. Databases of the thermodynamic model PHREEQC v3 were used to determine the saturated state of the brines with respect to principal chemical species. They are more oversaturated with calcite than with gypsum. The saturated state compared to the halite is no longer reached. Indeed, we use the thermodynamic model PHREEQC v3 to calculate the mineral/solutions equilibria on the basis of different assumptions. The test of these assumptions shows that each one has its limits to simulate and control the thermodynamic behaviour of the brines.

Keywords: Thermodynamic model; Brines; Chotts; Northeastern Sahara

Introduction

The most famous works devoted to the study of the mineral / solutions interactions and the geochemical evolutions of salt and the brines in lakes' ecosystems, chotts and sebkhas, are the works of Ref. [1-15] about the geochemistry and mineralogy of brine in chott Merouane located in Algerian northeastern Sahara, and the work of Ref. [16] about the application of the Pitzer equation in the geochemistry of brine in chott El Jerid located in southern Tunisia. The thermodynamic models used in this work are applicable for brines where the ionic strength is less than 1 mol/l, which is not the case for our region where the ionic strength of the brines is greater than 1 mol/l. Therefore, a simulation of the geochemical evolution of brine using mathematical models will be necessary to verify the applicability of these models. For these purposes, databases of the thermodynamic model PHREEQC v3 (version 3.0 recently issued by the US Geological Survey) [17] have been used to determine the activity coefficients and the saturated state of the brines from the main chemical species. Considering the high concentration of the dissolved salt in the brines, the treatment of the results of the chemical analysis acquired during this research requires significant work. In order to evaluate the quality of the analysis results, a simple method based on the calculation of the ionic balance [18], is used. The results are therefore corrected by accepting an error of imbalances of an ionic balance lower than 10%. Nevertheless, this ionic balance provides information on the analytical quality only for the dominant chemical species in the solution. The chemical nature of the brines is influenced by the chemical properties of the drainage water, the severe climatic conditions, and by the human impact that can influence or accelerate the natural processes.

Materials and Methods

Study site

The Algerian Northeastern Sahara is a vast depression (-37 to 300 m above the sea level), partly occupied by the Gran Erg Oriental. It is defined by the boundary: the Saharan Atlas in the North; Saharan ridge and the Grand Erg Occidental in the West; the plateaux of Tademaït and Tinrhert in the South. This depression is opened in the North East over the Gulf of Gabes. It extends between 34° or 35° lat. N (of Chott) and 28° or 29° lat. N (the Tademaït) and for the longitude, 3°E (the Saharan Ridge) at 10° or 11°E (Figure 1). Chotts and sebkhas constitute the lowest level. The chott that corresponds to the edge of the sebkha is composed of siliceous and/or gypseous sand and gypseous soils with gypseous crust at surface and subsurface. The shallow is characterized by the presence of a permanent water table, very little deep (1 to 5 m) in the chott, which is flush with the surface in the center of the sebkha.

Sampling protocol for solutions

The choice of sampling site is based on:

- their situation in the chott (1 to 3 m from the periphery)
- their contact with discharges (drainage water, for example)
- their depth with respect to the surface (without contact with the air)

Water samples were collected and analyzed per the protocol developed by A. Probst at the Laboratory of Mechanisms and Transfers in Geology (LMTG) in Toulouse [19]. The polypropylene sampling bottles were cleaned with hydrochloric acid (HCl) and then rinsed with distilled water and dried at 45°C in an oven. During sampling, the sample bottles were rinsed with a little sample before filling. The amount required for the analysis of the anions is taken without acidification. But, the dosage of the cations requires acidification beforehand. All samples once collected are stored in a cooler, and within a maximum of 12 hours placed in a fridge at 5°C. All manipulations are systematically performed with sterilized gloves (Figure 1).

Thermodynamic modelling

In the development of geochemical models, we can distinguish, with Ref. [20], the models that calculate the speciation of aqueous solutions and minerals. The theory of the thermodynamic equilibria

***Corresponding author:** Abdelaziz B, Laboratoire Géologie du Sahara, Université de Kasdi Merbah, BP 511, Ouargla 30000, Algérie
E-mail: bellaoueuraz@gmail.com

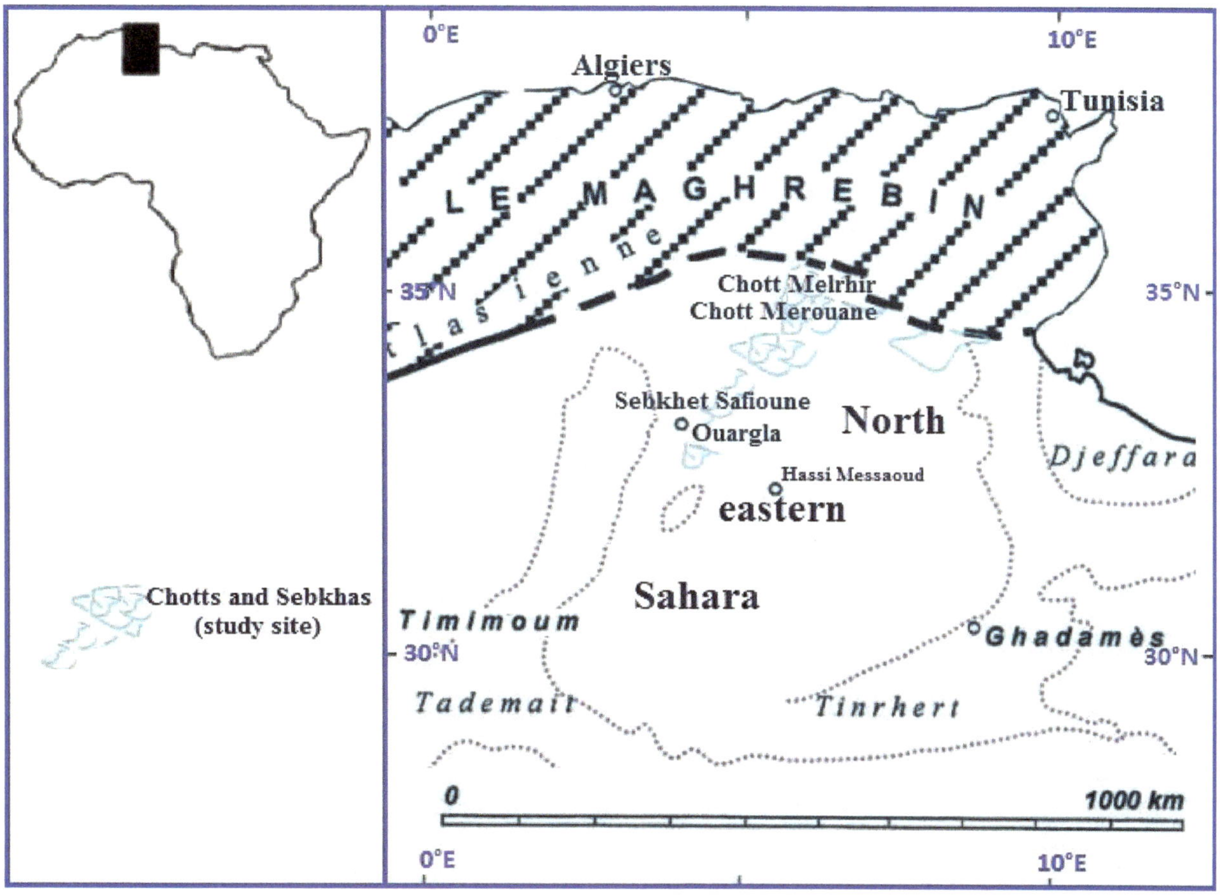

Figure 1: Study site localisation. Document opened: Geological Map of Northwestern Africa with 1/5.000.000 [29].

is the most adequate concept to facilitate the identification of several variables in the determination of the relationship between the solution and the minerals [21]. The work of Debye and Hückel [22] is a major opening in the evolution of the thermodynamics of the electrolyte solutions. Indeed, we use the model PHREEQC v3 to calculate the activity coefficients and the mineral/solutions equilibria based on various assumptions:

1) The ion association model based on the extended Debye-Hückel function: the extent expression of the Debye-Hückel law is written for union i solution (Equation 1):

$$\text{Log } \gamma i = -(A\ zi2\ \sqrt{I})/(1+ai^{\circ}\ B\sqrt{I})+b\gamma I \quad (1)$$

With: $b\gamma$=0.2 or 0.3: constant of error from the Debye-Hückel law; ai°: the hard-core diameter, which is specific to each aqueous species i (in cm); z: the electric charge of i; I: ionic strength; A=0.5092 and B=0.3283 in t=25°C: universal constants of the Debye-Hückel law [22].

2) The Ion-specific interaction model based on Pitzer's equations where the term $b\gamma I$ is replaced by the sum of: $\Sigma\beta ij$ mj; where mj: the molality of the ion j; βij: the characteristic coefficient of species pair i and j [23-25].

3) The Specific-ion Interaction Theory (SIT): The expression for a SIT parameter is the same as for a Pitzer's parameter [26].

In the case of the brines, the activity coefficients that take into account the influence of the concentration of various species in the solution are easily calculated by the model of the specific ion interaction drawn up by Ref. [27]. In the conception of this model, ions remain unassociated and the ionic strength I of the solution is directly calculated from the molality of salt (Equation 2).

$$I=0.5\ \Sigma\ [Ei]\ zi2 \quad (2)$$

Where [Ei] is the molality of the constituent Ei in the solution and zi is the electric charge of Ei. The saturated state of the solution with respect to the minerals is expressed by the Saturation Index (SI) defined as the logarithm of the ratio between the Ionic Activity Product (IAP) and the solubility constant of the solid phase considered (Ks) (Equation 3):

$$SI=\text{Log}\ (\Omega)=\text{Log}\ (IAP/Ks) \quad (3)$$

The sub-saturation of the solution relative to the mineral is characterized by a ratio less than 1 and therefore the mineral tends to dissolve. The saturation (or thermodynamic equilibrium) is characterized by a ratio equal to 1 and no change in the net quantity of the dissolved mineral precipitated and is therefore expected. Oversaturation is characterized by a ratio greater than 1, and consequently, the mineral tends to precipitate.

Results and Discussion

Chemical composition and facies

The Chemical analysis of the different ions allows to specify the geochemical facies of the brines. They generally include simple ions in the solution, complex aqueous species and mineral species. The Table 1 summarizes the concentration of the dissolved salt in these brines under natural conditions. At average, they typically contain both chloride (6.04 mol/kgw) as the dominant anion and sodium (5.36 mol/kgw) as the dominant cation. The other dissolved elements have less significant concentrations relative to sodium and chloride; they are at average about 0.53 mol/kgw for sulphate and magnesium, 0.26 mol/kgw for calcium and of about 0.41 mol/kgw for potassium. The sulphate concentration is associated with evaporated formations (gypseous soils with gypseous crust). The total concentration of the dissolved elements can sometimes approach many hundreds of grams per liter (400 g/l); however, the water of the sea contains only 19 g/l of chlorides and an average of salinity of about 35 g/l. The Chemical facies of these brines are mainly chlorinated-sodium and rarely sodic-sulfated: (Equation 4 and 5)

$$r\%Cl>r\%SO_4>r\%HCO_3=r\%Na>r\%Mg>r\%Ca>r\%K \quad (4)$$

$$r\%Cl>r\%SO_4>r\%HCO_3=r\%Na>r\%Ca>r\%Mg>r\%K \quad (5)$$

r%: quantity in reaction in percent as milliequivalents per liter (meq/l).

Saturated state

Indexes of saturation in calcite, gypsum, anhydrite, sylvite and thenardite were calculated by using three thermodynamic models (Table 2). Regarding the effect of the ionic strength on the solubility of the mineral salt, there is a slight increase in the solubility of calcite, gypsum and anhydrite because the ionic strengths are less than 6 mol/l. Apart from this ionic strength, the solubility of these minerals will augment with the increase of the ionic strength. Contrary to the halite and thenardite, the increase in the solubility of these two mineral salts is directly proportional with the ionic strength, except for the sylvite which has more than an ionic strength superior to 6 mol/l where the solubility becomes inversely proportional with the increase of the ionic strength. The results of the calculation of the mineral saturation indexes presented by the above three thermodynamic models are almost identical. The Thermodynamic modelling indicates that the brines are substantially saturated with respect to calcite than with respect to gypsum. Indeed, the calcite precipitation is controlled by the concentration of calcium ions. In the case of low concentrations of Ca, the precipitation kinetics of calcite control the concentration of the brines in Ca. In addition to this, when the gypsum dissolution is important, the precipitation of the kinetics of calcite has no significant influence on the concentrations of Ca. The state of thermodynamic equilibrium observed with halite is not reached. For the other minerals, the brines are more or less under-saturated with respect to sylvite and thenardite. The databases of three thermodynamic models used in the calculations show more or less significant differences between the measured values of the mineral saturation indexes. Taking into account the limited applicability of the

Sample	t (°C)	pH	Density	Ca (mol/kgw)	Mg	Na	K	Cl	SO_4	HCO_3
1	29.3	8.3	1.37	0.04	0.21	4.48	1.11	4.12	0.68	0.001
2	30.8	8.43	1.36	0.04	0.2	4.38	1.16	4.24	0.57	0.001
3	24.1	7.68	1.46	0.03	0.53	5.3	1.24	5.49	0.88	0.003
4	30.1	7.79	1.5	0.11	0.95	5.24	0.83	5.25	1.37	0.01
5	13.5	8.07	1.65	0.68	0.47	8.17	0.06	10.42	0.51	0.02
6	14	8.13	1.67	0.74	0.84	8.23	0.12	10.12	0.69	0.03
7	16	8.38	1.17	0.15	0.52	1.39	0.13	1.79	0.52	0.02
8	23.7	7.28	1.46	0.04	0.65	4.75	1.88	6.37	0.3	0.003
9	20.3	7.25	1.39	0.02	1.2	3.85	0.55	5.81	0.46	0.003
10	19.4	7.88	1.2	0.07	0.08	2.61	0.54	3.08	0.08	/
11	21.4	7.36	1.37	0.02	0.56	4.57	0.59	5.61	0.3	0.003
12	22.4	7.25	1.44	0.04	0.64	4.82	1.52	6.32	0.32	0.003
13	17.4	7.45	1.47	0.04	0.83	4.98	1.8	6.38	0.35	0.003
14	16.5	7.6	1.27	0.03	0.32	3.22	0.55	3.96	0.24	0.005
15	16	7.85	1.54	0.01	0.96	6.86	0.13	9.03	0.37	0.02
16	18	7.95	1.52	0.01	0.82	6.74	0.11	8.84	0.3	0.01
17	19	8.04	1.44	0.51	0.5	5.98	0.04	5.75	0.65	0.01
18	19.7	8.26	1.69	0.89	0.48	8.39	0.12	10.2	0.82	0.02
19	19.8	8.12	1.79	1	0.84	8.94	0.12	9.48	1.87	0.02
20	19.9	8.74	1.14	0.33	0.17	1.31	0.02	1.31	0.42	0.02
21	20	8.76	1.24	0.45	0.46	2	0.27	1.53	1.04	0.02
22	19	9.4	1.06	0.03	0.08	0.77	0.01	0.53	0.24	0.004
23	13.7	8.3	1.54	0.1	0.58	7.79	0.09	7.18	0.92	0.02
24	14.3	8.36	1.3	0.07	0.44	4.06	0.05	3.07	0.86	0.02
25	14.4	9.04	1.34	0.07	0.97	3.33	0.29	3.14	1.2	0.03
26	20	8.22	1.15	0.21	0.27	1.48	0.06	1.53	0.41	0.02
27	14	7.89	1.46	0.55	0.44	6.15	0.07	7.35	0.22	0.01
28	15	7.81	1.5	0.56	0.53	6.74	0.07	8.13	0.22	0.01
29	20.8	8.02	1.5	0.68	0.46	6.62	0.07	8.02	0.26	0.01
30	21.2	8.07	1.52	0.76	0.51	6.88	0.08	8.39	0.23	0.01
31	15.7	7.85	1.46	0.11	0.48	7.01	0.07	7.08	0.27	0.01
32	15.9	8.01	1.46	0.11	0.48	7.1	0.07	7.32	0.15	0.01
33	19	7.35	1.56	0.53	0.64	7.25	0.11	9.13	0.28	0.01

(mol/kgw: mole per kilogram of water)

Table 1: Chemical compositions of brines in Algerian Northeastern Sahara (2009-2010).

Sample	I (mol/l)	SI Calcite			SI Gypsum			SI Anhydrite			SI Halite			SI Sylvite			SI Thenardite	
		IA	M-Pitzer	SIT	IA	M-Pitzer	SIT	IA	M-Pitzer	SIT	IA	M-Pitzer	SIT	IA	M-Pitzer	SIT	M-Pitzer	SIT
1	4.93	0.85	0.79	0.48	0.7	0.63	0.40	0.61	0.52	0.43	-0.42	-0.48	-0.71	-0.78	-0.67	-0.71	-0.48	-0.67
2	4.81	0.83	0.81	0.56	0.61	0.57	0.35	0.53	0.47	0.39	-0.41	-0.47	-0.69	-0.76	-0.64	-0.68	-0.55	-0.72
3	6.08	0.47	0.81	0.42	0.72	0.78	0.36	0.62	0.68	0.37	-0.08	-0.08	-0.48	-0.51	-0.35	-0.49	-0.13	-0.53
4	7.02	1.56	1.75	1.52	1.29	1.39	0.88	1.25	1.38	0.95	-0.12	-0.02	-0.61	-0.78	-0.57	-0.79	0.08	-0.49
5	7.68	2.62	2.76	2.69	1.92	2.69	1.38	1.85	2.78	1.47	0.83	1.08	0.36	-1.3	-0.96	-1.23	0.60	-0.56
6	8.26	2.64	2.12	2.87	2.04	2.73	1.46	1.98	2.87	1.55	0.87	1.17	0.30	-1.03	-0.65	-1	0.77	-0.46
7	3.46	2.39	2.36	2.20	1.23	1.06	1.09	0.89	0.66	0.89	-1.47	-1.49	-1.60	-1.96	-1.91	-1.87	-1.62	-1.47
8	5.83	-0.01	0.82	0.35	0.37	0.55	0.22	0.29	0.47	0.26	0.03	0.00	-0.40	-0.23	-0.05	-0.23	-0.62	-0.90
9	6.08	-0.76	0.17	-0.37	0.05	0.37	-0.05	-0.12	0.22	-0.09	-0.15	-0.08	-0.61	-0.78	-0.58	-0.82	-0.57	-0.82
10	2.97	/	/	/	0.24	0.17	0.23	-0.03	-0.15	0.10	-0.89	-0.96	-1.00	-1.11	-1.07	-1.01	-1.80	-1.65
11	5.21	-0.23	0.51	-0.03	0.01	0.25	-0.08	-0.15	0.09	-0.10	-0.15	-0.15	-0.48	-0.79	-0.64	-0.76	-0.67	-0.84
12	5.78	-0.08	0.77	0.25	0.37	0.57	0.22	0.26	0.47	0.23	0.02	0.01	-0.39	-0.32	-0.14	-0.31	-0.57	-0.84
13	6.15	-0.03	0.80	0.38	0.46	0.63	0.26	0.32	0.48	0.23	0.10	0.10	-0.37	-0.17	0.03	-0.18	-0.50	-0.83
14	3.98	0.57	0.94	0.49	0.21	0.23	0.14	-0.06	-0.09	0.01	-0.61	-0.66	-0.81	-0.95	-0.89	-0.89	-1.18	-1.09
15	6.92	0.02	0.42	0.65	-0.46	0.36	-0.62	-0.56	0.36	-0.59	0.54	0.71	0.04	-1.12	-0.83	-1.12	0.22	-0.51
16	6.64	0.15	0.45	0.73	-0.54	0.25	-0.65	-0.63	0.24	-0.61	0.48	0.63	0.02	-1.20	-0.93	-1.19	0.07	-0.58
17	6.4	2.38	2.81	2.30	1.78	1.91	1.38	1.63	1.77	1.37	0.07	0.16	-0.28	-1.90	-1.68	-1.82	-0.12	-0.59
18	8.16	2.85	/	2.83	2.20	2.78	1.56	2.20	2.96	1.72	0.84	1.10	0.33	-1.07	-0.73	-1.01	0.76	-0.42
19	9.34	2.57	/	2.62	2.53	2.86	1.76	2.54	3.10	1.89	0.82	1.22	0.18	-1.11	-0.72	-1.11	1.20	-0.21
20	2.79	2.91	2.89	2.66	1.52	1.31	1.37	1.22	0.94	1.2	-1.67	-1.70	-1.74	-2.84	-2.80	-2.72	-1.72	-1.57
21	4.68	3.07	2.85	2.74	1.88	1.55	1.6	1.60	1.21	1.45	-1.39	-1.40	-1.55	-1.77	-1.69	-1.67	-1.17	-1.08
22	1.28	1.81	1.68	1.6	0.49	0.34	0.41	0.14	-0.08	0.2	-2.34	-2.38	-2.37	-3.55	-3.54	-3.44	-2.10	-1.94
23	6.95	2.11	2.09	2.11	1.24	1.62	0.76	1.08	1.53	0.73	0.38	0.54	-0.06	-1.38	-1.13	-1.34	0.52	-0.33
24	4.87	2.19	2.13	1.88	1.03	0.87	0.72	0.74	0.52	0.56	-0.65	-0.67	-0.89	-2.11	-2.04	-2.03	-0.55	-0.63
25	5.68	2.43	1.93	2.12	1.10	0.96	0.76	0.80	0.63	0.6	-0.71	-0.67	-1.03	-1.32	-1.21	-1.3	-0.57	-0.7
26	2.89	2.53	2.57	2.31	1.29	1.12	1.17	0.99	0.76	1	-1.54	-1.57	-1.63	-2.37	-2.33	-2.26	-1.62	-1.48
27	6.31	2.08	2.84	2.14	1.44	1.81	1.09	1.26	1.67	1.07	0.31	0.40	-0.04	-1.44	-1.21	-1.35	-0.43	-1
28	6.71	1.98	2.79	2.15	1.43	1.91	1.06	1.30	1.85	1.08	0.48	0.60	0.08	-1.38	-1.11	-1.29	-0.25	-0.99
29	6.75	2.32	2.88	2.41	1.56	1.99	1.16	1.48	1.99	1.24	0.45	0.56	0.06	-1.50	-1.22	-1.4	-0.22	-0.97
30	7	2.34	2.79	2.47	1.57	2.02	1.13	1.51	2.07	1.23	0.54	0.67	0.12	-1.41	-1.12	-1.31	-0.19	-1.03
31	6.04	1.48	2.10	1.69	0.76	1.21	0.52	0.61	1.08	0.51	0.30	0.39	-0.05	-1.50	-1.27	-1.43	-0.16	-0.72
32	6	1.65	2.19	1.91	0.51	1.01	0.32	0.37	0.90	0.32	0.35	0.43	-0.01	-1.48	-1.24	-1.4	-0.38	-0.94
33	7.14	1.59	2.72	1.92	1.49	2.13	1.12	1.43	2.19	1.22	0.63	0.79	0.17	-1.21	-0.90	-1.14	0.07	-0.83

IA: Ion-association method; M-Pitzer: Pitzer specific-ion-interaction model; SIT: Specific ion Interaction Theory; I: Ionic strength.

Table 2: Ionic strength and saturation indexes (SI) of mineral salts in brines of chotts and sebkhas in Algerian Northeastern Sahara (2009-2010). calculated by three thermodynamic models: IA. M-Pitzer and SIT.

ionic association model for the highly concentrated solutions [28]. It is estimated: 21, 24, 9 and 12% of aberantes values under or overestimated SI Calcite, SI Gypsum, SI Anhydrite and SI Sylvite calculated by the SIT method against respectively 45, 24, 57 and 63% calculated by the Pitzer's method except for halite, where it is estimated that the values of aberantes are 0% under or overestimated SI Halite calculated by the method of Pitzer against 84% calculated by the SIT method. This leads us to deduce that the calculation with the specific ion interaction theory SIT is satisfactory.

The saturation indexes of thenardite are not calculated by the ionic association model; however, in the activity diagrams, the equilibrium state simulated by different thermodynamic models shows that:

i) For the saturation with respect to the gypsum and with respect to calcite, points are organized, more or less in accordance with the straight equilibrium solubility product of gypsum Ksp (gypsum) and of calcite Ksp (calcite) drawn up by the method of Pitzer against that of the other methods. When saturation is reached, the concentrations of Ca and SO_4 are conditioned by their solubility products (Figures 2 and 3) [29-31].

ii) For the saturation with respect to halite, it will not be well reached if the concentrations of sodium and chloride increase; therefore, their precipitation can only be obtained for higher concentration factors (Figure 4).

The equilibrium straight line of the solubility product of gypsum:

Log Ksp (gypsum)=log aCa+log aSO_4+2log aw=-4.58

The equilibrium straight line of the solubility product of calcite:

Log Ksp (calcite)=log aCa+log aCO_3=-8.36

The equilibrium straight line of the solubility product of halite:

Log Ksp (halite)=log aNa+log aCl=1.56

Conclusion

The results obtained during this research project show that the brines of chotts and sebkhas in the Algerian Northeastern Sahara are characterized by a high concentration of about 290 g/l at average, a state of oversaturation with respect to calcite and more or less with respect to gypsum. Speciation and thermodynamic modelling of these brines differ according to the used model. This model depends mainly on the thermodynamic database with which the calculations are performed.

Taking into account the limited applicability of the ionic association model for highly concentrated solutions, the specific ionic

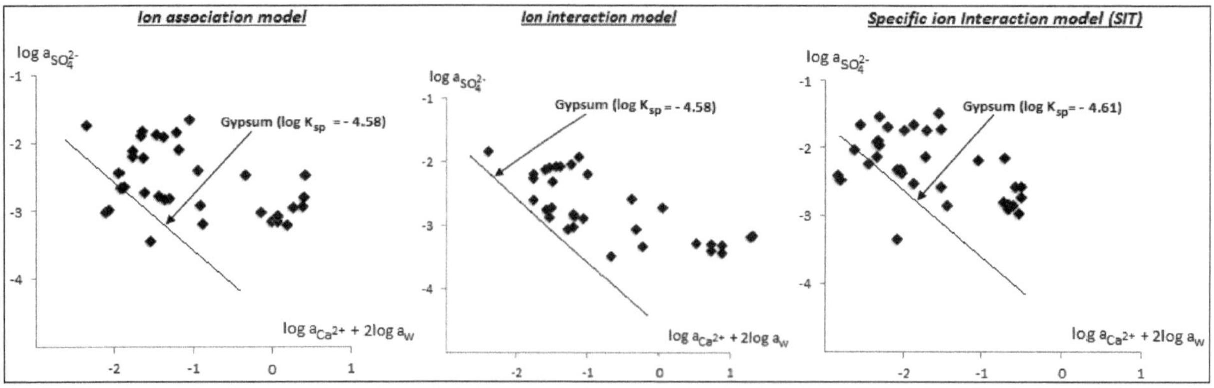

Figure 2: Brine equilibrium relative to gypsum.

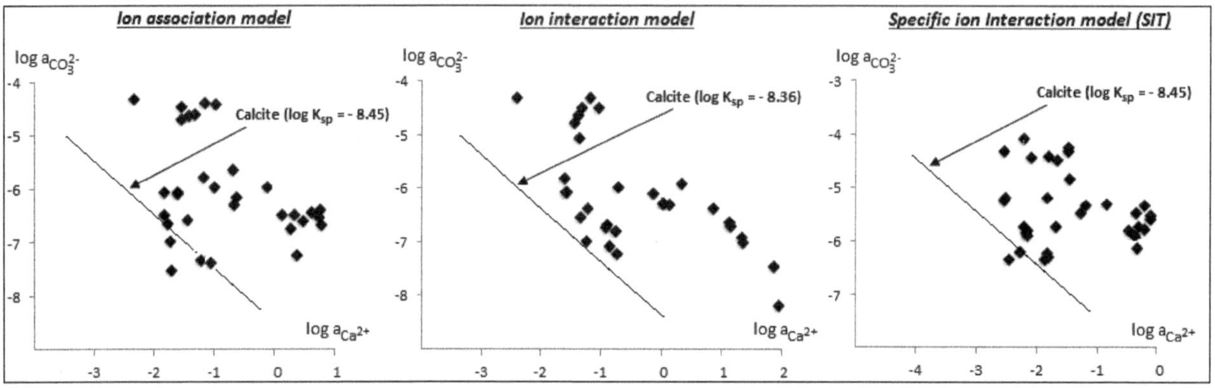

Figure 3: Brine equilibrium relative to calcite.

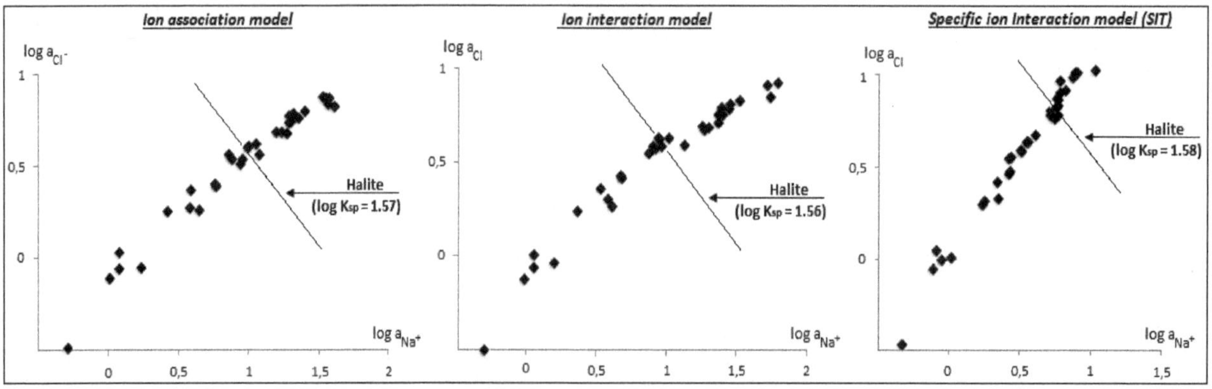

Figure 4: Brine equilibrium relative to halite.

interaction model drawn up by Pitzer was compared with that of the Specific ion Interaction Theory. It is generally observed that the SIT model involves more parameters and it is more or less satisfactory, in its field of application, compared to that of Pitzer. However, the diagrams of the activity of halite show that there are points over the equilibrium straight line, which is important, i.e., none of these three models will work. This paragraph highlights the fact that there is no more ideal modelling tool. Each one has its advantages and limitations, but still modelling remains a valuable tool to simulate and control the thermodynamic behaviour of the brines. Therefore, the modeler should primarily select the best model relative to the study site, depending on the data available, the objectives of the modelling, etc.

References

1. Jones BF, Deocampo DM (2004) Geochemistry of Saline Lakes. US Geological Survey, Reston, VA, USA 5: 393-424.
2. Moshood NT (2004) Evolution of saline waters and brines in the Benue-Trough, Nigeria. Geochemistry 19: 1355-1365.
3. Risacher F, Alonso H, Salazar C (2003) The origin of brines and salts in Chilean salars: a hydrochemical review 63: 249-293.

4. Risacher F, Clement A (2001) A computer program for the simulation of evaporation of natural waters to high concentration. Computers and Geosciences 27: 191-201.

5. Garcés I (2000) Geochemistry of Huasco salar, Chile. Origin of solutes and brine evolution. In: Geertman RM (ed.), Proceedings of the 8th World Salt Symposium 2: 1159-1160.

6. Bein A, Dutton LR (1993) Origin, distribution and movement of brine in the Permian Basin (USA): a model for displacement of connate brine. Geol Soc Am Bull 105: 695-707.

7. Weare JH (1987) Models of mineral solubility in concentrated brines with application to feild observations. Rev Mineral 17: 143-176.

8. Brantley SL, Crerar DA, Moller NE, Weare JH (1984) Geochemistry of a modern marine evaporate: Bocana De Virrila, Peru. J Sediment Petrol 54: 447-462.

9. Frape SK, Fritz P (1982) The chemistry and isotopic composition of saline groundwaters from Sudbury Basin, Ontario. Can J Earth Sci 19: 645-661.

10. Land LS, Prezbindowski DR (1981) The origin and evolution of saline formation water, Lower Cretaceous carbonates, South-central Texas, USA. J Hydrol 54: 51-74.

11. Rettig SL, Jones BF, Risacher F (1980) Geochemical evolution of brines in the salar of Uyuni, Bolivia. Chemical Geology 30: 57-79.

12. Eugster HP, Jones BF (1979) Behavior of major solutes during closed-basin brine evolution. Am J Sci 279: 609-631.

13. Clayton R, Friedman I, Graf D, Mayeda P, Meets W, et al. (1966) The origin of saline formation waters: isotopic composition. J Geophys Res 71: 3869-3882.

14. Hacini M, Oelkers EH, Kherici N (2006) Retrieval and interpretation of precipitation rates generated from the composition of the Merouane Chott ephemeral lake. J Geochem Explor 88: 284-287.

15. Hacini M, Oelkers EH, Kherici N (2008) Mineral precipitation rates during the complete evaporation of the Merouane Chott ephemeral lake. Geochim Cosmochim Acta 72: 1583-1597.

16. Gueddari M, Monnin C, Perret D, Fritz, B, Tardy Y (1983) Geochemistry of brines of the Chott El Jerid in southern Tunisia–Application of Pitzer's equations. Chem Geol 39: 165-178.

17. Parkhurst DL, Appelo CAJ (2013) Description of input and examples for PHREEQC version 3-A computer program for speciation, batch-reaction, one-dimensional transport, and inverse geochemical calculations. US Geological Survey Techniques and Methods A43: 497.

18. Freeze RA, Cherry JA (1979) Groundwater. Prentice-Hall Inc., Englewood Cliffs, NY, USA, p: 604.

19. Savary V (2002) Rapport de fin de contrat. LMTG-Université Paul Sabatier, Toulouse.

20. Coudrain-Ribstein A (1988) Transport d'éléments et réactions chimiques dans les aquifères. Th. d'Etat, Univ. L. Pasteur, Strasbourg, p: 381.

21. Stumm W, Morgan J (1970) Aquatic Chemistry. An introduction emphasizing chemical equilibria in natural waters. Wiley Interscience, p: 583.

22. Helgeson HC, Kirkham DH (1974) Theoretical prediction of the thermodynamic behavior of aqueous electrolytes at high pressures and temperatures. II. Debye-Hückel parameters for activity coefficients and relative partial molar properties. Am J Sci 274: 1199-1261.

23. Millero FJ (2001) The Physical Chemistry of Natural Waters. Wiley, NY, p: 654.

24. Plummer LN, Parkhurst DL, Fleming GW, Dunkle SA (1988) A computer program incorporating Pitzer's equations for calculation of geochemical reactions in brines: US Geological Survey Water-Resources Investigations Report 88: 4153.

25. Harvie CE, Weare JH, Hardie LA, Eugster HP (1980) Evaporation of seawater: calculated mineral sequences. Science 208: 498-500.

26. Grenthe I, Plyasunov AV, Kastriot S (1997) Estimations of medium effects on thermodynamic data. In: Aquatic chemistry. Chapter IX: Paris, OECD Nuclear Energy Agency, pp: 325-426.

27. Pitzer KS (1973) Thermodynamics of electrolytes -1. Theoretical basis and general equations. Journal of Physical Chemistry 77: 268-277.

28. Michard G (1989) Equilibres chimiques dans les eaux naturelles. Paris, p: 357.

29. Bronner G, Fabre J (1978) Carte géologique du Nord-Ouest de l'Afrique au 1/5.000.000. SNED ed. Alger.

30. Lewis GN, Randall M (1921) The activity coefficient of strong electrolytes. J Am Chem Soc 43: 1112-1154.

31. Wolf M, Breitkopf O, Puk R (1989) Solubility of calcite in different electrolytes at temperatures between 10 and 60°C and at CO_2 partial pressures of about 1 kPa. Chemical Geology 76: 291-301.

The Kinetics of the *Cis*-to-*Trans* Thermal Isomerization of 4-Anilino-4'-Nitroazobenzene are Highly Influenced by Solvent Polarity

Spurgeon Smith and Fadi Bou-Abdallah*

Department of Chemistry, State University of New York at Potsdam, Potsdam, New York, USA

Abstract

The thermally stable *trans* isomers of donor-acceptor substituted azobenzene dyes can be easily transformed into unstable *cis*-form by light. Although azobenzene and its derivatives have been extensively investigated over the years, many aspects of the isomerization reaction kinetics are still unclear or controversial at best. Here, we investigate the kinetics of the thermal *cis-trans* isomerization of 4-anilino-4'-nitroazobenzene (4A4NAB) in six solvents of different polarities by means of flash photolysis. Using the transition state theory, we were able to determine a number of thermodynamic quantities including the standard enthalpy of activation $\Delta^{\ddagger}H^{o}$, the standard entropy of activation $\Delta^{\ddagger}S^{o}$, the Gibbs free energy of activation $\Delta^{\ddagger}G^{o}$, and the quasi-equilibrium constant $K^{\ddagger}$ of the isomerization reaction. Higher rates of isomerization and lower activation energies were observed in polar media, consistent with the formation of a charged transition state complex and the presence of solute−solvent interactions such as intermolecular hydrogen bonding. The data suggest that the *cis-trans* thermal isomerization of 4A4NAB is highly influenced by solvent polarity and the isomerization mechanism is likely to proceed via rotation about the azo double bond.

Keywords: Isomerization; Photo-isomerization; Electron-donor; Cis-trans

Introduction

Azobenzenes are common organic dyes that have been extensively studied both experimentally and theoretically owing to their potential applications in material science, medicinal chemistry, molecular switches and other devices [1-15]. They are photo-reactive molecules that undergo reversible photo-isomerization from the more stable *trans*-isomer to the less stable *cis*-isomer. Despite decades of research, the mechanism of the *cis-trans* isomerization is not unequivocally established and depends largely on the experimental conditions (i.e., solvent polarity and viscosity), as well as molecular substitutions on the basic azobenzene molecule (i.e., azobenzene derivatives), both of which have a dramatic effect on the spectroscopic properties of the molecule and the kinetics of isomerization.

Photo-isomerization mechanisms

The 4-Anilino-4'-nitroazobenzene (4A4NAB) is a "push-pull" type molecule that can undergo a resonance structure due to the presence of an electron-donor (the anilino group) and an electron-acceptor (the nitro group) at the *p* and *p'* position of the conjugated aromatic system in the same molecule (Figure 1).

Two possible mechanisms have been proposed for the reversible photo-isomerization of azobenzenes. The inversion mechanism is suggested to proceed *via* a linear transition state in which the N=N double bond remains intact, whereas the rotation mechanism is proposed to occur *via* a twisted transition state in which the N=N π-bond is broken. Upon photo-excitation of the *trans* form, an electron is excited from its ground-state (S_0) orbital to its first singlet excited state (S_1) or second singlet excited state (S_2) in which the electron retains its spin under an n-π* excitation condition or a π-π* excitation condition, respectively[1,2]. Azo groups are reported to photo-isomerize via two distinct mechanisms: the π-π* transition with an out-of-plane rotation mechanism in which the nitrogen-nitrogen π bond is ruptured heterolytically and a dipolar transition state is involved (Scheme 1), or the n-π* electronic transition with an inversion of one sp^2 hybridized nitrogen atom through an sp hybridized linear transition state in which the double bond is retained (Scheme 2) [2-6]. The rate of isomerization for the inversion mechanism is relatively rapid and mostly independent of the polarity of the medium or the electronic nature of substituents on the azobenzene, but the rate for the rotation mechanism increases rapidly with increasing solvent polarity [3].

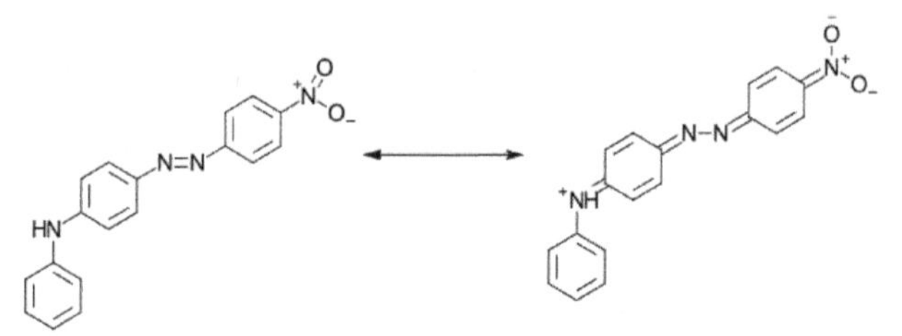

Figure 1: Resonance structures of *trans*-4-anilino-4'-nitroazobenzene.

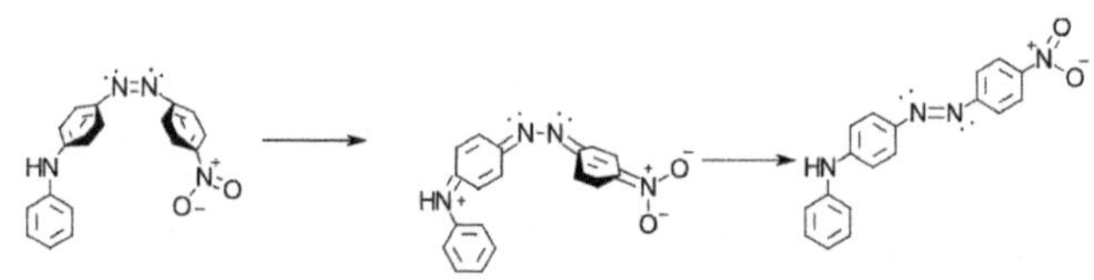

Scheme 1: Rotation mechanism for the dark cis-to-trans isomerization reaction of azobenzene following a photo-conversion of the trans-stable isomer to the cis-unstable isomer. Note that the electronic transitions discussed above pertain to the trans to cis reaction and not the cis to trans reaction.

***Corresponding author:** Fadi Bou-Abdallah, Department of Chemistry, State University of New York at Potsdam, Potsdam, New York, USA
E-mail: bouabdf@potsdam.edu

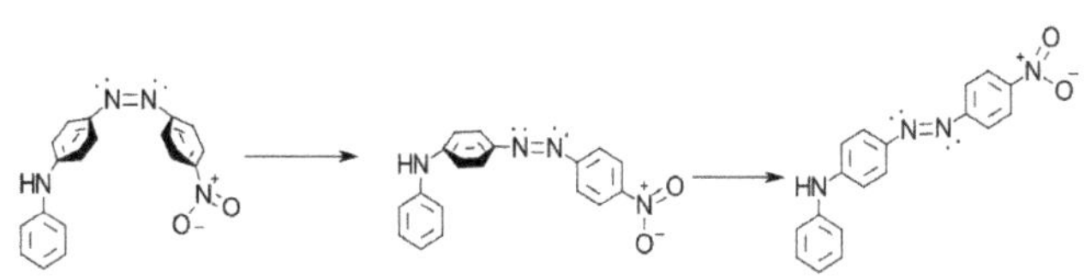

Scheme 2: Inversion mechanism for the dark *cis*-to-*trans* isomerization reaction of azobenzene following a photo-conversion of the *trans*-stable isomer to the *cis*-unstable isomer.

A conceptual diagram of the reaction coordinate for the *cis-trans* isomerization of disperse orange 1 is shown in Figure 2 whereby the absorption of a photon of visible light causes the promotion of the stable ground-state form (*trans*-disperse orange 1) into an electronically excited state. The excited molecule quickly relaxes back to the ground electronic state of the *cis*-form. The ground-state *cis* molecule will then convert back to the *trans* isomer with the activation energy for the isomerization process shown as E_a in Figure 2.

Taking advantage of the unique properties of azobenzenes, this study explores important aspects of photochemistry and the effect of six different solvents (cyclohexane, toluene, benzene, tetrahydrofuran, acetone, and 3-pentanol) of different polarity on the kinetics of the *cis-trans* isomerization of 4-anilino-4'-nitroazobenzene using a camera flash and a UV/Vis spectrophotometer. The camera flash initiates a photochemical isomerization of 4A4NAB (*trans* → *cis*) in which the stable *trans*-isomer is initially promoted to an electronically excited species (step 1 in Figure 2) followed by a decay to the ground-state *cis* isomer of lower energy level (step 2 in Figure 2). Subsequently, the *cis* isomer undergoes a thermal isomerization to the more stable *trans* isomer (step 3 in Figure 2). The *trans* → *cis* → *trans* isomerization process is completely reversible and could be carried for several cycles. Due to the steric hindrance of the two phenyl rings, the *cis* isomer decays down to the more energetically favored *trans* isomer (Figure 2). In this experiment the absorbance of a disperse orange 1 solution will be monitored at a wavelength where only the *trans* isomer absorbs. Upon exposure to the camera flash, a certain portion of the *trans* molecules will be converted to the *cis* isomer and the absorbance will drop before it starts to rise back to its original value as the *cis* molecules convert back to the *trans* isomer. This absorbance value is labeled A_∞ (the absorbance at infinite time). A typical experimental scan of a flash photolysis experiment is shown in Figure 3.

Integrated rate law and the transition state theory thermodynamic equations

The overall isomerization reaction can be written according to the following scheme:

$$Trans \xrightarrow{hv} Trans^{*}_{excited} \rightarrow Cis \xrightarrow{k} Trans$$

The rate of the thermal reaction *Cis* → *Trans* can be written as:

$$rate = \frac{d[trans]}{dt} = -\frac{d[cis]}{dt} = k[cis] \qquad (1)$$

where k is the first order rate constant, and [*trans*] and [*cis*] are the concentrations of the *cis* and *trans* isomers, respectively at any time *t*. Knowing that the total number of molecules is constant and assuming that at t=∞ all the molecules have converted back to the *trans* isomer gives:

$$[trans]_\infty=[trans]+[cis] \qquad (2)$$

so that

$$[cis]=[trans]_\infty-[trans] \qquad (3)$$

Substituting equation (3) into equation (1) yields:

$$rate=k\,([trans]_\infty-[trans]) \qquad (4)$$

which upon integration gives:

$$\ln([trans]_\infty-[trans]) = -kt + \ln([trans]_\infty-[trans]_0) \qquad (5)$$

where $[trans]_0$ is the concentration of the *trans* isomer at time *t*=0. Because the concentration of the *trans* isomer at any time *t* is proportional to the absorbance of the solution (via. Beer's Lambert

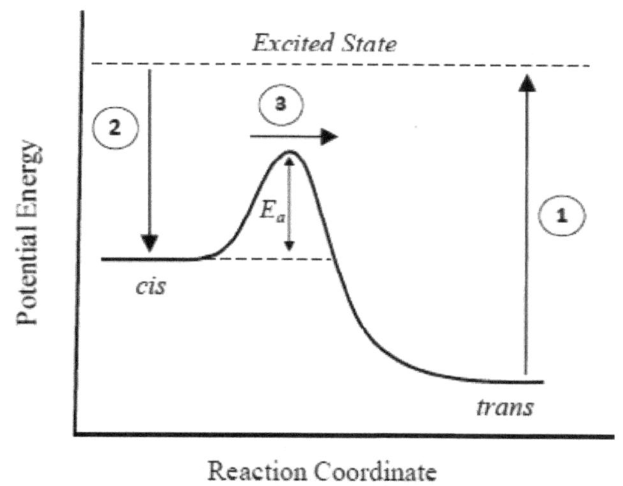

Figure 2: Energy profile for the photo-induced *trans*→ *cis*→ *trans* isomerization of disperse orange 1. Initially, the camera flash excites the *trans* molecules (step 1) which then relax to the ground electronic state (step 2) to form the *cis* isomer before converting back to the *trans* isomer (step 3).

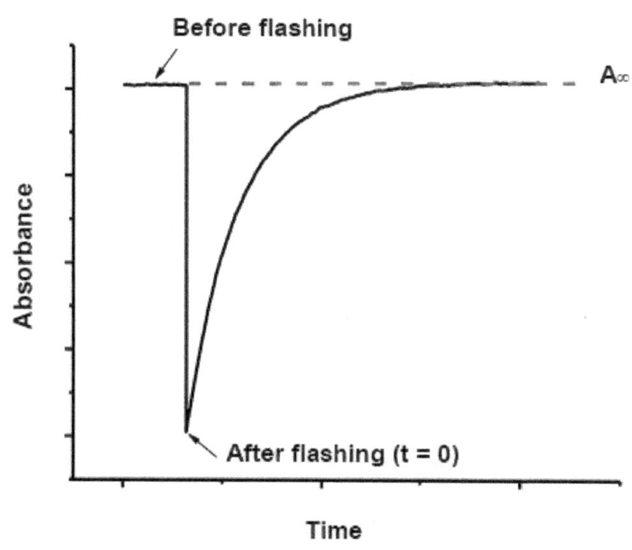

Figure 3: Typical change in the absorbance of *trans*-disperse orange 1 as a function of time in a flash photolysis experiment. The time *t*=0 is taken to be the time immediately after the flash occurs.

law), the integrated rate law may then be written as:

$$\ln(A_\infty - A) = -kt + \ln(A_\infty - A_0) \quad (6)$$

where A, A_0, and A_∞ are the absorbances at time t, t_0, and t_∞, respectively.

The dependence of the rate constants on temperature can be used to determine the reaction's activation energy and the Arrhenius constant A according to Arrhenius equation 7:

$$\ln k = -\frac{E_a}{R}\frac{1}{T} + \ln A \quad (7)$$

Using the transition state theory, the activation energy Ea and Arrhenius constant A can be used to determine the standard enthalpy of activation $\Delta^\ddagger H^0$, the standard entropy of activation $\Delta^\ddagger S^0$, the Gibbs free energy of activation $\Delta^\ddagger G^0$, and the quasi-equilibrium constant $K^\ddagger$ using the following equations:

$$E_a = \Delta^\ddagger H^0 + RT \text{ (for a condensed phase reaction)} \quad (8)$$

$$A = \frac{k_B Te}{h}\exp\left(\frac{\Delta^\ddagger S^o}{R}\right) \quad (9)$$

$$\Delta^\ddagger G^0 = \Delta^\ddagger H^0 - T\Delta^\ddagger S^0 \quad (10)$$

$$\Delta^\ddagger G^0 = -RT\ln K^\ddagger \quad (11)$$

where k_B is the Boltzmann constant, e is the Euler number 2.718..., and h is Planck's constant.

Experimental

All solvents employed in this study were used without further purification. The dye, 4-anilino-4'-nitroazobenzene was purchased from MP Biomedicals Inc. under the common name disperse orange 1, and is typically 15-25% by mass 4-anilino-4'-nitroazobenzene. There was no detailed safety information available on the manufacturer's website other than to wear suitable protective clothing, gloves and eye/face protection. In order to minimize degradation, the dye was freshly prepared in the appropriate solvent using a pinch of the red coloured dye and the stock solution was either wrapped in aluminium foil or stored in an amber bottle away from direct light. Any salts that did not dissolve in the organic solvents were decanted before the dye solutions, whose absorbance values ranged between 0.4 and 0.5, were used. In case of higher absorbance values, the dye solutions were appropriately diluted in the corresponding solvent. The kinetic data were essentially identical using a freshly prepared dye solution or one that was stored overnight in the dark for complete equilibration. However, it is important to keep all solutions tightly sealed to prevent evaporation.

The source of the solvents used in this study and their purity are as follow: cyclohexane (Acros Organics), ACS Grade, >99% (GC grade); toluene (Fischer Scientific), ACS certified, >99.5%; benzene (Mallinckrodt Pharmaceuticals), ACS Certified; tetrahydrofurane (EMD Millipore Corp.), ACS/USP Grade, >99.9%; acetone (Pharmco-Aaper), ACS/USP grade, >99.9%; 3-pentanol (Sigma-Aldrich), ≥ 98% (GC grade). A Varian Cary 50 Bio UV-Visible spectrophotometer with a Northwest TC 125 Quantum temperature control unit (ΔT ± 0.1°C) were used. The excitation light flash was produced by a Quantaray MS-1 camera flash.

The solvent-dependent maximum wavelength at which the *trans* isomer $\pi^* \leftarrow \pi$ transition occurs is shown in Table 1. Once the required temperature is reached, the sample was quickly removed from the spectrometer cell holder and flashed with the camera flash (by placing the flash against the quartz cuvette and depressing the flash button) and then immediately placed back into the spectrometer. The sample absorbance was monitored over time until its initial absorbance value before excitation was reached.

Results and Discussion

The absorbance spectra of the 4A4NAB solutions in six different solvents are shown in Figure 4. The data show a bathochromic shift in 3-pentanol compared to cyclohexane. This pronounced red-shift of ~60 nm can be rationalized by hydrogen bond formation between the solute and the solvent and a more stable zwitterionic form of the dye in the more polar solvents. Additionally, the polarity of the solvent may affect the N-N-C angle and the energy of electronic transitions of the N=N bond [16,17].

Figure 5 shows the typical absorbance change *vs.* time when a 4A4NAB dye solution is excited with a camera flash. Before excitation, the steady absorbance indicates a stable solution (mostly molecules in the *trans*-form) at equilibrium. Once excited, a sudden drop in absorbance is observed followed by a return to the initial values as the 4A4NAB molecules transition from the unstable *cis* orientation to the more stable *trans* orientation. For each solvent, the experiment was repeated two to three times to ensure reproducibility, and at each temperature the same dye solution was excited/flashed two to three times.

The rate constants for the isomerization of 4A4NAB in cyclohexane, toluene, benzene, tetrahydrofurane, acetone and 3-pentanol are determined from a plot of $\ln(A_\infty - A)$ *vs.* time as shown in Figure 6. In

Solvent	λ/nm	Relative Polarity	1st order Rate Constant ($k \times 10^3$/s^{-1})	Log(k)
Cyclohexane	431	0.006	1.38 ± 0.09	- 2.86 ± 0.07
Toluene	445	0.099	1.89 ± 0.08	- 2.72 ± 0.04
Benzene	445	0.111	1.87 ± 0.05	- 2.73 ± 0.03
THF	463	0.207	25.9 ± 0.8	- 1.38 ± 0.02
Acetone	468	0.355	95.5 ± 4.1	- 1.02 ± 0.04
3-Pentanol	485	0.463	449.2 ± 2.8	- 0.35 ± 0.01

Table 1: First-order rate constants for the *cis-trans* isomerization of 4A4NAB at 25°C in six different solvents. The relative polarity and maximum wavelength of each solvent are also listed. The reported uncertainties are standard errors determined from replicate measurements.

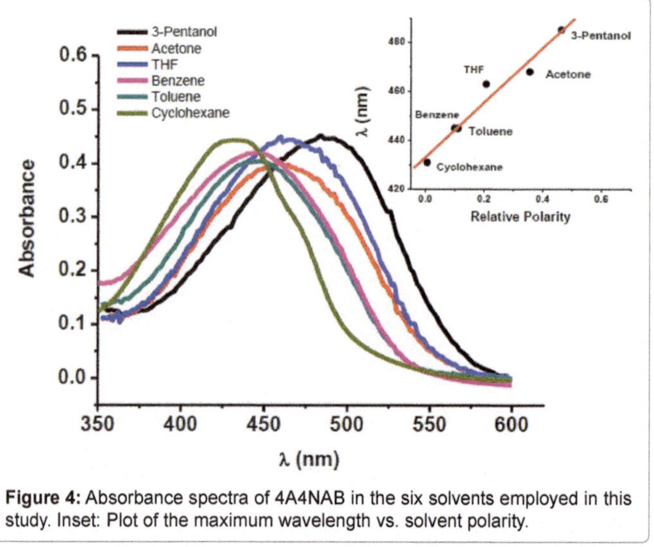

Figure 4: Absorbance spectra of 4A4NAB in the six solvents employed in this study. Inset: Plot of the maximum wavelength vs. solvent polarity.

each case, the reaction followed the expected first order kinetics with the slope of the line equals to *k*. Clearly, the results of Figure 6 show a pronounced effect of solvent polarity on the rate of isomerization.

Table 1 summarizes the kinetics data obtained with six solvents whose relative polarity values are defined by the solvent polarity parameter as proposed by Reichardt [18]. The data in Table 1 show that the rate constants for the isomerization reactions increase significantly with an increase in the solvent relative polarity.

A plot of log(k) *vs.* relative polarity (Figure 7) exhibits a linear relationship consistent with earlier literature data for similar isomerization reactions involving other azobenzenes [18-22]. The increase in the rate constants of isomerization is accompanied by a decrease in activation energy which is typically the case with *para*-donor, *para*-acceptor type azobenzenes [21,22].

To determine the activation energy of the reaction, a plot of the natural logarithm of the isomerization rate versus temperature was created (Figure 8). The slopes and Y-intercepts of these plots represent the activation energy $E^{\ddagger}_a$ and the Arrhenius frequency factor *A* for the isomerization reactions, respectively. All other thermodynamic parameters including the standard enthalpy of activation $\Delta^{\ddagger}H^0$, the standard entropy of activation $\Delta^{\ddagger}S^0$, the Gibbs free energy of activation

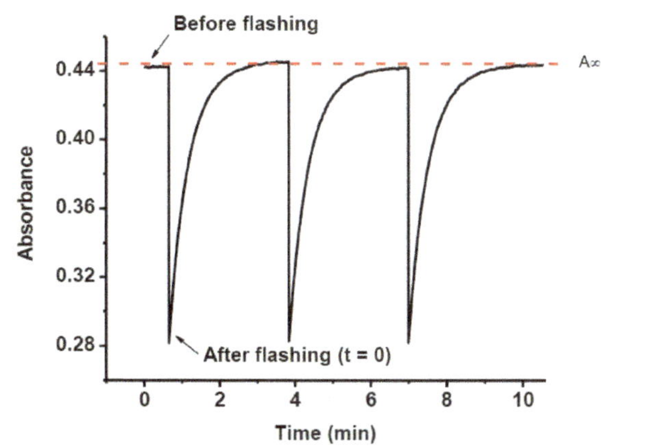

Figure 5: Plot of the change in absorbance vs. time for 4A4NAB in THF at 30°C before and after 3 flashes are applied to the same solution.

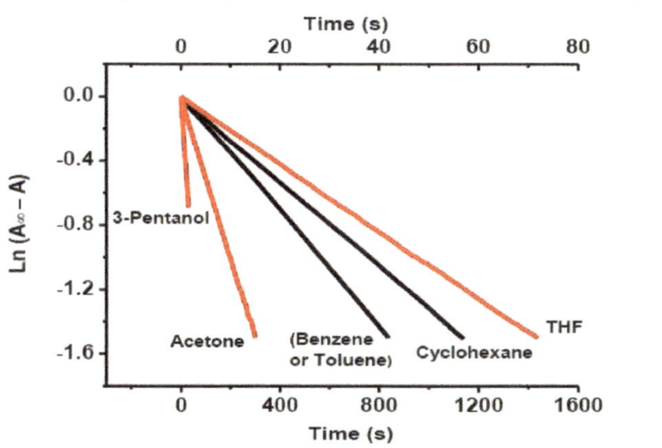

Figure 6: Plot of ln(A_{∞}-A) vs. time for 4A4NAB in different solvents at 25°C. The red lines for 3-pentanol, acetone, and tetrahydrofurane follow the time scale of the upper X-axis. Only the initial linear portion of these plots is shown.

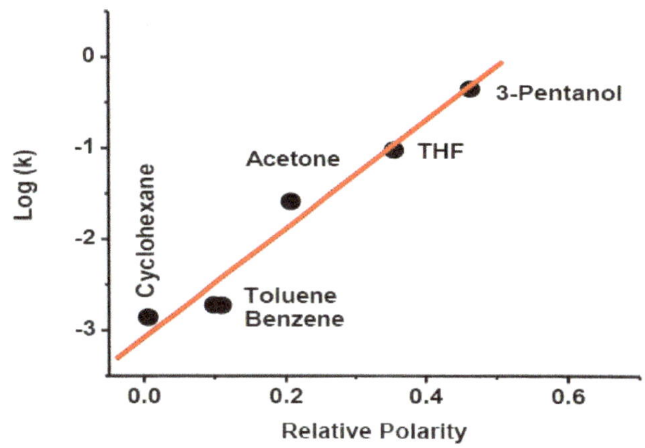

Figure 7: Plot of the log of the rate constants vs. the solvents relative polarity for 4A4NAB. The values of the rates constants were taken at 25°C.

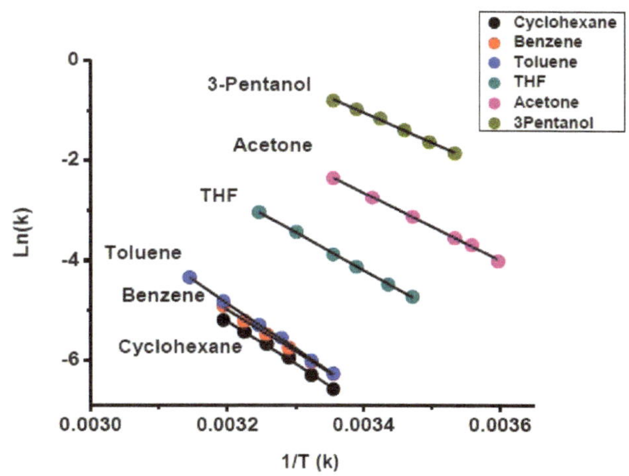

Figure 8: Arrhenius plots for the isomerization of 4A4NAB in six different solvents.

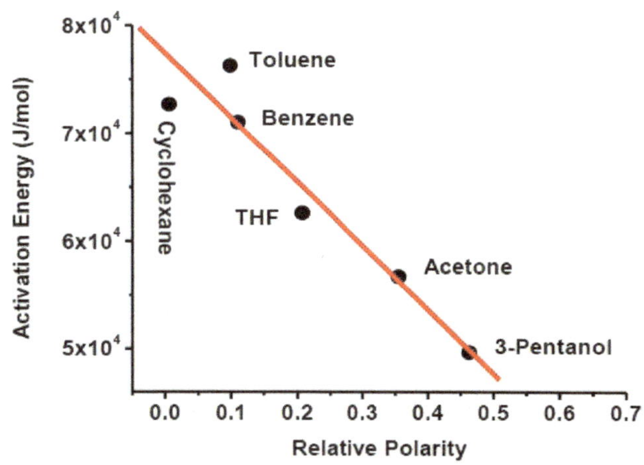

Figure 9: Plot of the activation energy of 4A4NAB as a function of solvent relative polarity.

Solvent	Relative Polarity	Isomerization Energy ($E^{\ddagger}a \times 10^{-3}$ mol/J)	Arrhenius Constant, Ln(A)	Enthalpy of isomerization ($\Delta^{\ddagger}H^{0} \times 10^{-3}$ mol/J)	Entropy of isomerization ($\Delta^{\ddagger}S^{0}$/K.mol/J)	Gibbs Free Energy $\Delta\,^{\ddagger}G^{0} \times 10^{-3}$ mol/J)	Equilibrium Constant $K^{\ddagger}$
Cyclohexane	0.006	72.7 ± 0.2	22.7 ± 0.7	70.2 ±1.9	- 64.0 ± 6.3	89.3 ± 2.7	(2.26 ± 0.07) ´10^{-16}
Toluene	0.099	76.3 ± 1.1	24.6 ± 0.4	73.8 ±1.1	49.5 ± 3.5	88.5 ± 1.5	(3.01 ± 0.05) ´10^{-16}
Benzene	0.111	71.0 ± 1.5	22.4 ± 0.6	68.5 ±1.5	- 67.4 ± 4.9	88.6 ± 2.1	(2.94 ± 0.07) ´10^{-16}
THF	0.207	62.6 ± 0.7	21.4 ± 0.3	60.1 ±0.7	- 77.4 ± 2.4	83.2 ± 1.1	(2.65 ± 0.04) ´10^{-15}
Acetone	0.355	56.7 ± 0.9	20.4 ± 0.4	54.2 ± 0.9	- 82.4 ± 3.5	78.8 ±1.4	(1.55 ± 0.03) ´10^{-14}
3-Pentanol	0.463	49.7 ± 1.2	19.3 ± 0.5	47.2 ± 1.2	-89.2 ± 4.1	73.8 ± 1.7	(1.17 ± 0.02) ´10^{-13}

Table 2: Thermodynamic values for the isomerization of 4A4NAB obtained in each of the six solvents. The reported uncertainties are standard errors determined from replicate measurements.

$\Delta^{\ddagger}G^{0}$, and the quasi-equilibrium constant K‡ for the *cis-trans* isomerization process are summarized in Table 2.

The activation energies are then plotted against the relative polarity of the solvents (Figure 9). In all cases, the rate of isomerization increases as the solvent polarity increases with a concomitant decrease in the activation energy, enthalpy, entropy and free energy of activation for the *cis-trans* isomerization process (Tables 1 and 2) [20,23].

Overall, the data show that the rate of transition from *cis* to *trans* is strongly solvent dependent. Typically, solvent polarity has very little effect on the rate of isomerization when considering an inversion-based mechanism. A strong relationship between rate of reactions and solvent polarity points to an intermediate transition state that is considerably more polar than the *cis* conformation. This polar transition state is indicative of a rotational mechanism for the isomerization of 4A4NAB [21,22,24]. This result is expected considering the strong electron withdrawing group (the nitro group) on 4A4NAB and is corroborated by computational studies [22,25,26]. The isomerization mechanism of 4A4NAB can thus be defined as a rotation around the azo bond assisted by a linearization of the angle of the ring attached to the electron-withdrawing substituent. The extent of such assistance depends on the polarity of the solvent, being less important as the polarity increases [8]. Indeed, as the solvent polarity increases, a noticeable decrease in the activation energy, entropy, enthalpy and Gibbs free energy of activation for the *cis-trans* isomerization process is observed, a result typical of *para*-donor, *para*-acceptor type azobenzenes. However, for some azobenzenes, a deviation from linearity of the rate constants or activation energy versus solvent polarity, particularly at low polarity, may indicate a change in the isomerization mechanisms, from rotational for polar solvents to inversion mechanisms for non-polar solvents [20,27]. While structural differences and types of substituents play an important role in the isomerization mechanisms of azobenzenes, a detailed mechanistic study is required to delineate the different steps involved.

Acknowledgments

The authors wish to thank the Collegiate Science and Technology Entry Program (CSTEP) at SUNY Potsdam for funding this project and the Chemistry department for providing some of the materials used in the experiments.

References

1. Lednev IK, Ye TQ, Abbott LC, Hester RE, Moore JN (1998) Photoisomerization of a Capped Azobenzene in Solution Probed by Ultrafast Time-Resolved Electronic Absorption Spectroscopy. J Phys Chem A 102: 9161-9166.
2. Fujino T, Tahara T (2000) Picosecond time-resolved Raman study of trans-azobenzene. The Journal of Physical Chemistry A 104: 4203-4210.
3. Satzger H, Spörlein S, Root C, Wachtveitl J, Zinth W, et al. (2003) Fluorescence spectra of trans-and cis-azobenzene–emission from the Franck–Condon state. Chemical physics letters 372: 216-223.
4. Conti I, Garavelli M, Orlandi G (2008) The different photo isomerization efficiency of azobenzene in the lowest nπ* and ππ* singlets: the role of a phantom state. Journal of the American Chemical Society 130: 5216-5230.
5. Ootani Y, Satoh K, Nakayama A, Noro T, Taketsugu T (2009) Ab initio molecular dynamics simulation of photoisomerization in azobenzene in the n π□ state. The Journal of chemical physics 131: 194306.
6. Shao J, Lei Y, Wen Z, Dou Y, Wang Z (2008) Nonadiabatic simulation study of photoisomerization of azobenzene: Detailed mechanism and load-resisting capacity. The Journal of chemical physics 129: 164111.
7. Yuan S, Dou Y, Wu W, Hu Y, Zhao J (2008) Why does trans-azobenzene have a smaller isomerization yield for ππ* excitation than for nπ* excitation?. The Journal of Physical Chemistry A 112: 13326-13334.
8. Corchado JC, Sanchez ML, Fdez Galván I, Martin ME, Muñoz-Losa A, et al. (2014) Theoretical study of solvent effects on the ground and low-lying excited free energy surfaces of a push-pull substituted azobenzene. The Journal of Physical Chemistry B 118: 12518-12530.
9. Rau H, Lueddecke E (1982) On the rotation-inversion controversy on photoisomerization of azobenzenes. Experimental proof of inversion. Journal of the American Chemical Society 104: 1616-1620.
10. Fujino T, Arzhantsev SY, Tahara T (2001) Femtosecond time-resolved fluorescence study of photoisomerization of trans-azobenzene. The Journal of Physical Chemistry A 105: 8123-8129.
11. Schultz T, Quenneville J, Levine B, Toniolo A, Martínez TJ, et al. (2003) Mechanism and dynamics of azobenzene photoisomerization. Journal of the American Chemical Society. 125: 8098-8099.
12. Cembran A, Bernardi F, Garavelli M, Gagliardi L, Orlandi G (2004) On the mechanism of the cis− trans isomerization in the lowest electronic states of azobenzene: S0, S1, and T1. Journal of the American Chemical Society 26: 3234-3243.
13. Crecca CR, Roitberg AE (2006) Theoretical study of the isomerization mechanism of azobenzene and disubstituted azobenzene derivatives. The Journal of Physical Chemistry A 110: 8188-8203.
14. Serra F, Terentjev EM (2008) Effects of solvent viscosity and polarity on the isomerization of azobenzene. Macromolecules 41: 981-986.
15. Joshi NK, Fuyuki M, Wada A (2014) Polarity Controlled Reaction Path and Kinetics of Thermal Cis-to-Trans Isomerization of 4-Aminoazobenzene. The Journal of Physical Chemistry B 118: 1891-1899.
16. Hair SR, Taylor GA, Schultz LW (1990) An easily implemented flash photolysis experiment for the physical chemistry laboratory: The isomerization of 4-anilino-4'-nitroazobenzene. J Chem Educ 67: 709.
17. Bahrenburg J, Sievers CM, Schönborn JB, Hartke B, Renth F, et al. (2013) Photochemical properties of multi-azobenzene compounds. Photochemical & Photobiological Sciences 12: 511-518.
18. Reichardt C (1998) Solvents and Solvent Effects in Organic Chemistry. VCH: Weinheim, Federal Republic of Germany, p: 418.

19. Matczyszyn K, Bartkowiak W, Leszczynski J (2001) Influence of the environment on kinetics and electronic structure of asymmetric azobenzene derivatives—experiment and quantum-chemical calculations. Journal of Molecular Structure 565: 53-57.

20. Luner P, Winkler CA (1952) Solvent Effects in Cis-trans Isomerization. Canadian Journal of Chemistry 30: 679-686.

21. Marcandalli B, Liddo LPD, Fede CD, Bellobono IR (1984) J Chem Soc, Perkin Trans 4: 589.

22. Douglas P, Clarke D. Substituent and solvent effects on the thermal isomerisation of 7H-pyrazolo [5, 1-c]-1, 2, 4-triazole azomethine dyes. Journal of the Chemical Society, Perkin Transactions 2: 1363-1367.

23. Joshi NK, Fuyuki M, Wada A (2014) Polarity Controlled Reaction Path and Kinetics of Thermal Cis-to-Trans Isomerization of 4-Aminoazobenzene. The Journal of Physical Chemistry B 118: 1891-1899.

24. Surampudi SK, Patel HR, Nagarjuna G, Venkataraman D (2013) Mechano-isomerization of azobenzene. Chemical Communications 49: 7519-7521.

25. Wazzan NA, Richardson PR, Jones AC (2010) Cis-Trans isomerisation of azobenzenes studied by laser-coupled NMR spectroscopy and DFT calculations. Photochemical & Photobiological Sciences 9: 968-974.

26. Whitten DG, Wildes PD, Pacifici JG, Irick Jr G (1971) Solvent and substituent on the thermal isomerization of substituted azobenzenes. Flash spectroscopic study. Journal of the American Chemical Society, pp: 2004-2008.

27. Angelini G, Canilho N, Emo M, Kingsley M, Gasbarri C (2015) Role of Solvent and Effect of Substituent on Azobenzene Isomerization by Using Room-Temperature Ionic Liquids as Reaction Media. The Journal of Organic Chemistry 80: 7430-7434.

30

Statistical Analysis of the Catalyst and Activator Level on *In Situ* Polymerization of Laurolactam during Reactive Melt Blending

Faridirad F, Ahmadi S, Barmar MD* and Naderi G

Iran Polymer and Petrochemical Institution, Pazhouhesh Blvd., Pazhouhesh Science Park, Tehran-Karaj Highway, Iran

Abstract

In the present paper, *in situ* ring opening anionic polymerization of laurolactam towards polyamide 12 (PA12) was investigated, during reactive melt blending. The reaction was taken place in an internal mixer, in the presence of sodium caprolactam (catalyst), and toluene diisocyanate (activator). The effect of catalyst and activator contents on initiation time of the reaction and the residual monomer was carried out in this study. Solvent extraction method was used in order to remove the residual monomer. The experimentation is further undergone to statistical analysis to extract the absolute optimal values of activator- catalyst content in synthesized polyamide 12. Two-factor design in Response Surface Method (RMS) is suitably applied for validation analysis of experimental data. The optimum content of catalyst and activator was obtained 2.53% and 1.15%, which offers the polyamide 12 with minimum residual monomer equal to 1.26%.

Keywords: Ring opening anionic polymerization; Polyamide12; Laurolactam; Sodium caprolactam; Toluene diisocyanate

Introduction

Polyamides are well-known engineering thermoplastic materials which have high mechanical strength, high melting point, low density and high potential of forming complicated structures from molding injection [1]. They widely used in industrial applications for their remarkable mechanical and thermal properties [2-5].

Polyamides are prepared by applying three technique [6], Hydrolytic [7], Solid-phase [6] and Anionic polymerization. PA12 can be obtained through anionic polymerization of laurolactam. This method must be carried out strictly in the absence of water [8].

The presence of an anionically activated monomer and an activator is essential for starting the anionic polymerization. Only the correct ratios of monomer, activator, and catalyst will complete the reaction so that it will give the least residual monomer [8,9]. The initiators of anionic lactam polymerization are the lactamate anions being formed by the reaction of a lactam with a strong base [10]. Schwartz and Paul proposed the use of the alkaline and earth alkaline metals, especially sodium and potassium, and their oxides, hydroxides, hydrides, carbonates, and other reactive compounds of these metals, including organometallic reagents as catalyst [10].

Luisierv et al. worked on improving impregnation prior to *in-situ* polymerization of Polyamide 12 by exploiting a very low monomer viscosity. The polymerization kinetics of anionic PA12 was studied and the viscosity changes during polymerization were measured to determine how the processing temperature influences the impregnation time [11]. Also there are some researches on investigation the effect of catalyst- activator content in the formulation.

Ahmadi et al. investigated the effect of catalyst and activator on the residual monomer and other characteristics of PA6. The experimental results showed that the addition of 3% catalyst and 3% activator to the formulation gave the best properties. These conditions led to the lowest residual monomer and better mechanical properties as well [12].

Experimental Design and Response Surface Methodology (RSM) has been used for a long time to develop descriptive–predictive models to fulfill optimization purposes in different areas of dosage form design [13-15]. Its popularity stems from its ability to locate the combination of formulation and/or manufacturing parameter values that are most likely to produce the desired release profile.

In this study anionic polymerization of Laurolactam was studied to produce polyamide 12, and then analytical tests were applied to investigate the effect of sodium caprolactam (NaCL), as catalyst and toluene diisocyanate (TDI) as an activator, on reaction time and the residual monomer, in different concentrations. Moreover an optimization study and statistical analysis of these properties was also carried out using Design Expert Software. The optimization study was evaluated in order to validate the experimental results and to identify the best-fit values.

Experiment

Materials and methods

The monomer used in this project was laurolactam supplied by Evonik Co., Germany. Toluene diisocyanate (TDI), used as an activator, was obtained from Karoon Co., Iran. The catalyst, sodium Caprolactam was obtained from Karangin Co.

Polymerization of laurolactam was carried out in a 50 mL chamber of a Haake type internal mixer model Sys 9000 for the sample preparation and the subsequent evaluations. The rotor speed was fixed at 60 rpm. The temperature was set at 180°C under the nitrogen atmosphere. By changing the catalyst and activator concentration at three levels (1, 2 and 3%), nine different formulations where synthesized which were named as I_1C_1, I_1C_2, I_1C_3, I_2C_1, I_2C_2, I_2C_3, I_3C_1, I_3C_2 and I_3C_3 respectively. in these formulations "I" refers to initiators (activator) and "C" refers to catalyst. also subscripts 1, 2 and 3 refers to the concentration percentage of each material.

***Corresponding author:** Mohammed Barmar, Iran Polymer and Petrochemical Institution, Pazhouhesh Blvd., Pazhouhesh Science Park, Tehran-Karaj Highway, Iran
E-mail: m.barmar@ippi.ac.ir

Residual monomer characterization

Different methods have been proposed to determine the conversion of monomers to polymer. Among these methods the gravimetric method was applied here. This was based on the mass ratio of polymerized products of which the residual monomers were removed by solvent extraction. Toluene was used as a solvent for Soxhlet extraction of un-reacted laurolactam. Monomer extraction was continued for 40 h at 120°C. Then, the samples were dried in a vacuum oven for 30 h at 70°C.

In order to confirm the monomer extraction results, TGA was performed for detecting and also determining the residual monomer content of the PA12 samples. TGA tests were conducted under nitrogen flow from 25°C up to 600°C and then continued up to 1000°C under oxygen. The heating rate was 10°C/min.

Statistical analysis

The Statistically designed experiments play an important role in Science, Engineering, and Industry. The experimentation is a part of scientific method, which requires observing and gathering the information about how the process and system works. In an experiment, some inputs transform into an output that have one or more observable response variables. Therefore, useful results and conclusions can be drawn by experiment. Experimenter needs to plan and design the experiment, and analyze the results, in order to obtain an objective conclusion. There are many types of experiments used in real-world situations and problems. Statistical analysis (regression and ANOVA analysis) of the responses are carried out to estimate the coefficients of the polynomial equation of the response by regression and to check the significance of the regression coefficients of independent variables and interaction variables by ANOVA. Analysis of Variance (ANOVA) table is used to determine the significance of the different degrees, and cross-product terms of the polynomial.

The present study is conducted to find the optimum catalyst and activator contents validate the obtained experiment results; Response Surface Methodology (RSM) is adopted to determine the functional relationships between the process variables and initiation reaction time and also residual monomer. In this case, non-linear trends in the response are likely and hence a second order polynomial model could be considered to fit adequately the experimental results. Here there were only two factors (catalyst and activator %) to be analyzed.

Results and Discussion

Synthesis of polymer

The Figure 1 shows the torque versus mixing time for different formulations. In all cases, the torque started to increase at a time, which is called reaction time. Duration time differed, depending on the formulation. Due to the lower activation energy required for the nucleophilic attack of the anionic caprolactam to its activated lactam group, after the addition of only 1% of TDI, the polymerization took place. For all samples except samples I_1C_3 and I_3C_1, the torque curves increased very rapidly after about 50-70 seconds of mixing, indicating the starting point of the polymerization. Diagrams reached to its maximum after about 120 seconds and then decreased, which indicated the end of polymerization [16].

Catalyst and activator content had a considerable effect on polymerization. Catalyst led the reaction to be occurred faster and activator increased the anionic attack speed. The outcome of these two factors together is shown in Figure 2. It is obvious from Figure 2, polymerization reaction time decreased, by increasing the amount of catalyst (NaCL) and the activator (TDI), but the effect of NaCL is more significant. Although lower reaction time seemed to be better but it may limit the polymerization and reduce the molecular weight too.

Characterization of the synthesized polymers

Residual monomer determination: Monomer conversion is an important parameter, because a little content of monomer sometimes can remarkably influence the properties of specimen. Generally, the loss weight of sample from 200°C up to 320°C was corresponding to Laurolactam content [17]. The Figure 3 shows the effect of catalyst and activator concentration on the residual monomer content of the samples. This figure clearly shows that in all of the concentrations of the activator, by increasing the amount of catalyst, residual monomer percentage decreased, while activator had an inverse effect.

The maximum conversion of Laurolactam was reported to be around 97.3% and 96% (2.7%-4%residual monomer) by Libo Du et al. and Ha and White respectively [18]. However, the residual monomer content (Figure 3) in this work was reduced to 1.86% for PA12 containing 1% catalyst and 1% activator; which was lower than the cited references. The improvement in the polymerization efficiency is believed to be caused by the use of a double functional cyanate (TDI). Because in this case the reaction proceeded from two sides of the activator, and the monomers were consumed more rapidly. The results of TGA confirmed the Soxhlet results and both of them showed the same procedure as it is shown in Table 1.

Statistical analysis of experimental results: Second-order polynomial equation was developed using RSM including quadratic terms and two-factor interactions that explained the nonlinear nature of the response [19]. The second-order polynomial equations are expressed generally in the following form:

$$y=a_0+a_1X_1+a_2X_2+a_3X_1X_2+a_4X_1^2+a_5X_2^2+E \quad (1)$$

Where a_1-a_5 are the coefficients of the respective variables and their interaction terms, and E is an error term. Results of analysis are usually considered significantly if their corresponding P values are less than 0.05. A two-factor, multiple-level full factorial design was used to construct the response surface and the polynomial model for the polyamide12 polymerization, optimization. The catalyst (NaCL) (X1) and also the activator percentage (X2) were studied at three levels: 1, 2 and 3%.

In this study, the effect of catalyst and activator on the residual monomer and reaction time of the synthesized polyamide 12 was investigated using two-factor design which is one of the response surface methodology designs. The results were analyzed by using Design Expert Software Version 8 to establish the mathematical functional relations as well as a number of statistics to confirm the variables of the models.

Analysis of the reaction time

The results of the ANOVA, two-factor design experiments on the effect of the process variable over the reaction time are tabulated in Tables 2 and 3. A suitable model was selected from these results. Through the estimation of all regression coefficients, the experimental response could be modeled as a polynomial equation that showed the effect of process variable (activator and catalyst content) on the reaction time of the synthesized polyamide 12. The quadratic function obtained is given in Equation (2). This equation was used to generate

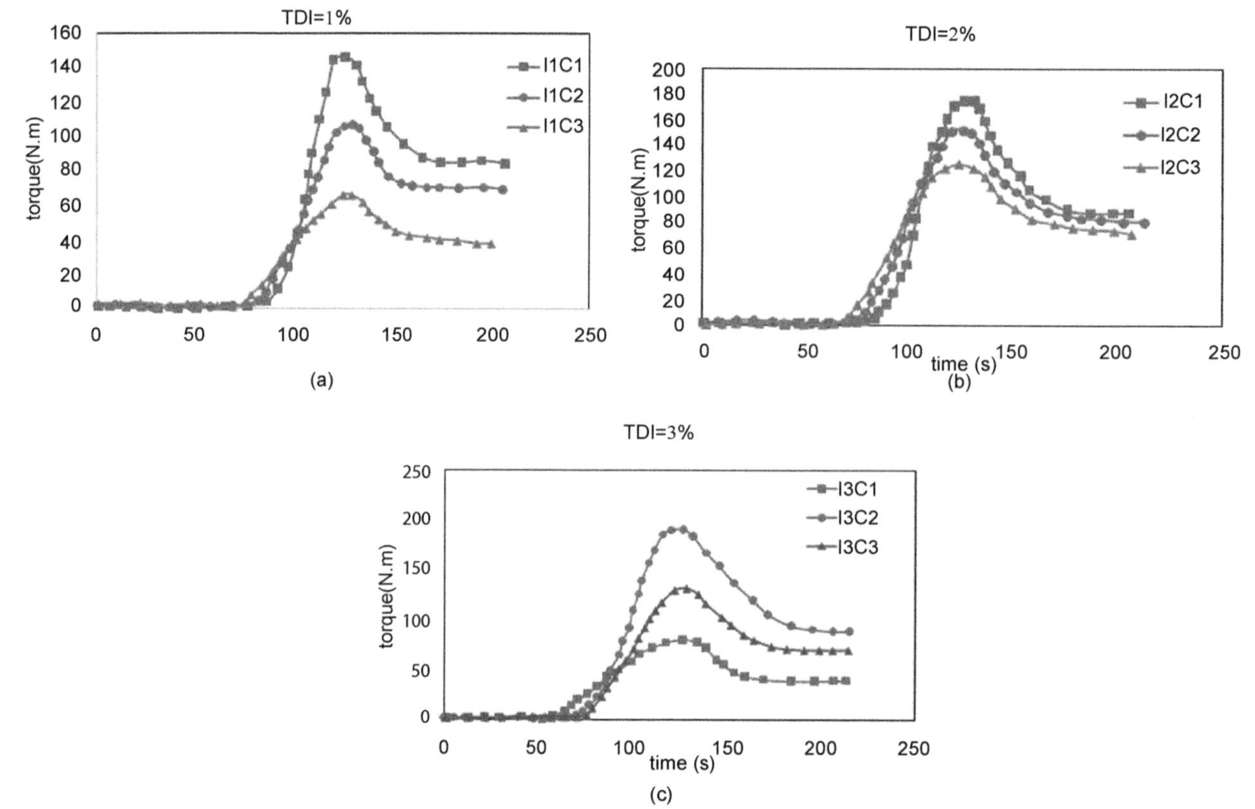

Figure 1: Torque-time curves at 150 rpm and 200°C for different formulations (TDI and NaCL contents) of PA12 in a Haake mixer, (a) TDI=1%, (b) TDI=2% and (c) TDI=3%.

Sample	TDI	NaCl	T (peak1)	T (peak2)	T (10%wt)	T (30%wt)	T (50%wt)	Residual monomer (TGA)%	Residual monomer (Soxhlet)%
I_1C_1	1	1	467.25	357.27	373.562	449.776	459.946	3.6	3.6
I_1C_2	1	2	458.8	359.27	377.128	446.616	457.883	1.9	1.8
I_1C_3	1	3	462.55	359.3	384.768	449.402	453.926	1.8	1.8
I_2C_1	2	1	464.026	238.96	241.585	429.848	452.685	12.2	11.9
I_2C_2	2	2	465.092	348.3	360.211	443.515	455.865	2.7	2.7
I_2C_3	2	3	463.12	348.31	355.331	441.107	455.865	2.2	2.3
I_3C_1	3	1	460.94	244.56	213.101	251.874	426.57	33.5	33.5
I_3C_2	3	2	464.29	338.34	343.933	441.083	454.782	3.8	3.7
I_3C_3	3	3	464.27	343.404	342.708	438.934	453.792	3.4	3.5

Table 1: TGA and Soxhlet results of residual monomer for different formulations.

the predicted values using the software as shown in Table 4. A graph is plotted between the predicted and experimental (actual) values as shown in Figure 4. Here the transform function was chosen as ($y'=(y+k)^3$).

The Model F-value of 10.81 implied the significancy of model. There was only a 0.26% chance that a "Model F-Value" this large could occur due to noise.

Values of "Prob>F" less than 0.0500 indicate model terms are significant. In this case A, B, AB were significant model terms. Values greater than 0.1000 indicate the model terms are not significant. If there are many insignificant model terms (not counting those required to support hierarchy), model reduction may improve the model. Thus the effect of activator, catalyst and the cross product of activator and catalyst were more prominent (Figures 5 and 6).

A negative "Pred R-Squared" implied that the overall mean was a better predictor of present response than the current model. "Adeq Precision" measured the signal to noise ratio. A ratio greater than 4 is desirable. Here the ratio of 11.262 indicated an adequate signal. This model can be used to navigate the design space.

Final equation in terms of coded factors: $(\text{reaction time})^3 = +3.809E+005 - 41967.65 \times A - 71692.01 \times B + 59081.85 \times A \times B - 40356.60 \times B^2$ (2)

Final equation in terms of actual factors: $(\text{reactiontime})^3 = +6.83148E+005 - 1.60131E+005 \times \text{catalyst} - 28429.32392 \times \text{activator} + 59081.84642 \times \text{catalyst} \times \text{activator} - 40356.59568 \times \text{activator}^2$ (3)

An interaction occurs when the response is different depending on the settings of two factors. Plots in Figure 7 make it easy to interpret

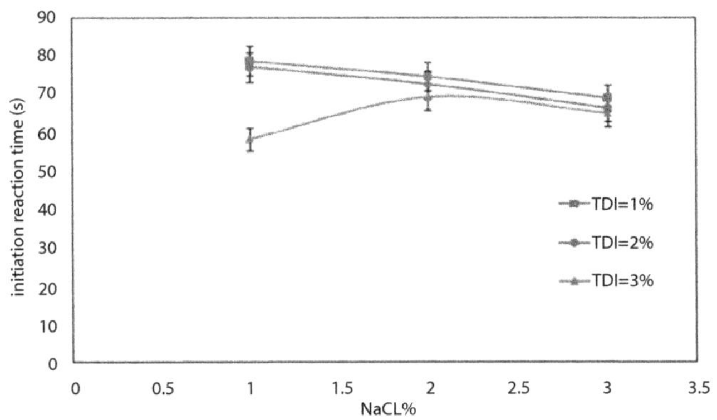

Figure 2: Effects of the catalyst (NaCl) and the activator (TDI) on the initiation reaction time.

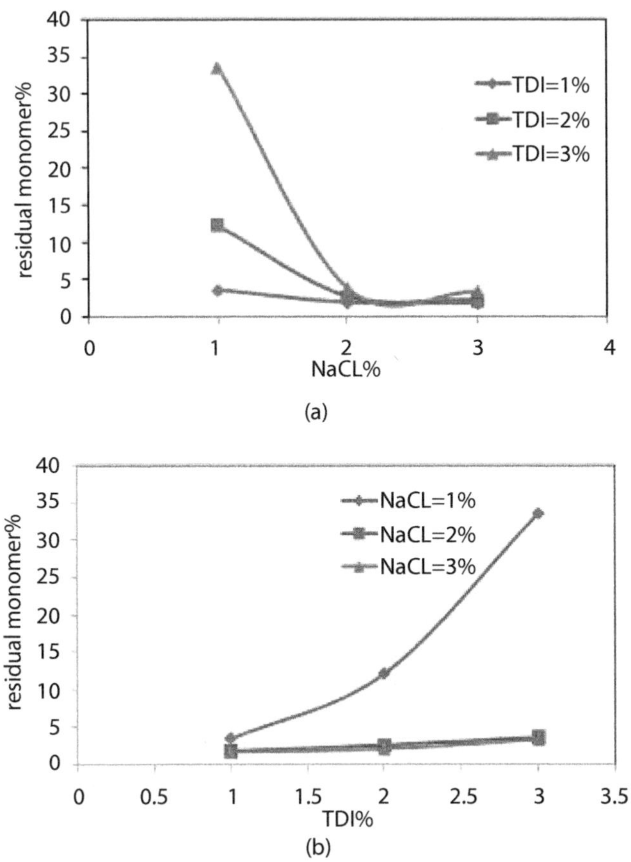

Figure 3: Effect of the catalyst (a) and the activator (b) on the residual monomer determined from TGA analysis.

two factor interactions. They will appear with two non-parallel lines, indicating that the effect of one factor depends on the level of the other.

Analysis of the residual monomer

The results of the ANOVA, two-factor design experiments on the effect of the process variable over residual monomer are tabulated in Table 5, part (a). Through the estimation of all regression coefficients, the experimental response was modeled as a polynomial equation which showed the effect of process variable on the residual monomer of synthesized polyamide 12. The quadratic function obtained is given in Equation (4). This equation was used to generate the predicted values using the software as shown in Table 6. A graph is plotted between the predicted and experimental (actual) values as shown in Figures 8-11.

The Model F-value of 14.05 implied the significancy of the model. There was only a 0.16% chance that a "Model F-Value" this large could occur due to noise. Due to "Prob>F" values which were less than 0.0500 model terms were significant. In this case A, B, AB, A^2 were significant model terms.

The "Pred R-Squared" of 0.1033 is not as close to the "Adj R-Squared" of 0.8447 as is normally expected. This may indicate a large

Source	Sum of squares	df	MeanSquare	F - value	P- value Prob>F
Model	6.063E+010	4	1.516E+010	10.81	0.0026
A-catalyst	1.057E+010	1	1.057E+010	7.54	0.0252
B-activator	3.084E+010	1	3.084E+010	21.99	0.0016
AB	1.396E+010	1	1.396E+010	9.96	0.0135
B^2	5.262E+009	1	5.262E+009	3.75	0.0887
Residual	1.122E+010	8	1.402E+009		
Total	7.185E+010	12			

Table 2: Analysis of variance (ANOVA) results for reaction time of PA12 polymerization.

Std. Dev	37446.20	**R-Squared**	0.8439
Mean	3.623E+005	**Adj R-Squared**	0.7658
C.V.%	10.34	**Pred R-Squared**	-0.0553
PRESS	7.582E+010	**Adeq R-Squared**	11.262

Table 3: Modeling parameters for raction time of PA12 polymerization.

Sample	Reaction time (ACTUAL), (s)	(Reaction time)^3 (ACTUAL), (s)	(Reaction time)^3 (PREDICTED), (s)
I1C1	78.839	490030.7361	513300
I1C2	74.754	417737.3512	412300
I1C3	69.0359	329022.0265	311200
I2C1	77.149	459188.3947	422900
I2C2	72.612	382846.9547	380900
I2C3	66.424	293072.5039	339000
I3C1	58.43	199483.8121	251800
I3C2	69.268	332351.7308	268900
I3C3	65.014	274802.4882	286000

Table 4: Predicted Vs Actual values of reaction time.

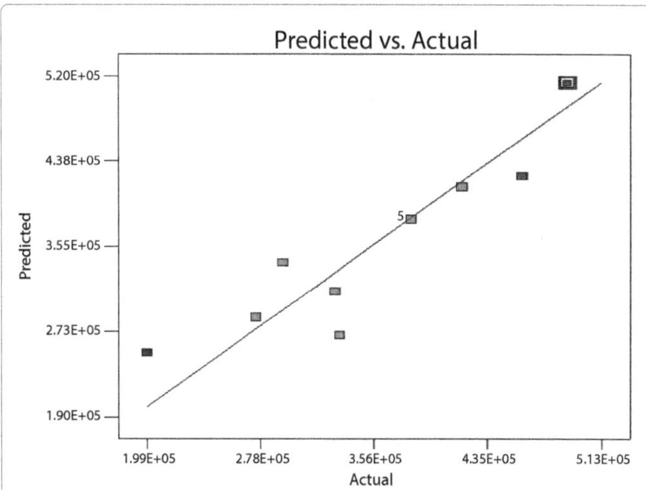

Figure 4: Graph showing the predicted reaction time vs actual values (s).

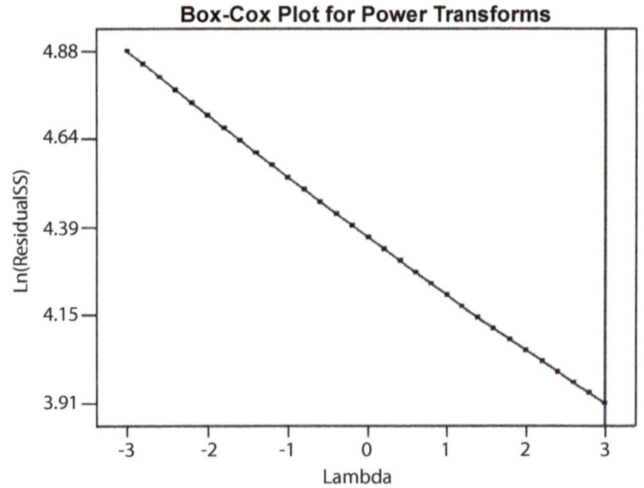

Figure 5: Graph showing the Box-Cox plot for power transform of the reaction time.

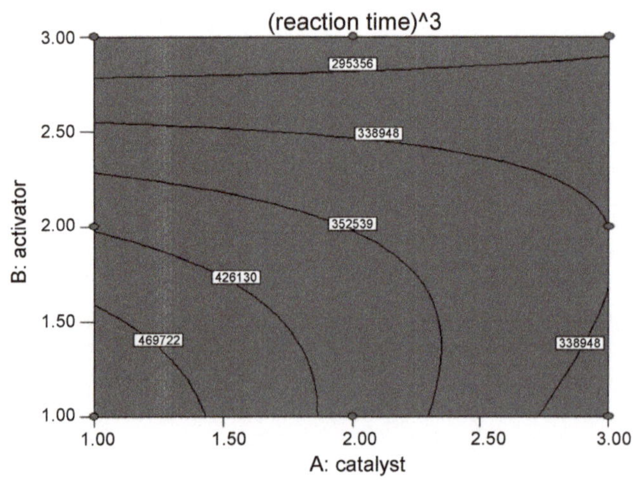

Figure 6: Graph showing the effect of catalyst and activator simultaneously on the reaction time.

block effect or a possible problem with the model and/or data. Here the ratio of 13.373 for "Adeq Precision" indicated an adequate signal.

Final equation in terms of coded factors: The residual monomer $=+2.27-6.96\times A+5.56\times B-7.08\times A\times B+6.08\times A^2+1.70\times B^2$ (4)

Final equation in terms of actual factors: The residual monomer $=+7.87140-17.13644\times$ catalyst $+12.93044\times$ activator $-7.07703\times$ catalyst $\times$ activator $+6.08306\times$ catalyst$^2+1.69681\times$ activator^2 (5)

As it is clear from Table 6, this model did not present appropriate prediction for the residual monomer, therfore the transform function was chosen to be $(y+k)^{-1}$ and again the quadradic model was applied. The results of the new model is presented in Table 5 (part b). In this case A, B, A^2 were significant model terms.

Final equation in terms of coded factors: (residual monomer)$^{-1}=+0.37+0.15\times A-0.13\times B+4.160E\text{-}005\times A\times B-0.11\times A^2+0.022\times B^2$ (6)

Final equation in terms of actual factors: (residual monomer)$^{-1}=-0.012639+0.57570\times$catalyst$-0.21507\times$activator$+4.16047E\text{-}005\times$ catalyst $\times$ activator$-0.10733\times$ catalyst$^2+0.022340\times$ activator^2 (7)

Since the variation in catalyst and activator content between the experimental value and predicted value was not that large, and the desirability factor was 82%, the data obtained from the experiment can be considered as the best-fit value, as it is shown in Tables 6-8. The

(a) Linear trasnform function					
Source	Sum of squares	Df	Mean Square	F - value	P- value Prob>F
Model	830.82	5	166.16	14.05	0.0016
A-catalyst	290.50	1	290.50	24.56	0.0016
B-activator	185.72	1	185.72	15.70	0.0054
AB	200.34	1	200.34	16.94	0.0045
A^2	102.20	1	102.20	8.64	0.0217
B^2	7.95	1	7.95	0.67	0.4393
Residual	82.79	7	11.83		
Total	913.62	12			
(b) Transform function=$(y+k)^{-1}$					
Source	**Sum ofSquares**	**df**	**Mean Square**	**F - value**	**P- value Prob>F**
Model	0.26	5	0.051	106.35	<0.0001
A-catalyst	0.13	1	0.13	266.99	<0.0001
B-activator	0.095	1	0.095	196.45	<0.0001
AB	6.92×10^{-9}	1	6.92×10^{-9}	1.43×10^{-5}	0.9971
A^2	0.032	1	0.032	66.01	<0.0001
B^2	1.378×10^{-3}	1	1.378×10^{-3}	2.66	0.1347
Residual	3.374×10^{-3}	7	4.820×10^{-4}		
total	0.26	12			

Table 5: Analysis of variance (ANOVA) results for the residual monomer, (a) Linear trasnform function and (b) Transform function=$(y+k)^{-1}$.

Sample	Residual monomer (actual), (%)	Residual monomer (predicted by Linear trasnform function), (%)	Residual monomer (predicted by $(y+k)^{-1}$ trasnform function), (%)
I_1C_1	3.6	4.3	3.8
I_1C_2	1.9	-1.5	1.9
I_1C_3	1.9	4.6	1.8
I_2C1	12.2	15.1	8.7
I_2C_2	2.7	2.2	2.7
I_2C_3	2.2	1.4	2.4
I_3C_1	33.5	29.6	83.3
I_3C_2	3.8	9.5	3.6
I_3C_3	3.4	1.6	3.3

Table 6: Predicted residual monomer vs the actual values.

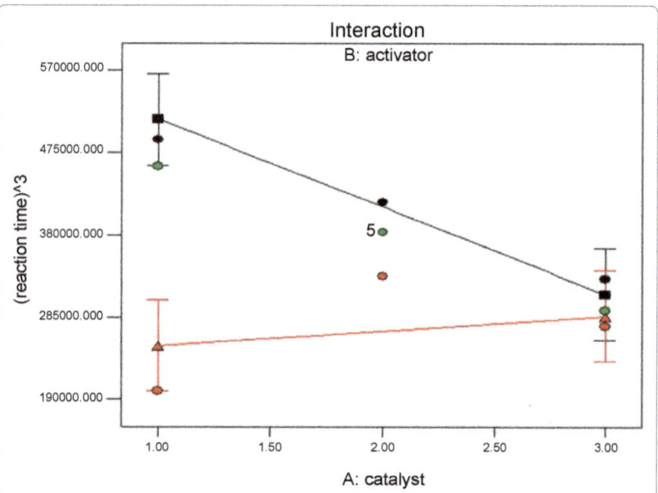

Figure 7: Graph showing the interaction of catalyst and activator on the reaction time.

validation checks using response surface method (two-factor design), gave 1.15% activator and 2.53% catalyst as the optimum values for the minimum residual monomer. This formulation led to the monomer content equal to 1.26%. Therefore by investigating a set of random formulations and applying statistical analysis, a prediction of optimum condition was simply done. According to this prediction, a large number of possible formulations, which should be synthesized in order to determine the best catalyst and activator content, would be omitted.

Conclusion

In this study, the effect of catalyst and activator on reaction time and residual monomer during the ring opening polymerization of polyamide 12 was investigated. The experimental results are validated using Response Surface Method (RMS). The following facts are concluded from the present research:

- The response surface method along with ANOVA technique is found to be an effective approach for optimization of decreasing the residual monomer of synthesized PA12.
- Evidently, activator and the cross product of activator and catalyst are detected to be the significant factors which effects on reaction time.
- According to the above results, it can be concluded that the catalyst (2.53%) and activator (1.15%) is the optimal combination for achieving the required objectives (the least residual monomer content).

Name	Goal	Lower Limit	Higher Limit
catalyst	is in range	1	3
activator	is in range	1	3
(reaction time)^3	is in range	199484	490031
(residual monomer)^-1	maximize	0.0298531	0.537634

Table 7: Objectives of combined effect – reaction time and residual monomer.

Catalyst%	Activator %	(reaction time)^3	(residual monomer)^-1	Desirability
2.53	1.15	374738.106	0.79	1.000

Table 8: Final Result for combined effect - reaction time and residual monomer.

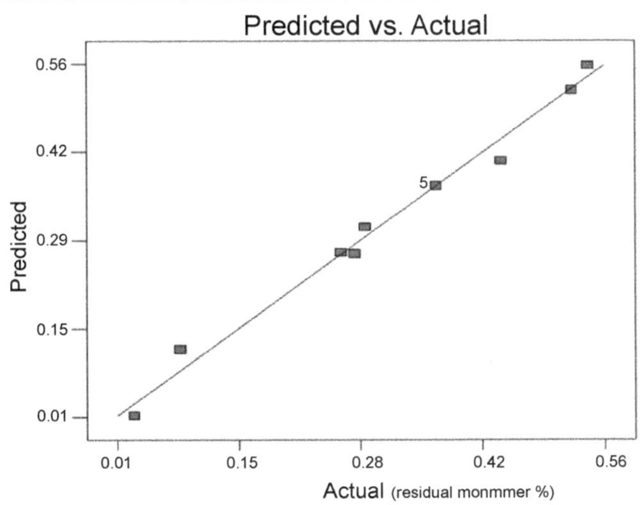

Figure 8: Graph showing the predicted residual monomer vs actual values.

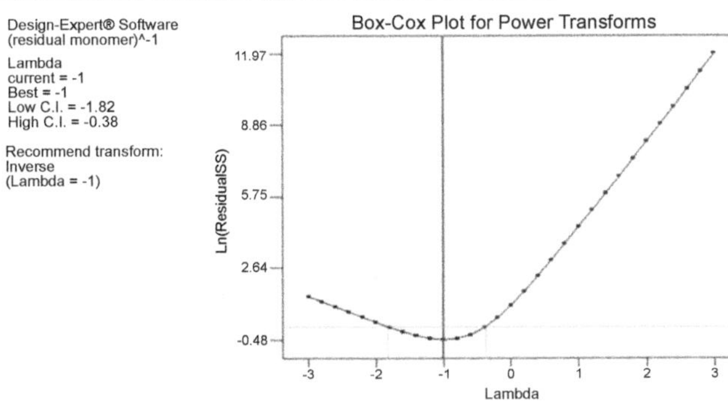

Figure 9: Graph showing the Box-Cox plot for power transform of residual monomer.

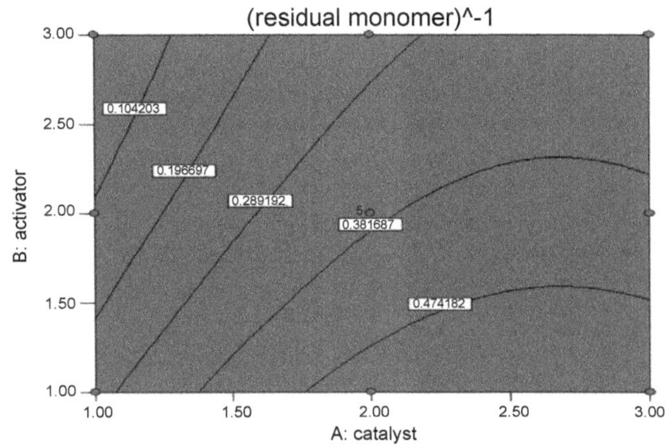

Figure 10: Graph showing the effect of catalyst and activator simultaneously on residual monomer.

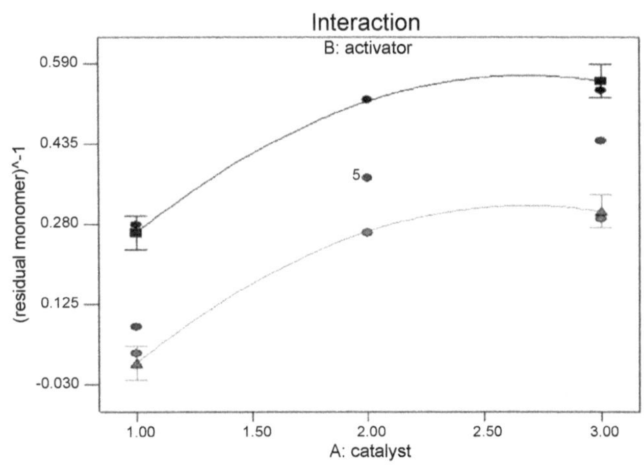

Figure 11: Graph showing the interaction of catalyst and activator on residual monomer.

References

1. Kohan MI (1995) Nylon plastics handbook. Hanser, New York.
2. Erdmann E, Dias M, Pita VJRR, Destéfanis H, Monasterio F, et al. (2007) Characterization of HDPE/Polyamide 6/Nanocomposites Using Scanning-and Transmission Electron Microscopy. Macromolecular symposia 258: 82-89.
3. Gatos K, Apostolov A, Karger-Kocsis J (2005) Compatibilizer effect of grafted glycidyl methacrylate on EPDM/organoclay nanocomposites. Materials Science Forum 482: 347-350.
4. Liu XQ, Yang W, Xie BH, Yang MB (2012) Influence of multiwall carbon nanotubes on the morphology, melting, crystallization and mechanical properties of polyamide 6/acrylonitrile–butadiene–styrene blends. Materials & Design 34: 355-362.

5. Taghizadeh E, Naderi G, Razavi-Nouri M (2011) Effects of organoclay on the mechanical properties and microstructure of PA6/ECO blend. Polymer Testing 30: 327-334.
6. Melvin I (1995) Nylon Plastic Handbook. Hanser: New York, USA.
7. Cheremisinoff NP (1989) Handbook of polymer science and technology. CRC Press.
8. Griehl W, Ruestem D (1970) Nylon-12-preparation, properties, and applications. Industrial & Engineering Chemistry 62: 16-22.
9. Ahmadi S, Morshedian J, Hashemi SA (2010) Effects of molecular weight on the dynamic mechanical and rheological properties of anionically polymerized polyamide 6 containing nanofiber. Journal of Vinyl and Additive Technology 16: 152-160.
10. Wollny A, Nitz H, Faulhammer H, Hoogen N, Mulhaupt R (2003) In situ formation and compounding of polyamide 12 by reactive extrusion. Journal of applied polymer science 90: 344-351.
11. Luisier A, Bourban P, Manson J (1999) In-situ polymerisation of polyamide 12 for thermoplastic composites. Proceedings of ICCM-12, Paris.
12. Ahmadi S, Morshedian J, Hashemi SA, Carreau PJ, Leelapornpisit W (2010) Novel anionic polymerization of e-caprolactam towards polyamide 6 containing nanofibrils. Iran Polym J 19: 229-240.
13. Sastry SV, Reddy IK, Khan MA (1997) Atenolol gastrointestinal therapeutic system: optimization of formulation variables using response surface methodology. Journal of controlled release 45: 121-130.
14. Hamed E, Sakr A (2001) Application of multiple response optimization technique to extended release formulations design. Journal of controlled release 73: 329-338.
15. Turkoglu M, Sakr A (1992) Mathematical modelling and optimization of a rotary fluidized-bed coating process. International journal of pharmaceutics 88: 75-87.
16. Whelan T (1994) Consultant, Polymer Technology Dictionary. Chapman & Hall, New York, USA.
17. Ha SK, White JL (1998) Continuous polymerization and copolymerization of lauryl lactam in a modular corotating twin screw extruder. International Polymer Processing 13: 136-141.
18. Du L, Yang G (2010) A super-toughened nylon 12 blends via anionic ring-opening polymerization of lauryllactam in a twin screw extruder. Preparation, morphology, and mechanical properties. Polymer Engineering & Science 50: 1178-1185.
19. Myers RH, Montgomery DC, Anderson-Cook CM (2016) Response surface methodology: process and product optimization using designed experiments. John Wiley & Sons, USA.

Thermodynamic Description of the Gd-Sb and Gd-Bi-Sb Systems

Jinming Liu[1,2*], Xiaoyang Chen[2], Yinghui Zhang[1], Ming Zhang[1], Lan Zhang[1], Tao Zhong[1], Juan Lin[1] and Caifang Cao[3]

[1]*School of Material Science and Engineering, Jiangxi University of Science and Technology, Ganzhou, 341000, PR China*
[2]*Xingle Group Co. Ltd., Yueqing, Zhejiang, 325604, PR China*
[3]*School of Metallurgy and Chemical Engineering, Jiangxi University of Science and Technology, Ganzhou, 341000, PR China*

Abstract

The thermodynamic modeling and optimization of the Gd-Sb and Gd-Bi-Sb systems were critically carried out by means of the CALPHAD (CALculation of PHAse Diagram) technique. The solution phases, liquid, bcc, rhomb and hcp(Gd) were described by the substitutional solution model. The compounds, β-$Gd_5(Bi, Sb)_3$, γ-$Gd_4(Bi, Sb)_3$, δ-Gd(Bi, Sb), δ'-Gd(Bi, Sb), ε-$Gd_{16}Sb_{39}$ and $GdBi_2$ were treated as the formulae $(Gd)_{m/m+n}(Gd, Bi, Sb)_{n/m+n}$ using two sublattice model in the Gd-Bi-Sb system. A self-consistent thermodynamic description of the Gd-Sb and Gd-Bi-Sb system were developed. The isothermal section at 300 K for the Gd-Bi-Sb system in the literature were reproduced in the present work.

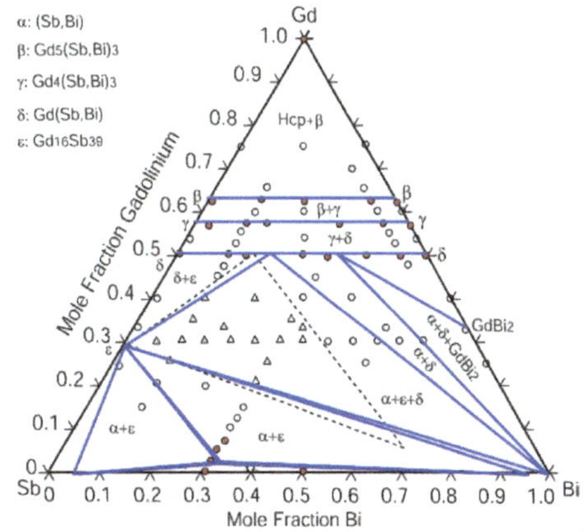

Keywords: Gd-Bi-Sb system; Phase diagram; CALPHAD technique; Thermodynamic properties

Introduction

In past years, the advanced magnetic materials and its potential as an energy savings technology greatly stimulated the interest of researchers and promoted the rapid development of magnetic refrigeration. Magnetic refrigeration could be realized by utilizing the heat release or absorption caused by the magnetic entropy change DSM of a magnetic material due to a magnetic field change DH [1]. The recent discovery of the giant magnetocaloric effect in $Gd_5(Si_2Ge_2)$ [1], gave further impulse towards the development of new materials [2]. Gd_4Sb_3 was one of the candidates for magnetic refrigerant near room temperature (266 K). Substitution of Bi for Sb in Gd_4Sb_3 increased its Curie temperature T_c up to 330 K [3]. The calculation of phase diagrams (CALPHAD) method, which was a powerful approach to save cost and short time during development of materials, effectively provided a clear guideline for material design. So in order to better understand the interactions of Gd and Sb with Bi and design high performance thermoelectric materials, it was important to study the thermochemical properties and the phase equilibria concerning the Gd-Bi-Sb system and to obtain the thermodynamic parameters of the system.

Binary Systems

To obtain a thermodynamic description of a ternary system, the thermodynamic description of each involved binary system was necessary.

Gd-Bi system

The partial phase diagram of the composition range 0–60 at % Bi of the Gd-Bi binary system was firstly measured by Gambino [4], and three intermetallics, Gd_5Bi_3, Gd_4Bi_3 and GdBi, were reported in this system. In 1993, the phase diagram of the Gd-Bi system was re-determined and a new phase, $GdBi_2$ was discovered [5]. The four intermetallics, Gd_5Bi_3, Gd_4Bi_3, $GdBi_2$ and GdBi, were reported in the Gd-Bi binary system. All of the information of the Gd-Bi binary system was accompiled by Wang et al. [6]. The Gd-Bi binary system was assessed by Wang et al. [6]. Figure 1 presented the calculated phase diagram of Gd-Bi system using the thermodynamic description of Wang et al. [6]. The thermodynamic parameters obtained by Wang et al. [6] were adopted in the present work.

Bi-Sb system

The Bi-Sb system was very simple and only two phases, liquid and

***Corresponding author:** Jinming Liu, School of Material Science and Engineering, Jiangxi University of Science and Technology, Ganzhou, 341000, PR China
E-mail: liujm2011@sina.com

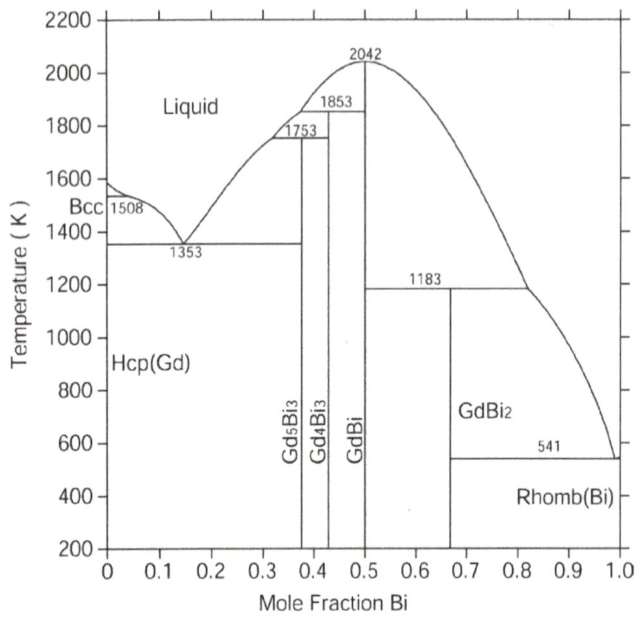

Figure 1: Calculated Gd-Bi phase diagram by Wang et al. [6].

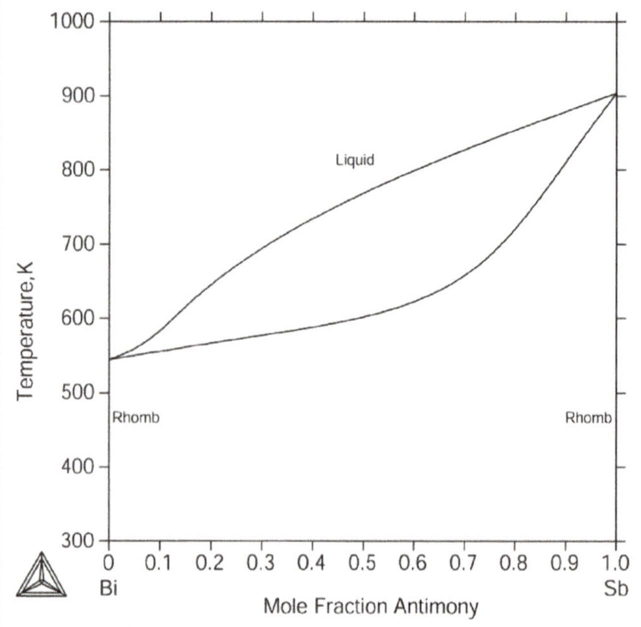

Figure 2a: Calculated Bi-Sb phase diagram by Dinsdale et al. [7], reproduced using the Thermo-Calc software [10].

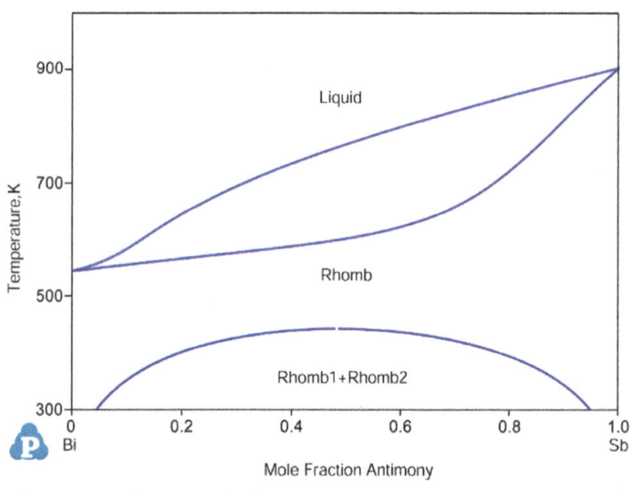

Figure 2b: Calculated Bi-Sb phase diagram by Dinsdale et al. [7], reproduced using the Pandat software [11].

rhomb(Bi, Sb) in this system, which was calculated by Dinsdale et al. [7]. The calculated the thermodynamic parameters [7] were accepted in the assessment of the Bi-Sb-Sn system [8] and the Ag-Bi-Sb system [9]. So the thermodynamic parameters obtained by Dinsdale et al. [7] were adopted in the present work. Figure 2 presented the calculated phase diagram of Bi-Sb system using the thermodynamic description of Dinsdale et al. [7]. Figure 2a presented the Bi-Sb system reproduced using the Thermo-Calc software [10] and Figure 2b presented the Bi-Sb system reproduced using the Pandat software [11].

Gd-Sb system

The Gd-Sb system was firstly determined in the concentration range 0-60 at% Sb by Gambino [4]. Later, Gerasimov [12] updated all the invariant equilibria temperatures. The whole composition and temperature ranges in the Gd-Sb phase diagram were established by Abdusalyamova et al. [13]. The four intermediate compounds, Gd_5Sb_3, Gd_4Sb_3, $GdSb_2$, and GdSb were reported in this system. Recently, the previously called "$GdSb_2$" phase was determined to be $Gd_{16}Sb_{39}$ [14], which was confirmed in the Gd-Bi-Sb ternary system [15]. So the Gd-Sb phase diagram was revised by Borzone et al. [14]. All of the information of the Gd-Sb system was compiled by Li et al. [16]. The updated phase diagram was assessed by Li et al. [16]. The thermodynamic parameters of the Gd-Sb system were well reproduced [16], but the parameters of Bcc(Gd) were difficult to reproduce the temperature of the Bcc(Gd). In the present work, to assess the Gd-Bi-Sb ternary system, the model of the five intermediate compounds, Gd_5Sb_3, Gd_4Sb_3, $Gd_{16}Sb_{39}$, α-GdSb (GdSb-low temperature as GdSb-LT) and β-GdSb(GdSb-high temperature as GdSb-HT), was re-built to cosistent with other thermodynamic parameters of the relative compounds in the Gd-Bi system [6]. So the thermodynamic parameters of the five intermediate compounds were re-assessed in the present work. Figure 3 presented the calculated phase diagram of Gd-Sb system using the thermodynamic parameters of the five intermediate compounds calculated in this work and the thermodynamic parameters of the liquid assessed by Li et al. [16].

Experimental Information on the Gd-Bi-Sb System

The information about the Gd-Bi-Sb system was very scarce. Recently, the phase diagram of the Gd-Bi-Sb ternary system had been determined at room temperature using X-ray powder diffraction and differential scanning calorimetry analysis by Jian et al. [15]. There were composed of six single-phase regions, six two-phase regions and one three-phase region. No ternary compound was reported in the Gd-Bi-Sb ternary system. The compounds, Gd_5Sb_3, Gd_4Sb_3 and GdSb (α-GdSb and β-GdSb) in the Gd-Sb system and their relative compounds, Gd_5Bi_3, Gd_4Bi_3 and GdBi in the Gd-Bi system were formed the continuous solid solutions, respectively. So the six single-phase regions are (Gd, Bi, Sb) solid solution (α phase), $Gd_5(Bi, Sb)_3$ solid solution (β phase), $Gd_4(Bi, Sb)_3$ solid solution (γ phase), Gd(Bi, Sb) solid solution (δ and δ' phases), $Gd_{16}Ga_{39}$ (ε phase) phase and Gd (Hcp phase) phase in the Gd-Bi-Sb system. The maximum solid solubility of Gd in (Gd, Bi, Sb) at room temperature was determined to be 7.5 at %. The solid solubility of Gd in the other phases could not be detected. The above information was described in detail in the Ref. [15].

Thermodynamic Models

Unary phases

The Gibbs energy function for the element *i* (*i*=Gd, Bi, Sb) in the phase φ (φ = liquid, hcp, rhomb) was described as follows,

$$G_i^{\varphi}(T) = {}^0G_i^{\varphi}(T) - H_i^{SER}(298.15K)$$
$$= a + bT + cT\ln T + dT^2 + eT^3 + fT^{-1} + gT^7 + hT^{-9} \quad (1)$$

Where H_i^{SER} (298.15 K) was the molar enthalpy of the element *i* at 298.15 K in its standard element reference (SER) state, hcp for Gd, rhomb for Bi and Sb. The Gibbs energy of the element *i*(*i*=Gd, Bi, Sb), $G_i^{\varphi}(T)$, in its SER state, was denoted by GHSER_i, i.e.,

$$\mathrm{GHSER}_i = {}^0G_i^{\phi}(T) - H_i^{\mathrm{SER}}(298.15\ \mathrm{K}) \quad (2)$$

In the present work, the Gibbs energy functions were taken from the SGTE(Scientific Group of Thermodata Europe) pure elements database compiled by Dinsdale [17] and listed in Table 1.

Solution phases

In the Gd-Bi-Sb system, there are three solution phases, liquid, hcp, and rhomb. Their molar Gibbs energies are described by the following expression:

$$G_{\mathrm{m}}^{\phi} = x_{\mathrm{Gd}}G_{\mathrm{Gd}}^{\phi}(T) + x_{\mathrm{Bi}}G_{\mathrm{Bi}}^{\phi}(T) + x_{\mathrm{Sb}}G_{\mathrm{Sb}}^{\phi}(T) + RT(x_{\mathrm{Gd}}\ln x_{\mathrm{Gd}} + x_{\mathrm{Bi}}\ln x_{\mathrm{Bi}} + x_{\mathrm{Sb}}\ln x_{\mathrm{Sb}}) + {}^{\mathrm{E}}G_{\mathrm{m}}^{\phi} \quad (3)$$

where *R* is the gas constant; x_{Gd}, x_{Bi} and x_{Sb} were the mole fractions of the pure elements Gd, Bi and Sb, respectively; ${}^{\mathrm{E}}G_{\mathrm{m}}^{\phi}$ was the excess Gibbs energy, expressed by the Redlich-Kister polynomial [18].

$${}^{\mathrm{E}}G_{\mathrm{m}}^{\phi} = x_{\mathrm{Gd}}x_{\mathrm{Bi}}\sum_j {}^jL_{\mathrm{Gd,Bi}}^{\phi}(x_{\mathrm{Gd}} - x_{\mathrm{Bi}})^j + x_{\mathrm{Gd}}x_{\mathrm{Sb}}\sum_j {}^jL_{\mathrm{Gd,Sb}}^{\phi}(x_{\mathrm{Gd}} - x_{\mathrm{Sb}})^j + x_{\mathrm{Bi}}x_{\mathrm{Sb}}\sum_j {}^jL_{\mathrm{Bi,Sb}}^{\phi}(x_{\mathrm{Bi}} - x_{\mathrm{Sb}})^j$$
$$+ x_{\mathrm{Gd}}x_{\mathrm{Bi}}x_{\mathrm{Sb}}L_{\mathrm{Gd,Bi,Sb}}^{\phi} \quad (4)$$

where ${}^jL_{\mathrm{Gd,SB}}^{\phi}$, ${}^jL_{\mathrm{Gd,SB}}^{\phi}$ and ${}^jL_{\mathrm{Bi,Sb}}^{\phi}$ were the binary interaction parameters between elements Gd and Bi, Gd and Sb, and Bi and Sb, respectively. Its general form was

$$L^{\varphi} = a + bT + cT\ln T + dT^2 + eT^3 + fT^{-1} \quad (5)$$

but in most case only the first one or two terms were used according to the temperature dependence of the experimental data. $L_{\mathrm{Gd,Bi,Sb}}^{\phi}$ was the ternary interaction parameter expressed as:

$$L_{\mathrm{Gd,Bi,Sb}}^{\phi} = x_{\mathrm{Gd}}\,{}^0L_{\mathrm{Gd,Bi,Sb}}^{\phi} + x_{\mathrm{Bi}}\,{}^1L_{\mathrm{Gd,Bi,Sb}}^{\phi} + x_{\mathrm{Sb}}\,{}^2L_{\mathrm{Gd,Bi,Sb}}^{\phi} \quad (6)$$

where ${}^jL_{\mathrm{Gd,Bi,Sb}}^{\phi} = a_j + b_jT$ (j = 0, 1, 2). a_j and b_j were the parameters to be optimized in this work.

Intermetallic compounds

The compounds, Gd_5Sb_3, Gd_4Sb_3 and GdSb (GdSb-high temperature named as GdSb-HT and GdSb-low temperature named as GdSb-LT) in the Gd-Sb system and their relative compounds, Gd_5Bi_3, Gd_4Bi_3 and GdBi in the Gd-Bi system were formed the continuous solid solutions, respectively. So the three continuous solid solutions single-phase (named as $Gd_m(Bi, Sb)_n$), β-$Gd_5(Bi, Sb)_3$, γ-$Gd_4(Bi, Sb)_3$, δ-Gd(Bi, Sb) and δ'-Gd(Bi, Sb) in the Gd-Bi-Sb system were treated as two-sublattice model $(Gd)_{m/m+n}(Gd, Bi, Sb)_{n/m+n}$. In the Gd-Bi-Sb system, δ-Gd(Bi, Sb) and δ'-Gd(Bi, Sb) were the relative compounds, GdSb-high temperature (as GdSb-HT) and GdSb-low temperature (as GdSb-LT) in the Gd-Sb system. The Gibbs energy per mole of formula unit $Gd_m(Bi, Sb)_n$ was given by the following expression:

$$G_m^{\mathrm{Gd}_m(\mathrm{Bi,\,Sb})_n} = y'_{\mathrm{Gd}}\,y''_{\mathrm{Bi}}\,G_{\mathrm{Gd:Bi}}^{\mathrm{Gd}_m(\mathrm{Bi,\,Sb})_n} + y'_{\mathrm{Gd}}\,y''_{\mathrm{Sb}}\,G_{\mathrm{Gd:Sb}}^{\mathrm{Gd}_m(\mathrm{Bi,\,Sb})_n}$$
$$+ \frac{n}{m+n}RT(y''_{\mathrm{Bi}}\ln y''_{\mathrm{Bi}} + y''_{\mathrm{Sb}}\ln y''_{\mathrm{Sb}})$$
$$+ y'_{\mathrm{Gd}}\,y''_{\mathrm{Bi}}\,y''_{\mathrm{Sb}}\,{}^jL_{\mathrm{Gd:Bi,Sb}}^{\mathrm{Gd}_m(\mathrm{Bi,\,Sb})_n} \quad (7)$$

The parameter y''_i was the site fractions of Bi or Sb (*i* = Bi or Sb) on the second sublattices; the parameter $G_{\mathrm{Gd:*}}^{\mathrm{Gd}_m(\mathrm{Bi,\,Sb})_n}$ represented the Gibbs energies of the compound $Gd_m(Bi, Sb)_n$ when the second sublattice was occupied by only one element Bi or Sb, respectively, which were relative to the enthalpies of pure rhomb for Bi and Sb in their SER state. ${}^jL_{\mathrm{Gd:Bi,Sb}}^{\mathrm{Gd}_m(\mathrm{Bi,\,Sb})_n}$ represented the *j*th interaction parameters (*j*=0) between the elements Bi or Sb on the second sublattice. The other binary intermetallic compounds, $GdBi_2$ and $Gd_{16}Sb_{39}$ in the Gd- Bi and Gd-Sb system were treated as stoichiometric compounds, and were used as the two-sublattice model, which was consistented with the relative binary system [6,16].

Assessment procedure

A general rule for selection of the adjustable parameters was that only those coefficients determined by the experimental values should be adjusted [19]. The assessment was carried out by means of the optimization module PARROT of the thermodynamic software Thermo-Calc [10], which could deal with various kinds of experimental information. A careful examination of thermodynamic descriptions of the Gd-Bi [6] and the Bi-Sb [7] systems were made. The thermodynamic optimization of the Gd-Sb and Gd-Bi-Sb system were carefully performed in this work. The thermodynamic parameters for the Gd-Sb system were optimized on the basis of the experimental information available in the experimental data. The thermodynamic parameters of liquid were taken from the assessed data [16]. The compounds, Gd_5Sb_3, Gd_4Sb_3, $Gd_{16}Sb_{39}$, α-GdSb and β-GdSb in the Gd-Sb system were assessed in the present work. The experimental results of Jian et al. [15] were given more weight during the process of optimization. The thermodynamic parameters for the Gd-Bi-Sb system were optimized on the basis of the experimental information available in the experimental data [15]. The experimental results of Jian et al. [15] were given more weight during the process of optimization. The thermodynamic parameters of liquid, bcc, rhomb and hcp in the Gd-Bi-Sb system, are obtained by a combination of the corresponding Gibbs energy functions from the assessments of the binary systems using Muggianu interpolation of binary excess terms [20]. The interaction parameters of the Gd-Bi and Bi-Sb system and the liquid of the Gd-Sb system were taken from the calculated data assessed by Wang et al. [6], Dinsdale et al. [7] and Li et al. [16]. The binary parameters of the compounds of the Gd-Sb system and the ternary parameters in the Gd-Bi-Sb system were optimized according to the experimental data [15].

Phase	Thermodynamic parameters		Reference
	Temperature (K)	$GHSER_{Gd}=$	[17]
	200–1300	$-11600.525+151.111948T-32.5013T\ln(T)+0.00281265T^{2}$	
		$-1.081237\times10^{-6}T^{3}+421363T^{-1}$	
	1300–1535	$-8106.163+115.953605T-27.459\ T\ln(T)$	
		$-2.76123\times10^{-4}T^{2}-7.37325\times10^{-7}T^{3}$	
	1535–3600	$-125618.511+711.994197T-103.403757T\ln(T)$	
		$+0.015817949T^{2}-6.72548\times10^{-7}T^{3}+30183452T^{-1}$	
		$GHSER_{Bi}=$	[17]
	298.15–544.54	$-7817.776+128.418925T-28.4096529T\ln(T)+0.012338888T^{2}$	
		$-8.381598\times10^{-6}T^{3}$	
	544.54–800	$+30208.022-393.650351T+51.8556592T\ln(T)-0.075311163T^{2}$	
		$+1.3499885\times10^{-5}T^{3}-3616168T^{-1}+1.66145\times10^{25}T^{-9}$	
	800–1200	$-11045.664+182.548971T-35.9824T\ln(T)+0.0074266T^{2}$	
		$-1.046\times10^{-6}T^{3}+1.66145\times10^{25}T^{-9}$	
	1200–3000	$-7581.312+124.77144T-27.196T\ln(T)+1.66145\times10^{25}T^{-9}$	
		$GHSER_{Sb}=$	[17]
	298.15–903.78	$-9242.858+156.154689T-30.5130752T\ln(T)$	
		$+7.748768\times10^{-3}T^{2}-3.003415\times10^{-6}T^{3}+100625T^{-1}$	
	903.78–2000	$-11738.83+169.485872T-31.38T\ln(T)+1616.8491024T^{-9}$	
liquid		model: $(Gd, Bi, Sb)_{1}$	
		$G(\text{liquid, Gd})=$	[17]
	200–1000	$+13557.615-8.612727T+GHSER_{GD}$	
	1000–1535	$+80721.646-609.764875T+74.8351133\ T\ln(T)-0.061538012T^{2}$	
		$+6.061728\times10^{-6}T^{3}-10325075T^{-1}$	
	1535–3600	$-3485.133+182.116258T-37.1539\ T\ln(T)$	
		$G(\text{liquid, Bi})=$	[17]
	298.14–544.540	$+11246.066-20.636509T-5.9549\times10^{-19}T^{7}+GHSER_{Bi}$	
	544.540–800	$+11336.26-20.810418T-1.66145\times10^{+25}T^{-9}+GHSER_{Bi}$	
	800–1200	$+11336.26-20.810418T-1.66145\times10^{+25}T^{-9}+GHSER_{Bi}$	
	1200–3000	$+3754.947+103.961021T-27.196\ T\ln(T)$	
	1234.93–3000	$-3587.111+180.964656T-33.472T\ln(T)$	
		$G(\text{liquid, Sb})=$	[17]
	298.14–903.78	$+19822.328-21.923164T-1.74847\times10^{-20}T^{7}+GHSER_{SB}$	
	903.78–2000	$+8175.359+147.455986T-31.38\ T\ln(T)$	
		${}^{0}L^{\text{liq.}}_{\text{Gd,Bi}}=-200000-2.0294T$	[6]
		${}^{1}L^{\text{liq.}}_{\text{Gd,Bi}}=-3168.5926+4.7319T$	[6]
		${}^{3}L^{\text{liq.}}_{\text{Gd,Bi}}=-30000+2.8846T$	[6]
		${}^{3}L^{\text{liq.}}_{\text{Gd,Bi}}=-30000+2.8846T$	[6]
		${}^{0}L^{\text{liq.}}_{\text{Bi,Sb}}=+2230.0+0.06T$	[7]
		${}^{0}L^{\text{liq.}}_{\text{Gd,Sb}}=-267653.636-25.5971477T$	[16]
		${}^{1}L^{\text{liq.}}_{\text{Gd,Sb}}=+13256.1962-30.01213T$	[16]
		${}^{2}L^{\text{liq.}}_{\text{Gd,Sb}}=+42899.3741+30.0T$	[16]
		${}^{1}L^{\text{liq.}}_{\text{Gd,Bi,Sb}}=-115000.0$	This work
		${}^{1}L^{\text{liq.}}_{\text{Gd,Bi,Sb}}=-115000.0$	This work
		${}^{2}L^{\text{liq.}}_{\text{Gd,Bi,Sb}}=+5000.0$	This work
bcc		model: $(Gd, Bi, Sb)_{1}$	
	298.14–3000	$G(\text{bcc, Gd})=$	[17]
	200–1000	$+3572.085-2.304997T+GHSER_{Gd}$	
	1000–1535	$+86879.486-747.960507T+95.1961984T\ln(T)-0.072542756T^{2}$	

		$+7.116157\times10^{-6}T^{3}-12664161T^{-1}$	
	1535−1587	$-14097.626+192.503674T-37.656T\ln(T)$	
	1587−3600	$-18620.027+217.622313T-40.9284427\,T\ln(T)$	
		$+9.53751\times10^{-4}T^{2}-4.1967\times10^{-8}T^{3}+1107440T^{-1}$	
		$G(\text{bcc, Bi})=$	[17]
	298.14−3000	$+11297-13.9T+\text{GHSER}_{\text{Bi}}$	
		$G(\text{bcc, Sb})=$	
	298.14−2000	$+19874-15.1T+\text{GHSER}_{\text{Sb}}$	[17]
		${}^{0}L^{\text{fcc}}_{\text{Gd,Sb}}=-267618.599$	This work
		${}^{0}L^{\text{bcc}}_{\text{Gd,Bi,Sb}}=+0.0$	This work
		${}^{2}L^{\text{bcc}}_{\text{Gd,Bi,Sb}}=+0.0$	This work
		${}^{2}L^{\text{bcc}}_{\text{Gd,Bi,Sb}}=+0.0$	This work
hcp_A3		model: $(\text{Gd, Bi, Sb})_1$	This work
	200−3600	$G(\text{hcp\_A3, Gd})=+\text{GHSER}_{\text{Gd}}$	[17]
	298.14−3000	$G(\text{hcp\_A3, Bi})=+9900-11.8T+\text{GHSER}_{\text{Bi}}$	[17]
	298.14−2000	$G(\text{hcp\_A3, Sb})=+19874-13T+\text{GHSER}_{\text{Sb}}$	[17]
		${}^{0}L^{\text{hcp\_A3}}_{\text{Gd,Sb}}=-269106.632$	This work
		${}^{0}L^{\text{hcp\_A3}}_{\text{Gd,Bi,Sb}}=+0.0$	This work
		${}^{1}L^{\text{hcp\_A3}}_{\text{Gd,Bi,Sb}}=+0.0$	This work
		${}^{2}L^{\text{hcp\_A3}}_{\text{Gd,Bi,Sb}}=+0.0$	This work
α(Gd,Bi,Sb) (rhomb)		model: $(\text{Gd, Bi, Sb})_1$	
	298.14−6000	$G(\text{rhomb, Gd})=+\text{GHSER}_{\text{Gd}}+5000$	This work
	298.14−2000	$G(\text{rhomb, Bi})=+\text{GHSER}_{\text{Bi}}$	[17]
	298.14−2000	$G(\text{rhomb, Sb})=+\text{GHSER}_{\text{Sb}}$	[17]
		${}^{0}L^{\text{rhomb}}_{\text{Bi,Sb}}=+10150.0-6.3000T$	[7]
		${}^{1}L^{\text{rhomb}}_{\text{Bi,Sb}}=+150$	[7]
		${}^{0}L^{\text{rhomb}}_{\text{Gd,Bi,Sb}}=+0.0$	This work
		${}^{2}L^{\text{rhomb}}_{\text{Gd,Bi,Sb}}=-2202320.0-20.03232T$	This work
		${}^{2}L^{\text{rhomb}}_{\text{Gd,Bi,Sb}}=-2202320.0-20.03232T$	This work
β-$Gd_5(Bi, Sb)_3$		model: $(\text{Gd})_{0.625}(\text{Bi, Sb})_{0.375}$	This work
		$G^{\beta}_{\text{Gd:Bi}}=+0.625\text{GHSER}_{\text{Gd}}+0.375\text{GHSER}_{\text{Bi}}$ $-75332.61+4.1T$	[6]
		$G^{\beta}_{\text{Gd:Sb}}=+0.625\text{GHSER}_{\text{Gd}}+0.375\text{GHSER}_{\text{Sb}}$ $-104159.14+3.1036T$	This work
		${}^{0}L^{\beta}_{\text{Gd:Bi,Sb}}=-100200.0$	This work
γ-$Gd_4(Bi, Sb)_3$		model: $(\text{Gd})_{0.5714}(\text{Bi, Sb})_{0.4286}$	This work
		$G^{\gamma}_{\text{Gd:Bi}}=+0.5714\text{GHSER}_{\text{Gd}}+0.4286\text{GHSER}_{\text{Bi}}$ $-83701.2+5.1T$	[6]
		$G^{\gamma}_{\text{Gd:Sb}}=+0.5714\text{GHSER}_{\text{Gd}}+0.4286\text{GHSER}_{\text{Sb}}$ $-118710.41+5.8351T$	This work
		${}^{0}L^{\gamma}_{\text{Gd:Bi,Sb}}=-100200.0$	This work
δ(GdSb-HT)		model: $(\text{Gd})_{0.5}(\text{Bi, Sb})_{0.5}$	This work

		$-119418.209+1.4792T$	This work
		$G^{å}_{Gd:Sb} = +0.5GHSER_{Gd} + 0.5GHSER_{Sb}$ $-119418.209+1.4792T$	This work
		${}^{0}L^{å}_{Gd:Bi,Sb} = -100200.0$	This work
δ'(GdSb-LT)		model: $(Gd)_{0.5}(Bi, Sb)_{0.5}$	This work
		$G^{å}_{Gd:Bi} = +0.5GHSER_{Gd} + 0.5GHSER_{Bi} - 88053.58+3.8T$	[6]
		$G^{å}_{Gd:Sb} = +0.5GHSER_{Gd} + 0.5GHSER_{Sb}$ $-119399.189+1.4702T$	This work
		${}^{0}L^{å}_{Gd:Bi,Sb} = -100200.0$	This work
$GdBi_2$		model: $(Gd)_{0.333}(Bi)_{0.667}$	
		$G^{GdBi_2}_{Gd:Bi} = +0.333GHSER_{Gd} + 0.667GHSER_{Bi}$ $-68103.64+4.84T$	[6]
ε($Gd_{16}Sb_{39}$)		model: $(Gd)_{0.29}(Sb)_{0.71}$	This work
		$G^{å}_{Gd:Bi} = +0.29GHSER_{Gd} + 0.71GHSER_{Sb}$ $-87952.125+17.1823T$	This work

Table 1: Thermodynamic parameters of the Gd–Bi–Sb system (in J $mole^{-1}$ of the formula units).

Results and Calculations

The Gd-Bi and the Bi-Sb systems were assessed by Wang et al. [6], and Dinsdale et al. [7], respectively. Their thermodynamic descriptions were accepted in the present work and listed in Table 1. Figures 1 and 2 presented the Gd-Bi and Bi-Sb phase diagrams using the thermodynamic description of Wang et al. [6], and Dinsdale et al. [7], respectively. There was some different at low temperature using the Thermo-Calc Software [10] and the Pandat software [11] to reproduce the Bi-Sb phase diagram, as shown in Figures 2a and 2b.

Figure 3 presented the Gd-Sb phase diagram using the thermodynamic description of liquid assessed by Li et al. [16] and the calculated result of the compounds in the present work, and comparison with the experimental data [13]. Figure 4 presented the calculated standard enthalpies of formation in The Gd-Sb system at 300 K and comparison with the experimental data [14]. The reference states are hcp for Gd and rhomb for Sb. Reasonable agreement was obtained between the calculated results and the experimental data [14]. Figure 5 was the calculated enthalpies of mixing of liquid in the Gd-Sb system at 6000 K.

The calculated standard enthalpies of formation in the Gd-Sb system and comparison with the experimental data [14,21-25] were shown in Figure 4 and Table 2. The calculated invariant equilibria in the Gd–Sb system were listed in Table 3. As shown in the tables, a very good agreement was obtained between the calculated results and the experimental data [14]. The Gd-Bi-Sb system were optimized on the basis of the available experimental data [15]. The thermodynamic description of the Gd-Bi-Sb system obtained in the present work was shown in Table 1. Figure 6 were the calculated isothermal section at 300 K using the present thermodynamic description in comparison with experimental data [15] in the Gd-Bi-Sb system. Satisfactory agreements were obtained between the calculated results and the experimental data [15] in Figure 6. For the Bi-Sb phase diagram at low temperature (under 300 K), the rhomb was broken down into two component parts rhomb1 and rhomb2, as shown in Figure 2b. So there was some different at the rhomb phase side from the experiment data [15].

Compound	Temperature, K	$\Delta_f H$	References
Gd_5Sb_3	300	-110	[16]
Gd_5Sb_3	673	-141.9	[23]
Gd_5Sb_3	300	-103.664	This work
Gd_5Sb_3	673	-93.623	This work
Gd_4Sb_3	300	-116	[16]
Gd_4Sb_3	673	-145.1	[23]
Gd_4Sb_3	300	-118.254	This work
Gd_4Sb_3	673	-108.228	This work
GdSb	300	-120	[16]
GdSb	300	-131	[25]
GdSb	693	-102.5	[21]
GdSb-LT	300	-118.994	This work
GdSb-LT	753	-106.672	This work
GdSb-LT	693	-108.429	This work
$Gd_{0.52}Sb_{0.48}$	≈753	-137.1	[22]
$Gd_{16}Sb_{39}$ [a]	300	-78	[16]
$Gd_{16}Sb_{39}$	300	-87.697	This work
$Gd_{16}Sb_{39}$	803	-73.972	This work
$GdSb_2$	803	-96	[24]

[a]The nominal composition of $GdSb_2$ was determined by Borzone et al. [14] to be $Gd_{16}Sb_{39}$ (70.91 at %Sb).

Table 2: Calculated formation enthalpy (kJ mole of atoms).

Conclusions

The thermodynamic parameters in the Gd-Sb and Gd-Bi-Sb ternary system were critically evaluated from the experimental information available in the literature. A set of self-consistent thermodynamic

Reaction	*T* (K)	*x*(Sb)			Reference
Liq. + bcc(Gd) → hcp(Gd)	1505	0.065	0.025	0.02	[13]
	1501	0.058	0.0191	0.0183	[16]
	1505	0.0821	0.0408	0.0387	This work
Liq. + hcp(Gd) → Gd_5Sb_3	1393	0.135	0.025	0.375	[13]
	1390	0.1018	0.0402	0.375	[16]
	1393	0.1329	0.0685	0.375	This work
Liq. + Gd_5Sb_3 → Gd_4Sb_3	1913	0.35	0.375	0.4286	[13]
	1917	0.3029	0.375	0.4286	[16]
	1913	0.3016	0.375	0.4286	This work
Liq. + Gd_4Sb_3 → βGdSb	2043	0.389	0.4286	0.50	[13]
	2048	0.3517	0.4286	0.50	[16]
	2043	0.3515	0.4286	0.50	This work
Liq. → GdSb-HT	2403	0.50	0.50	-	[13]
	2401	0.50	0.50	-	[16]
	2403	0.50	0.50	-	This work
GdSb-HT → GdSb-LT	2113	0.50	0.50	-	[13]
	2113	0.50	0.50	-	[16]
	2114	0.50	0.50	-	This work
Liq. + GdSb-LT → $Gd_{16}Sb_{39}$	1053	0.958	0.7091[a]	0.50	[13]
	1054	0.9665	0.709	0.50	[16]
	1053	0.9514	0.709	0.50	This work
Liq. + $Gd_{16}Sb_{39}$ → rhomb	897	>0.99	0.7091[a]	1.0	[13]
	903	>0.99	0.709	1.0	[16]
	903	0.9984	0.709	1.0	This work

[a]The nominal composition of $GdSb_2$ was determined by Borzone et al. [14] to be $Gd_{16}Sb_{39}$ (70.91 at %Sb).

Table 3: Invariant reactions of the Gd–Sb system.

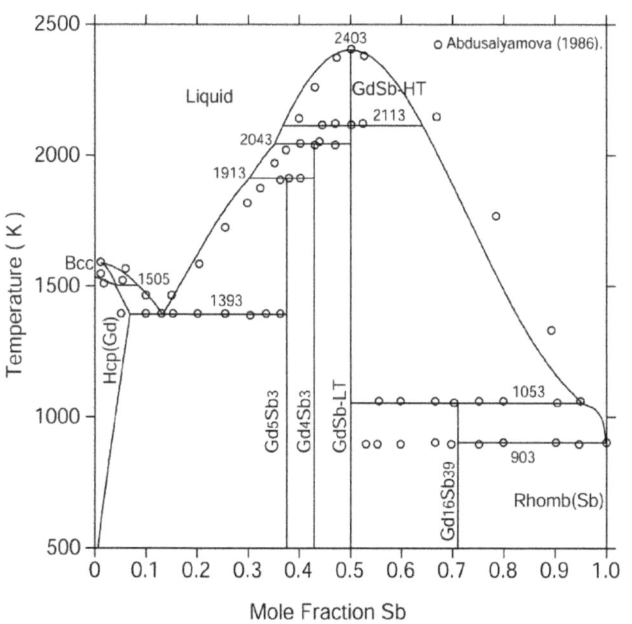

Figure 3: Calculated Gd-Sb phase diagram in the present work and using the parameters of liquid by Li et al. [16], and comparison with the experimental data [13].

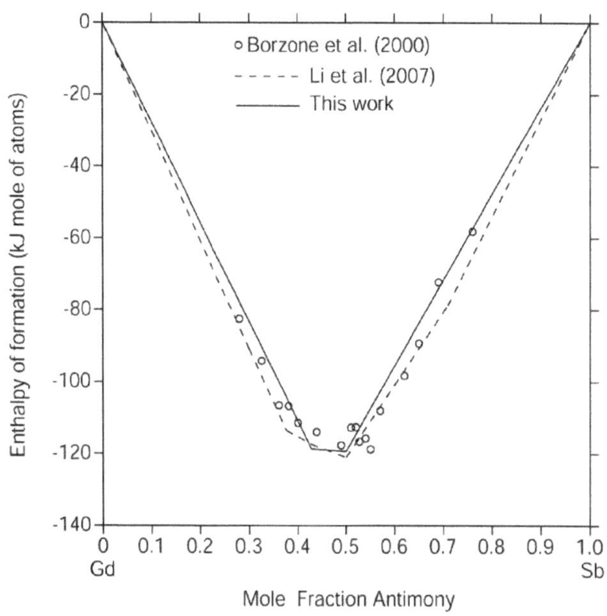

Figure 4: Calculated enthalpies of formation at 300 K in the Gd-Sb system and comparison with the experimental data [14] and the calculated data [16]. The reference states for the elements were hcp for Gd and rhomb for Sb.

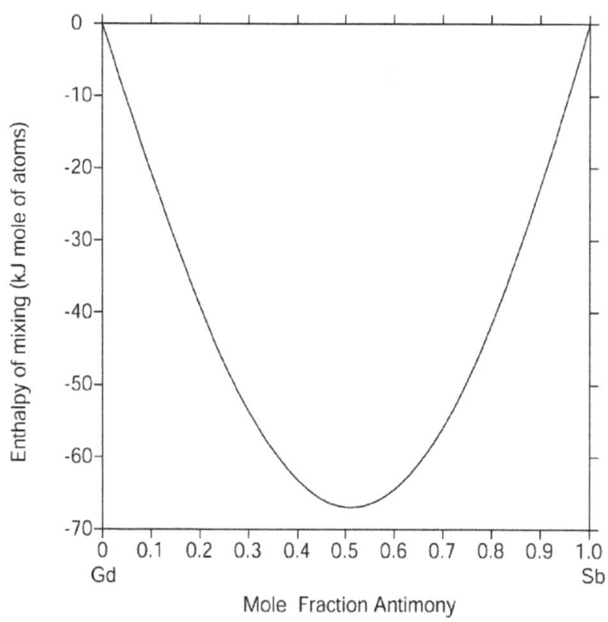

Figure 5: Calculated enthalpies of mixing of liquid at 6000 K in the Gd-Sb system. The reference states were liquid for Gd and Sb.

parameters describing the Gibbs energy of each individual phase as a function of composition and temperature was derived.

Acknowledgements

This work was supported by National Natural Science Foundation of China (NSFC) (Grant No. 51264012), Jiangxi Province Science and Technology Support Program (Grant No. 20141BBE50006) and Jiangxi Province Natural Science Foundation (Grant No. 20142BAB206015), and the China Postdoctoral Science Foundation (Grant No. 2014M551712).

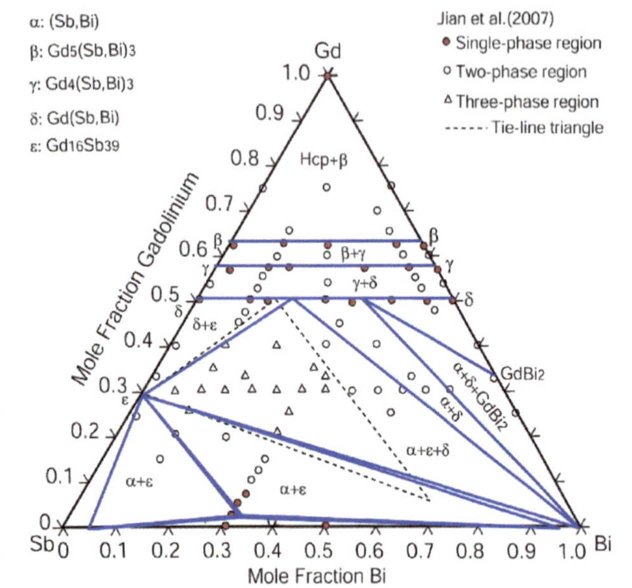

Figure 6: Calculated isothermal section of the Gd-Bi-Sb system at 300 K by the present thermodynamic description and comparison with the experimental data [15].

References

1. Pecharsky VK, Gschneidner KA (1997) Giant Magnetocaloric Effect in $Gd_5(Si_2Ge_2)$. Phys Rev Lett 78: 4494-4497.
2. Gschneidner KA, Pecharsky VK (1997) Rare Earth: Science, Technology and Application III. In: Bautista RC, Bounds CO, Ellis TW, Kilbourn BT (eds.), The Minerals, Metals & Materials Society.
3. Niu XJ, Gschneidner KA, Pecharsky AO, Pecharsky VK (2001) Crystallography,magnetic properties and magnetocaloric effect in $Gd_4(B_xSb_{1-x})_3$ alloys. J Magn Magn Mater 234: 193-206.
4. Gambino RJ (1967) Rare-earth Sb and Bi compounds with the Gd_4Bi_3 (anti-Th_3P_4) structure. J Less-Common Met 12: 344-352.
5. Abulkhaev VD (1993) Gd-Bi phase diagram. Izv Ross Akad Nauk Met 1: 187-190.
6. Wang J, Li C, Guo C, Du Z, Wu B (2013) Thermodynamic assessment of the Gd-Bi and the Ho-Bi systems. Calphad 41: 1-5.
7. Dinsdale AT, Kroupa A, Vizdal J, Vrestal J, Watson A, et al. (2006) COST 531 Database for Lead-free Solders. Version 2.0.
8. Manasijevic D, Vrestal J, Minic D, Kroupa A, Zivkovic D, et al. (2007) Phase equilibria and thermodynamics of the Bi-Sb-Sn ternary system. J Alloys Compd 438: 150-157.
9. Liu J, Guo C, Li C, Du Z (2012) Thermodynamic description of the Ag-Bi-Sb system. Thermochim Acta 539: 44-50.
10. Sundman B, Jansson B, Andersson JO (1985) The Thermo-Calc databank system. Calphad 9: 153-190.
11. Chen SL, Zhang F, Daniel S (2003) Calculating phase diagrams using PANDAT and PanEngine. JOM 55: 48-51.
12. YI Gerasimov (1971) Dokl Phys Chem 197: 258-261.
13. Abdusalyamova MN, Burnashev OR, Miranov KY (1986) The Alloy Systems Ln-Sb. J Less-Common Met 125: 1-6.
14. Borzone G, Fornasini ML, Parodia N, Ferro R (2000) Gd-Sb system: standard enthalpies of formation of solid alloys and crystal structure of $Gd_{16}Sb3_9$. Intermetallics 8: 189-194.
15. Jian YX, Li JQ, Liu FS, Ao WQ (2007) Phase equilibria in the Gd–Bi–Sb ternary system at room temperature. J Alloys Compd 433: 125-128.
16. Li JQ, Jian YX, Chen HL, Liu YQ (2007) Thermodynamic assessment of the Gd-Sb system. Intermetallics 15: 1332-1336.
17. Dinsdale AT (2006) SGTE Pure elements (unary) database. Version 4.5.
18. Redlich O, Kister AT (1948) Algebraic representation of thermodynamic properties and the classification of solutions. Ind Eng Chem 40: 345-348.
19. Lukas HL, Fries SG (1992) Demonstration of the use of BINGSS with the Mg-Zn system as example. J Phase Equilib 13: 532-542.
20. Muggianu YM, Gambino M, Bros JP (1975) Enthalpies of formation of liquid Bi–Ga–Sn tin alloys at 723 K--The analytical representation of the total and partial excess functions of mixing. J Chim Phys 72: 83-88.
21. Pratt JN, Chua KS (1970) Thermodynamic studies of some rare-earth alloys. Final report min tech agreement no. 2027/048/RL, University of Birmingham.
22. Goryacheva VI, Nikol'skaya AV, Gerasimov Yal (1971) Dokl Akad Nauk SSSR 197: 389-392.
23. Goryacheva VI, Messauden N, Nikol'skaya AV, Gerasimov YI (1977) Vestn Mosk Univ. Ser 2, Khim 18: 749.
24. Kober VI, Nichkov IF, Raspopin SP, Kuzminnykh VM (1983) Splavy Redk Met. Osobymi Fiz. Svoistvami: Redkozem Blagorodn Met, pp: 132-135.
25. Sh G, Viksman SP, Gordienko I, Akad Nauk SSSR (1984) Neorg Mater 20: 1441-1444.

32

Stellate MSN-based Dual-enzyme Nano-Biocatalyst for the Cascade Conversion of Non-Food Feedstocks to Food Products

Hao-Hsin Chi[1], Daniela R Radu[1,2], Geraldine Dezayas[1,3], Miranda Penney[1,3] and Cheng-Yu Lai[1*]

[1]*Department of Chemistry, Delaware State University, 1200 N. DuPont Highway, Dover, DE, 19901, USA*
[2]*Department of Materials Science and Engineering, University of Delaware, Newark, DE, 19716, USA*
[3]*Department of Biological Sciences, Delaware State University, 1200 N. DuPont Highway, Dover, DE, 19901, USA*

Abstract

The successful demonstration of a tandem enzymatic catalyst which utilizes stellate macroporous silica nanospheres (Stellate MSN) platform as dual-enzyme host is reported herein. Upon simultaneous loading of beta-glucosidase and glucose isomerase inside their porous structure, Stellate MSNs-featuring a hierarchical pore arrangement and large surface area, show capability to perform a cascade reaction that converts cellobiose, a cellulosic hydrolysis product, into glucose and further to fructose. The silica platform provides a modality for substrate channelling which involves the transfer of the cascade intermediate, glucose, to the next enzyme without first diffusing to the bulk. A key aspect to this proof-of-concept is the two-enzyme system working in an optimized pH domain to fit the modus operandi for both enzymes. The concept could be extrapolated to other enzyme tandems, with potential to impact dramatically enzymatic processes which require multi-catalyst, one-pot transformations.

Keywords: Catalyst; Macromolecules; Glucose; Enzymatic catalysis

Introduction

Biotechnology is at the core of the environmentally friendly chemical processes of the 21st century. Enzymes, macromolecules capable to accomplish chemical transformation with high selectivity, are key components of the biotechnology toolbox. Applications, such as new industrial processes, creating new functional foods, and contributing to medical treatments are a few areas where innovative approaches toward reducing waste and harmful catalytic processes are required, thus driving the desire to replace existing processes with enzymatic catalysis. However, to fulfill current demands of chemical industry, enzyme productivity, recyclability and increased shelf life are required to enable the replacement of existing processes at scales and in an economical fashion. Food security is one of paramount importance in the wake of increased Globe population. As such, potential sources of food basic ingredients such as glucose and fructose from non-food biomass renewable resources such as switch grass, corn stover, and other lignocellulosic waste, could significantly supplement the ever-growing contemporary food needs. The principal constituent of lignocellulosic biomass is cellulose, a glucose-based polymer. The conversion of cellulose to glucose involves a mixture of enzymes termed cellulase. In microorganisms such as *Trichoderma viride*, the conversion is step-wise process: first, beta-1,4 glucanase breaks the glyosidic linkage to cellobiose—a glucose dimer (Figure 1) which is subsequently converted to glucose by beta-glucosidase (β-glucosidase). However, assuming that the enzymatic catalysis research efforts ultimately aim to product commercialization, high-fructose corn syrup (HFCS) is a more desired outcome, given its sweetness, equating 1.3 times more than sucrose, and the amenability of this product to be used by diabetics. Conversion of glucose to fructose is accomplished by Glucose isomerase (GI) enzyme, catalyzing the reversible isomerization of D-glucose (glucose) to D-fructose (fructose). This enzyme, with high occurrence in prokaryotes, is subjected to research efforts that could improve the catalytic ability and application in industry [1]. The process of converting cellulose to fructose is illustrated in Figure 1.

In the process of converting cellulose to glucose, β-1,4,-glucanase enzyme leads to a glucose dimer, called cellobiose. The enzyme β-glucosidase is necessary to accomplish further conversion to glucose. β-glucosidase, known also as cellobiose, is a ubiquitous component of cellulases [2]. Given that cellobiose inhibits the reaction, cellulose hydrolysis is greatly impaired. Without sufficient β-glucosidase present in the enzymatic process of cellulose, little glucose is formed; the main product being cellobiose. Published work has showed that supplementation of commercially produced cellulases with β-glucosidase increases the rate and extent of glucose production [3,4]. The native level of β-glucosidase activity in cellulase is, therefore, insufficient for the maximum rate and extent of glucose production to be reached [2]. Since the substrate of β-glucosidase, cellobiose, is water soluble, an immobilized (water-insoluble) enzyme preparation could be used to supplement commercial cellulase/cellulose mixtures, and such immobilized enzyme could be subsequently recovered and reused. Enzyme immobilization provides an excellent opportunity increase turnover over a considerable period of time, enzyme availability to the substrate is greater. Several natural and synthetic supports have been assessed for their efficiency for enzyme immobilization [5].

The typical approach is one enzyme per support. Several examples of enzyme immobilization onto porous silica have been reported [6-12]. Co-immobilization of enzymes on various platforms is an emerging field, and involves considerations of various factors: spanning from facile access of substrate to the next enzyme in the cascade, optimization of catalytic activity to allow performance of all components, and the stability of enzyme in the solid support [13]. Porous silica materials emerged as scaffolds due to their resilient nature, capacity to protect the encapsulated enzyme, easy access of reactants in a confined space, and large surface area [14]. Among these platforms,

***Corresponding author:** Cheng-Yu Lai, Department of Chemistry, Delaware State University, 1200 N. DuPont Highway, Dover, DE, 19901, USA
E-mail: cylai@desu.edu

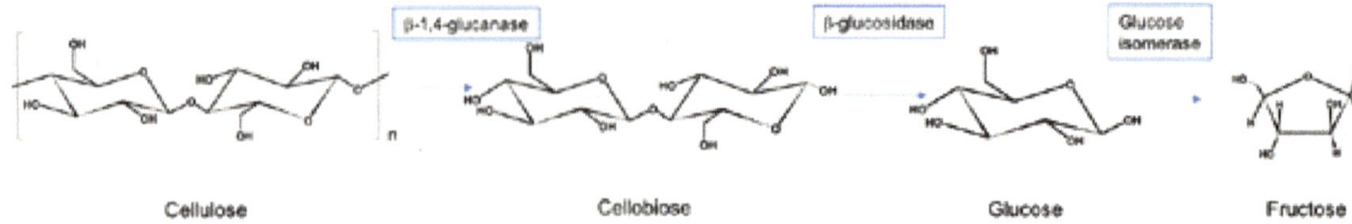

Figure 1: Schematic representation of the step-wise transformation process from cellulose to fructose.

Stellate MSN, a macroporous silica nanospheres platform, presents the advantage of facile and reproducible synthesis, narrow particle size distribution, hierarchical pores and significant surface area. To the best of our knowledge, this is the first report of dual-enzyme immobilization on Stellate MSN showing a functioning cascade enzymatic reaction, in optimized reaction conditions that are suitable for both enzyme activities.

In addition, present work aims to demonstrate that immobilization of two cascade reaction enzymes on a Stellate MSN platform would control the direct transfer of a reactant from one catalytic site to another without first diffusing it to the bulk environment. The process of intermediates control along a pathway is termed substrate channeling, and is key to efficient one-pot multi-step catalysis. In addition to channeling, another aspect of multi-step enzymatic catalysis is the significant overlap of enzymatic reaction conditions. Enzymes optimal function require a defined set of conditions, including reaction pH, and participating ions concentration, and temperature. Present work achieved optimal reaction conditions for β-glucosidase and glucose isomerase tandem system on Stellate MSN platform while demonstrating channelling effect occurring in the conversion of cellobiose to fructose.

Materials and Methods

Materials

Tetraethyl orthosilicate (TEOS), cetyltrimethylammonium tosylate, triethanolamine, anhydrous ethanol, D-(+)-Cellobiose, D-(+)-glucose, D-(-)-fructose, Tris (hydroxymethyl) aminomethane hydrochloride, magnesium sulphate, 4-Morpholinepropanesulfonic acid, Cobalt(II) chloride hexahydrate, acetonitrile (HPLC grade), Immobilized Glucose Isomerase from *Streptomyces murinus* (Sweetzyme T, Novazyme) and β-glucosidase from almonds lyophilized powder were purchased from Sigma Aldrich without further purification. *E. coli* strains DH5α, IPTG (isopropylthio-β-galactoside), B-PER™ Bacterial Protein Extraction Reagent, and Ni-NTA Agarose and Kanamycin were purchased from Thermo-Fisher. BL21 (DE3) Chemically Competent *E. coli* was purchased from Agilent. Nanopure water was used in all experiments.

Methods

The experiments carried out in this work involve the synthesis of the silica platform, preparation of enzymes, and the fabrication and validation of the enzyme catalyst platforms (both dual-enzyme and single enzyme immobilization were prepared for the control experiments).

Synthesis and characterization of stellate MSN platform

The synthetic procedure of MSNs with stellate morphology was modified from previous literature reports [15,16]. In a typical experiment, a mixture of 1.920 g of cetyl-trimethylammonium tosylate (CTATos), 0.4 g of triethanolamine (TEA) and 100 mL of nanopure water was stirred at 80°C for 1 h, and then 14.58 g of tetraethyl-orthosilicate (TEOS, 6.998 mol) was quickly added into the surfactant solution. The molar composition of the precursors was 1.0 SiO_2: 0.06, CTATos: 0.026, TEA: 80.0 H_2O. The mixture was stirred at 80°C for another 2 h. The solid product resulting upon reaction was filtered, washed copiously with nanopure water and methanol, and dried in under vacuum at 60°C for 12 h. To remove the surfactant template (CTATos), 1.50 g of as-synthesized Stellate MSN was refluxed for 24 h in a methanolic solution of hydrochloric acid (9.00 mL of HCl (37.4%) in 160.00 mL of methanol) followed by extensive washes with nanopure water and methanol. The resulting surfactant-free Stellate MSN material was placed under high vacuum to remove the remaining solvent in pores of the material. Nitrogen adsorption-desorption isotherms of the sorbents were obtained on a Nova 4200e (Quantachrome, Boynton Beach, FL) surface area and porosity analyzer) in the relative pressure range of 0.05-0.95 at 77K. Stellate MSN was degassed at 110°C for 3 h under high vacuum. The total pore volume was calculated as the adsorbed volume of liquid nitrogen at the relative pressure of 0.95. Specific surface areas were calculated with the Brunauer-Emmett-Teller (BET) method. Pore volume was determined by Non-Local Density Functional Theory model. High-resolution TEM images were obtained on a JEM-2100F model operated at 200 kV. SEM images were obtained on a Hitachi S4700 instrument.

Glucose isomerase enzyme: plasmid construction and bacterial expression in *E. coli*.: The Glucose Isomerase enzyme is not commercially available in pure form and therefore it was produced in-house by gene cloning. The gene encoding hyperthermostable D-glucose-isomerase from *Thermotoga neapolitana*-5068 (*xlyA* gene) was chemically synthesized (GenScript, Piscataway, NJ). The synthetic gene was cloned as an NdeI-XhoI insert into plasmid pET28a (+), thereby creating pTNXI [17]. The pTNXI plasmid was transformed into chemically competent *E. coli* DH5α according to the manufacturer's instructions.

Recombinant protein purification: The plasmid carrying the Thermotoganeapolitana glucose/xylose isomerase gene (PET28a-TNXI) was transformed into BL21 (DE3) bacteria for protein expression. Bacterial cells were grown in 50 mL of LB media with 50 μg/mL kanamycin to an A600 of 0.6-0.8 prior to induction with isopropyl-β-D-thiogalactopyranoside (IPTG, 0.5 mM). After 12 h, the cells were collected and cells pellets resuspended in 20 mL of Bugbuster (Novagen) protein extraction reagent supplemented with Benzonase (20 U/mL, Novagen), 1.0 mg/mL of lysozyme (Sigma), and one tablet of EDTA free protease inhibitor cocktail (Roche). After 60 min rocking at 4°C, the suspension was subjected to centrifugation at 7000 rpm for 30 min at 4°C to remove cell debris. Recombinant proteins were purified on nitrilotriacetic acid-Ni^{2+} agarose columns (Thermo-Fisher). After extensive washing, the recombinant proteins were eluted with 10 mL of 50 mM Tris, 300 mM NaCl, 500 mM imidazole, pH 8 and then immediately dialyzed at 4°C into M Buffer, consisting of 50 mM MOPS, 5 mM $MgSO_4$, 0.5 mM $CoCl_2$, with pH 7.

Protein quantification, electrophoresis, and molecular mass determination: Protein concentration was determined according to the Bradford method using a BSA standard curve and a prefabricated assay solution (Bio-Rad). SDS-PAGE was carried out using WedgeWell Tris-Glycine precast gradient gels (4–12%, ThermoFisher) and Tri-color Prestained Protein Marker II (Bioland Scientific). SimplyBlue™ safe stain was used for visualization of the protein bands (Invitrogen) (The gel images are included in Supporting Information). The theoretical mass was derived from the ExPASy ProtParam tool (http://web.expasy.org/protparam/).

Enzyme immobilization on stellate MSN

Dual-enzyme immobilization: A volume of 2 mL recombinant glucose isomerase (0.2 mg/mL in in M buffer, 0.4 mg) was mixed with 2mL β-glucosidase solution (1 mg/mL in M buffer, 2 mg). Enzyme immobilization was performed by mixing the enzymes solution with 50 mg of the mesoporous supports in 4mL M buffer, at 4°C under stirring for 12 h at 200 rpm. The suspension was then centrifuged, and the enzyme immobilized on Stellate MSN was washed with buffer. To quantify the immobilization efficiency, the remaining protein in supernatant was measured by a previously described method, showing that enzyme was completely immobilized. The resulting dual-enzyme Stellate MSN was resuspended in 4mL buffer and stored at 4°C.

Single-enzyme immobilization: Single-enzyme immobilization of recombinant GI enzyme on Stellate MSN, and separately, of the β-glucosidase on Stellate MSN were also performed for control experiments in the same manner as described for the dual enzyme immobilization.

Enzymatic assays

Specific substrate conversion assay by HPLC analysis: The products of enzymatic analyses such as glucose, fructose or mixtures were performed on a Shimadzu Nexera UHPLC/HPLC System with a Restek Ultra Amino column. After enzyme conversion reaction, the reaction products were purified by syringe filtration, possible products were identified, and their calibration curves were developed. An isocratic run at 20% water and 80% acetonitrile was used for the quantitative analyses of all possible analytes. Flow rate of mobile phase was 1.5 mL/min. The column temperature was set at 35°C and the injection volume of samples was 25 μL for all analyses.

Dual-enzyme MSN enzymatic conversion of cellobiose to fructose: A volume of 2 mL of immobilized dual-enzyme platform was mixed with 1 mL of 2 % Cellobiose in the M buffer and maintained at 20 h at 55°C to obtain cellobiose to fructose conversion.

Conversion of cellobiose to glucose catalysed by free β-glucosidase enzyme: For cellobiose to glucose conversion, 2.0 mL of 1 % Cellobiose (0.02 g of cellobiose) in the M buffer was mixed with β-glucosidase solution in the M buffer (1 mg/mL, 1 mL) and the mixture was converted to glucose for 20 h at 55°C.

Conversion of glucose to fructose catalysed by free recombinant GI enzyme: For glucose to fructose conversion, 2 mL of 1 % Glucose (0.02 g of glucose) in the M buffer was mixed with the recombinant glucose isomerase 1mL (0.2 mg/mL, 0.2 mg) and the mixture was converted to D-Fructose for 20 h at 55°C.

Conversion of glucose to fructose catalysed by commercially available immobilized GI enzyme (sigma-aldrich): Immobilized Glucose isomerase from *Streptomyces murinus* (50 mg/1 mL) was used for enzyme activity comparison. In a typical experiment, 2 mL of Sigma-Aldrich GI, as provided by vendor, was mixed with 2 mL of 1% Glucose and the experiment was conducted for 20 h at 55°C.

Conversion of cellobiose to fructose catalysed by a mixture of β-glucosidase and recombinant GI: For cellobiose to fructose conversion, 1 mL of 2% cellobiose in the M buffer was mixed with β-glucosidase solution in the M buffer (1 mg/mL, 1 mL) and recombinant glucose isomerase 1mL (0.2 mg/mL). Same experiment was conducted for a mixture of β-glucosidase and Immobilized Glucose Isomerase from Streptomyces murinus (Sigma-Aldrich GI, 50 mg/1mL) and whole mixture was converted to glucose for 20 h at 55°C.

Effects of the temperature, ph, and metal ions on recombinant enzyme activities: The optimum pH for β-glucosidase is 5 whereas the commercially available GI operates at pH 7 (courtesy of Sigma Aldrich website). To optimize the operating conditions toward enabling the one-pot reaction, the effects of temperature, pH, and metal ion were investigated for the dual system. The effect of the temperature on enzyme activity was analyzed by assaying the enzymatic conversion of D-Glucose to D-Fructose catalyzed by the recombinant GI over a temperature range of 4–65°C. To determine the influence of metal ions on enzyme activities, the recombinant enzyme was dialyzed separately in two different customized buffers: M buffer (50 mM MOPS, 5 mM MgSO4, 0.5 mM Co^{2+}, pH 7.0) and C buffer (50 mM citric acid, 5 mM MgSO4, 0.5 mM Co^{2+}, pH 5.0) (Table 1) [17]. The optimal pH for both β-glucosidase and recombinant GI were determined using the aforementioned assay conditions for the conversion of cellobiose to fructose in the two buffer systems. The working conditions for the recombinant GI enzyme and the GI-BG couple are illustrated in Table 2.

Catalyst reusability

Following the completion of the enzymatic reaction, the dual-enzyme Stellate MSN, termed Stellate (BG+GI) was separated from the reaction mixture by centrifugation, washed copiously with water and stored in M buffer. The enzymatic reaction was repeated within 24 h. This sequence was repeated four consecutive times.

Results

The preparation of Stellate MSN, as previously reported by us and others, [16] renders materials with average surface area of 580 m^2/g, as illustrated in Table 3. The macroporous Stellate materials, as illustrated by SEM and TEM images (Figure 2), present spherical morphology and hierarchical porosity, suitable for encapsulation of large macromolecules. The GI in pure form is not commercially available. As such, this product could not be used for immobilization experiments. GI in pure form was produced in-house by genetic engineering and

Buffer Components and Conditions	C Buffer (citric)	M Buffer (MOPS)
Dialysis Buffer pH	5	7
50 mM Buffer Salt	Citric	MOPS
5 mM $MgSO_4$	+	+
0.5 mM $CoCl_2$	+	+

Table 1: Buffer Components.

Enzyme	C Buffer	M Buffer
Recombinant Enzyme GI*	pH Too close to enzyme's pI no reaction	Optimal conditions
BG+GI Cellobiose to D-Fructose	Trace amount of fructose	optimal conditions

Table 2: Optimal reaction conditions.

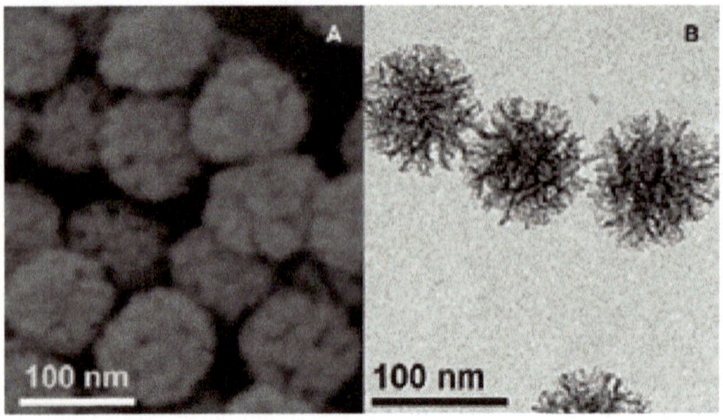

Figure 2: SEM (A) and TEM (B) of Stellate MSN.

Material	Average BET (m²/g)	Average Pore Size Distribution (nm)
Surfactant–free Stellate MSN	580.0	12.6

Table 3: Surface Area (BET) and Average Pore Size.

molecular cloning, followed by protein expression and purification techniques. Preparation of GI along with characterization is described in detail in supporting information. Enzyme macromolecules loading in Stellate MSN was performed by an impregnation method which typically does not differentiate between external and internal surface of the nanospheres. According to the protein assay used upon the immobilization of two enzymes, β-glucosidase and GI on Stellate-MSN, the residual amount of protein remaining in solution after encapsulation by physical entrapment was negligible. We assumed therefore complete immobilization of the two enzymes in the Stellate MSN. The experimental evidence of complete disappearance of the enzyme in the loading solution upon impregnation, corroborated with the enzyme sizes (Table 4) suggest that the enzymes are primarily located in the pores of Stellate MSN, given the amount of material used and the reported porosity and surface area. Optimization of enzymatic reaction conditions resulted in choosing the M buffer for all experiments performed, while the time and temperature were chosen as 20 h and 55°C, respectively. The recombinant GI activity was assessed in comparison with the commercially available GI. Figure 3 shows the performance of the free recombinant enzyme in the time interval reported active for the Sigma GI. The performance of the immobilized dual-enzyme catalyst in M buffer was compared with the activity exhibited by the free enzymes in solution. It is important to note that commercially available GI represents a cellular extract containing GI which is immobilized by encapsulation (Sigma GI). Upon characterization of enzymatic activity, it was determined that the activity of recombinant GI is similar with the commercially available GI and the pure form could be used for the immobilization experiments. Figure 4 shows the relative activity of the free and immobilized enzymes (Stellate MSN and controls) used in conversion of cellobiose to fructose. The controls used included direct conversion of cellulose to fructose using the Stellate-MSN-encapsulated GI and the free GI respectively. Controls for the dual system included the free enzymes, as follows: (BG+recombinant GI) and (BG+commercially available GI from Sigma Aldrich). Enzyme recyclability, critical to economic feasibility of the dual-enzyme immobilization process, has been evaluated. The experiments were conducted four consecutive times, showing gradual decrease in activity as illustrated in Figure 5.

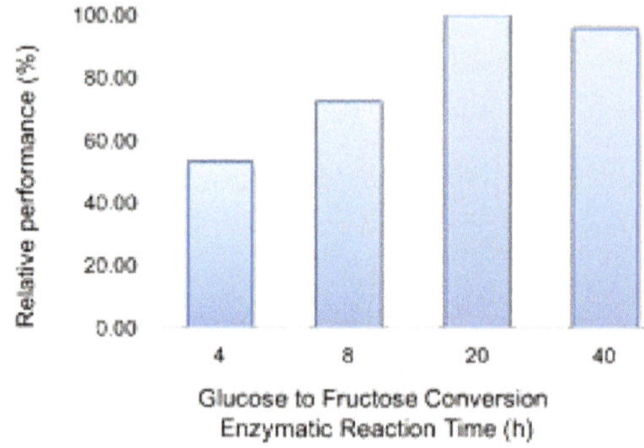

Figure 3: Timeline optimization of recombinant GI activity.

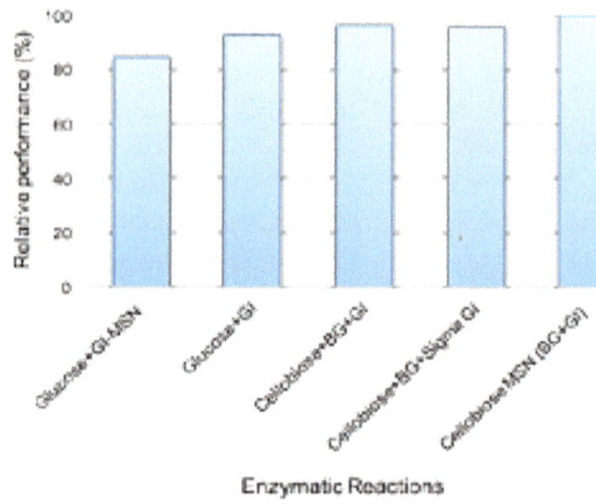

Figure 4: Relative performance of free and immobilized enzymes.

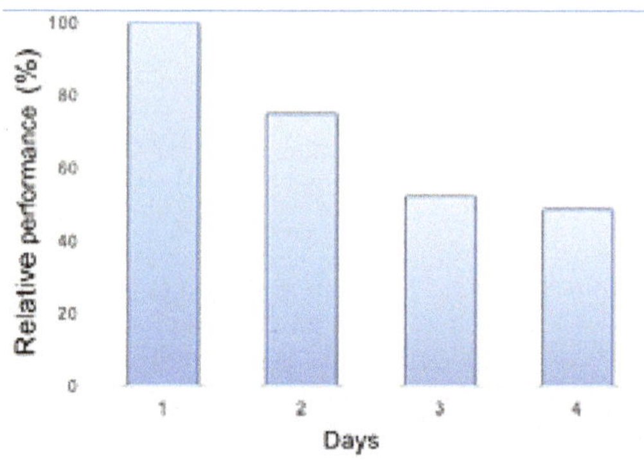

Figure 5: Enzyme recyclability.

Enzyme	Size (kDa)
β-Glucosidase	135
Recombinant Glucose Isomerase	52.7

Table 4: Size of Loaded Enzymes.

Discussion

Prior work reporting enzyme encapsulation in porous silica materials with application to the cellulose to glucose to fructose conversion sequence, has been accomplished by two single enzyme nanospheres, where the two separate platforms were utilized sequentially [8,18]. Therefore, reaction conditions were not unified but changed for each sequential step to match the enzyme working

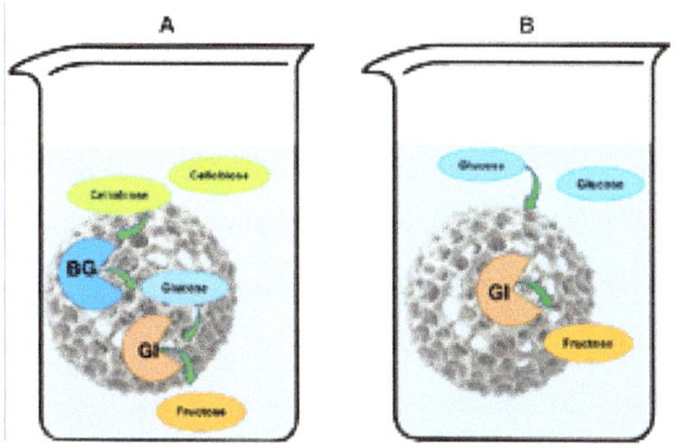

Figure 6: Schematic of channelling effect in the dual enzyme system: A. Glucose proximal access to the enzymatic site *vs*. B. Bulk diffusion to the enzymatic site.

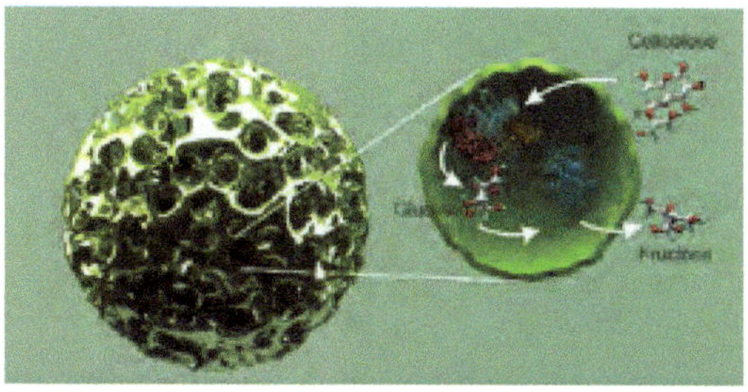

Figure 7: Schematic representation of the cascade reaction occurring in the Stellate MSN pore.

conditions. The significant contribution of the dual-enzyme system accomplished in this work resides with finding a set of conditions that enables working condition adequate for both enzymes function. While the common reaction conditions need to be identified and optimized for any other tandem enzyme potentially explored in the future, this work established the proof-of-concept for the specific GI-β-glucosidase (BG) tandem.

In addition, the encapsulation of two enzymes on the same nanosphere provides ground for demonstrating the channelling effect derived from easy access of the first product of the cascade, glucose, to the next catalytic site, without diffusion in solution. When cellobiose is reacted with Stellate MSN (BG+GI), the two-step cascade reaction enables direct access of Glucose produced by β-glucosidase to GI immobilized on the material, as schematically showed in Figure 6. In contrast, Stellate-MSN immobilized GI (single enzyme) is accessed by glucose from bulk solution, involving a longer diffusion path. The conversion data showed in Figure 4, in the same reaction conditions, the enzymatic reaction in the cascade case is more efficient then in the single enzyme case and support this hypothesis. Further investigations will be directed to understand the effect of the enzymes location in respect to each other and how the relative location impacts the reaction rates and the channelling effect. An important advantage of enzyme immobilization is represented by the stability typically conferred to the encapsulated enzymes; thus, it is expected that the enzyme would be recyclable. To demonstrate that the Stellate MSN host structure provides stability to degradation, we have performed four cycles of the cascade reaction. While a gradual decrease of enzyme activity was observed, potentially due to extensive washing and centrifugation, the enzymes are still active upon the four cycles. Further work aims to mitigate the potential detachment of the enzyme from the support by host functionalization and covalent attachment of the enzymes to the silica support.

Conclusion

Stellate MSN proves to be an excellent platform for multi-site cascade reactions as demonstrated here for the dual-enzyme catalyst produced by immobilization of β-glucosidase and glucose isomerase on the silica nanospheres (Figure 7). This is the first demonstration of biomacromolecules encapsulation in Stellate MSN and while this application was targeted to enzymatic catalysis, a plethora of other applications are envisioned. Future work will undertake mechanistic studies of to prove the channelling hypothesis. Additional focus will be on elucidating the exact location of the enzymes in the Stellate MSN, by enzyme labelling with chromophores toward enabling localization of enzymes in the porous platform and in respect with each other. Stellate MSN proves to be an excellent platform for multi-site cascade reactions as demonstrated here for the dual-enzyme catalyst produced by immobilization of β-glucosidase and glucose isomerase on the silica nanospheres.

Acknowledgements

The work was funded by U.S. Department of Agriculture (Award number: USDA-2014-38821-22459) and National Science Foundation (Award number NSF 1435716) for funding of this research. The authors thank Mr. Nicholas Pizzi for the porosity measurements of Stellate MSN.

References

1. Bhosale SH, Rao MB, Deshpande VV (1996) Molecular and industrial aspects of glucose isomerase. Microbiol Rev 60: 280-300.
2. Woodward, JKoran Jr LJ, Hernandez LJ, Stephan LM (1993) Use of Immobilized β-Glucosidase in the Hydrolysis of Cellulose, in β-Glucosidases. Am Chem Soc pp: 240-250.
3. Woodward J, Bales JC (1989) The Adsorption of Trichoderma reesei C30 Cellulase Protein to Protein Adsorbents: Application to Cellulase Recovery and Reuse. pp: 87-101.
4. Sternberg D, Mandels GR (1980) Regulation of the cellulolytic system in Trichoderma reesei by sophorose: induction of cellulase and repression of beta-glucosidase. J Bacteriol 144: 1197-1199.
5. Datta S, Christena LR, Rajaram YR (2013) Enzyme immobilization: an overview on techniques and support materials. 3 Biotech 3: 1-9.
6. Yiu HH, Wright PA, Botting NP (2001) Enzyme immobilisation using siliceous mesoporous molecular sieves. Microporous Mesoporous Mater 44: 763-768.
7. Peng WH, Lee YY, Wu C, Wu KC (2012) Acid–base bi-functionalized, large-pored mesoporous silica nanoparticles for cooperative catalysis of one-pot cellulose-to-HMF conversion. J Mater Chem 22: 23181-23185.
8. Lee YC, Chen CT, Chiu YT, Wu KC (2013) An Effective Cellulose-to-Glucose-to-Fructose Conversion Sequence by Using Enzyme Immobilized Fe3O4-Loaded Mesoporous Silica Nanoparticles as Recyclable Biocatalysts. Chem Cat Chem 5: 2153-2157.
9. Ispas C, Sokolov I, Andreescu S (2009) Enzyme-functionalized mesoporous silica for bioanalytical applications. Anal Bioanal Chem 393: 543-554.
10. Hartmann M, Kostrov X (2013) Immobilization of enzymes on porous silicas-benefits and challenges. Chem Soc Rev 42: 6277-6289.
11. Dutta S, Wu KC (2014) Enzymatic breakdown of biomass: enzyme active sites, immobilization, and biofuel production. Green Chem 16: 4615-4626.
12. Chang RH, Jang J, Wu KC (2011) Cellulase immobilized mesoporous silica nanocatalysts for efficient cellulose-to-glucose conversion. Green Chem 13: 2844-2850.
13. Storey KB, Chakrabarti AC (1990) One-step conversion of cellulose to fructose using coimmobilized cellulase, β-glucosidase, and glucose isomerase. Appl Biochem Biotechnol 23: 139-154.
14. Betancor L, Luckarift HR (2010) Co-immobilized coupled enzyme systems in biotechnology. Biotechnol Genetic Eng Rev 27: 95-114.

15. Zhang K, Xu LL, Jiang JG, Calin N, Lam KF, et al. (2013) Facile Large-Scale Synthesis of Monodisperse Mesoporous Silica Nanospheres with Tunable Pore Structure. J Am Chem Soc 135: 2427-2430.

16. Radu DR, Pizzi NA, Lai CY (2016) Functionalized stellate macroporous silica nanospheres for CO_2 mitigation. J Mater Sci 51: 10632-10640.

17. Sriprapundh D, Vieille C, Zeikus JG (2003) Directed evolution of Thermotoga neapolitana xylose isomerase: high activity on glucose at low temperature and low pH. Protein Eng Design Selec 16: 683-690.

18. Tran DN, Balkus Jr KJ (2011) Perspective of Recent Progress in Immobilization of Enzymes. ACS Catal 1: 956-968.

Thermogravimetric Investigation of Normal and Pathological Human Meniscus

Gality H[1*], Aigner Z[2], Toth K[1], Szabo-Revesz P[2] and Sohar G[1]

[1]Department of Orthopaedics, Faculty of Medicine, University of Szeged, Semmelweis u. 6, 6725 Szeged, Hungary
[2]Department of Pharmaceutical Technology, Faculty of Pharmacy, University of Szeged, Eötvös u. 6, 6720 Szeged, Hungary

Abstract

Meniscus degeneration is a common problem in the human knee. Meniscal diseases can significantly alter the quality of life. Thermal analysis of human meniscus extracted during surgery can provide insight into the degenerative process. The aim of present study was to characterise the water loss from the investigated biological samples caused by altered metabolism, and compare the thermogravimetric properties of normal, moderately, and severely degenerated meniscus samples. Although total water content was similar in normal, moderately and severely degenerated meniscal cartilages (70.26, 75.51 and 76.29% respectively), there was a significant difference in activation energy (44.92, 51.44 and 62.04 kJ M-1). This can be attributed to the difference in chemical composition of meniscus fibrocartilagenous matrix between the healthy and degenerated samples.

Keywords: Activation energy; Human meniscus; Osteoarthritis; Thermogravimetry

Introduction

In present study the authors demonstrate thermogravimetric analysis of healthy and degenerated human knee joint menisci. Meniscus mainly consists of robust fibrocartilage, it is an important shock absorber in the knee. It protects the hyaline cartilage of the joint, shielding it from physical damage and therefore contributing to prevention of early osteoarthritis [1]. A large proportion of body weight is distributed through the meniscus during walking, running, and jumping. Menisci contribute to the stability of the knee by helping the shape of the femur to conform to the tibia. Several factors can cause damage to the menisci. They can tear or rip from force, or get pinched between the femur and the tibia [2,3]. It does not always require a major fall or twist to cause a meniscal tear. Degenerative changes of knee menisci frequently result from trauma to the joint or associate with joint diseases; in other cases they are caused as a result of chronic overload, and in some cases even with no obvious reason. Degeneration of human knee joint menisci has been the subject of numerous morphological studies [4,5] but detailed studies on the chemical composition and the metabolism of menisci are still limited.

Several thermoanalytical techniques can be applied to measure the change of physical or chemical properties of a sample as a function of temperature. Thermogravimetric analysis (TG) is one of the oldest thermal analytical procedures; it has been used extensively in the study of polymeric systems and similarly in case of biological samples. The technique involves monitoring the mass loss of the sample in a specific atmosphere (usually nitrogen or air) as a function of temperature. The applied equipment – among others – is suitable for simultaneous recording of thermogravimetric, derivative thermogravimetric (DTG) and differential thermoanalytical (DTA) curves [6]. The usefulness of TG for analyzing complex systems was greatly enhanced by the introduction of the ability to record the first derivative of the mass loss simultaneously. The ability of TG to generate fundamental quantitative data from almost any class of materials, has led to its widespread use in virtually every scientific field. Compositional analysis is a key application: by careful choice of temperature programming and gaseous environment, complex materials or mixtures can be analyzed by selectively decomposing them, or removing their components. This approach is regularly used to analyze moisture content of many substances. TG is inherently quantitative technique, and it does not give direct chemical information. The ability to analyze the volatile products during a mass loss is also a valuable possibility, when available [7-10]. Previously differential scanning calorimetry (DSC) was used for the investigation of normal and degenerative human meniscus cartilage, and water content has not been measured [11-14]. A significant paper from this field was the study of Than et al. [15], further studies measured the difference between primary osteoarthritis and septic arthritis [16]. Most of the known changes in the extracellular matrix in OA come from animal models, because human samples are only available for experiment to a limited extent. Thermogravimetric study of human hyaline cartilage has been carried out previously [17], although meniscus has not been studied extensively. The purpose of this investigation was to gain insight on the role of water content in contributing to progression of fibrocartilage degeneration, and to reveal the kinetic character of water loss effect of heating on menisci in good health and in different stages of degeneration.

Experimental

Materials

The human knee joint meniscus specimens were received from the Orthopaedic Department, University of Szeged. After surgical removal, the menisci were dissected free of their attachments and a 5 mm thick disc was produced under sterile conditions from the middle segment of each meniscus. The disc was first washed in sterile saline, and then stored in 20 ml saline for transportation at room temperature. Mean storage time was 6 hours (Minimum: 1 hour, maximum: 26 hours), 10 samples out of 11 were studied within 6 h of preparation. One sample was stored

***Corresponding author:** Gality H, Department of Orthopaedics, Faculty of Medicine, University of Szeged, Semmelweis u. 6, 6725 Szeged, Hungary
E-mail: galityh@yahoo.com

over-night at 5°C. Pre-emptive control examinations did not show any change in the calorimetric and thermogravimetric properties after storage for 26 hours at 5°C. All tissues were managed in accordance to legal regulation, international ethical concerns, and patients' consent. Diagnosis of the patient was established on basis of the medical history, clinical signs and radiological findings preoperatively. In order to conduct the thermoanalytical study, 11 samples were collected. The state of the joint degeneration, and the meniscus was determined intraoperatively. All patients in the moderately osteoarthritic group (mean age: 60.47 years) were considered to be Osteoarthritis Research Society International (OARSI) [18] grade 1-3 articular degeneration. Severely osteoarthritic group (mean age: 70.88) were considered to be OARSI grade 4-6. Normal samples were extracted during arthroscopic surgeries of patients (mean age: 40.94 years) with knee injury, when ruptured but otherwise healthy meniscus had to be sacrificed for the partial meniscectomy procedure. The gender distribution was the same in both osteoarthritic groups.

Methods

We used a new protocol established by Sohár et al. [17] for sample extraction during live surgery using simple saline solution instead of the previously used phosphate buffer. This new method proved to be suitable for the thermoanalytical investigations. Thermal analysis was performed with a MOM Derivatograph-C (MOM, Budapest, Hungary), and the TG, DTG, and DTA curves were determined. TG allows the determination of the mass as a function of temperature. This thermal technique provides information concerning the thermal stability and composition of the sample and of any intermediate compound which may be formed. The heating was linear from 25°C to 150°C and the rate of heating was 5 K minute^{-1}. Platinum sample holders were used, Al_2O_3 was used as reference material, the purge gas was air. In the first step, the total water loss and kinetic parameters were calculated. The kinetic parameters calculated by the software are the following: the reaction order (n), the activation energy (E_{act}) and the pre-exponential factor (A). The value of n (reaction order) is allocated by the Kissinger method [19] and it is the first kinetic parameter calculated by the computer:

$$n = 1,26 \cdot S^{\frac{1}{2}} \qquad (1)$$

where S is the form factor which presents the absolute value of the gradients of DTG curves in the points of min/max. The activation energy (E_{act}) is determined according to the natural logarithmic form of the Arrhenius-equation

$$k(T) = A \cdot e^{\frac{E_{act}}{RT}} \qquad (2)$$

which is widely used in the literature [20,21]. The results of the investigations were statistically evaluated by Microsoft Excel. A Paired t-test was used to compare mean water loss of normal and degenerative meniscus during thermal analysis. The linear range of the first step of TG curve in water loss was analyzed to establish its slope. The determination of linear parts of TG curves and fitting linear straight line on the curve was performed by the statistical software SPSS.

Results and Discussion

TG curves of healthy meniscus, samples with no degeneration, and degenerated menisci are presented on Figure 1. The results show, that the total water content of healthy fibrocartilage was 70.26%, to remove the water content 44.92 kJ M^{-1} was needed. Fibrocartilage obtained from the moderately degenerated meniscus samples had a 75.51% water content, which differed significantly from the healthy tissue.

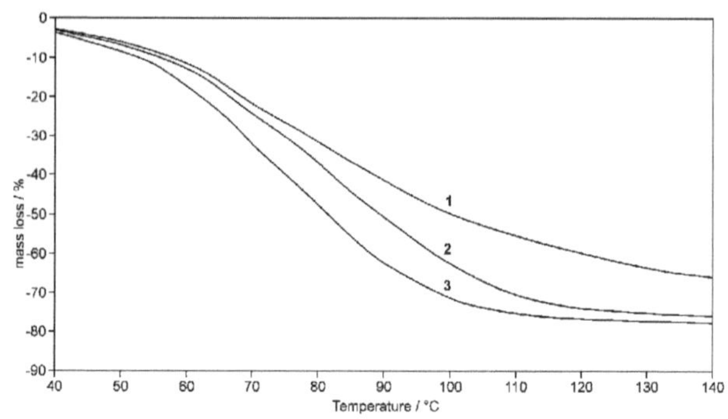

Figure 1: TG curves of normal (1), moderately (2) and severely (3) degenerated meniscus samples.

Extraction of water content needed 51.44 kJ M^{-1} energy on average. Fibrocartilage obtained from the severely degenerated meniscus samples had a 76.29% water content, which did not differ significantly from the moderately degenerated samples, but differed significantly from results of normal tissue. Extraction of water content needed 62.04 kJ M^{-1} energy on average. Loss of water content in all groups is presented with a sharp step on the TG curve, starting on average temperature of 44.65°C and ending at 137.2°C. Linear part of the TG curve begins at 63.53°C for normal meniscus, 63.73°C for moderately degenerated meniscus, and 62.65°C for severely degenerated meniscus; while ends at 96.73°C for normal meniscus, 97.20 °C for moderately degenerated meniscus, and 96.55°C for severely degenerated meniscus. A straight line was fitted to the curve in this section and the slope of this curve was calculated which represents the rate of the water content loss (Table 1). The slope of the linear region correlated in both groups. In case of the normal fibrocartilage 1.0% °C^{-1} mass loss was detected. In case of moderately degenerated samples 1.5% °C^{-1} decrease in mass was observed. In the severely degenerated samples also 1.5% °C^{-1} mass reduction was observed. The resulting amount of mass loss in the linear region was recounted from these results (Table 1).

Conclusion

The fibrocartilage of knee menisci is a functionally specialized tissue that plays a role in weight bearing, nutrition, and lubrication of the articular cartilage and in joint stability. Past biomechanical studies done under static loading have shown that the menisci, regardless of their condition, contribute to force [5] and stress distribution [1]. Meniscus degeneration in the knee is frequently a result from trauma to the joint or it is associated with joint diseases; in other cases it arises after chronic overload, or even spontaneously [4]. Our study has several potential limitations. First of all, the sample size was not large enough to draw definitive conclusions. Although the mass loss change of the process initiated by the temperature change showed marked difference between the normal and pathological groups (Table 2), statistical analysis did not show significant alteration at $p \leq 0.05$. Consequently, further measurements are needed to confirm the results of our study. Previous thermo analytical studies used cadaver samples for the investigation as normal human meniscus. All samples that were extracted for our studies were obtained during live surgeries and were macroscopically intact. Only full thickness meniscus was included in the analysis of normal group. More information will be needed on the events leading to meniscus degeneration and the interactions between different pathologic effects in order to discover new therapeutic targets. Several potential disease modifying agents are studied currently that

Sample group	Sample number	TG step linear region /°C			Mass loss /%	Reaction order/n SD	Slope of linear region
Normal	3	63.53	–	96.73	37.51	1.00	0.22
						0.44	
Moderate	4	63.73	–	97.20	45.45	1.00	0.23
						0.18	
Severe	4	62.65	–	96.55	46.88	1.13	0.21
						0.26	

Table 1: Reaction kinetic parameters of meniscus samples with no degeneration, moderate, and severe degree of degeneration.

Sample group	Sample number	TG step /°C			Total mass loss /% SD	E_{act} / kJ M^{-1} SD
Normal	3	30.50	–	130.20	70.26	44.92
					7.44	10.50
Moderate	4	27.88	–	133.05	75.51	51.44
					3.66	12.41
ssSevere	4	27.23	–	129.45	76.29	62.04
					4.29	17.12

Table 2: Mass loss and activation energy of normal, moderately, and severely degenerated meniscus samples.

may significantly contribute to changes in the treatment approach; therefore sensitive tools are needed to study biological matter in clinical trials. Meniscus pathology plays an important role in osteoarthritis pathophysiology. However, whether meniscus damage or cartilage degeneration is the first factor is unknown. Meniscal rupture seems to be a strong risk factor in the development and progression of knee osteoarthritis. Meniscectomy increases knee osteoarthritis risk twofold [1,22]. Modifiable risk factors are natural target for clinical efforts in managing knee degeneration. At present, there is a number of modifying therapeutic options that may be used to alter the rate of disease progression. While these therapies are currently underutilized, they might play more prominent role when modification of disease progression has been demonstrated. However, the role of these therapies is unclear, given the paucity of long-term, well-designed controlled trials [23]. In summary, we examined the thermal properties of human meniscus tissue of normal origin in adults and in patients with moderate and severe stages of degeneration. A newly established thermogravimetric protocol was used for our experiments. This method proved to be suitable for thermogravimetric analysis of normal and degenerative human meniscus samples.

Acknowledgements

Grant: This study was supported by OTKA T-047166.

Conflicts of Interest

No conflicts of interest to declare.

References

1. Hoshino A, Wallace WA (1987) Impact-absorbing properties of the human knee. J Bone Joint Surg Br 69: 807-811.
2. Brindle T, Nyland J, Johnson DL (2001) The meniscus: review of basic principles with application to surgery and rehabilitation. J Athl Train 36: 160-169.
3. Bonnie RB, Finlay JB, Papadopoulos P, Andreae P (1984) The effect of medial meniscectomy on strain distribution in the proximal part of the tibia. J Bone Joint Surg Am 66: 1431-1437.
4. Herwig J, Egner E, Eckhart B (1984) Chemical changes of human knee joint menisci in various stages of degeneration. Ann Rheum Dis 43: 635-40.
5. Walker PS, Erkman MJ (1975) The role of the menisci in force transmission across the knee. Clin Orthop Relat Res 109: 184-192.
6. Paulik J, Paulik F, Arnold M (1986) The Derivatograph-C. J Therm Anal 32: 375-378.
7. Gill PS, Sauerbrunn SR, Crowe BS (1992) High resolution thermogravimetry. J Therm Anal Calorim 38: 255-266.
8. Reading M (1992) Controlled Rate Thermal Analysis and Beyond, in E L Charsley and S B Warrington, Thermal Analysis - Techniques & Applications The Royal Society of Chemistry, Cambridge 127-155.
9. Riesen R (1998) Adjustment of Heating Rate for Maximum Resolution in TG and TMA. J Therm Anal Calorim 53: 365-374.
10. Brown ME (1998) Controlled Rate Thermal Analysis and Related Techniques, Handbook of Thermal Analysis and Calorimetry, Principles and Practice Elsevier Science B.V., Amsterdam 1: 423-443.
11. Sohár G, Aigner Z, Gálity H, Tóth K, Szabó-Révész P (2013) Calorimetric investigation of normal and pathological human meniscus. J Therm Anal Calorim 115: 2031-2035.
12. Wiegand N, Vámhidy L, Patczai B, Dömse E, Kereskai L, et al. (2009) Differential scanning calorimetric examination of the human skeletal muscle in a compartment syndrome of the lower extremities. J Therm Anal Cal 98: 177-182.
13. Wiegand N, Vámhidy L, Patczai B, Dömse E, Than P, et al. (2009) Differential scanning calorimetric examination of the degenerated human palmar aponeurosis in dupuytren disease . J Therm Anal Cal 95: 797-800.
14. Wiegand N, Vámhidy L, Lőrinczy D (2010) Differential scanning calorimetric examination of ruptured lower limb tendons in human. J Therm Anal Cal 101: 487-492.
15. Than P, Vermes C, Schäffer B, Lőrinczy D (2000) Differential scanning calorimetric examination of the human hyaline cartilage. Thermochim Acta 346: 147-151.
16. Sillinger T, Lőrinczy D, Kocsis B, Kereskay L, Nöt LG, et al. (2014) Differential scanning calorimetric measurement of cartilage destruction caused by Gram-negative septic arthritis. J Therm Anal Calorim 116: 747-752.
17. Sohár G, Pallagi E, Szabó-Révész P, Tóth K (2007) New thermogravimetric protocol for the investigation of normal and damaged human hyaline cartilage. J Therm Anal Calorim 89: 853-856.
18. Pritzker K, Gay S, Jimenez S, Ostergaard K, Pelletier JP, et al. (2005) Osteoarthritis cartilage histopathology: grading and staging. Osteoarthritis Cartilage.
19. Kissinger HE (1957) Reaction Kinetics in Differential Thermal Analysis. Anal Chem 29: 1702-1706.

20. Arnold M, Somogyvári P, Paulik J, Paulik F (1987) Derivatograph-C-a microcomputer-controlled simultaneous TG, DTG, DTA, TD and EGA apparatus. J Therm Anal Calorim 32: 679-683.

21. Doyle CD (1965) Series Approximations to the Equation of Thermogravimetric Data. Nature 207: 290-291.

22. Johnson RJ, Kettelkam DB, Clark W, Leaverton P (1974) Factors effecting late results after meniscectomy. J Bone Joint Surg Am 56: 719-729.

23. Brandt KD (2004) Non-surgical treatment of osteoarthritis: a half century of "advances". Ann Rheum Dis 63: 117-122.

Thermophysical Properties of Binary Liquid Mixtures of Acetronitrile with 2- Ethoxyethanol and 2- Butoxyethanol at Temperatures 293K, 303K and 313K: Experimental and Computational Study

Rahul Singh[1], Sudir Kumar[2], Sangeeta Sagar[2] and Manisha Gupta[2]*

[1]Department of Physics, Ambalika Institute of Technology, Charbagh, Lucknow, Uttar Pradesh, India
[2]Department of Physics, University of Lucknow, Lucknow, Uttar Pradesh, India

Abstract

With the aim to study the viscosity and density of acetronitrile with 2- ethoxyethanol and 2- butoxyethanol at temperatures 293K, 303K and 313K over the whole composition range. The experimental results were used to calculate deviation in viscosity ($\Delta\eta$), excess internal pressure (π_i^E) and excess enthalpy (H_m^E). These results are used to quantitatively discuss the nature of interaction taking place between the components when mixed in terms of hydrogen-bond effects, dipole-dipole interaction weak vanderwaals type dispersion force. The calculated quantities have been fitted to Redlich-Kister equation.

Keywords: Viscosity; Internal pressure; Excess enthalpy; Redlich Kister equation

Introduction

Mixed solvents rather than single pure liquids, are extensively used in industrial process as they provide a wide range of solvent with variable physical properties. The physical property data on mixed solvents are important for the theoretical and applied areas of research and are frequently used in many chemical and industrial processes such as design of new process and process equipment (fluid flow, mass transfer or heat transfer calculation). Acetonitrile is used as a polar aprotic solvent in organic synthesis and in the purification of butadiene. In the laboratory, it is used as a medium-polarity solvent that is miscible with water and a range of organic solvents, but not saturated hydrocarbons. It has a convenient liquid range and a high dielectric constant of 38.8. With a dipole moment of 3.92 D. Acetonitrile plays a significant role as the dominant solvent used in the manufacture of DNA oligonucleotides from monomers. Industrially, it is used as a solvent for the manufacture of pharmaceuticals and photographic film. 2- ethoxyethanol is a versatile organic liquid used in chemical industries as a solvent for lacquers, oils, and resin and as an antifreeze for explosive. It is miscible with a number of organic solvents as well as water. 2-butoxyethanol is a solvent in paint and surface coatings, as well as cleaning products and inks. Other products that contains 2- butoxyethanol include acrylic resin formulation, asphalt release agents, firefighting foam, leather protectors, oil spill dispersants, lane degreases and photographic strip solutions. Other products containing 2-butoxyethanol as a primary ingredient include some whiteboard cleaners, liquid soaps, cosmetics, dry cleanings solutions, lacquers, varnishes, herbicides and latex paints.

Viscosity and density data are required in many chemical engineering calculations involving fluid flow, heat and mass transfer. Moreover, the knowledge of dependence of viscosity and density on mixture composition and temperature is of considerable interest in understanding the intermolecular interactions [1-6] Internal pressure has been a subject of active interest and attempts have been made to show the significance of internal pressure as a fundamental property [7]. The present study reports the experimental values of viscosity (η) for the binary mixture of acetonitrile with 2 ethoxyethanol and 2-butoxyethanol at temperatures 293K, 303K and 313K over the whole composition range. The experimental results were used to calculate deviation in viscosity ($\Delta\eta$), excess internal pressure (π_i^E) and excess enthalpy (H_m^E). These results are used to quantitatively discuss the nature of interaction taking place between the components when mixed in terms of hydrogen-bond effects, dipole-dipole interaction weak vanderwaals type dispersion force. The calculated quantities have been fitted to Redlich – Kister equation [8].

Experimental Methods

Materials

The analytical grade chemicals Acetronitrile (Fluka), 2-ethoxyethanol (Sigma- Aldrich, ACS reagent, ≥ 99.5%) and 2-Butoxyethanol (Sigma- Aldrich, ACS reagent, ≥ 98.6%) have been obtained from Sigma Aldrich Chemicals Pvt. Ltd. All the chemicals were purified by standard procedure discussed by Perrin and Armarego [9]. Solutions have been prepared by mass, using air tight bottles and have been measured on electronic balance OHAUS- AR 2104 (Ohaus Corp. Pine Brook, NJ, USA) with an accuracy of 1×10^{-4} g. The possible error in the estimation of mole fraction is less than ± 0.0001.

Apparatus and procedures

The viscosities have been measured using Brookfield LVDV-II+ Pro Programmable cone and plate viscometer (Brookfield Engineering Laboratories, Inc., USA). The experimental assembly allows measurement of viscosities in the range of 0.15 cP to 3065 cP (with spindle CPE-40) and 4.6 cP to 92130 cP (with spindle CPE-52) with an accuracy of ± 1.0% of full scale range and repeatability of ± 0.2%. The apparatus requires only 0.5 ml of the mixture for measurement of viscosity. The apparatus measures fluid absolute

***Corresponding author:** Manisha Gupta, Department of Physics, University of Lucknow, Babuganj, Hasanganj, Lucknow-226 004, Uttar Pradesh, India
E-mail: guptagm@rediffmail.com

viscosity directly in cP. The apparatus was calibrated by two viscosity standards Polydimethylsiloxane with viscosity 4.6 and 485 cP provided by Brookfield Engineering Laboratories. The viscosity standards are Newtonian, and therefore, have the same viscosity regardless to the spindle speed.

Circulating water bath with programmable temperature controller (TC-502, Brookfield Engineering Laboratories, Inc., USA), having variable pump speeds, has been used for water circulation in water jackets of the apparatuses. The temperature controller covers the temperature measurement range of -20°C to 200°C, with temperature stability of ± 0.01°C.

Structure of chemical compounds

The Figure 1 gives the structural information of the chemical compounds Acetronitrile (Fluka), 2- ethoxyethanol (Sigma- Aldrich, ACS reagent, ≥ 99.5%) and 2- Butoxyethanol (Sigma- Aldrich, ACS reagent, ≥ 98.6%).

Results and Discussion

Thermophysical parameters deviation in viscosity, excess internal energy and excess enthalpy were calculated from equations

$$\Delta\eta=\eta_m-(x_1\eta_1+x_2\eta_2) \tag{1}$$

$$\pi_{im}^{E}=\frac{bRTk^{\frac{1}{2}}\eta_1^{\frac{1}{2}}\rho_1^{\frac{2}{3}}}{u_1^{\frac{1}{2}}M^{\frac{7}{6}}}\left(\frac{x_1bRTk^{\frac{1}{2}}\eta_1^{\frac{1}{2}}\rho_1^{\frac{2}{3}}}{u_1^{\frac{1}{2}}M^{\frac{7}{6}}}+\frac{x_2bRTk^{\frac{1}{2}}\eta_2^{\frac{1}{2}}\rho_2^{\frac{2}{3}}}{u_2^{\frac{1}{2}}M^{\frac{7}{6}}}\right) \tag{2}$$

$$H_m^E=x_1\pi_{i1}V_1+x_2\pi_{i2}V_2-\pi_{im}V_m \tag{3}$$

Experimental values of viscosity are given in Table 1 for binary mixture of acetonitrile + 2- ethoxyethanol and acetonitrile + 2-butoxyethanol at temperature 293K, 303K and 313K respectively. Deviation in viscosity ($\Delta\eta$), excess internal pressure (π_i^E) and excess molar enthalpy (H_m^E) are given in Table 2, with the mole fraction of acetonitrile. Variations of respective parameters Δη , π_i^E and against mole fraction of acetonitrile (x_1) are shown in Figures 2-4 respectively.

Excess parameters

It has been reported by many workers [10-14] that three main types of contributions to excess thermodynamic properties of mixtures are:

(1) Physical: due to non-specific Vander Walls type interactions;

(2) Chemical: due to hydrogen bonding;

(3) Structural: due to changes of interstitial accommodation and free volume.

The study of excess parameters is very important in order to explain the various type of interaction occurring in binary mixtures.

Deviation in viscosity (Δη) is plotted against mole fraction of acetonitrile (x_1) for both mixtures. Generally the positive values of deviation viscosity (Δη) indicate the predominance of hydrogen bonding interaction between unlike molecules over the dissociation effects of the mixing components [15]. It is known that the strength of the molecular hydrogen bonding is not only the factor influencing viscosity deviation (Δη) in the liquid mixture, the molecular size and shape of the components and average degree of association is equally important factor [16]. In the absence of strong specific interaction leading to complex formations, negative deviation generally occur when dispersive forces are primarily responsible for interaction or even when they interact more strongly via dipole – dipole interaction [17]. This also means that the interaction between pairs of like molecules is stronger than between pairs of unlike molecules. When temperature is increased then the values of deviation in viscosity (Δη) becomes less negative in case of both mixtures. It has been noticed from Figure 2 that deviation in viscosity (Δη) found to be negative for both mixtures. This negative deviation shows dispersive force is acting between unlike molecules of component mixtures. More negative values of (Δη) in case of acetonitrile + 2-butoxyethanol mixtures support our results.

It has been seen from Figure 4 that the values of excess internal pressure (π_i^E) has found to be negative for both the systems at temperatures 293K, 303K and 313K over entire composition range. At around x_1=0.7 maximum value is found shows the complex formation

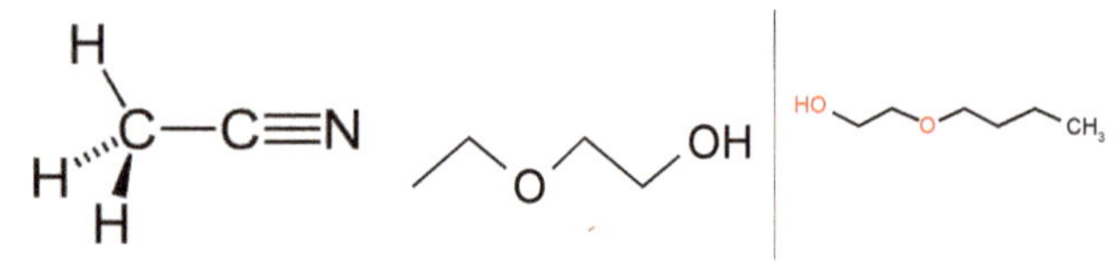

Acetronitrile 2-ethoxyethanol 2- butoxyethanol

Figure 1: Structure of Acetronitrile, 2- ethoxyethanol and 2- butoxyethanol.

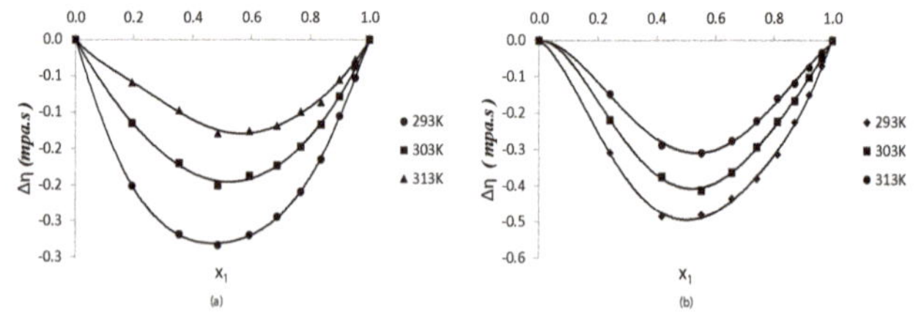

Figure 2: Viscosity deviation ($\Delta\eta$) vs. mole fraction of methyl cyanide (x_1) for binary mixtures (a) acetronitrile+2-ethoxyethanol and (b) acetronitrile+2-butoxyethanol at varying temperatures.

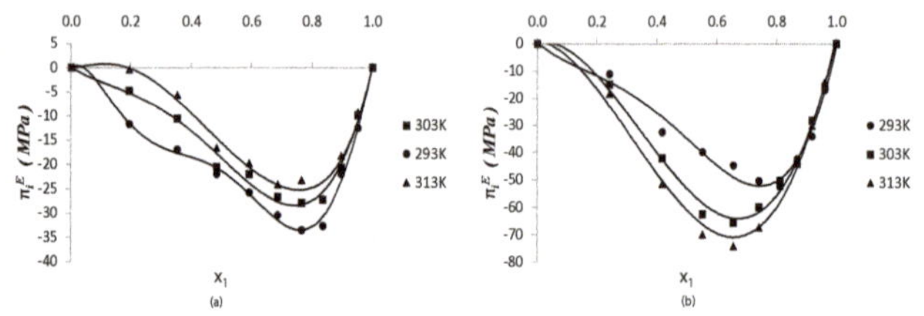

Figure 3: Excess internal pressure (π_i^E) vs. mole fraction of methyl cyanide (x_1) for binary mixtures (a) acetronitrile+2-ethoxyethanol and (b) acetronitrile +2-butoxyethanol at varying temperatures.

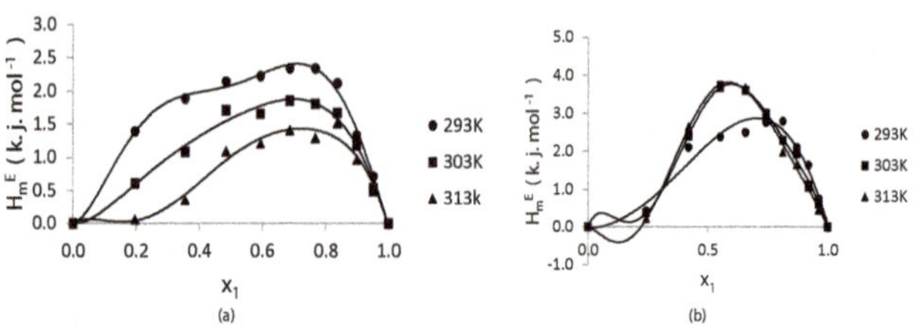

Figure 4: Excess molar enthalpy (H_m^E) vs. mole fraction of methyl cyanide (x_1) for binary mixtures (a) acetronitrile+2-ethoxyethanol and (b) acetronitrile+2-butoxyethanol at varying temperatures.

at this concentration. Excess negative values for both mixture show long range dispersive force operating between components mixture suggest that rupture of hydrogen bonding chain of 2- ethoxyethanol and 2- butoxyethanol is stronger than the hydrogen bonding between unlike molecules.

Values of excess molar enthalpy (H_m^E) were found to be negative in the case of specific complex formation and positive in the case of dissociation. In our present study the values of excess molar enthalpy (H_m^E) is found to be positive for both mixtures. Larger positive values in both mixtures suggest specific weak interaction between molecules of components mixtures. More positive values in case of acetonitrile +2-butoxyethanol suggest that long range dispersive force is prominent in this case. The trend of excess molar enthalpy (H_m^E) supports our earlier result of excess internal pressure (π_i^E).

Redlich-Kister polynomial equation data

The composition dependence of the excess properties is correlated by the Redlich- Kister polynomial equation. The values of excess parameters for each mixture were fitted to the Redlich- Kister polynomial equation of the type,

$$Y^E = x_1(1-x_1)\sum_{i=1}^{5} a_i(2x_1-1)^{i-1} \tag{4}$$

Where a_i is the polynomial co-efficients.

The values of the co-efficients a_i were obtained by the least squares method with all points weighted equally. In each case, the optimum number of co-efficients was ascertained from an examination of the variation of the standard deviation $\sigma(Y^E)$ with no. of co-efficients (p).

$$\sigma(Y^E) = \left[\frac{\sum(Y_{expt}-Y_{cal})^2}{n-p}\right]^{1/2} \tag{5}$$

The values of co-efficient ai of Redlich – Kister polynomial equation evaluated using the method of least squares for the mixtures are given in Table 3 along with the standard deviation σ(YE).

Internal pressure

Internal pressure has been evaluated using empirical relations. A brief discussion of each empirical relation used is being given here.

Thermo dynamical relation: The relation among applied pressure (P), molar volume (V), temperature (T) and molar internal energy (U) is given by thermodynamic relation.

$$\left(\frac{\partial U}{\partial V}\right)_T = T\left(\frac{\partial P}{\partial T}\right)_V - P$$

The isothermal internal energy-volume coefficient $(\partial Y/\partial V)_T$ is often called internal pressure πi. Since $\alpha = 1/V\ (\partial V/\partial T)_P$ and $k_T = -1/V\ (\partial V/\partial P)_T$, the above equation, for the pressure limiting to zero, can be written as

$$\pi_i = \frac{\alpha T}{k_T} \tag{6}$$

Relation using empirical equations for α and k_T: From the empirical equations for thermal expansion coefficient α and isothermal compressibility k_T [17] we get

$$\pi_i = 44.2T^{4/3}u^{3/2}\rho\,\mathrm{b} \tag{7}$$

Relation based upon Buehler-Hirschfelder-Curtiss equation of state: With the help of equation proposed by Buehler et al. [18,19] the

X_1	η_m (mPa.s)	$\Delta\eta$(mPa.s)	π_i^E (MPa)	H_m^E (kJ.mol^{-1})
		293K		
0	2.084	0	0	0
0.1961	1.551	-0.201	-11.6365	1.3954
0.3544	1.216	-0.2681	-16.8876	1.8896
0.4848	0.979	-0.2832	-21.9262	2.141
0.5941	0.808	-0.2693	-25.7399	2.224
0.6871	0.676	-0.2444	-30.4327	2.341
0.7671	0.575	-0.2091	-33.4705	2.3421
0.8367	0.502	-0.1644	-32.6073	2.1159
0.8978	0.458	-0.105	-21.9008	1.3324
0.9518	0.419	-0.0525	-12.4416	0.7158
1	0.39	0	0	0
		303K		
0	1.67	0	0	0
0.1961	1.296	-0.1148	-4.7431	0.6106
0.3544	1.032	-0.1697	-10.4712	1.0882
0.4848	0.829	-0.2005	-20.4787	1.7154
0.5941	0.698	-0.1871	-21.9318	1.6606
0.6871	0.589	-0.1732	-26.5711	1.8587
0.7671	0.509	-0.1474	-27.8748	1.8126
0.8367	0.448	-0.117	-27.1662	1.6739
0.8978	0.406	-0.078	-20.5587	1.1868
0.9518	0.376	-0.0368	-9.815	0.5449
1	0.349	0	0	0
		313K		
0	1.341	0	0	0
0.1961	1.082	-0.059	-0.3634	0.0632
0.3544	0.882	-0.0972	-5.5194	0.3563
0.4848	0.716	-0.129	-16.5546	1.0958
0.5941	0.608	-0.126	-19.6978	1.2131
0.6871	0.52	-0.1186	-24.0431	1.4123
0.7671	0.458	-0.099	-23.2169	1.2942
0.8367	0.399	-0.086	-27.0576	1.5207
0.8978	0.368	-0.055	-18.2036	0.973
0.9518	0.341	-0.027	-9.308	0.4987
1	0.318	0	0	0

Table 1: Viscosity (η_m) deviation in viscosity ($\Delta\eta_m$), excess internal pressure (π_i^E) and excess molar enthalpy (H_m^E) for acetonitrile+2-ethoxyethanol mixtures.

X_1	η_m (mPa.s)	$\Delta\eta$ (mPa.s)	π_i^E (MPa)	H_m^E (kJ.mol^{-1})
		293K		
0	3.126	0	0	0
0.2423	2.155	-0.3081	-11.0081	0.425
0.4185	1.497	-0.4842	-32.4926	2.1097
0.5523	1.134	-0.4811	-39.8633	2.3839
0.6573	0.892	-0.4355	-44.7111	2.5005
0.7422	0.714	-0.3809	-50.4412	2.7707
0.812	0.591	-0.313	-52.4877	2.7966
0.8704	0.519	-0.2255	-42.3732	2.1053
0.92	0.458	-0.1504	-34.0228	1.6355
0.9628	0.421	-0.0705	-16.889	0.7586
1	0.39	0	0	0
		303K		
0	2.43	0	0	0
0.2423	1.707	-0.2186	-14.7708	0.3631
0.4185	1.185	-0.3746	-42.0453	2.4124
0.5523	0.869	-0.412	-62.5906	3.7278
0.6573	0.699	-0.3635	-65.7516	3.6031
0.7422	0.592	-0.2936	-60.0033	2.9971
0.812	0.517	-0.223	-50.3727	2.2897
0.8704	0.452	-0.1665	-43.8708	1.939
0.92	0.413	-0.1022	-28.1809	1.128
0.9628	0.376	-0.0501	-15.5975	0.6142
1	0.349	0	0	0
		313K		
0	1.8977	0	0	0
0.2423	1.3685	-0.1465	-18.1925	0.2078
0.4185	0.9488	-0.288	-51.542	2.6503
0.5523	0.7156	-0.3098	-69.9963	3.6835
0.6573	0.5827	-0.2769	-74.3164	3.673
0.7422	0.5041	-0.2215	-67.3879	3.0109
0.812	0.4559	-0.1594	-51.6844	1.9755
0.8704	0.4042	-0.1189	-44.08	1.6493
0.92	0.3696	-0.0751	-30.1247	1.0614
0.9628	0.3433	-0.0339	-13.9832	0.4332
1	0.3184	0	0	0

Table 2: Viscosity (η_m), deviation in viscosity ($\Delta\eta_m$) excess internal pressure (π_i^E) and deviation in molar enthalpy (H_m^E) for acetonitrile+2-butoxyethanol mixtures.

Parameters	Temp. (K)	α1	α2	α3	α4	α5	σ(Y^E)
acetonitrile+2-ethoxyethanol							
Δη (mPa.s)	293	-1.1346	-0.1909	-0.5333	0.3178	0.7479	0.0016
	303	-0.9303	-0.6333	-0.297	1.964	1.8851	0.0016
	313	-0.4816	0.1773	-0.0073	-0.0308	0.0794	0.0034
π_i^E (MPa)	293	-83.2904	81.3762	-234.9172	79.8204	216.5565	1.1721
	303	-520.1019	-1314.9	57.2027	3882.977	2908.0146	0.9995
	313	-67.2473	117.49	132.2675	-75.3104	-278.3652	1.8347
H_m^E (kJ.mol-1)	293	8.2279	-2.8301	14.2574	-3.901	-15.3335	0.0753
	303	29.8727	71.0567	-1.845	-208.4377	-159.4172	0.0794
	313	4.3963	-6.932	-12.3042	7.2786	22.6867	0.1216
acetonitrile+2- butoxyethanol							
Δη (mPa.s)	293	0.0757	5.1174	-4.3184	-18.1271	-10.9172	0.0087
	303	-1.6466	0.3884	1.6367	-0.288	-2.003	0.0028
	313	0.1883	5.2001	-1.7859	-17.401	-11.7185	0.0163
π_i^E (MPa)	293	-145.8562	81.8381	-196.6476	492.6593	410.6228	1.6683
	303	-224.8594	293.2934	222.2356	-239.4692	-470.4946	1.5923
	313	-250.5386	250.2075	36.5039	33.3339	104.8389	2.1697
H_m^E (kJ.mol-1)	293	9.0466	-1.6854	9.5056	-40.8999	-40.101	0.1112
	303	13.4362	-17.7698	-27.2281	17.5737	35.9649	0.1082
	313	13.318	-13.4031	-16.5041	-4.4572	-3.2247	0.1418

Table 3: Adjustable parameters α_i with the standard deviations σ(Y^E) for deviation in viscosity (Δη) excess internal pressure (π_i^E) and excess molar enthalpy (H_m^E) for both the binary mixtures at varying temperatures.

internal pressure can be given as

$$\pi_i = \frac{2^{1/6} RT}{2^{1/6} V - dN^{1/3} V^{2/3}} \qquad (8)$$

where V is the molar volume, d is the molecular diameter, T is the absolute temperature and R is the gas constant

Relation based upon Flory statistical theory: Knowing the thermal expansion coefficient α and isothermal compressibility k_T of pure liquid components, the values of α and k_T for liquid mixture can be evaluated as

$$(\alpha)_{Flory} = \frac{3\left(\tilde{V}^{1/3} - 1\right)}{\left\{1 - 3\left(\tilde{V}^{1/3} - 1\right)\right\} T}$$

And

$$(k_T)_{Flory} = \frac{(\alpha)_{Flory} T \tilde{V}^2}{P^*}$$

where V is the reduced volume, T is the absolute temperature and P* is the characteristic pressure. Using the values of α and k_T thus obtained, one can evaluate the internal pressure of the liquid mixture using the relation [20].

$$\pi_i = \frac{(\alpha)_{Flory} T}{(k_T)_{Flory}}$$

These relations have been applied to both the mixtures for the evaluation of internal pressure. The calculated values of internal pressure are given in Tables 4 and 5 respectively with average percentage deviation (APD) values. Here we have taken thermodynamic relation as reference because in this relation internal pressure has been evaluated from experimental data's. By comparing APD values of each empirical formula we can say that empirical relations giving the best results for both binary mixtures at all three temperatures.

Conclusion

The observed values and their excess parameters suggest weak dipole-dipole interaction and dispersive force may contribute in result of viscosity, internal pressure and molar enthalpy and their excess values.

Acknowledgements

One of the authors (Sangeeta Sagar) thankful to UGC for providing financial support through Rajiv Gandhi National Fellowship.

X_1	Thermodynamic relation	Empirical relation	F. S. T. relation	B.H.C. relation
		T=293K		
0	387.1	387	387.1	380.3
0.1961	379.6	379.5	380.3	419
0.3544	372.4	372.3	373.4	455.5
0.4848	364.5	364.4	366.2	489.4
0.5941	356.5	356.4	358.9	521.5
0.6871	348.8	348.7	351.6	552.2
0.7671	341.4	341.3	344.2	581.6
0.8367	334.3	334.2	336.9	610
0.8978	327.4	327.3	329.6	637
0.9518	321.1	321.1	322.4	663.6
1	315.4	315.3	315.4	689.4
	APD	**-0.024**	-0.466	58.41
		T=303K		
0	382	381.9	402.2	388.5
0.1961	375.8	375.7	395.4	428.3
0.3544	368.8	368.7	388.4	465.4
0.4848	360.8	360.7	381	499.8
0.5941	353.1	353.1	373.6	532.4
0.6871	345.9	345.9	366.2	563.9
0.7671	338.4	338.4	358.7	593.8
0.8367	331.4	331.4	351.2	622.6
0.8978	324.5	324.5	343.8	649.8
0.9518	318.2	318.2	336.5	676.4
1	311.9	311.9	329.2	701.9
APD		**-0.024**	**5.657**	**63.2**
		T=313K		
0	375.8	375.7	375.8	396.1
0.1961	370.9	370.8	369.6	437.5
0.3544	363.7	363.6	363	474.7
0.4848	356.3	356.2	356.2	509.9
0.5941	349	348.9	349.3	543.3
0.6871	341.6	341.5	342.3	574.8
0.7671	334.5	334.4	335.3	605.1
0.8367	327.8	327.7	328.4	634.6
0.8978	321	320.9	321.5	662.3
0.9518	314.8	314.7	314.7	689.2
1	307.8	307.7	307.8	714
	APD **0.024**	**0.024**		**68.42**

Table 4: Internal pressure of acetonitrile+2-ethoxyethanol mixture from various methods at T=293, 303, 313 K.

X_1	Thermodynamic relation	Empirical relation	F. S. T. relation	B.H.C. relation
		T=293K		
0	372.8	372.8	372.8	282.3
0.2423	366	365.9	368.3	329.9
0.4185	359.4	359.3	363.3	375.7
0.5523	353	352.9	358	419.8
0.6573	346.6	346.5	352.3	461.9
0.7422	340.8	340.7	346.4	503
0.812	335.2	335.1	340.3	542.5
0.8704	329.8	329.7	334.2	580.8
0.92	324.8	324.7	327.9	618
0.9628	319.9	319.9	321.7	654.2

1	315.4	315.3	315.4	689.4
APD	**-0.024**	**0.9833**	**47.42**	
		T=303K		
0	372.5	372.4	372.5	288.5
0.2423	365	364.9	368.6	425.4
0.4185	357.9	357.8	364.2	470.6
0.5523	351.2	351.1	359.3	510.9
0.6573	344.7	344.6	354	547.2
0.7422	338.6	338.5	348.1	580.2
0.812	332.8	332.8	341.9	609.7
0.8704	327.2	327.1	335.1	636.4
0.92	322.1	322	327.9	660.5
0.9628	317	316.9	320.2	682.4
1	311.9	311.9	311.9	701.9
APD		**-0.024**	**1.683**	**65.83**
		T=313K		
0	375.8	375.7	375.8	396.1
0.2423	370.9	370.8	369.6	437.5
0.4185	363.7	363.6	363	474.7
0.5523	356.3	356.2	356.2	509.9
0.6573	349	348.9	349.3	543.3
0.7422	341.6	341.5	342.3	574.8
0.812	334.5	334.4	335.3	605.1
0.8704	327.8	327.7	328.4	634.6
0.92	321	320.9	321.5	662.3
0.9628	314.8	314.7	314.7	689.2
1	307.8	307.7	307.8	714
APD		**-0.024**	**1.923**	**55.36**

Table 5: Internal pressure of acetonitrile+2-butoxyethanol mixture from various methods at T=293, 303, 313 K.

References

1. Motin MA, Ali MA, Sultana S (2007) Density and excess molar volumes of binary mixtures of sulfolane with methanol, n-propanol, n-butanol, and n-pentanol at 298.15-323.15 K and atmospheric pressure. J Phys Chem Liq 45: 221-229.
2. MT Zafarani Moattar, S. Hosseinzadeh (2006) Refractive Index, Viscosity, Density, and Speed of Sound of Aqueous Sodium Tartrate Solutions at Various Temperatures. J Chem Eng Data 51: 1190-1193.
3. Jui-Tang Chen, Hsiao-Pei Chu (2007) Densities and Viscosities for Binary Mixtures of Ethyl Lactate with Methacrylic Acid, Benzyl Methacrylate, and 2-Hydroxyethyl Methacrylate at (298.15, 308.15, and 318.15) K. J Chem Eng Data 52: 650-654.
4. Gonzalez B, Dominguez A, J Tojo (2004) J Chem Thermdyn 36: 267.
5. Ali A, Abida, S Hyder, Nain AK (2002) Ultrasonic, volumetric and viscometric study of molecular interactions in binary mixtures of 2,2,4-trimethyl pentane with n-hexane and cyclohexane at 308 K. Ind J Phys 76B: 661-667.
6. Vesovic V (2002) Predicting the viscosity of halogenated hydrocarbon mixtures. Fluid Phase Equilibria 199: 295-306.
7. Chen JT, Chang WC (2005) J Chem Eng Data 50: 753.
8. Redlich O, Kister A (1948) Ind Eng Chem 40: 345.
9. Valles C, Eduardoperez C, Cardoso M, Dominguez M, Mainar AM (2004) Excess Enthalpy, Density, Viscosity, and Speed of Sound for the Mixture Tetrahydropyran + 1-Butanol at (283.15, 298.15, and 313.15) K. J Chem Eng Data 49: 1460-1464.
10. Ali A, Nain AK (2002) Ultrasonic study of molecular interaction in binary liquid mixtures at 30°C. J of Phys 58: 695.
11. Nath J (2002) J Chem Thdym 34: 1857-1872.
12. Pal A, Kumar A, Arbad BR, Takale AB (2002) Ind J of Chem 41: 1857.
13. Mehta SK, Sharma AK, Bhasin KK, Prakash R (2002) Fluid Phase Equilibria 201: 203.
14. Narendra K, Narayanamurthy P, Srinivasu C (2010) Evaluation of Excess Thermodynamic Parameters in a Binary Liquid Mixture (Cyclohexane+O-Xylene) at Different Temperatures. E-Journal of Chemistry 7: 927-934.
15. Oswal SL, Desai HS (2001) Fluid phase Equilib 186: 81.
16. Grunberg L, Nissan AH (1949) Mixture law for viscosity. Nature 164: 799.
17. Pandey JD, Dubey GP, Tripathi N (1997) J Int Acad Phys Sci 1: 117.
18. Buehler RJ, Wentorff RH, Hirschfelder JO, Curtiss CF (1951) The Free Volume for Rigid Sphere Molecules. J Chem Phys 19: 61.
19. Hirschfelder JO, Curtiss CF, Bird RB (1954) Molecular Theory of Liquids. Wiley, New York, USA.
20. Flory PJ (1965) J Am Chem Soc 87: 1833.

Permissions

All chapters in this book were first published in JTC, by OMICS International; hereby published with permission under the Creative Commons Attribution License or equivalent. Every chapter published in this book has been scrutinized by our experts. Their significance has been extensively debated. The topics covered herein carry significant findings which will fuel the growth of the discipline. They may even be implemented as practical applications or may be referred to as a beginning point for another development.

The contributors of this book come from diverse backgrounds, making this book a truly international effort. This book will bring forth new frontiers with its revolutionizing research information and detailed analysis of the nascent developments around the world.

We would like to thank all the contributing authors for lending their expertise to make the book truly unique. They have played a crucial role in the development of this book. Without their invaluable contributions this book wouldn't have been possible. They have made vital efforts to compile up to date information on the varied aspects of this subject to make this book a valuable addition to the collection of many professionals and students.

This book was conceptualized with the vision of imparting up-to-date information and advanced data in this field. To ensure the same, a matchless editorial board was set up. Every individual on the board went through rigorous rounds of assessment to prove their worth. After which they invested a large part of their time researching and compiling the most relevant data for our readers.

The editorial board has been involved in producing this book since its inception. They have spent rigorous hours researching and exploring the diverse topics which have resulted in the successful publishing of this book. They have passed on their knowledge of decades through this book. To expedite this challenging task, the publisher supported the team at every step. A small team of assistant editors was also appointed to further simplify the editing procedure and attain best results for the readers.

Apart from the editorial board, the designing team has also invested a significant amount of their time in understanding the subject and creating the most relevant covers. They scrutinized every image to scout for the most suitable representation of the subject and create an appropriate cover for the book.

The publishing team has been an ardent support to the editorial, designing and production team. Their endless efforts to recruit the best for this project, has resulted in the accomplishment of this book. They are a veteran in the field of academics and their pool of knowledge is as vast as their experience in printing. Their expertise and guidance has proved useful at every step. Their uncompromising quality standards have made this book an exceptional effort. Their encouragement from time to time has been an inspiration for everyone.

The publisher and the editorial board hope that this book will prove to be a valuable piece of knowledge for researchers, students, practitioners and scholars across the globe.

List of Contributors

Márcio José Da Silva and Abiney Lemos Cardoso
Chemistry Department, Federal University of Viçosa, Viçosa, Minas Gerais State, Brazil

Kamau JM, Mbui DN, Mwaniki JM, Mwaura FB and Kamau GN
Department of Chemistry, University of Nairobi, PO Box 30197-00100, Nairobi, Kenya

Derun Hua, Jian Li, Yongrong Xie and Xinning Lu
Chemical Institute of Chemical Industry, Gannan Normal University, Jingxi, 341000, China

Zheng Zhou
Natural Gas Processing Plant of Sinopec Zhongyuan Oilfield Branch, Henan, 457000, China

Mahendra Nath Roy, Partha Sarathi Sikdar and Pritam De
Department of Chemistry, University of North Bengal, Darjeeling 734013, India

Xianlong Zhang, Shuangshuang lv, Xiaobin Jia, Xueping Wu and Fanyue Meng
School of Chemistry and Chemical Engineering, Hefei University of Technology, Hefei, 230009, PR China

Mahendra Kumar Trivedi, Rama Mohan Tallapragada, Alice Branton, Dahryn Trivedi and Gopal Nayak
Trivedi Global Inc., 10624 S Eastern Avenue Suite A-969, Henderson, NV 89052, USA

Snehasis Jana and Rakesh Kumar Mishra
Trivedi Science Research Laboratory Pvt. Ltd., Hall-A, Chinar Mega Mall, Chinar Fortune City, Hoshangabad Rd., Bhopal, Madhya Pradesh, India

Jacques Thierie
Aquaphotomix Brussels Study Center (ABSC), 248 Beyseghem Street / 1120 NOH Brussels, Belgium

Jacques Ricard
Jacques Monod Institute CNRS, University Paris, Paris, France

Elin Moe, Netsanet Gizaw Assefa, Ingar Leiros, Arne O Smalås and Nils Peder Willassen
The Norwegian Structural Biology Centre, University of Tromsø, N-9037 Tromsø, Norway

Kathrin Torseth
Department of Laboratory Medicine, Children's and Women's Health, Norwegian University of Science and Technology, N-7489 Trondheim, Norway

Mohamed Ben Yahia, Fatma Aouaini and Abdemottaleb Ben Lamine
Department of Physics, Faculty of Science, King AbdulAziz University, Rabigh, Saudi Arabia

Manel Ben Yahia
Department of Physics, Faculty of Science, King AbdulAziz University, Rabigh, Saudi Arabia
Laboratory of Quantum Physics, Faculty of Science, Monastir University, Monastir, Tunisia

Korniy S, Pokhmurskii V and Kopylets V
Karpenko Physico-Mechanical Institute of the National Academie of Sciences of Ukraine, 5 Naukova Str., 79060 Lviv, Ukraine

Lei Wang, Hui Wang, Ruiyi Yan and Suojiang Zhang
Beijing Key Laboratory of Ionic Liquids Clean Process, Key Laboratory of Green Process and Engineering, State Key Laboratory of Multiphase Complex Systems, Institute of Process Engineering, Chinese Academy of Sciences, Beijing 100190, PR China

Lilong Zhou and Yunli Cao
Beijing Key Laboratory of Ionic Liquids Clean Process, Key Laboratory of Green Process and Engineering, State Key Laboratory of Multiphase Complex Systems, Institute of Process Engineering, Chinese Academy of Sciences, Beijing 100190, PR China
Chinese Academy of Sciences, University of Chinese Academy of Sciences, Beijing 100049, PR China

Afshin Davarpanah
Department of Petroleum Engineering, Science and Research Branch, Islamic Azad University, Tehran, Iran

Mala A Sainna
Manchester Institute of Biotechnology and School of Chemical Engineering and Analytical Science, The University of Manchester, 131 Princess Street, Manchester M1 7DN, United Kingdom

Abba MK
Department of Petroleum Engineering, Faculty of Engineering and Engineering Technology, Abubakar Tafawa Balewa University Bauchi, PMB 0248, Yelwa Road, Bauchi State, Nigeria

Rumki N, Aniruddha G, Susanta M and Bidyut Saha
Homogeneous Catalysis Laboratory, Department of Chemistry, The University of Burdwan, Golapbag, Burdwan, Pin 713104, WB, India

Josef Taucher, Sascha Baer, Philipp Schwerna, Dominik Hofmann, Magdalena Hümmer, Rainer Buchholz and Anna Becker
Institute of Bioprocess Engineering, Friedrich-Alexander-Universität Erlangen-Nürnberg, Erlangen, Germany

Guerdouh A and Barkat D
Laboratory of chemical molecular and environment, Department of Industrial Chemistry, Faculty of Science and Technology, Biskra University, 07000 Biskra, Algeria

Sobko AA
Department of Physics and Chemistry of New Materials AM Prokhorov Academy of Engineering Sciences 19 Presnensky Val, Moscow 123557, Russia

El-Bindary AA, Diab MA, El-Sonbati AZ and Serag LS
Chemistry Department, Faculty of Science, Damietta University, Damietta 34517, Egypt

Ghoneim MM
Chemistry Department, Faculty of Science, Tanta University, Tanta, Egypt

Startsev AN
Pensioner, Novosibirsk 630090, Russia

El Batouti M
Chemistry Department, Faculty of Science, Alexandria University, Egypt

El Mossalamya EH
Chemistry Department, Faculty of Science, Benha University, Egypt

Nawmosyc G
Chemistry Department, Faculty of Science and Arts, Al Qassim University, Saudi Arabia

Mohamad Javad Kamali and Gholamhossein Vatankhah
Department of Chemical Engineering, Islamic Azad University, Bushehr branch, Bushehr, Iran

Zakarya Kamali
National Iranian Gas Company (NIGC), Fajr-e Jam Gas Company, Bushehr, Iran

Prodyut Dhar, Sai Phani Kumar Vangala, Pankaj Tiwari, Amit Kumar and Vimal Katiyar
Department of Chemical Engineering, Indian Institute of Technology Guwahati, Assam, India

Wangdan Zhao, Wenge Yang, Qirun Guo, Shimin Fan and Jianfeng Hao
School of Pharmaceutical Sciences, Nanjing Tech University, No. 30, South Puzhu Road, Nanjing 211816, PR China

Yonghong Hu
State Key Laboratory of Materials-Oriented Chemical Engineering, Nanjing Tech University, No.200, North Zhongshan Road, Nanjing 210009, PR China

Hums E
Consulting Environmental Catalysis, PO Box 1848, 91008 Erlangen, Germany

Yoshihiro Kudo and Shoichi Katsuta
Graduate School of Science, Chiba University, Chiba 263-8522, Japan

Yuuki Ohsawa and Kohei Nozaki
Department of Chemistry, Faculty of Science, Chiba University, Chiba 263-8522, Japan

Abdelaziz B and Messaoud H
Laboratoire Géologie du Sahara, Université de Kasdi Merbah, BP 511, Ouargla 30000, Algérie

Spurgeon Smith and Fadi Bou-Abdallah
Department of Chemistry, State University of New York at Potsdam, Potsdam, New York, USA

Faridirad F, Ahmadi S, Barmar MD and Naderi G
Iran Polymer and Petrochemical Institution, Pazhouhesh Blvd., Pazhouhesh Science Park, Tehran-Karaj Highway, Iran

Jinming Liu
School of Material Science and Engineering, Jiangxi University of Science and Technology, Ganzhou, 341000, PR China
Xingle Group Co. Ltd., Yueqing, Zhejiang, 325604, PR China

Yinghui Zhang, Ming Zhang, Lan Zhang, Tao Zhong and Juan Lin
School of Material Science and Engineering, Jiangxi University of Science and Technology, Ganzhou, 341000, PR China

Xiaoyang Chen
Xingle Group Co. Ltd., Yueqing, Zhejiang, 325604, PR China

Caifang Cao
School of Metallurgy and Chemical Engineering, Jiangxi University of Science and Technology, Ganzhou, 341000, PR China

Hao-Hsin Chi and Cheng-Yu Lai
Department of Chemistry, Delaware State University, 1200 N. DuPont Highway, Dover, DE, 19901, USA

Daniela R Radu
Department of Chemistry, Delaware State University, 1200 N. DuPont Highway, Dover, DE, 19901, USA
Department of Materials Science and Engineering, University of Delaware, Newark, DE, 19716, USA

Geraldine Dezayas and Miranda Penney
Department of Chemistry, Delaware State University, 1200 N. DuPont Highway, Dover, DE, 19901, USA
Department of Biological Sciences, Delaware State University, 1200 N. DuPont Highway, Dover, DE, 19901, USA

Gality H, Toth K and Sohar G
Department of Orthopaedics, Faculty of Medicine, University of Szeged, Semmelweis u. 6, 6725 Szeged, Hungary

Aigner Z and Szabo-Revesz P
Department of Pharmaceutical Technology, Faculty of Pharmacy, University of Szeged, Eötvös u. 6, 6720 Szeged, Hungary

Rahul Singh
Department of Physics, Ambalika Institute of Technology, Charbagh, Lucknow, Uttar Pradesh, India

Sudir Kumar, Sangeeta Sagar and Manisha Gupta
Department of Physics, University of Lucknow, Lucknow, Uttar Pradesh, India

Index